Cities in Competition

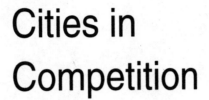

Cities in Competition

Productive and Sustainable Cities for the 21st Century

Editors:

John Brotchie, Mike Batty, Ed Blakely, Peter Hall and Peter Newton

LONGMAN AUSTRALIA

Longman Australia Pty Ltd.
Longman House
Kings Gardens
95 Coventry Street
Melbourne 3205 Australia

Offices in Sydney, Brisbane, Adelaide, Perth and associated companies throughout the world.

Designed by Lauren Statham
Set in 10/12 Plantin
Cover designed by Lauren Statham
Cover illustration by Lauren Statham
Produced by Longman Australia Pty Ltd
through Longman Malaysia

National Library of Australia
Cataloguing-in-Publication data

Cities in Competition: productive and sustainable cities for the 21st century.

  Bibliography.
  Includes index.
  ISBN 0 582 80106 0.

  1. Cities and towns - Forecasting. 2. Twenty-first century - forecasts. 3. Social change. 4.
  Urban ecology. I. Brotchie, J. F. (John F.).

307.76

Contents

Foreword

MAJOR CITIES OF THE WORLD are becoming increasingly linked—by global networks of telecommunications, computers and air transport. On land, highways and, in Europe and Japan, fast rail links further complement these. These networks are enabling the development of global and regional markets. The major cities at the network nodes are market centres and centres for corporate control and associated services. The networks are allowing these cities to have global as well as regional roles. They are also enabling the spatial expansion of organisations and systems for production of goods and services. They are facilitating a shift in resources from production of goods to provision of services, particularly information-based services. This is also enabling the increased participation of women in the workforce. The communication networks are increasingly carriers of information in its various forms. The fast travel networks carry expertise and high value to weight goods between production centres. Cities are increasingly the centres of population, skills and innovation and of production of goods and services. They are the new engines of the modern economy.

The lowering of national barriers through international trade agreements, formation of trading blocs and interregional alliances, and through the deregulation of markets within and between nations, is further increasing the importance of cities as economic entities. It is increasing the competition between cities for provision of goods and services and for the attraction of new industries, particularly company headquarters or regional offices, and associated services, and new technology, knowledge- and information-based industries. This increasing competition among cities both increases the opportunities for growth and prosperity but also increases their vulnerability to competition from other cities.

The transition to an information economy is changing the relative importance of cities—reinforcing the largest as global or supra-regional centres, at major network nodes; yet providing opportunities for others to ascend the urban hierarchy, displacing others in the process. It is also changing the pattern of development of these cities and their urban forms. The majority of employment is now located in the suburbs—in suburban centres of various types—metro centres at urban transport nodes, multi-centres for services (including retail), edge cities, special use zones including technology centres, development corridors along fast intercity rail routes, and some teleworking by firms and some individuals, thereby creating virtual communities. Commuting and daily activity patterns and lifestyles are changing accordingly.

This book plots the trajectory of these changes and their impacts. The introduction outlines a general theory of the factors and processes underlying this urban change and their different outcomes for cities.

Part 1 describes the global expansion of markets through technology diffusion, decreasing cost and increasing speed of transport and communication, increasing diffusion of information and declining barriers to trade. The urban impacts of this globalisation are also defined and include the concentration of corporate control and financial and other service functions in the largest cities.

Part 2 outlines the increasing competition generated among firms and between cities and the impacts on the urban hierarchy including the creation of winners and losers. Winner cities are those with a global orientation and a post industrial base, including corporate headquarters and financial and other producer service activities, and are well located at global network nodes.

Part 3 considers the creation of technopoles or technology cities as centres for innovation and development of new technologies and industries. High technology firms are concentrated in a few metropolitan environments which are rich in intellectual resources, and these in turn tend to congregate in high amenity urban areas with good connections to global networks. The planned creation of these centres is also discussed.

Part 4 analyses changing patterns of living and working at the metropolitan level including dispersal of employment into the suburbs and beyond, and the consequences for commuting patterns, public transport and sustainability of development. Urban consolidation policy is brought into question and telecommuting experiments are described.

Part 5 considers the nature of sustainability of urban systems and of ecosystems for their support and the implications for urban system simulation and planning.

The Epilogue outlines the continuing international study of which this volume is a part, and summarises the factors, processes, and urban products and systems generated in the transition to a global information economy.

List of Contributors

Mr Miles Anderson, Division of Building, Construction and Engineering, CSIRO, Australia

Professor Michael Batty, National Center for Geographic Information & Analysis, State University of New York, USA

Professor Edward Blakely, School of Urban and Regional Planning, University of California, USA

Professor Michael Breheny, Department of Geography, University of Reading, UK

Dr John Brotchie, Division of Building, Construction and Engineering, CSIRO, Australia

Dr John Carey, Director of Greystone Communications, New York, USA

Professor Robert Cervero, City and Regional Planning, University of California, USA

Professor Paul Cheshire, Dept. of Economics, University of Reading, UK

Professor Peter Daniels, Service Sector Research Unit, University of Birmingham, UK

Professor John Dickey, University Center for Innovation Research and Support, Virginia Polytechnic Institute, USA

Dr Martin Frost, Department of Geography, Kings College, London, UK

Professor John Goddard, Center for Urban and Regional Development, University of Newcastle Upon Tyne, UK

Professor Ian Gordon, Dept. of Geography, University of Reading, UK

Professor Peter Gordon, School of Urban and Regional Planning, University of Southern California, USA

Professor Peter Hall, Bartlett School of Architecture and Planning, University College London, UK

Professor Stephen Hamnett, Regional and Urban Planning, University of South Australia, Australia

Professor Britton Harris, City and Regional Planning, University of Pennsylvania, USA

Jean Monnet Professor Klaus Kunzmann, European Spatial Planning, University of Dortmund, Germany

Ms Cheryl McNamara, Division of Building, Construction and Engineering, CSIRO, Australia

Professor Richard Meier, City and Regional Planning, University of California, USA

Professor Mitchell Moss, Wagner Graduate School of Public Service, New York University, USA

Dr Dai Nakagawa, College of Socio-Economic Planning, University of Tsukuba, Japan

Dr Peter Newton, Division of Building, Construction and Engineering, CSIRO

Dr Kevin O'Connor, Geography and Environmental Science, Monash University, Australia

Professor Harry Richardson, School of Urban and Regional Planning, University of Southern California, USA

Dr Nigel Spence, Department of Geography, London School of Economics, UK

Professor Saskia Sassen, Urban Planning Program, Columbia University, USA

Professor Thomas Stanback, The Eisenhower Center for the Conservation of Human Resources, Columbia University, USA

Professor Robert Stimson, Brisbane City Council Chair in Urban Studies, Queensland University of Technology, Australia

Dr Mamoru Taniguchi, College of Socio-Economic Planning, University of Tsukuba, Japan

Dr Tsunekazu Toda, College of Socio-Economic Planning, University of Tsukuba, Japan

Dr Michael Wegener, Institute of Spatial Planning, University of Dortmund, Germany

Dr Kelvin Willoughby, David Eccles School of Business, The University of Utah, USA

Acknowledgements

This book is the latest from an international study of technological change and its effects on cities convened (by the Commonwealth Scientific and Industrial Research Organisation, CSIRO) on behalf of the International Council for Building Research Studies and Documentation (CIB) as its Working Commission 72. The book reports the findings of the study's Fourth International Workshop on Technological Change and Urban Form held at the University of California, Berkeley, in April 1993.

The editors are pleased to acknowledge the financial support of CIB and CSIRO and are grateful for assistance in organising the workshop and in producing the report to Kate Blood, Barbara Hadenfeldt (Manager. Institute of Urban and Regional Development), Carey Pelton, Martha Conway and the team at the Institute of Urban and Regional Development at UC, Berkeley, and to Cheryl McNamara, Cathy Bowdich, Rhyll Reed, Jill Connor and Richard Figar at the Division of Building, Construction and Engineering, CSIRO, Melbourne.

Publisher's Note

We are grateful to the following for permission to reproduce copyright material:

AGPS, reproduced by permission of Commonwealth of Australia copyright, for figure 3.8, p. 75, from *Internal Migration in Australia: 1981-86*, by M. Bell, Bureau of Immigration Research, 1992;

The Banker, Financial Times Magazine, for table 12.2, p. 229;

Belhaven Press, an imprint of Pinter Publishers for table 5.1, p. 118, from *Local Economic Development: Public-Private Partnership Initiatives in Britain and Germany* by R.J. Bennett & G. Krebs, 1991;

F. Bruinsma & P. Reitveld for figure 7.7, p. 147, from *Stedelijke Agglomeraties in Europese Infrastruktuurnetwerken*, 1992;

Kerstin Cederlund, Ulf Erlandsson & Gunnar Törnqvist for figure 7.4, p. 144, from 'Svenska kontaktvägar i det europeiska stadslandskapet' in *Svensk Geografisk Årsbok*, 67, Lund, 1991;

ECOTEC Research and Consulting Ltd. for tables 12.2, p. 407, 21.3 & 21.4, p. 409, from *Reducing Transport Emissions through Planning*, London, 1993;

Financial Times for figure 12.1, p. 226;

J. Kenworthy & P.W.G. Newman for table 12.1, p. 405, from 'Gasoline consumption & cities—a comparison of US cities with a global survey' in *Journal of the American Planning Association*, vol. 55, no. 1, 1989;

Klaus R. Kunzmann & M. Wegener for figures 7.1, p. 141, 7.2, p. 142, 7.8, p. 149 & 7.11, p. 159 from *The Pattern of Urbanisation in Western Europe: 1960-90*, 1991;

Kevin O'Connor for table 3.2 , p. 67, from 'Economic Activity in Australian Cities: National and Local Trends and Policy' in *Urban Futures*, February 1992 and table 3.4, p. 73, from 'The Geography of Research and Development Activity in Australia' in *Urban Policy & Research*, December 1993;

Office of Official Publications of the EC for figure 7.5, p. 145;

Office of Population Censuses & Surveys for figures 21.1, p. 417 & 21.2, p. 426 and tables 21.6, p. 418 & 21.7, p. 419;

Ohkurasho Insatsukyoku (Bureau of Printing, Ministry of Finance, Japan) for figure 9.2, p. 196;

RECLUS for figure 7.6, p. 146;

Shutoken Seibi Kyokai (Institute for Tokyo Metropolitan Improvement) for table 9.2, p. 195;

Klaus Spiekermann and Michael Wegener for figures 7.9 and 7.10, p. 155, from *Zietkarten fur die Raumplanung*, 1991;

David Wilmouth for table 3.6, p. 79, from 'Metropolitan Planning for Sydney' in *Urban Australian Planning Issues & Policies*, eds. D. Hamnett & R. Bunker, 1987.

While every effort has been made to trace and acknowledge copyright, in some cases copyright proved untraceable. Should any infringement have occurred, the publishers tender their apologies and invite copyright owners to contact them.

Introduction

1 Towards a General Urban Theory

Peter Hall

THERE is a general sense, in recent urban literature, that a set of forces—economic, technological, social—are operating together, to influence profoundly the global system of cities. These changes express themselves both in the relationship of individual cities to each other and to the system of which they form a part; and also in the internal structure of these cities.[1] In this introductory chapter, drawing heavily on insights in the chapters that follow, I shall first seek to summarise the nature and effect of the forces; and then to speculate on the resulting changes to the urban system. Inevitably, that will mean a degree of repetition; but this may be helpful in seeking to trace the connections—sometimes subtle ones—between causes and results.

The contributing forces

Globalisation

It seems to be generally accepted that, economically and thereby politically, nations and regions are being increasingly brought into close relationships with one another; often competitive, sometimes co-operative.

The most evident basic expression is a New International Division of Labour. Since about 1960, an increasingly wide group of middle-income countries (newly industrialising countries) have shown themselves capable of accelerated industrialisation, achieving in about one generation a process that formerly took a century or more. The most spectacular examples are all in eastern Asia (the Republic of Korea, Taiwan, Hong Kong and Singapore, now in process of being joined by Malaysia, Thailand and the southern provinces of the People's Republic of China, with the Philippines as a possible candidate); but the most advanced core regions of several Latin American countries (Mexico, Venezuela, Brazil, Argentina, Chile) also belong in this group. These countries (more precisely, their leading cities) have shown themselves variously capable of competing not only in older staple industries, such as textiles and shipbuilding, but also in technologically sophisticated industries like automobile manufacturing and electronics, where they may form increasingly complex arrangements with manufacturers in the older core industrial countries. The result is the familiar process of de-industrialisation in these countries, as their basic industries find it increasingly difficult to compete with the combination of modern equipment, highly qualified but cheap labour, and competent management in the newer industrial regions.

All this is familiar. A major question now however concerns the countries of eastern Europe, which potentially could join the newly industrialising group as soon as they reinvest in modern industrial plant and equipment and acquire western management techniques. Given that these countries have long industrial traditions, that their workforces are traditionally well educated and well trained technologically, and that some of them have reasonably good (if under-maintained) transport and urban infrastructure, they seem well placed to exploit their advantages as soon as the transitional period (including the current recession) is over. They seem likely to develop a relationship to western Europe similar to that of Mexico to the United States; at least for a time, they will become specialised low-cost industrial platforms and tourist resort areas. Much of the resulting development will be concentrated in the larger cities; a point to which we shall return.

Associated with this, especially in the last decade, is the development of multinational trading blocks, notably the European Single Market and the (somewhat uncertain) European Monetary Union, and the North American Free Trade Area. It seems certain that these blocks will result in some degree of concentration in a few key core cities, especially those that are chosen as the sites of international institutions (such as the projected European Monetary Institute, its proto-central bank); hence here, too, there is intense political competition for the location of such institutions.

The informational economy

The second fundamental feature is the shift of all advanced economies from goods production and goods handling to the processing of information. It appears that this is a fundamental economic shift, perhaps as momentous as the transition from an agrarian to a manufacturing economy in the 18th and 19th centuries (Castells 1991). It is however a shift of long duration, observable since at least the start of the 20th century; in advanced countries an actual majority of the population is now involved in producing services, while some 40% of the entire workforce in the US and Britain, 30% in Germany and Japan, are involved in processing information (Clark 1940; Hall 1987; Hall 1989a). In such an economy traditional location factors, like coal and iron and water, are no longer very significant; the question is: what takes their place?

Here, one trend is highly significant: as a combined result of globalisation and informationalisation, the production of services becomes increasingly detached from that of production (Sassen 1991). Though producer services may depend in some sense on production (Gershuny & Miles 1983; Cohen & Zysman 1987), the financial services sector, notably, has no direct relationship to production, leading to a locational disarticulation: as production disperses worldwide, services increasingly concentrate; and they do so in places different from the old production locations (Sassen 1991). They cluster in a relatively few trading cities, both the well-known 'global cities' and a second rung of about 20 cities immediately below these, which we can distinguish as 'sub-global'. These cities are global centres for financial services (banking, insurance) and headquarters of major production companies; most are also seats of the major world-power governments (King 1990; Sassen 1991). They also attract clusters of specialised business services like commercial law and accountancy, advertising and public relations services and legal services, many of which are themselves becoming globalised, and which relate not to production locations but to the controlling headquarters locations; in Britain, for instance, they are overwhelmingly concentrated in and around London (Hall 1991a; Sassen 1991).

In turn this attracts business tourism and real estate functions, with effects on the transportation, communication, personal services and entertainment-cultural sectors. Since these cities have tended to be traditional centres for culture and recreation, they also attract a large share of the growing leisure tourism market. There is intense competition between cities both at a given level in the hierarchy and also between levels in the hierarchy, especially in Europe where the position is most complex; but also a great deal of stability because of forces of historic inertia.

The new locational logic is governed by access to information. This is obtained either by face-to-face communication or by electronic transfer. The first encourages agglomeration in major cities, which are the traditional points of concentration for specialised information-generating and information-exchanging activities (government, universities, the financial sector, the media) and which also serve as key nodes for national and international transportation (railways and ports in the 19th century, motorways and airports in the 20th, high speed train systems in the 21st). Thus they continue to attract informational activities by a process of massive locational inertia (Hall 1991b).

Electronic communication, it might be thought, should work in an opposite direction: as telecommunications costs have dropped dramatically, informational activities should have been increasingly free to locate away from the old face-to-face exchange locations. But evidently they have not, for several reasons: telecommunications will never be entirely costless, and (unlike the 19th century postal tradition) there is still a distance-deterrent effect, even nationally; major cities create a built-in demand for new and improved telecommunications services (such as all-digital systems), which gets satisfied first there. Even within the European Community, one of the most advanced industrial regions of the world, there are marked centre–periphery differences, even in basic telephone service; the diffusion of personal computers has been far more rapid in and around London than elsewhere in Britain (Goddard & Gillespie 1987, 1988; Batty 1988). Paradoxically, therefore, in the informational economy the largest global cities retain their primacy.

Technical change

Technical change impacts on the urban system in two distinct ways: through the generation of new basic industries, and through changes in the underlying transportation and communication system.

New Industries

There has been a huge literature in the past decade about the forces that have generated new technologies serving as the basis of new industrial complexes, including the bunching of such innovative forces in time (so-called long waves) and geographical space. No general agreement has resulted on the question of long waves; though many observers include that such waves have in fact generated related bunches of new industries at approximately half-century intervals, there is much controversy about the precise triggering mechanisms. What does seem to be generally agreed is that in general, new industrial complexes tend to have developed in locations somewhat different from those of older 'sunset' industries. The reasons are complex and may include freer entrepreneurial attitudes, the availability of venture capital, the location of technologically oriented universities, the arrival of key technological developments that have transformed geographical space and opened up new regions, and government policy (particularly government defence policies after World War II), and more recently the development by some governments of deliberate 'Technopole' policies. The outcome in the last half century has been the development of new industrial regions in the older industrial countries (the Western Crescent around London; the south west sector of Paris, and the south of France; the Munich region; Silicon Valley; Los Angeles) and in some of the newly industrialising countries (Seoul, Singapore, Guangdong), as well as the rarer case of the older industrial metropolis that successfully transforms itself into a high-tech centre (Munich, Tokyo).

Recent analyses of these new industrial complexes suggest that their industrial organisation is 'post Fordist': like industrial districts of the traditional kind, analysed by Alfred Marshall a century ago, they consist of highly networked small—and medium sized—enterprises, many of them of fairly recent foundation, and connected not only by demand–supply relationships for specialist components and services, but also more subtly by a general climate of innovative ideas within a system that is part competitive, part collaborative; the benefits from these interrelationships outweigh the transaction costs that they impose. Through these relationships, such areas seem to avoid the progression from small scale to large scale production, which has characterised other industries and other industrial areas (Saxenian 1991; 1992; 1993; Scott 1988a; 1988b; 1989; Scott & Angel 1987).

Such intensely networked industrial complexes appear to be typical not only of high-tech, technologically very advanced industries (electronics in Silicon Valley; biotechnology in the San Francisco Bay Area and Cambridge, England), but also of 'high-touch' industries dependent for their success on style (clothing, fashion accessories, the media); the archetype is the 'Third Italy' of Emilio-Romagna, which may combine high-tech and high-touch in the same product, and which also uses very advanced technologies (computer-aided design; computerised production of short runs of customised products) to produce high-touch goods and services. Traditionally high-touch industries (clothing, the media) have clustered in metropolitan cores with very intense and rapid access to unprogrammed information, much of it conveyed by personal contact; the injection of advanced technology seems to have had little impact here, except in allowing certain processes (e.g. the physical production of print media) to be decentralised to lower-cost locations; at most, the location may shift to just-off-centre locations (e.g. the displacement of daily newspaper editing and production to Stockholm's western tangential motorway in the 1960s and London's docklands in the 1980s). Interestingly, also, shifts to post-Fordist production methods tend simply to confirm and intensify existing agglomerative tendencies; notice that the shift of television production in Britain to small independent firms in the early 1990s has left the industry even more firmly tied to London, just as an earlier shift of movie and TV production had the same effect in Los Angeles.

Some observers see the so-called high-touch industries—in an extended definition that includes high-touch services such as education, health care, museums, galleries, libraries, and the media—as a major source of urban economic growth, perhaps *the* major source, in the current period and for the immediate future. (In London, for instance, cultural and entertainment services rivalled financial services as a source of employment growth during the 1980s, and seems to have been more resilient than the latter during the 1990s recession). There is no formal theory of location for these industries; but it is clear that they tend to cluster strongly in established major metropolitan cities, particularly those that historically were seats of royal or aristocratic patronage (the major European capitals, including former capitals such as Edinburgh, Munich, Milan and St Petersburg; Kyoto in Japan is a similar case) or that were major concentrations of the *haute bourgeoisie* (New York, Chicago). They may also cluster disproportionately in specialised cities such as university towns (Oxford, Cambridge, Heidelberg, Lund, Uppsala, Boston-Cambridge, New Haven, Princeton, Berkeley) and places of unique historic character, which may overlap with the last category (York, Nimes, Bruges, Vicenza, Verona, Bologna, Firenze, Sevilla, Kraków). They thus have a close symbiotic relationship with the more specialised non-mass segments of the tourist industry, notably business tourism and cultural tourism. Only rarely, and only in the mass tourist sector, can entirely new urban spaces be created for this complex of industries (Eurodisney, La Grande Motte, Torremolinos; Anaheim, Orlando); and this sector is the most vulnerable to globalised third world competition exploiting the potential offered by long-haul jets and lower labour costs (as the problems of Spain now demonstrate). Otherwise, a very powerful principle of urban inertia applies; so intense is the concentration of these industries in the older cities, that they create considerable planning problems in the form of traffic congestion and pollution, leading to the possibility that eventually they might require regulation either by price or by rationing.

The major question concerns the future: does the present recession constitute the start of a Kondratieff downswing, as forecast by some observers, to be followed by another set of key innovations; and, if so, what these might be and where they might occur (Hall & Preston 1988; Hall 1988; Berry 1991)? Some observers for instance have identified high-temperature superconductivity, new industrial materials, the convergence of telecommunications and computing, and biotechnology as future growth manufacturing sectors. What does appear certain is that all these will depend very heavily on the commercialisation of basic research and so will be most likely to appear around major technologically oriented research universities on the model of Cambridge (England), Cambridge (Massachusetts) or Palo Alto (California). Indeed, some observers have suggested that the next round of innovation may simply confirm the advantages held by key research core regions such as the San Francisco Bay Area or the Tokyo region. This could be especially true if a major source of growth proved to be the injection of high technology into high-touch, for instance in the electronic media. In France Jack Lang, the former Minister for Culture, was reported as promoting a Centre for Virtual Reality. Major advances are now occurring in knowledge-based systems (applied artificial intelligence); both these depend on research now taking place in this field in a relatively few research laboratories, which again happen to be concentrated in 'high-tech nests' like the Paris region and the San Francisco Bay area.

There is one open question here. Japan in its 26 Technopolis sites, and some other countries through individual promotions (the Multi Function Polis in Adelaide, South Australia; the Cartuja '93 project in Sevilla, Spain) are deliberately attempting to implant new innovative complexes in urban areas and in regions that previously lacked them, but which appear to have some basis for their development (especially, one or more technological universities or faculties; an existing industrial base; a trained labour force). Some of these (Technopolis, Cartuja) stress traditional advanced technology; the Australian MFP, the most experimental, also builds on a variety of high-touch activities, apparently reflecting the fact that its Japanese sponsors believe that the future lies in a more spontaneous style of innovation than they have previously been able to achieve at home. These attempts are relatively new, and indeed two of them (Cartuja, MFP) do not yet have a physical manifestation; so it is too early to say whether any of them will prove successful. The provisional verdict from Japan seems to be that relatively few of the Technopolis sites seem to be really successful in attracting inward investment; that a disproportionate number of these are relatively close to the existing Tokaido corridor; and that more distant sites seem to attract a branch-plant kind of high-technology development, with basic innovation still highly concentrated in the Tokyo and Osaka–Kyoto core regions (Castells & Hall 1993). In Adelaide, Hamnett (pp.310–326) reports an encouraging development of new initiatives which essentially use informational technology to build value-added services, especially in the educational, health and environmental fields; this feature, he suggests, may prove more important than the physical concept of the 'polis'.

A related question is whether the nature of innovation may be changing with the transition from the manufacturing to the informational economy; for the future, innovation may consist not so much in the development of new products and processes as in the formulation of new modes of organisation. Of this, there is however, as yet, not much sign. Rather, it appears that during the early 1990s technological innovation is belatedly impinging on the informational economy, improving productivity and thereby reducing demands for labour in the same ways as were long observable for manufacturing. This may well lead to a long period of jobless growth in advanced industrial (or post-industrial) economies as the service industries at last absorb the revolutionary technological changes of the 1980s. The precise effects are as yet difficult to gauge, but they could well impinge very seriously on the 'white-collar proletariat' which expanded in so many service sectors, notably financial services, during the 1980s boom.

Transportation and Communication

Colin Clark (1957) truly wrote that transport was the 'maker and breaker of cities'. Over centuries, its profound influence has come in two ways: first, through opening up new flows of people, goods and information at national and international scales, thus affecting the fortunes of different cities; secondly through affecting accessibility within cities. Of the first effect there are numerous historic examples (the use of the inland seas, above all the Mediterranean and Black Seas, for early commerce; the development of the great medieval overland trade routes from the Mediterranean to the North Sea and from the Baltic to the Atlantic; the opening up of the Atlantic ports of Western Europe and Eastern North America between the 16th and 19th centuries; the development of the 19th century trading economy between the European manufacturing heartland and the new primary producing regions; the development in North America first of the river systems and then the transcontinental railroads). Most, though not all, of these were associated with technological innovations (better sailing ships, navigational aids, steam, iron ships). In the 20th century, paradoxically, there have been very few technological breakthroughs in transportation; the most important have been the diffusion of the internal combustion engine through automobile and truck traffic on improved highways from the 1920s but above all from the 1950s, the development of commercial jet traffic from the 1950s, and the containerisation of freight from the 1960s. These, coupled with rural electrification, have had profound effects in opening up new regions which were hitherto inaccessible (such as the American South and

West), in vastly accelerating the growth of international passenger traffic, and in reorienting freight traffic between ports.

The precise effects on cities have depended, in many cases, on entrepreneurial adaptation. Some cities benefited greatly from prescient investment in new facilities such as airports (Los Angeles, Atlanta) or container ports (Seattle-Tacoma, Harwich, Felixstowe); others languished because they did not act in time (the ports of London and San Francisco). Some benefited because they were the home base of either major national flag airlines (London, Paris, Amsterdam, Frankfurt) or vigorous commercial carriers (Singapore, Hong Kong, Sydney, Atlanta, Chicago); some others suffered because their base airlines were insufficiently entrepreneurial (Miami). Some seem to have enjoyed an advantage because national highway systems made them into natural foci for manufacturing and distribution (Paris, Frankfurt, Atlanta, Denver). Technological developments played a role here: though airplane speeds remained unchanged after the first generation of jets in 1957–58, their range steadily increased, allowing them to overfly former hubs, particularly on the long trans-Pacific and trans-Asia hauls, and this is underlined by the latest generation of long-range 747–400s. An increasingly difficult constraint in air travel is limited capacity at major hub airports, accompanied in a number of cases by resistance on environmental or other grounds to expansion (London Stansted, Frankfurt, Chicago, Tokyo Narita). O'Connor (pp.88–104) notices that technology was a dominant factor in changes in worldwide air travel patterns during the post-World War II period and will continue to be so.

Within cities, as is well known, the impacts have been considerable. Hub airports acted as major magnets for new activity, not merely directly generated (hotels, service facilities), but also drawn to the airport because of accessibility (business parks, high-tech belts). The most spectacular cases include Los Angeles' Aerospace Alley (though this, ironically, resulted from the letting of airport sites in the 1930s when the airport was a failure); the Dulles Corridor in Washington, DC; the Heathrow M4 Corridor; and Stockholm Arlanda's E4 Corridor (Markusen *et al.* 1991; Castells & Hall 1993). Containerisation has led to a major reorientation of port activity, taking it out of older constricted dockland sites and into estuarial locations (New York-New Jersey; London; Bristol; Tokyo); rationalisation of rail freight operations had similar effects (Hoyle 1988). This has left large areas of prime space, close to city centres, available for redevelopment (London Docklands, Cardiff Bay, Salford Quays, Rotterdam Oude Haven, San Francisco Mission Bay, Toronto Harbourfront, Sydney Darling Harbour and numerous other cases).

The most spectacular impacts have of course come from the growth of automobile and truck traffic. This led not only to the development of automobile-based suburbs beyond the range of mass transit systems, first in the US but now increasingly in western Europe, but also to the growth of new activity centres in the suburbs, originally in manufacturing and warehousing which located close to the new arterial highway interchanges, but more recently in office-based service activities (Dillon et al. 1989;

Garreau 1991). (These are the 'Edge Cities' or 'New Downtowns' in American metropolitan areas, the largest of which now rival traditional downtown areas in their concentrations of commercial space.) The result is a new and complex pattern of non-radial travel to work, almost exclusively by the private car, over a highway network that was typically built in the 1950s or 1960s for quite different purposes, and which is overwhelmed as a result; the condition described by Robert Cervero (1986a; 1989a; 1991) as 'suburban gridlock'.

The question is whether, given the general lack of major technical progress in the present century, we are soon likely to see major innovations in transport. One is already evident: the high-speed train, an increasingly important mode first in Japan (since 1964) and now in Europe (in Great Britain since 1976, France since 1981 and Germany since 1991). High-speed trains appear to enjoy a clear competitive advantage over air for journeys less than approximately 500–900 km (300–500 miles), in length, typically up to about three hours in length depending on the speeds that can be achieved, in densely urbanised regions; they function most efficiently in serving linear corridors of urbanisation in such regions, such as the Tokaido corridor of Japan, the north west European *dorsale* and the Eastern Seaboard of the US. Taniguchi *et al.* (pp.191–199) confirm these competitive advantages for Japan, and show that cities with access to Shinkansen or major airports performed better than cities lacking such access. The precise impacts of high-speed trains on urbanisation are not yet completely clear. They appear to benefit end-cities, especially the largest cities, at the expense of smaller intermediate cities (Tokyo–Osaka versus Nagoya, Paris–Lyon at the expense of Machine); they may assist second-order cities to exploit markets in first-order ones (Lyon at the expense of Paris), and it is certain that they aid cities on their routes at the expense of those that are bypassed (Lille at the expense of Amiens) (Hall 1991b).

Within cities, high-speed trains may provide the platforms for major urban investments. Though generally the new trains serve the old 19th century stations (Tokyo, Nagoya, Sendai; Paris, Marseille, Nantes, Bordeaux; Hamburg, Hanover, Frankfurt, Stuttgart, Munich), in a few cases they have been placed in new city-centre stations some distance from the old (Shin–Osaka; Lyon Part–Dieu; Kassel Wilhelmshöhe; Lille Euralille); in most of these cases, they appear in effect to have created new city subcentres, though in most cases this was deliberately planned, and French planners now talk of the planned creation of 'Gareovilles' (Savel & Rabin 1992). Most interesting are a very few instances where new 'Edge Cities' have been planned around greenfield or edge-of-city suburban stations (Shin–Yokohama; Massy; Ebbsfleet). Only the first of these has been in existence long enough to be able to judge its impact; it appears to have been considerable, though the location (in the middle of the Kanagawa high-tech belt south of Tokyo) was exceptionally propitious.

The other most likely development in ground transportation is the injection of information technology. This is already evident in both offboard (centrally controlled highway systems giving information about traffic conditions) and onboard (traffic-responsive in-car direction systems) forms. The next critical advance would consist in automatic guidance of specially equipped vehicles to provide acceleration and braking, allowing optimal spacing on specially equipped highway lanes; this is a logical development of the control systems that are already widely used in recent urban rapid transit systems (Lille; Vancouver; London Docklands; Kobe), and considerable progress has already been made here in controlled laboratory conditions. The final stage, which would involve remote control of lateral (between-lane) movements, is much more complex and still some time away. Automatic guidance in any form will raise difficult issues of legal responsibility, maintenance of the vehicles and problems of system failure. Its first applications may well come in controlled and segregated conditions on public transport vehicles, such as guided busways.

Improvements in communications have a more subtle but nevertheless a potentially large impact. They arise from the fact that the informational services are voracious consumers of high-quality communications technology—especially digital telecommunications for voice, and above all data transmission. This creates first a distinction between those countries whose telecommunications systems invest most aggressively in new technology and which manage, either through competition or through regulation, to maintain low charges, and those that fail to do so. There are major differences in this respect between European countries, including members of the European Community, which may be reduced under the present Community proposals for open access.

Secondly, a distinction may emerge between particular cities that enjoy investments in the most advanced applications (ISDN; teleports) and those that do not. It is notable that the major global cities appear to enjoy a continuing advantage in this respect: London for instance has just achieved the distinction of being the first city in the world to have a 100% advanced digital telephone system; teleports tend to be created in major new developments in these same cities, such as Battery Park City or London Docklands. Particularly notable however are the attempts of some Japanese cities, encouraged by the country's Ministry of Telecommunications, to develop themselves as 'intelligent cities'. Kawasaki, an old industrial city just outside Tokyo which was a victim of de-industrialisation, has been notably successful at this, including the attraction of a major IBM office facility and the development of the Kanagawa Science Park (KSP) in an old industrial area.

For the future, it appears the current trend will soon result in the appearance of highly personalised, miniaturised technologies which will combine communication and information-processing functions in the same set, constantly available. Rapid progress is already being made in this direction. This will in turn create a demand for very sophisticated cellular or alternative remote technology, available nationally. Equipping nations with this capacity is likely to prove a very large piece of investment, comparable with the original creation of the telephone network. It is likely to come first in the most advanced industrial countries, and then to spread progressively from the major centres to smaller cities and only later to the deep countryside.

Related to this is the question of telecommuting. In the earliest discussions of this topic, about a decade ago, the general assumption was that this would be an agent of liberation, reducing the strain of the daily commute, and allowing flexible time management (for instance, by parents with young children). Limited empirical research, mainly coming from the major experiment among public sector workers in California in the late 1980s, seems to show that these expectations are indeed realised: commute travel is reduced; non-commute trips do not increase; indeed, they actually decrease, and in some cases trip making by household members has also declined. It has been suggested that within 20 years, a potential 50 million people could be telecommuting in the US alone. But other summaries are more cautious, pointing out that though 20% of all urban trips and as many as 50% of white-collar trips might be replaced by telecommuting, only 5–10% of the work force is able and willing to substitute telecommuting for part or all of the commute. And this potential reduction does not directly translate into reduced traffic congestion, since not all the relevant trips are on congested roads, and a good deal of congestion depends on random events like accidents. If work locations become less dependent on local labour supplies, people may choose new more scattered work locations, less accessible to public transport. A key question is whether telecommuting saves more miles of travel than may be created by these resulting shifts in location (Garrison & Deakin 1988; Mokhtarian 1991; Nilles 1991).

However, in the current recession, telecommuting may also be a response by employers to cut costs, not only by saving city centre office rentals, but also by placing workers on unfavourable freelance contracts which deprive them of pension and other benefits (and also in destroying the last vestiges of union organisation). In other words, it might prove to be the equivalent of homework in manufacturing a century ago: a highly exploitative relationship.

The demographic imperative

A marked trend, during the 1980s and early 1990s, has been an accelerating global migration from the less developed countries to the more developed (and more urbanised) regions. With few exceptions, the flows are emanating in areas with high rates of natural increase, while the importing areas have much lower rates. The major targets for immigration are North America and, above all, Australasia; the European Community, not an important recipient area before 1970, began to receive accelerating numbers and had reached quite unprecedented levels of immigration by the end of the 1980s. Down to the early 1970s, countries on the EC's southern periphery (Greece, Spain and Portugal and even Italy) were exporters of population; by the 1980s that was reversed, though in Ireland, a major exporter on the western margin, it resumed. Within the urban-industrial core of Europe, Germany has consistently been a major importer of people, save for a few years in the 1970s; France and Belgium in contrast have not been major net importers save at the end of their colonial era in the 1960s; the Netherlands have taken in more people, particularly at the end of their own colonial episode; the UK has generally been a net exporter. United Nations data shows a fairly consistent rising trend of immigration in Europe during the 1980s, but—rather surprisingly—a downturn for the US until the late 1980s, when the Reagan amnesty to illegal immigrants brought a big recorded inflow.

Both in Europe and North America, this immigration came overwhelmingly from the less developed countries. Generally, the movements have been regional and short-distance. Thus in the late 1980s the great bulk of all long-term migration into Germany came from Europe, and no less than half of that was from Poland; virtually all the immigration into France was from Algeria; the US drew migrants from Mexico, the Philippines and Korea. This represents a remarkable shift: in the 1960s nearly 40% of all migrants into the US came from Europe, by the 1980s less than 10%; Asia, the mirror image, provided only 14% in the 1960s, but close on 40% in the 1980s; Mexico's share has risen from 14 to more than 22%. Though political persecution has undoubtedly played a role in some cases, overwhelmingly the motive for these migration trends has been economic. These migrants have been overwhelmingly young: UN data shows that in the late 1980s typically about 20% of long-term migrants were under 15; between 20 and 30% were young adults of student age, 15–24; around 40% were young workers in the 25–44 group, and less than 20%, usually less than 15%, were over 45.

For the future, the migratory pressures on the developed world are unlikely to ease. The European Community provides a case: the countries to the south and east, in the Balkans and on the Mediterranean fringes, demonstrate relatively high fertility rates and high proportions of under-15s. What could be called the economic gradient is as steep as the demographic slope, and a good deal less stable; the Mediterranean could well represent the European equivalent of the Rio Grande. The pressures are already coming from this direction: in 1992, 64% of all those claiming political refugee status in the EC came from Europe, mostly from Rumania, Bulgaria and the former Yugoslavia; total asylum seekers from eastern Europe almost doubled between 1991 and 1992. Germany bore the brunt, but will no longer do so now that it has changed its own rules. And this represented a general trend: at their Copenhagen meeting in June 1992, EC Ministers agreed to a much stronger control of the common border, the so-called Fortress Europe policy, which involves the return of immigrants and agreements with neighbouring countries to take them back (Carvel 1993). The International Labour Office estimates that in 1993 there are already 2.5 million illegal immigrants in Europe, and perhaps double that (Marshall 1993).

The immigrants, whether legal or illegal, overwhelmingly concentrate in a few major urban areas. That is clear in the US, where four out of six main urban destinations of immigrants (in 1990) are at the southern periphery of the country, three are within 500 km (300 miles) of the Mexican border. These six metropolitan areas were among those with the highest concentrations of Hispanic and/or Asian population; ports of entry become ports of permanent call. Similarly, in 1991 Britain's black and Asian population were overwhelmingly concentrated in London and (in the case of the Asian population only) a few counties of the Midlands and North, which contain industrial cities (Leicester, Bradford) in which the immigrants have found jobs. Overall, in 1991, more than 25% of the Inner London population consisted of racial minorities, by far the highest relative concentration in Britain. Paris, Amsterdam and Berlin probably display similar features.

This reflects not only social factors—the presence, in these cities, of existing concentrations of compatriots with linguistic and cultural ties, plus associated institutions and businesses—but economics: in these larger cities are found the major concentrations of lower-paid entry level service jobs which tend to have been rejected by the native-born population but which are attractive to newcomers. And, given the facts of demography in the host community—most notably, the virtual certainty of an increase in the dependency ratio as the proportion of old people rises—there will inevitably again be pressures, perhaps not immediately but early in the 21st century, again to admit the immigrants to fill shortages of labour in the personal service industries.

An urban underclass?

The question is bound to arise, because it has been so insistently put in the recent literature: do these selective migrations threaten to develop a deprived urban underclass, characterised by persistent poverty, in the cores of major agglomerations of the western world? It has been strongly suggested that such a phenomenon can be observed in some American cities (Wilson 1987). Other American observers have questioned this, suggesting that there is not a single underclass, but many different deprived populations depending on the definition that is used (Jencks & Peterson 1991).

In any case, the American phenomenon arises from circumstances that probably do not repeat themselves in European and certainly not in Japanese cities: the migration, during and after World War II, of large numbers of relatively very poor and under-educated Afro-Americans from the American south, displaced from the land by technological change and attracted by the expansion of manufacturing jobs in the north at a time when traditional immigrant flows had dried up, followed by the ghettoisation of the migrants (a product of poverty and racial prejudice) and then by collapse of the urban manufacturing base in the 1970s; all leading to large concentrations of very deprived males without job prospects, in turn precipitating the flight of middle-class white-collar black Americans to the suburbs, thus leading to further ghettoisation. The result of all these forces has been the development of extraordinary concentrations of poor people, who then find that society discriminates against them because of their home origins: the phenomenon of so-called Zip Code discrimination.

No one has yet suggested that there is any parallel elsewhere, at least to the problem on the scale that it is manifested in Chicago or Detroit. European cities do have concentrations of lower-income people and of high unemployment in social housing areas in some of their cities, though these do not demonstrate any clear spatial pattern: in Britain they may be in inner area or outer area locations, in France generally on the urban periphery. But these concentrations are by no means as large, or as ethnically homogeneous, as their American equivalents; nor does it appear that they are so geographically removed from job markets. The phenomena are certainly different in degree, and may well be different in kind. Nevertheless, there is some common element: a degree of job mismatch, whereby part of the urban work force is under-educated and under-trained for the new urban informational economy, so that high urban unemployment levels para-doxically exist side-by-side with job vacancies (during upturns in the trade cycle) and with large-scale commuter movements from the suburbs to centre-city jobs.

Environment

There is a relatively new set of considerations that have loomed larger during the last decade, culminating in the 1992 Rio conference: the concern for environmental sustainability at the national and the international levels. These concerns are expressing themselves in policy initiatives at both national and more local levels, which promise to have quite profound impacts on the functioning and long-term prospects of cities.

At the national level, there is increased willingness on the part of governments to impose taxes on consumption of non-renewable and/or polluting fuels, in the form of higher energy taxes or carbon taxes. There is also an increased interest in the idea of charging for the use of road space, both on inter-urban highways (proposed by both Britain and Germany in early 1993) and in cities (Singapore and Oslo with operational schemes; programmed for Stockholm; under consideration for London and the Western Netherlands). The position is complicated by the fact that in some cases (Britain, Norway) charges are also seen as a way of financing road construction and as a way of part-privatising the road system; but, once the principle is admitted it can readily be extended to include environmental considerations. Additionally, the US has pioneered the development of fuel consumption norms (the so-called CAFE standards).

Locally, apart from road pricing initiatives, the US has also seen the development of local air quality norms, most notably in Los Angeles by the South Coast Air Quality Control District. These norms will make alternative energy sources for the private car mandatory from 1997 onwards; this is expected to give a huge fillip to the mass production of electric cars, with both domestic and overseas producers actively engaged in the development of commercial models. Another notable local initiative in the US has been the development of Transportation Systems Management and Transportation Demand Management plans, the thrust of which is to produce incentives and disincentives to discourage the single-occupant vehicle and encourage the use of both public transport and shared automobiles (car-sharing and van-sharing). The means are very varied: they include:

- high-occupancy vehicle lanes on both surface streets and highways (notably in Houston, but widely in other cities),
- preferential peak-hour access to highways (Los Angeles, Silicon Valley, San Francisco Bay Bridge),
- reservation of highways at certain hours to high-occupancy vehicles (Interstate 66 in Washington, DC),
- preferential parking places for shared vehicles,
- obligatory management plans for employers, and
- incentives to employers to provide shared vehicles.

An associated approach, though it tends to have been introduced as part of general policies to promote public transport without specific environmental objectives in mind, has been the development of public transport innovations, specifically to deal with the problem of dispersed origins and destinations in suburban living and working areas. Some of the more notable developments here have been:

- segregated busways, both guided (Essen, Germany; Adelaide, South Australia; Leeds, England) and unguided (Ottawa, Canada; Nagoya, Japan; île-de-France; Curitiba, Brazil);
- use of para-transit (minibus or van), widely used in British cities since bus deregulation;
- dual-mode vehicles, enabling buses to be run in city-centre tunnels (Essen; Seattle; planned for Ottawa); and
- 'hub-and-spoke' systems of bus scheduling (Edmonton, Canada; Houston).

It should be said that most of these examples are still based primarily on traditional radial route patterns, but some (Edmonton, Ottawa) break away from that mould. The most ambitious attempt to grapple with the problem so far is in île-de-France, where plans include an inner orbital public transport system (*ORBITALE*), in course of implementation, comprising segments of automated light rail, traditional light rail (tramway) and unguided busway, and an outer orbitale heavy rail ring (*LUTECE*) connecting the region's five new towns. Likewise, the Rhine-Main region of Germany plans a magnetically levitated, automated route through the suburbs, with frequent connections to the main radial lines. The development of automated technologies, which save labour costs and thus offer the potential of frequent service by small vehicles on relatively lightly trafficked routes, is germane here (Hall *et al.* 1993).

However, there is a limit to this process; such technologies are likely to involve high front-end capital costs in both track and vehicles (though the development of new electronic control technologies could bring a radical change in this regard). So there will continue to be a premium in developing land-use policies which concentrate activities, as far as possible, in relatively high-density zones.

However, it needs stressing that experts have had a great deal of difficulty in agreeing on a precise, operational definition of a sustainable urban environment. Though all would agree that it would consist (among other things) in a reduction of travel by motorised modes, especially those involving single occupancy, there is no agreement at all on the urban structure that would produce this.

- Newman and Kenworthy's well-known work (1989a; 1989b; 1992) has been criticised by Gordon and Richardson (Gordon *et al.* 1988; Gordon & Richardson 1989b) on the basis that dispersal of homes and jobs leads to shorter journeys;
- Breheny (1992) has criticised the European Commission's endorsement of high urban densities as lacking in any empirical foundation;
- Owens (1984; 1986; 1990; 1992a; 1992b) has suggested that compact medium-sized settlements, with dispersed but clustered employment, may represent an optimum form.

The major problem is to handle the complex adjustments that occur in the transition from a traditional single-core, radially focused metropolis with a high dependence on public transport to a polycentric structure based (at least so far) on automobile dependence.

Environmental politics

This is another, quite separate aspect of the environment. During the boom years of the 1980s, strong growth control politics influenced the development of the more dynamic metropolitan areas of the western world. Alike in south-east England and the San Francisco Bay Area, there was intense opposition to the idea of further ex-urban growth, often coming from the very people who had moved into the area a few years before. Nimbyism became yet another aspect of single-issue community politics, causing local elected bodies to resist the idea of further growth.

In south-east England this brought the demise of several proposals for new community development; in both Britain and California, it boosted the tendency to strike developer agreements (or planning agreements) to provide infrastructure or other mitigation, to such a degree that they became almost universal in southern England in the early 1990s. Generally, it had the effect of rolling the development pressures further out, beyond an invisible 'Nimby frontier' which by the early 1990s tended to be 60–70 miles (100–120 km) from the urban cores in the case of the Bay Area, up to 100 miles (160 km) distant in the case of London. Finally, it brought a great deal of pressure to redevelop inner urban areas that, as already noted, were becoming vacant for a variety of reasons (London Docklands; Rotterdam Oude Haven; Mission Bay in San Francisco).

During the recession-wracked early 1990s, an evident contradiction has begun to develop in western urban policies: between demands for wealth creation policies to regenerate economies, and demands for ever more stringent environmental protection policies, often embedded in regulatory codes with legal force. These may come to a head, for instance, over issues of inner-city regeneration, where the land may be classed as degraded by toxic waste; or in transportation policies, where new highways or rail systems pass through protected or precious habitats; or development generally, where plans for new shopping centres or suburban housing may be questioned on the ground that they are not sustainable environmentally. Very often, the problem with such issues is that there does not yet exist any precise way of evaluating environmental benefits and disbenefits, such that they may be compared with direct economic gains. This is one major unresolved question for urban planning policies in the 1990s.

Privatisation and deregulation

This last problem is further complicated by another development: during the 1980s and 1990s, in varying degrees, virtually every advanced industrial country has moved toward privatisation or deregulation of a range of urban services ranging from education to heath, from public transport to refuse collection. These services have been taken out of the control of local government (or public authorities) and given to a variety of direct providers, often private for-profit companies, sometimes competing non-profit organisations (e.g. individual schools, hospital trusts). Almost invariably it has been necessary to set up centralised regulatory or standards regimes in order to monitor the performance of these bodies and to guarantee that they fulfil certain obligatory standards. Thus deregulated transport services may be franchised or even thrown completely open to free competition, but with payments to provide certain socially necessary or desirable services; deregulated schools are subjected to national standardised tests; private prisons compete with public ones in tendering for prison services, against mandatory service standards.

This means that private profit again becomes a criterion in providing public services, as it was a century ago. The question is how this can be reconciled with the need to maintain these services at a level that will ensure the continued efficient and effective running of the city's economy. This is particularly true where these services compete with other, already privatised alternatives (public transport with automobiles; public hospitals with private ones; public schools with private ones).

An interesting result, observed in Great Britain: the thrust to privatise and deregulate urban transport, first through deregulation of bus services, then through privatisation of rail services, has led to a demand for road pricing in order to provide a level playing field. The prospect thus exists of a textbook model economic system, in which services are provided competitively at close to their marginal social costs.

There is, however, another possible outcome: as in previous eras of private infrastructure investment (American 19th century transcontinental railways; turn of the century urban rail systems in the US, Britain and Japan), such investments come to be evaluated only partly on their revenue-generating capacity, and much more on their capacity to generate profitable land development. Coupled with the other considerations, discussed earlier, this might mean for instance that new types of rail service (regional long distance express services, as in Paris, Frankfurt, Munich, Berlin and planned for London; or high speed rail as in Japan and France) become platforms for major urban developments. Particularly significant, in this regard, are the points at the edges of major metropolitan areas, where these two types of service interconnect (Shin–Yokohama, Omiya, Roissy, Eurodisney, Massy, Ebbsfleet).

Competition among places

Partly because the traditional force of locational factors is eroded, partly because the *Zeitgeist* of the 1980s and 1990s is so heavily pro-competitive, cities tend to market themselves rather like competing consumer goods. And, according to traditional rules of marketing, this leads to the imperative to establish a market niche: city administrations find themselves impelled to establish some unique quality for their city, some magic ingredient that no other city can precisely match. Since the sources of the new economic growth are so various and finally perhaps so fickle, the possibilities are endless. But one central element is quality of life. It is no accident that, as never before, rankings of cities dominate the media—just as in education, schools and universities compete for rankings on the basis of scholarship or teaching. Cities and smaller towns, that traditionally offered an attractive way of life and an attractive physical image, find little difficulty in this; older industrial cities, or more accurately ex-industrial cities, find it harder and may have to expend large sums to transform the picture that the public holds.

One fortunate accident is that manufacturing and goods handling themselves become exotic activities from a near-forgotten past, easy to sanitise and market to curious tourists; thus old mills can be transformed into museums (as in Lowell, Massachusetts) or even retail outlets (as in Saltaire, England, another model industrial community of the early Industrial Revolution). In England, classic industrial cities like Manchester and Bradford deliberately rehabilitate their old industrial buildings; Manchester even reopens its 19th century sewers as a museum; nearby, the authorities transform the Wigan Pier, a music hall joke immortalised by George Orwell, into a tourist complex; 100 miles to the south in Birmingham the 18th century canal system, which provided the arteries of the first Industrial Revolution, is transformed into leisure cruiseways as a backdrop to the new internal Convention Centre. These English examples drew on the consumer boom of the late 1980s, and may not readily be replicated in the recessionary 1990s; but the process is likely to continue, subject only to the vagaries of boom and bust.

The urban system

To understand how these changes impinge on the urban system, we need a different taxonomy from the one we have inherited from Christaller and Lösch. To be sure, there is nothing wrong with their basic insights: goods and services still have their distinct ranges, but these are now increasingly global; urban hierarchies may array themselves in transport-rich, city-rich sectors around major nodes, but these nodes may be global, or at least extra-national.

Specifically, we can distinguish at least two significant levels of a global hierarchy, though with considerable doubt as to where many cities belong, as well as a number of other categories that have significant relations to these two.

Global cities

These are cities that transact a substantial part of their business at a global scale, both with other global cities and with lower-order cities. This global business consists mainly in performance of specialised services, such as financial services, media services, educational and health services, and tourism (including business tourism). They are invariably seats of central banks, major clusters of clearing banks (including substantial representation of overseas banks), stock exchanges, insurance companies, headquarters of major corporations (including transnational), television stations, newspapers and magazines, publishers, major universities and hospitals, and leading international airports. Their relative importance is measured in data about flows of information and people: overseas banking transactions, airline passengers, audiences.

However, following the Christaller scheme, they also perform lower-order functions for more restricted areas. The most important of these is the national level, since—everywhere outside continental-scale countries like the US, Russia, China, Mexico and Brazil—the existence of nation states still constrains the globalisation of services. Western and central Europe provide the most spectacular case, with a score of central banks, national TV networks and newspapers, and major international airports. Even here, however, it is clear that a hierarchy exists: London is indisputably a global city; Paris, Brussels, Amsterdam and Milano aspire to that title and in some respects may qualify; while other capitals clearly restrict themselves to a national function.

These cities exported certain functions during the 1970s and 1980s, either to other locations in their own national space (including their own metropolitan peripheries) or in some cases to other countries. They have therefore demonstrated the paradoxical situation of sometimes large employment losses in traditional sectors (manufacturing, goods-handling, routine services) and simultaneously large gains in others (particularly financial services and specialised business services during the 1980s boom; though large losses are now occurring in these sectors in some global cities) (Noyelle & Peace 1991). Since the lost jobs tended to be of the skilled and junior clerical middle-income type, while the gains tended to be either at the upper-income and lower-income extremities (financial services; casual personal service jobs), it has been suggested that the result may have been an increasing polarisation of the labour force (Mollenkopf & Castells 1991).

National and regional cities

At the next level of the hierarchy it is therefore necessary to consider a complex phenomenon, since the European national capital cities immediately below the global level (Bonn, Amsterdam, Madrid, Rome, Copenhagen, Stockholm, Oslo, Vienna) can only be compared with major regional cities in the US and Japan (and indeed in some parts of Europe, such as Germany). In particular, nations with a federal structure or a history of late union, or both (Spain, Italy, Germany, the US, Mexico, Brazil) tend to have a small number of very important provincial cities which not only challenge their national capitals, splitting functions with them and effectively serving as regional capitals for wide areas, sometimes with great regional distinctiveness, but also in some cases perform global functions (Milano, Frankfurt, Los Angeles). Even in more centralised countries, there are certain provincial capitals so large and important that they belong in this category (Birmingham, Manchester; Lyon, Marseille).

This group of medium-sized city regions (typical population range: 500 000–2.5 million), serving as administrative and higher-level service centres either for a prosperous rural region within a larger national unit, or as the capital city of a small semi-industrialised nation state, has therefore shown great dynamism, rising per capita incomes and considerable economic prosperity during the late 20th century. This is so even though many have exported both manufacturing industry and population, a situation that Cheshire & Hay (1989) refer to as 'healthy decline'. Europe contains many examples in both categories, including all capitals below the global rank (Amsterdam, Brussels, Luxembourg, Copenhagen, Oslo, Stockholm, Dublin), some major 'sub-global' provincial capitals (Barcelona, Milano, Zürich, Frankfurt), and a much larger series of regional capitals (Bristol, Norwich; Bordeaux, Toulouse; Hanover, Stuttgart, Munich; Bologna, Napoli; Bilbao, Valencia, Sevilla). The US contains only regional examples (Atlanta, Dallas-Fort Worth, Minneapolis-St Paul, Denver, Seattle, Portland), as does Japan (Osaka, Nagoya, Sendai, Kumamoto). Similar regional cities exist in Latin America, such as Monterey and Guadalajara, Rio de Janeiro and Sao Paulo.

The top end of this group, especially in Europe and some eastern Asian city states (Hong Kong, Singapore), are potential global cities, even though in Europe some of them are not national capitals; they can perhaps best be described as sub-global, an appellation that applies also to major American regional centres like Boston, Atlanta, San Francisco and Los Angeles, and to Sydney and perhaps Melbourne. These places have proved outstandingly successful because they have enhanced their roles as high-level service centres for their spheres of influence, attracting the most dynamic information–service industries both in the public and the private sectors: local government, higher education, health care; banking, insurance, accountancy, legal services; airline operations, business tourism (Stanback 1985; Cheshire & Hay 1989).

There are two related problems concerning this group of cities. The first involves remote regional cities: places at the periphery of their national (or continental) space-economies, especially those dominated by older declining industries (Newcastle-upon-Tyne, Belfast, Glasgow; Clermont-Ferrand, Saarbrücken, Cadiz; Buffalo, Duluth; Sapporo, Nagaoka). They face special problems:

- their economies are based on industrial resource concentrations which have been exhausted or superseded;
- their labour force may be highly organised, resistant to change, and dominated by large industrial firms;
- they may lack a tradition of small firm entrepreneurship;
- they are too distant to benefit from the 'deconcentration wave' emanating from the global and sub-global cities (see below);
- they may be challenged by nearby regional cities for remaining regional service functions;
- they may suffer from the 'peripheralisation of the periphery': as better transport and communications effectively shrink the core urbanised regions on each continent, the peripheries may come to appear relatively more remote (Cheshire & Hay 1989).

Cheshire and Gordon (pp.108–126) stress that major cities near the core of the European Community showed improving performance during the 1980s, even demonstrating an improvement against lower-order cities as compared with the previous decade.

This poses a clear problem for the remote cities. They may try to counter these disadvantages by exploiting their accumulated labour skills and educational investments, as Glasgow did in attracting inward electronics investment after World War II or remote Japanese cities did through the Technopolis programme in the 1980s. Or they may seek totally new directions, as Glasgow did in establishing itself as a European cultural city.

Sometimes remoteness can actually encourage regional service economies: in Glasgow, distance from London and a separate Scottish banking system were significant factors in the city's adjustment from a manufacturing to a service-oriented city (Lever & Moore 1986).

The second problem concerns the major cities of eastern Europe. Their remoteness has been artificial, a product of the Iron Curtain and the development of the separate COMECON system from the 1950s; though these cities do remain distant from the north-western heartland of the European Community, they are no more so geographically than Italy or Spain.

As already suggested, this region seems certain to enjoy a major economic revival during the next decade, and the parallel with western Europe in the 1950s and 1960s suggests that much of this development will concentrate in the major cities, with big shifts of population from the countryside.

The service sector has been particularly under-represented in these cities, above all the informational sector; they have very limited retailing and office and hotel floor space compared with equivalent west European cities, and their airports are similarly under developed. We should therefore expect a particularly rapid development here, with big expansions of the central business districts, peripheral business parks, and airport-related business. Where older quarters were not destroyed in World War II they are likely to be reconstructed to become prestige residential areas; there is likely to be considerable low density individual home construction at the periphery, particularly since planning systems in the peripheral communities are unlikely to be professionally equipped to withstand the development pressures. Therefore the prospect is one of considerable low density sprawl and resulting transport problems, since car ownership levels in the more advanced of these countries are already relatively high (between half and two-thirds of their west European equivalents) and rising rapidly.

In addition, many of the peripheral high density high-rise public housing areas of the 1960s and 1970s are likely to become quite problematic; they will require expensive maintenance and many of them may become relatively undesirable areas to live, and will be occupied by low income new arrivals from the countryside or from abroad. Therefore, these areas are likely to replicate the history of some of their western equivalents (Kirkby, Sarcelles, Bijlermeer).

Patterns of internal shift in global and sub-global cities

The boom of the 1980s, coupled with the increasing tendencies to globalisation and tartarisation of economies, brought intense city-centre office development to all cities that could claim global status, and above all to the indisputable market leaders: London, New York and Tokyo. But many of the sub-global cities have grown similarly and have similar problems of congestion and deconcentration, with the largest reaching 3–5 million population (and Osaka even bigger) combined with downtown employment concentrations of 500 000 to 1 million. (This however depends in part on their age and hence their tendency to deconcentration. Some of the newest, in the American west and in Australasia, are characterised by transport planner Michael Thomson as highly deconcentrated, 'full motorisation' cities.) The difference usually consists in the scale over which development impacts extend: in the global cities they may extend to 100 miles (160 km) from the city centre, but here the effective radius would seldom be more than 30–40 miles (50–65 km) and generally a good deal less. A few of them, in the highly decentralised American west (Los Angeles, San Francisco) however have development impacts which stretch fully as far as those of the global centres; and the same applies, of course, to the mega-cities of Latin America, Mexico City and São Paulo.

The early phase of urban deconcentration, which can be traced in many of the world's largest cities from the late 19th century, was predominantly of residence from central cities to their suburbs, dependent on public transport technologies and in particular on electric commuter railways and metro systems from the end of the 19th century on (Clark 1951; 1967). A

distinctly new phase, beginning after World War II, involved decentralisation outside the traditional limits of effective public transport, and
thus dependent on the use of the private car for most trips including the
journey to work. This pattern of decentralisation was observable in the US
from the 1950s onward (and indeed perhaps from the 1930s, though
interrupted by the war); in Europe it spread progressively from Britain and
the Benelux countries to involve all the countries of the European Community by the early 1980s (Hall & Hay 1980; Cheshire & Hay 1989).

However, at first it remained predominantly a residential movement, with
some relocation of residentiary services. Employment, in contrast, remained
concentrated in central complexes in the downtown metropolitan cores: the
City, Whitehall and West End in London, Downtown and Midtown
Manhattan in New York, Kasumigaseki and Marunouchi and Otemachi in
Tokyo. Within these were specialist sub-agglomerations: finance, government, and business services, attesting to the strength of traditional
agglomeration forces. But, beginning in the US during the post-war boom
of the late 1950s and 1960s, and then progressively gaining pace, a parallel
decentralisation of employment began to be observable.

Three kinds of activities showed themselves particularly prone to
relocation.

1 Manufacturing and associated warehousing was drawn to suburban
   locations because of access to large areas of land suitable for efficient
   single-floor flow production next to the new national motorway systems,
   under construction everywhere during the 1960s, and also because the
   skilled blue collar work force was also moving to the suburbs.
2 Research and development (R&D) and associated high-technology manufacturing was attracted to high-amenity areas near the rural fringe, such as
   Santa Clara and Orange Counties in California, parts of Connecticut and
   New Jersey around New York, Berkshire in England, the Munich region of
   Germany, and southern Kanagawa prefecture in Japan.
3 Large offices engaged in electronic processing of standardised information, such as insurance or credit card operations, tended to disperse to
   major suburban nodes with easy access, lower office rents and supplies
   of suitable clerical labour: Reading west of London, Stamford in
   Connecticut, or Omiya and Kawasaki near Tokyo.

One key question is where this process might end. Where central rents are
particularly high, it may extend widely:

- around London, back offices have now diffused very widely across south-
  east England and even outside that region altogether;
- British Telecom have dispersed their London information service to
  remote regions;
- in the US Omaha (Nebraska) has managed to create a specialised service
  niche, telemarketing, around top-quality telecommunications networks
  and a supply of the right kind of labour (Brooks 1991; Feder 1991), and

- Salt Lake City has attracted American Express's telephone-based traveller's cheque operation, on the somewhat extraordinary ground that its ex-Mormon missionaries are highly motivated and proficient in languages (Sellers 1990; Donnelly 1991; Johnson 1991).

These instances, admittedly anecdotal, may presage a future trend for such functions to disperse to relatively remote places with low rents, low wages and/or a high-quality work force; but up to now the process seems constrained by the need for quick face-to-face communication between main office and back office. There is similar anecdotal evidence that some cities may be losing their attraction even for top-level headquarters functions: in California, some major banks and utility companies (Chevron; Pacific Bell; Bank of America) are decentralising from downtown San Francisco to suburban locations in the wider San Francisco Bay Area.

But these are far from conclusive examples. The evidence so far is that this pattern of deconcentration naturally tends to extend only into the surrounding metropolitan sphere of influence—effectively, to the outer limits of the commuter field. Here, it reconcentrates around existing medium-sized urban centres (population range, 50–200 000), typically county market towns serving as administrative and service centres for surrounding regions (counties, départements, Kreise, shi) within a radius of about 15–20 miles (25–35 km); these are distributed in a more or less regular Christaller like fashion (modified, perhaps, by the Lösch transport-rich principle) within the metropolitan sphere of influence. Perhaps the most spectacular example is London, where between 20 and 30 such centres can be identified in the so-called Outer Metropolitan Area, between 15 and 50 miles (20–80 km) from Central London, especially along the major road and rail corridors, about a dozen in number, which radiate from the city. The region around Tokyo presents a similar picture, while in the New York the development is more highly clustered along the historic routes which parallel the coast.

Interesting exceptions to this rule are Paris, where the corresponding pattern of outward diffusion is historically much weaker, and Berlin, where an earlier diffusion trend was completely halted during 40 years of communist control. A fascinating question for students of urbanism is the likely pattern of development, during the 1990s, in the artificially sterilised strip on either side of the ring motorway—itself an historic anomaly from the 1930s, built (for reasons of military aggression and national prestige) decades before its equivalents in the world's other great cities.

Apart from these exceptions, the medium-sized cities have thus proved very dynamic economically wherever they have been favourably located to receive the outward flows of people and jobs from neighbouring mega-city regions (south-east England, the lower Rhine, New Jersey, the I-680 corridor in California) (Stanback 1985; Cheshire & Hay 1989). Even in declining older industrial regions, many have been successful in preserving and enhancing their reputations by removing the scars of industrialisation.

Durham and Preston (England), Lille (Belgium), Essen (Germany) and Lowell (Massachusetts) provide examples.

A particularly interesting sub-group has prospered through specialisation on service functions that have shown exceptionally rapid growth: local government, educational services, health services, defence contracting (Noyelle & Stanback 1984). Examples include Reading, Oxford and Cambridge; Rennes and Aix-en-Provence; Marburg and Heidelberg; Lund and Uppsala; Padua and Bologna; Columbus, Austin, Salt Lake City and Sacramento; Kyoto and Nara. The success of these places has little to do either with metropolitan location; but if well-located in that regard also, they may exhibit exceptional dynamism, to the degree indeed that they may exhibit symptoms of overgrowth (Reading, Heidelberg, Kyoto).

However, downtown cores remained highly attractive to a wide range of informational services which required central agglomeration economies, as witnessed by the office construction boom that took place in all major metropolitan cores worldwide during the 1980s. Daniels (Chapter 12) points out that in London this exceeded any previous such construction boom, and emphasises that—despite the obvious importance of modern offices well-equipped for information technology—traditional location considerations, including prestige, remained important. Similarly, Stanback (Chapter 11) reviews his own and other work to show that in American cities, despite increasing competition from suburban sub-centres, the downtown cores retained their attraction for key high-level services. As a result, by the 1980s metropolitan dynamics entered a new phase: the pressures on downtown space for traditionally agglomerative activities, both in the global and the largest sub-global cities, were so intense that they led developers and planners to explore new, non-central locations for major activity concentrations within the metropolitan envelope. At least four types can be distinguished:

1 new commercial sub-centres, sometimes constituting downtown extensions, on reclaimed or redeveloped dock or freight-handling areas (London, Amsterdam, New York, Toronto, Hong Kong, Tokyo, Osaka-Kobe);

2 metropolitan sub-centres dependent on major inner- or middle-suburb public transport interchanges, sometimes serving as 'new downtowns' (Croydon, La Défense, Lyon-Part Dieu, Lille-Euralille, Newark Waterfront, Rosslyn, Ballston, Shinjuku, Shin-Osaka, Paulista);

3 'edge cities', speculatively developed around highway access and lacking public transport access (the New Jersey Zip Strip, Tysons Corner, Aurora between Denver and Fort Worth, Mesa outside Phoenix, San Ramon-Dublin-Pleasanton in the San Francisco Bay Area) (Dillon *et al.* 1989; Garreau 1991);

4 a special case, 'edge cities' developed around public transport interchanges: generally in redeveloped older cores of formerly free standing cities (the county town case, considered above), in course of incorpora-

tion into the wider metropolitan sphere of influence (Reading, Berkshire; Stamford, Connecticut; Omiya, Saitama Prefecture; Kawasaki, Kanagawa Prefecture); occasionally on new sites, either as planned new community centres (Cergy-Pontoise, St Quentin-en-Yvelines, Massy, Lille Val d'Ascq, Ebbsfleet) or as spontaneous developments (Walnut Creek, Shin-Yokohama). Even 'full motorisation' cities have begun to try to build urban rail systems and to densify areas or corridors around the new stations (Los Angeles).

The forms are very varied, and not easy to bring into a single schema because of the great variation in form from one metropolitan area to another; a bizarre case is the Paulista in Sao Paulo, a downtown 'edge city'. But overall, very large metropolitan areas have become steadily more polycentric in form. Even where this has been most carefully planned and coordinated with new public transport provision (as in the classic cases, the Stockholm region and the Region île-de-France), this has been associated with a steady shift away from use of public transport and toward reliance on the private automobile: in the Paris region as in the San Francisco Bay Area, suburb-to-suburb trips are made overwhelmingly by car, though Paris does manage to handle a respectable minority by RER and métro. Thus Cervero's 'suburban gridlock' tends to present a problem just as much for Europe's metropolitan centres as for their American equivalents.

The process of deconcentration has also been fundamentally affected by the fact that most of these metropolitan areas had made some serious attempt, by the early 1990s, to limit and guide their growth into planned channels, through a variety of devices such as green belts, green wedges, buffer zones, green hearts (in the special case of the Dutch Randstad), regional park systems and preferential development axes. These devices could profoundly affect the outward appearance of the metropolitan region: compare London (where they were used successfully) and Tokyo (where they were not). Given the basic dynamism of these regions, however, and coupled with general growth control and growth management policies even outside the defined green zones, the general effect of these controls was to channel growth pressures to even more distant locations, often further exacerbating the problems of automobile suburb-to-suburb commuting, as illustrated by the pressures on London's M25 orbital or California's I-580 connecting the new growth areas in the Central Valley with the Bay Area employment cores.

One critical question, in this complex process, was whether the decentralisation process would lead to some kind of re-equilibration, whereby employment would progressively move outward to the distant zones from which the commuters originated and, by reconcentrating there, lead to some relief. There was fairly clear evidence that this was happening in the outer London region in the 1980s, as the smaller centres in the outer ring became steadily more important not merely as local service centres but also for some manufacturing (high-technology industry along the M4 corridor and more generally in the entire 'Western Crescent' around London) and decentralised offices (of which Reading was only the most spectacular example) (Buck *et al.* 1986). It was not yet evident in the San Francisco Bay Area; there, it appeared that despite considerable 'edge city' deconcentration of employment, the frontier of new residential development was steadily moving even further out. Here however difficult terrain, which created natural barriers, may have exacerbated the phenomenon.

A critical question, addressed in a number of chapters for this volume, is the impact of these changes on journey patterns and thus on the achievement of sustainable urban development policies. The framework has essentially been set by the debate between Newman & Kenworthy (1989a; 1989b; 1992) and Gordon and Richardson (Gordon *et al.* 1988, Gordon & Richardson 1989b). Newman and Kenworthy argue for compact, high-density cities, Gordon and Richardson argue that suburban deconcentration of employment is essentially producing a form of market-based sustainability. Breheny (Chapter 21) comprehensively reviews the evidence, including a comprehensive new study of the British evidence by ECOTEC (Great Britain 1992). He concludes that the balance of evidence is in favour of compact, higher density forms with limited deconcentration, but that current trends in Britain run against these objectives. Similarly, Spence and Frost (Chapter 19) conclude for Britain that there has been an increase in longer-distance commuting and thus in median trip length. Cervero (Chapter 17) produces very similar conclusions in his review of trends during the 1980s in the US; he concludes that suburbanisation of homes and jobs leads to longer commute times and above all to almost complete dependence on single occupancy automobile commuting. The only positive note comes from Brotchie *et al.* (Chapter 20) who conclude that the marked suburbanisation of employment in the Melbourne metropolitan area has led to many more suburb-to-suburb trips and thus to shorter journeys; average journey length does increase marginally, because new residents are added at the periphery, but the rate of increase is lower than it would be if jobs remained concentrated in the centre. There are differences in interpretation here, but there is also some considerable agreement in these chapters that though trip lengths are increasing somewhat, average travel time is more stable, because people switch to faster routes or faster modes. Brotchie et al. conclude that 'There may be more than one pathway to future sustainability of urban development' (p.398): dispersal of employment close to homes, reducing average trip length and time, might be more effective than higher densities and rail-based transport. This suggests that further work will be needed to establish rigorous definitions of what we mean, in any particular spatial context, by sustainable urban development.

Growth corridors

A particular form of deconcentration, noticed in some of the largest metropolitan areas during the 1980s, was the growth axis or corridor: sometimes planned, sometimes spontaneous. Planned axial developments have been proposed recurrently throughout the 20th century, from Arturo Soria's *La Ciudad Lineal* onward, but few came to anything like complete realisation. Major exceptions were the 1948 Copenhagen Finger Plan, the parallel preferential axes in the 1965 *Schéma Directeur* for the île-de-France region, and the development corridors in the 1966 Stockholm regional plan—one of which, the E4 Corridor from Stockholm to Arlanda Airport, has become Sweden's preferred location for high-tech industry. This tendency is also observable in the M4 Corridor from Heathrow to Swindon west of London, in the so-called Dulles Corridor from the Pentagon and Washington National Airport to the Dulles Airport, and in Aerospace Alley from Los Angeles International Airport south-eastwards through Los Angeles and Orange Counties, all of which are also airport-related; unlike the Arlanda corridor, none of the others could be said to have been consciously planned, and indeed the British example was developed in defiance of national policies for regional development. All, too, seem to mimic the axial development of the original model, California's Silicon Valley, south and east from Palo Alto to San José, in the 1950s and 1960s; they correspond to the need for networking in a post-Fordist environment.

More broadly, there appears to be a clear tendency in some highly urbanised regions for deconcentration to follow linear corridors connecting major cities, which may be several hundred miles (or kilometres) long: thus the original 'Megalopolis' of the north eastern seaboard of the US, the Tokaido corridor in Japan, and the so-called *Dorsale* ('Blue Banana') in Europe from Birmingham via London, Brussels, Cologne, Frankfurt, Stuttgart and Zürich to Milano (Hall *et al.* 1973; Brunet *et al.* 1989). These corridors, which are extreme cases of the 'transport rich' sectors in the Lösch formulation, seem to have been propelled in the 1950s and 1960s by the development of key motorway links and then, from the 1960s in Japan and the 1970s in Europe, by the opening of high-speed train links. As national motorway plans have progressively extended high-quality accessibility from the original preferential corridors to the entire national system of cities (as in France, for instance, during the 1980s), the high speed trains have ironically had the opposite effect, re-emphasising the advantages of the old corridors.

There is so far only one attempt to create a planned development corridor along a new high speed rail line: the East Thames Corridor along the planned line from London to the Channel Tunnel, announced by the British government in March 1993. This is interesting because, like the earlier London Docklands development with which it will be spatially linked, it represents a conscious attempt to reverse a westward development trend that has long been observable in other major European cities (compare Paris, Berlin). The argument is that, just as airports have represented a major urban development trigger since World War II, so now will high speed train stations; and that, in the particular case of London, the train will logically take London's development in a new direction, eastwards toward the European mainland. The new line will also support long distance commuter services, mainly carrying passengers to work in central London but with the possibility of interception by major activity concentrations (which could be internationally oriented) at key points; between these points, stopping trains will provide commuter service to a series of intermediate planned communities. The new line will thus serve as one principal mechanism to regenerate an under-developed area which has suffered massively from polluting industries, waste disposal and general environmental degradation. Already, one major landowner has announced a major internationally oriented business park development at Ebbsfleet, the proposed edge-city station 25 miles (40 km) from London close to the M25 orbital highway.

There are obvious limits in trying to imitate this process: high speed lines must be efficiently routed to serve their primary purpose of connecting major cities (and thus to compete effectively with air transportation); diversions for strictly developmental purposes will thus have to be weighed against any resulting revenue losses.[2] But, insofar as cities do have unexploited development opportunities, East Thames may be a model for the future.

NOTES

[1] For convenience, throughout I shall use the term 'city' to mean a functional city region, in the familiar sense of the Metropolitan Statistical Area used in American analysis and similarly based work in other countries; this embraces a central city and its surrounding sphere of influence defined on the basis of daily urban flows of people, goods or information, commonly called the 'suburban ring' but not necessarily coterminous with the physical agglomeration.

[2] In the East Thames case, rather remarkably, there was no trade-off; the route is as direct as any alternative, and much cheaper than any costed alternative.

PART *1* Global Expansion of Markets

G LOBALISATION, the subject matter of this first main section, is a familiar theme in the recent urban literature. A number of forces are operating simultaneously to create a worldwide market in goods and services, in which competing suppliers find crumbling barriers but in which, conversely, there is virtually no protection from worldwide competition. The most important of these forces include:

- *technological diffusion*, which allows a growing club of middle- and even low-income (thereby, low-cost) countries to produce an increasing range of sophisticated products that were formerly monopolised by a few suppliers;
- *decreasing cost and increasing speed of transportation*, especially by air freight, allowing world markets to be reached within a few hours;
- *increasing diffusion of information*, which allows the development of complex networks of production across international frontiers (the Mexican *maquillodora* plants, Taiwanese and Hong Kong investment in the People's Republic of China, Western European branch plants in Eastern Europe); and
- *declining barriers to trade* through GATT agreements and the creation of regional trade blocs such as the European Community and the North American Free Trade area.

The most striking result of these processes is the development, sometimes with amazing rapidity, of new industrial spaces in what until recently were regarded as low-income agricultural areas, such as the Special Economic Zones of China. Such regions manufacture not only basic goods with high labour content, but also high-technology items like silicon chips and high-touch goods such as fashion shoes, which can be copied from the world's fashion centres and replicated a few days later. In consequence, a ferocious system of global competition begins to operate, in which the surviving advanced nations and regions find it increasingly difficult to compete save in a narrow range of very specialised items. But, hand in hand with this diffusion of manufacturing capacity across the world, another trend operates in the opposite direction: the increasing *information content* of production, coupled with a trend away from integration and mass production, and toward flexible networks of producers, means that command and control functions become ever more important, and these activities are apparently becoming concentrated in a decreasing number of cities and regions.

Among the most significant of these global informational functions are

- *financial services* (banking, insurance and finance, particularly the high end of these activities concerned with merchant and venture capital);
- specialised *producer services* (such as advertising, public relations, accountancy, legal services and management consultancy); associated with these,
- *the media* (publishing, TV and radio, multimedia);
- *research and development* functions of major corporations; and both
- *headquarters* and *back-office functions* of corporations both in the manufacturing and services sectors.

Several important research questions arise. What is the precise location of these activities as among the cities at the top of the global hierarchy? Do they concentrate in a very few global cities, as the work of Sassen (1991 and this volume Chapter 2) suggests? Or do these centres compete with a rather wider range of sub-global cities, embracing the bulk of the European capitals, the major regional cities of North America, the leading Latin American capitals, and the major cities of Eastern Asia and Australasia? Especially in Europe, with its long-established tradition of nation-states, what is the relationship of the global hierarchy to national hierarchies of cities? Is competition between cities affected by the quality of the information-exchange network, including international air travel, high speed trains (a new technological factor), and advanced telecommunications? What is the significance of international organisations such as UNO, UNESCO, OECD and the EC?

The chapters in this section begin to address some of these crucial research questions. To begin, Sassen reviews her own work on the global cities, through a comparison of New York and Tokyo. She postulates that global cities 'are a particular type of production site' for the production of specialised services, the production of financial innovations and the making of markets'. She shows that there is an increasing and powerful trend for cross-national investment in services, but also an increasing trend for geographical concentration in their supply. She poses the critical question of the effect of this concentration on the second level cities—a group that has until now received insufficient attention.

In the following chapter, Stimson takes up the same bundle of themes in an Australian context. Australia is rapidly becoming a service or informational economy, and both producer services and R&D are becoming concentrated in a few major metropolitan centres. There is increasing international investment in these centres, especially in real estate. Above all, 'Sydney stands out as the nation's international gateway and its only 'world city''. It may be in the process of becoming the core of a 'global city metropolis' stretching from Brisbane to Melbourne; these locations contrast with a 'rust belt' comprising other cities which have been badly hit by the decline of manufacturing.

O'Connor's chapter treats a specialised theme more or less ignored in the urban literature until now: the impact of shifts in airline service upon urban development worldwide, with a case study of Qantas. He shows that over the twenty years 1970–90 there have been remarkable shifts in the patterns of Qantas service worldwide, in part reflecting the technological move to bigger, longer range aircraft, in part the rise of dynamic metropolitan areas in Eastern Asia. Most importantly, the combined result is that services focus on fewer and fewer cities, and this both reflects the pattern of service sector concentration, and in turn reinforces it. For the future, more flexible mid-range aircraft might open up service to a wider range of cities, but market forces may well work in the opposite direction. High speed rail could also play a significant role in providing feeder service to major hub airports.

There is room for much more work on these themes. The most likely scenario seems to be one of even greater concentration in a limited number of first- and second-order global cities, aided and abetted both by technological change and by economies of scale and of scope. But there could be surprises, and the future appears far from predetermined.

2 Urban Impacts of Economic Globalisation

Saskia Sassen

THE specific forms assumed by globalisation over the last decade have created particular organisational requirements. The emergence of global markets for finance and specialised services, the growth of investment as a major type of international transaction, all have contributed to the expansion in command functions and in the demand for specialised services for firms. Transnational corporations and banks are major sites for international command functions and major consumers of specialised services. Yet much new global economic activity is not encompassed by the organisational form of the transnational corporation or bank. Nor is much of this activity encompassed by the power of such firms, a power often invoked to explain the fact of economic globalisation. Much of this activity involves, rather, questions of production and of place. The spatial and organisational forms assumed by globalisation and the actual work of running transnational operations have made cities one type of strategic place and producer services a strategic input.

The combination of geographic dispersal of economic activities and system integration which lies at the heart of the current economic era has contributed to new or expanded central functions and the complexity of transactions has raised the demand by firms for highly specialised services (Sassen 1991). Rather than becoming obsolete due to the dispersal made possible by information technologies, cities:

- concentrate command functions;
- are post-industrial production sites for the leading industries of this period, finance and specialised services; and
- are transnational marketplaces where firms and governments can buy financial instruments and specialised services.

This chapter examines how these processes of globalisation are actually constituted in cities. It does so by selecting a few key issues that have emerged from the specialised literature. The first section focuses on the intersection of the growth of services and globalisation. The second examines whether we are seeing the formation of a producer services complex. This includes an examination of the coordination and planning requirements of large transnational corporations, and a discussion of the locational patterns of major headquarters. The third very briefly discusses the impact of the international financial and real estate crisis beginning at the end of the 1980s. The fourth section discusses the intersection of globalisation and concentration by focusing on financial centres. The final section examines whether the rapid increase in cross-national business linkages among several major cities points to the formation of a trans-national urban system.

The intersection of service intensity and globalisation

The new or sharply expanded role of a particular kind of city in the world economy since the early 1980s is embedded in the intersection of two major processes. One is the sharp growth in the globalisation of economic activity. This has raised the scale and the complexity of economic transactions, thereby feeding the growth of top-level multinational headquarter functions and the growth of services for firms, particularly advanced corporate services. It is important to note that while globalisation raises the scale and complexity of these operations, these operations are also evident at smaller geographic scales and lower orders of complexity, e.g. firms that operate regionally. Thus, while such regionally oriented firms need not negotiate the complexities of international borders and regulations of different countries, they are still faced with a regionally dispersed network of operations that requires centralised control and servicing.

The second process is the growing service intensity in the organisation of all industries (Sassen 1991, pp.166–67; 1994, Chapter 4). This has contributed to a massive growth in the demand for services by firms in all industries, from mining and manufacturing to finance and consumer services. The components of this demand include financial, advertising, accounting, legal and consulting services. Cities are key sites for the production of services for firms. Hence the growing service intensity in the organisation of all industries has had a significant growth effect on cities in the 1980s. It is important to recognise that this growth in services for firms is evident in cities at different levels of a nation's urban system. Some of these cities cater to regional or sub–national markets; others cater to national markets and yet others cater to global markets. In this context, globalisation becomes a question of scale and added complexity. But the key process from the perspective of the urban economy is the growing demand for services by firms in all industries and the fact that cities are preferred production sites for such services, whether at the global, national or regional level. As a result we see in cities the formation of a new type of economic core of banking and service activities which comes to replace the older one, typically oriented towards servicing manufacturing and trade.

In cities that are major international business centres, the scale, power and profit levels suggest that we are seeing the formation of a new urban economy. This is so in at least two regards. On the one hand, while these cities have long been centres for business and finance, since the late 1970s there have been sharp changes in the structure of the business and financial sectors, and sharp increases in the overall magnitude of these sectors and their weight in the urban economy. On the other hand, the ascendance of the new finance and services complex, particularly international finance, engenders what I see as a new economic regime. That is to say, while this sector may account for only a fraction of the economy of a city, it imposes itself on that larger economy. Most notably, the possibility for super profits in finance has the effect of devalorising manufacturing insofar as manufacturing cannot generate the super profits typical of much financial activity.

This is not to say that everything in the economy of these cities has changed. On the contrary, there is much continuity and much similarity with cities that are not global nodes. It is rather that the implantation of global processes and markets has meant that the internationalised sector of the economy has expanded sharply and has imposed a new valorisation dynamic, often with devastating effects on large sectors of the urban economy. High prices and profit levels in the internationalised sector and its ancillary activities, e.g. restaurants and hotels, made it increasingly difficult in the 1980s for other sectors to compete for space and investments. Many of the latter have experienced considerable downgrading and/ or displacement. Illustrations are neighbourhood shops catering to local needs replaced by up-market boutiques and restaurants catering to new high income urban elites. The sharpness of the rise in profit levels in the international finance and service sector also contributed to the sharpness of the ensuing crisis.

Though at a different order of magnitude, these trends also became evident towards the late 1980s in a number of major cities in the developing world that have become integrated into various world markets: Sao Paulo, Buenos Aires, Bangkok, Taipei and Mexico City are some examples. Also central to the development of this new core in these cities were:

- the deregulation of financial markets,
- ascendance of finance and specialised services, and integration into the world markets,
- real estate speculation, and
- high-income commercial and residential gentrification.

The opening of stock markets to foreign investors and the privatisation of what were once public sector firms have been crucial institutional arenas for this articulation. Given the vast size of some of these cities, the impact of this new economic complex is not always as evident as in central London or Frankfurt, but the transformation has occurred.

Producer services

At the heart of the new urban economy are the producer services. There is by now a considerable literature on these services and their place in the urban economy (Stanback & Noyelle 1982; Noyelle & Dutka 1988; Daniels 1991; Leyshon et al. 1987 a & b, Hall, Chapter 1 this volume). Thus, here I will limit myself to a very brief review of recent growth figures in several developed countries and then move on to somewhat less familiar issues around the producer services. One of these concerns the production process in these services. Another is how these services fit in the work of running transnational operations.

The expansion of the producer services was a central feature of growth in developed countries in the 1980s, a pattern that continues though with significant slowdowns in the 1990s. In country after country we see a decline or stagnation in manufacturing alongside sharp growth in producer services. In my reading, the fundamental reason for this growth lies in the increased service intensity in the organisation of all industries (Sassen 1991, pp.166–68). Whether in manufacturing or in warehousing, firms use more legal, financial, advertising, consulting and accounting services. These services can be seen as part of the supply capacity of an economy in that they facilitate adjustments to changing economic circumstances (Marshall et al. 1986, p.16). These services are a mechanism that 'organises and adjudicates economic exchange for a fee' (Thrift 1987). They are part of a broader intermediary space of economic activity.[1]

Though disproportionately concentrated in the largest cities, producer services are actually growing at faster rates at the national level in most developed economies. In addition to the growing use of service inputs by firms in all industries, households have also raised their consumption of professional services. They have done so both directly, e.g. the growing use of accountants for preparation of tax returns, or indirectly via the reorganisation of consumer industries, e.g. buying flowers or dinner from franchises or chains which have contributed to the expansion of headquarters functions, in contrast to self-standing and privately owned 'family' shops. Services directly bought by consumers will tend to be located wherever there are population concentrations. In that regard they are far less concentrated than producer services, especially those catering to top firms. The demand for specialised services by households, from accounting to architects, may be a key factor contributing to the growth of these services at the national level.

National figures on employment trends clearly show how the producer services are the fastest growing sector in most developed economies. Thus in the USA, total employment grew from 76.8 million in 1970 to 116.9 million in 1991. But producer services grew from 6.2 to 16.4 million—an almost threefold increase; the largest single increase was in the category miscellaneous business services, and the second largest was in legal services. In contrast, manufacturing grew from 19.8 to 20.4 million. The other major growth sectors were the social services which grew from 16.9 to 29.8 million and personal services from 7.8 to 13.7 million; significant levels but not nearly the rate of producer services. Distributive services showed strong growth as well from 17.2 to 24 million.

We see parallel patterns in other developed economies. Thus, in Japan total employment grew from 52 million in 1970 to 61.7 million in 1990. Producer services more than doubled from 2.5 to 5.9 million; also here the largest single increase was in miscellaneous business services. Manufacturing grew from 13.5 to 14.5 million. Social services grew from 5.4 to 8.8 million, and personal services from 4.4 to 6.3 million. As in the US, distributive services showed considerable growth, going from 11.7 million to 15 million.

In France, total employment went from 20 million in 1968 to 21.8 million in 1989. Producer services doubled from 1 to 2.2 million, with the largest single increase also in miscellaneous business services. Personal services almost doubled, from 1.6 to 3 million in 1989, and social services grew from 3 to 4.3 million. Distributive services also showed strong growth, going from 3.7 to 4.5 million. Manufacturing fell from 5.4 to 4.6 million.

In the UK total employment fell from 23.4 million in 1970 to 21.3 million in 1992. Manufacturing lost almost 50% of its jobs, going from 9 to 4.5 million. Distributive services remained unchanged. But, as in other developed economies, the producer services more than doubled, from 1.2 million in 1970 to 2.6 million in 1992. Social services grew from 4.2 to 6.1 million and personal services from 1.9 to 2 million.

Finally, in Canada total employment grew from 8.4 million in 1971 to 13.9 million in 1992. But producer services tripled from 0.5 million to 1.6 million; miscellaneous business services accounted for two-thirds of this growth. All other service sectors also grew strongly. Distributive services almost doubled, social services grew from 1.8 to 3.1 million, and personal services grew from 0.6 to 1.9 million. Manufacturing grew from 1.6 to 2 million.

Over the last decade producer services have become the most dynamic, fastest growing sector in many cities. Particularly notable here is the case of the UK, where overall employment actually fell and there were severe losses in manufacturing. Yet in central London, there were sharp increases in producer services between 1984 and 1987; their share rose from 31 to 37% of all employment, reaching 40% by 1989 (Frost & Spence 1993). Central London saw both relative and absolute declines in all other major employment sectors. Similar developments can be seen in New York City: in 1987, at the height of the 1980s boom, producer services accounted for 37.7% of private sector jobs, up from 29.8% in 1977. There were high growth rates in many of the producer services during the period when economic restructuring was consolidated in New York City: from 1977 to 1985 employment in legal services grew by 62%, in business services by 42%, and in banking by 23%; in contrast, employment fell by 22% in manufacturing and by 20% in transport.

Accompanying these sharp growth rates in producer services was an increase in the level of employment specialisation in business and financial services in major cities throughout the 1980s. For example, over 90% of jobs in FIRE (finance, insurance, real estate) in New York City were located in Manhattan, as were 85% of business service jobs. If we consider only those components of producer services that respond to the description of information industries (even though so-called information industries can include a large share of jobs that have nothing to do with the handling of information), we can see that NYC has a significantly higher concentration than any other major city in the US. Information industries accounted for 31% of jobs in NYC in the mid 1980s, 17.8% in Los Angeles (county) and 20.3% in Chicago; all three cities show a higher incidence than the US average, which stood at 15.1%.

There is today a general trend towards high concentrations of finance and certain producer services in the downtowns of major international financial centres around the world: from Toronto and Sydney to Frankfurt and Zurich we are seeing growing specialisation in financial districts everywhere (see Sassen 1994, Chapter 5). It is worth noting that this trend is also evident in the multi-polar urban system of the US: against all odds, NYC has kept its place at the top in terms of concentration in banking and finance (see Tables 2.1 and 2.2).

Furthermore, finance and business services in the New York metro are are more concentrated in Manhatten today than they were in 1956 (Harris 1991a).[2]

These cities have emerged as important producers of services for export, with a tendency towards specialisation. New York and London are leading producers and exporters in accounting, advertising, management consulting, international legal services, and other business services. They are the most important international markets for these services, with New York the world's largest source of service exports. Tokyo is emerging as an important centre for the international trade in services, going beyond its initial role restricted to exporting the services required by its large international trading houses. Japanese firms are more likely to gain a significant share of the world market in certain producer services than others (Rimmer 1987). Construction and engineering services are examples of the first, advertising and international legal services of the second. As recently as 1978, the US accounted for 60 of the top 200 international construction contractors and Japan for ten (Rimmer 1988). By 1985, each accounted for 34 of such firms (see Sassen 1991, Tables 7.3 and 7.4).

There are also tendencies towards specialisation among different cities within a country. In the US, New York leads in banking, securities, manufacturing administration, accounting and advertising. Washington leads in legal services, computing and data processing, management and public relations, research and development, and membership organisations. New York is more narrowly specialised as a financial and business centre and cultural centre. Some of the legal activity concentrated in Washington

Table 2.1   US top ten concentrations of foreign banks, 1990–91

	Total offices	
	1990	*1991*
New York	463	464
Los Angeles	138	133
Chicago	83	80
Miami	59	65
San Francisco	57	62
Houston	55	42
Atlanta	31	29
Washington DC	15	15
Seattle	11	13
Boston	12	10
Total in top 10	924	913
Total all US offices	1 046	1 029

Source: *The American Banker*, 23 April 1992.

Table 2.2   Assets of 100 largest diversified service companies, by region (in millions of dollars)

City	Assets	Percentage of US Top 100
Total, top 100 US firms	366 844.6	–
New York City	123 733.0	34.66
NYC metro area	3 411.6	
Total	127 144.6	
Los Angeles	8 736.3	2.38
Chicago	6 323.5	8.45
Chicago metro area	24 659.1	
Total	30 982.6	
San Francisco	2 756.9	0.75
Total of above areas	169 620.4	46.24

Rank	City	Assets (in millions of $)
1	New York City (including metro area)	127 144.6
2	Chicago (including metro area)	30 982.6
3	Dallas	14 005.0
4	Houston	12 965.5
5	Cincinnati*	11 706.8
6	Burbank, CA*	10 861.7
7	Westwood, MO*	10 188.4
8	Washington, DC*	9 678
9	Kirkland, WA	9 632.5
10	Los Angeles	8 736.3
25	San Francisco*	2 756.9

* Denotes cities with only one company in top 100.
Source: 'The Service 500,' *Fortune*, 31 May 1993, pp.199–230.

is actually serving NYC businesses which have to go through legal and regulatory procedures, lobbying, etc. These are bound to be in the national capital. Many of these services are oriented to the national economy and to non-economic purposes. Furthermore, much of the research activity in Washington is aimed not at the world economy but at national medical and health research agendas. Thus it is obviously important to distinguish whether there is orientation to world markets and integration into the global economy or simply a focus on domestic demand.[3]

It is important to recognise that manufacturing remains a crucial economic sector in all of these economies, even when it may have ceased to be so in some of these cities. Indeed, several scholars have argued that there could be no producer services sector without manufacturing (Cohen & Zysman 1987; Markusen & Gwiasda 1993). In this context it has been argued, for instance, that the weakening of the manufacturing sector in the broader New York region is a threat to the city's status as a leading financial and producer services centre (Markusen & Gwiasda 1993). A key proposition for these and other authors is that producer services are dependent on a strong manufacturing sector in order to grow. There is considerable debate around this issue (Noyelle & Dutka 1988; Drennan 1989; Sassen 1991). Drennan (1992), a leading analyst of the producer services sector in NYC, argues that a strong finance and producer services sector is possible in New York notwithstanding decline in its industrial base, and that these sectors are so strongly integrated into the world markets that the articulation with the hinterland becomes secondary.

Sassen (1991), in a variant on both positions, argues that manufacturing indeed feeds the growth of the producer services sector, but that it does so whether located in the particular area, in another region, or overseas. While manufacturing, and mining and agriculture for that matter, feed the growth in the demand for producer services, their actual location is of secondary importance in the case of global level service firms: thus whether a manufacturing corporation has its plants offshore or inside a country may be quite irrelevant as long as it buys its services from those top-level firms. Secondly, the territorial dispersal of plants, especially if international, actually raises the demand for producer services because of the increased complexity of transactions. This is yet another meaning of globalisation: that the growth of producer service firms headquartered in New York or London or Paris can be fed by manufacturing located anywhere in the world as long as it is part of a multinational corporate network. Thirdly, a good part of the producer services sector is fed by financial and business transactions that either have nothing to do with manufacturing, as is the case in many of the global financial markets, or for which manufacturing is incidental, as in much of the merger and acquisition activity which was really centred on buying and selling rather than the buying of manufacturing firms as such.

Some of the figures on New York and London, two cities which experienced heavy losses in manufacturing and sharp gains in producer services, illustrate this point. New York lost almost 34% of its manufacturing jobs from 1969 to 1989 in a national economy that overall lost only 2% of such jobs, and there actually was manufacturing growth in many areas. The British economy lost 32% of its manufacturing jobs from 1971 to 1989; and the London region lost 47% of such jobs (Fainstein *et al.* 1992a; Buck *et al.* 1992). Yet both cities had sharp growth in producer services and raised their shares of such jobs in total city employment. Further, it is also worth noting the different conditions in each city's larger region: London's region had a 2% decline compared with a 22% job growth rate in the larger NY region. This points to the fact that the finance and producer services complex in each city rests on a growth dynamic that is somewhat independent from the broader regional economy—a sharp change from the past, when a city was presumed to be deeply articulated with its hinterland.

The formation of a new production complex

The rapid growth and disproportionate concentration of producer services in central cities should not have happened according to standard conceptions about information industries. As they are thoroughly embedded in the most advanced information technologies they could be expected to have locational options that by-pass the high costs and congestion typical of major cities. But cities offer agglomeration economies and highly innovative environments. Some of these services are produced in-house by firms, but a large share are bought from specialised service firms. The growing complexity, diversity and specialisation of the services required makes it more efficient to buy them from specialised firms rather than hiring in-house professionals. The growing demand for these services has made possible the economic viability of a free-standing specialised service sector.

There is a production process in these services which benefits from proximity to other specialised services. This is especially the case in the leading and most innovative sectors of these industries. Complexity and innovation often require multiple highly specialised inputs from several industries. One example is that of financial instruments. The production of a financial instrument requires inputs from accounting, advertising, legal expertise, economic consulting, public relations, designers and printers. Time replaces weight in these sectors as a force for agglomeration. That is to say, if there were no need to hurry, one could conceivably have a widely dispersed array of specialised firms that could still cooperate. And this is often the case in more routine operations. But where time is of the essence, as it is today in many of the leading sectors of these industries, the benefits of agglomeration are still extremely high—to the point that it is not simply a cost advantage, but an indispensable arrangement.

It is this combination of constraints that has promoted the formation of a producer services complex in all major cities. This producer services complex is intimately connected to the world of corporate headquarters;

they are often thought of as forming a joint headquarters-corporate services complex. But it seems to me that we need to distinguish the two. While it is true that headquarters still tend to be disproportionately concentrated in cities, many have moved out over the last two decades. Headquarters can indeed locate outside cities. But they need a producer services complex somewhere in order to buy or contract for the needed specialised services and financing. Further, headquarters of firms with very high overseas activity or in highly innovative and complex lines of business tend to locate in major cities. In brief, firms in more routinised lines of activity, with predominantly regional or national markets, appear to be increasingly free to move or install their headquarters outside cities. Firms in highly competitive and innovative lines of activity and/or with a strong world market orientation appear to benefit from being located at the centre of major international business centres, no matter how high the costs.

But it is clear, in my view, that both types of headquarters need a corporate services sector complex to be located somewhere. Exactly where is probably increasingly unimportant from the perspective of many, though not all, headquarters. From the perspective of producer services firms, such a specialised complex is most likely to be in a city rather than, for instance, a suburban office park. The latter will be the site for producer services firms, but not for a services complex. And it is only such a complex that can handle the most advanced and complicated corporate demands.

To illustrate some of these points let me focus on two issues. One of these is how these services fit in the work of running transnational operations; the other, headquarters location.

The servicing of transnational corporations

The sharp rise in the use of producer services has also been fed by the territorial dispersal, whether at the regional, national or global level, of multi-establishment firms. Firms operating many plants, offices and service outlets must coordinate planning, internal administration and distribution, marketing and other central headquarters activities. As large corporations move into the production and sale of final consumer services, a wide range of activities, previously performed by free-standing consumer service firms, are shifted to the central headquarters of the new corporate owners. Regional, national or global chains of motels, food outlets, flower shops, require vast centralised administrative and servicing structures. A parallel pattern of expansion of central high level planning and control operations takes place in governments, brought about partly by the technical developments that make this possible and partly by the growing complexity of regulatory and administrative tasks. Formally, the development of the modern corporation and its massive participation in world markets and foreign countries (Table 2.3) has made planning, internal administration, product development and research increasingly important and complex. Diversification of product lines, mergers, and transnationalisation of economic activities all require highly specialised services.

Table 2.3   Number of parent transnational corporations and foreign affiliates by area and country, 1990–92 (in numbers)

All developed countries

Year	Select countries	Parent corporations based in country	Foreign affiliates located in country
1992	Australia	1 036	695
1991	Canada	1 308	5 874
1990	France	2 056	6 870
1990	Germany (Fed. Rep.)	6 984	11 821
1992	Japan	3 529	3 150
1992	Netherlands	1 426	2 014
1991	Sweden	3 529	2 400
1985	Switzerland	4 000	4 000
1991	UK	1 500	2 900
1990	USA	3 000	14 900

All developing countries
Select countries

Year	Select countries	Parent corporations based in country	Foreign affiliates located in country
1992	Brazil	566	7 110
1989	China	379	15 966
1987	Colombia	–	1 041
1991	Hong Kong	500	2 828
1988	Indonesia	–	1 064
1989	Mexico	–	8 953
1987	Philippines	–	1 952
1991	Republic of Korea	1,049	3 671
1989	Saudi Arabia	–	1 461
1986	Singapore	–	10 709
1990	Taiwan Province of China	–	5 733
1991	Former Yugoslavia	112	3 900
	Central and Eastern Europe		
1992	Commonwealth of Independent States*	68	3 900

* Relates to the whole of the economic territory of the former USSR.
Sources: Based on UNCTAD on Transnational Corporations (1993, pp.20–21).

A brief examination of the territorial dispersal entailed by transnational operations of large enterprises can serve to illustrate some of the points raised here. For instance, Table 2.4 lists the numbers of workers employed abroad by the largest 25 non-financial transnational corporations world-wide. They are rather large numbers. Thus about half of each Exxon's and IBM's and about a third each of Ford Motors' and GM's total work force is employed outside the US. We know, furthermore, that large transnationals have very high numbers of affiliates. Thus in 1990 German firms had over 19 000 affiliates in foreign countries, up from 14 000; and the US had almost 19 000. Finally, we know that the top transnationals have very high shares of foreign operations: the top ten largest transnational corporations in the

Table 2.4 The 25 non-financial transnational corporations, ranked by foreign assets, 1990 (billions of dollars and number of employees)

Rank	Corporation	Country	Industry	Foreign assets	Total assets	Foreign sales	Total sales	Foreign employment	Total employment
1	Royal Dutch Shell	UK	Petroleum refining	69.2	106.4	47.1	106.5	99 000	137 000
2	Ford	US	Motor vehicles & parts	55.2	173.7	47.3	97.7	188 904	370 383
3	GM	US	Motor vehicles & parts	52.6	180.2	37.3	122.0	251 130	767 200
4	Exxon	US	Petroleum refining	51.6	87.7	90.5	115.8	65 000	104 000
5	IBM	US	Computers	45.7	87.6	41.9	69.0	167 868	373 816
6	British Petroleum	UK	Petroleum refining	31.6	59.3	43.3	59.3	87 200	118 050
7	Asea Brown Bovera	Switzerland	Industrial & farm equipment	26.9	30.2	25.6	26.7	200 177	215 154
8	Nestle	Switzerland	Food	–	28.8	35.8	36.5	192 070	199 021
9	Phillips Electronics	Netherlands	Electronics	23.3	30.6	28.8	30.8	217 149	272 800
10	Mobil	US	Petroleum refining	22.3	41.7	44.3	57.8	27 593	67 300
11	Unilever	UK/Netherlands	Food	–	24.7	16.7	39.6	261 000	304 000
12	Matsushita Electric	Japan	Electronics	–	62.0	21.0	46.8	67 000	210 848
13	Fiat	Italy	Motor vehicles & parts	19.5	66.3	20.7	47.5	66 712	303 238
14	Siemens	Germany	Electronics	–	43.1	14.7	39.2	143 000	383 000
15	Sony	Japan	Electronics	–	32.6	12.7	20.9	62 100	112 900
16	Volkswagen	Germany	Motor vehicles & parts	–	42.0	25.5	42.1	95 934	268 744
17	Elf Acquitaine	France	Petroleum refining	17.0	42.6	11.4	32.4	33 957	90 000
18	Mitsubishi	Japan	Trading	16.7	73.8	45.5	129.3	–	32 417
19	G.E.	US	Electronics	16.5	153.9	8.3	57.7	62 580	112 900
20	DuPont	US	Chemicals	16.0	38.9	17.5	37.8	36 400	124 900
21	Alcatel Alsthom	France	Electronics	15.3	38.2	13.0	26.6	112 966	205 500
22	Mitsui	Japan	Trading	15.0	60.8	48.1	136.2	–	9 094
23	News Corporation	Australia	Publishing & printing	14.6	20.7	4.6	5.7	–	38 432
24	Bayer	Germany	Chemicals	14.2	25.4	20.3	25.9	80 000	171 000
25	B.A.T. Industries	UK	Tobacco	–	48.1	16.5	22.9	–	217 373

Source: Based on UNCTAD 1993, pp.26–27.

world had 61% of their sales abroad. The average for the 100 largest corporations was almost 50%.

These figures show a vast operation dispersed over a multiplicity of locations. This generates a large demand for producer services, from international accounting to advertising. Operations as vast as these feed the expansion of central management, coordination, control and servicing functions. Some of these functions are performed in headquarters, some are bought or contracted for, thus feeding the growth of the producer services complex.

Corporate headquarters and cities

It is common in the general literature and in some more scholarly accounts to use headquarters concentration as an indication of whether a city is an international business centre. The loss of headquarters is then interpreted as a decline in a city's status. The use of headquarters concentration as an index is actually a problematic measure given the way in which corporations are classified.

Which headquarters concentrate in major international financial and business centres depends on a number of variables. First, how we measure or simply count headquarters makes a difference. Frequently, the key measure is size of firm in terms of employment and overall revenue. In this case, some of the largest firms in the world are still manufacturing firms and many of these have their main headquarters in proximity to their major factory complex, which is unlikely to be in a large city due to space constraints. Such firms are likely, however, to have secondary headquarters for highly specialised functions in major cities. Further, many manufacturing firms are oriented to the national market and do not need to be located in a city's national business centre. Thus, the much publicised departure of major headquarters from New York City in the 1960s and 1970s involved these types of firms. If we look at the Fortune 500 largest firms in the US (cf. 'Fortune Magazine 500 list'), many have left New York City and other large cities. If instead of size we use share of total firm revenue coming from international sales, a large number of firms that are not part of the Fortune 500 list come into play. For instance, in the case of NYC the results change dramatically: 40% of US firms with half their revenue from international sales have their headquarters in New York City.

Secondly, the nature of the urban system in a country is a factor. Sharp urban primacy will tend to entail a disproportionate concentration of headquarters no matter what measure one uses. Thirdly, different economic histories and business traditions may combine to produce different results. Further, headquarters concentration may be linked with a specific economic phase. For instance, unlike New York's loss of top Fortune 500 headquarters, Tokyo has been gaining headquarters. Osaka and Nagoya, the two other major economic centres in Japan, are losing headquarters to Tokyo. This is in good part linked to the increasing internationalisation of the Japanese economy and the corresponding increase in central command and servicing functions in major international business centres. In Japan, extensive government regulation over the economy is an added factor contributing to headquarter location in Tokyo, as all international activities have to go through various government approvals (see Sassen 1991, Chapters 1 & 7).

In brief, understanding the meaning of headquarters concentration requires differentiation along the above described variables. While headquarters are still disproportionately concentrated in major cities, the patterns evident today do represent a change from twenty years ago.

Impacts of the crisis on global city functions: the case of New York City

The high level of speculation and of profitability that fed this growth in the 1980s was clearly unsustainable. The financial 'crisis' that became evident by 1990 raises two types of possibilities. One is that we are seeing a crisis of an economic system; the other, that we are seeing a rather sharp readjustment to more sustainable levels of speculation and profitability. New York was the first of the major international financial centres to experience massive losses. Its post-1990 evolution may provide some useful insights into the interaction between crisis and readjustment in the dominant sector.

Employment in banking in the city fell from 169 000 in 1989 to 157 000 in 1991. Most of this loss (over 10 000 jobs) was in domestic banking. It should also be noted that some of these losses are due to the massive restructuring happening within the industry, including a tendency towards mergers among large domestic banks. Also, in the securities industry, which suffered some of the sharpest job losses in the last two years, New York City remains strong. It houses 6 of the top 10 securities firms worldwide as of late 1990. City firms and their overseas affiliates acted as advisers for almost 80% of the value of all international mergers and acquisitions at the height of the financial boom in the mid-1980s. As the securities industry is almost completely export oriented, it may well be less sensitive to the overall crisis in the US economy and in NYC.

Even after the financial crisis of the last few years, NYC continues to function as an important international centre and continues to be dominated by the financial and related industries. According to many analysts, the developments of the last year constitute a much needed adjustment to the excess of the 1980s. Within the US, NYC remains as the banking capital of the country, leading in total assets, number of banks, volume in various markets—currency markets, options trading, merchant banking. Furthermore, foreign banking is a growth sector in NYC and may well be a key factor in the continuing role of the city as a leading financial centre for the world. So, while Japanese and European banks have far surpassed US banks and dominate the international banking industry, they all have offices in NYC. Indeed, in 1990 NYC surpassed London in its number of international bank offices. New York is truly a platform for international operations, notwithstanding reductions in the domestic banking industry and major crises in several industry branches.

What emerges from these developments is that NYC may retain its central role as a financial centre but with a far greater participation of foreign firms making loans, selling financial services, assisting in mergers and acquisitions. The job losses and bankruptcies in the securities industry since 1987 point to the possibility of a major transformation in the role of Wall Street and other stock markets, most particularly the fact that large corporations can bypass stock markets in order to raise investment capital. But it does not spell the end of Wall Street. Thus the crisis on Wall Street may be partly a restructuring process from which it may emerge as a smaller market, catering to smaller firms, but without losing an international base, and continuing as a provider of the most specialised and complex services.

Globalisation and concentration: financial centres

All the major economies in the developed world display a similar pattern towards sharp concentration of financial activity in one centre: Paris in France, Milano in Italy, Zurich in Switzerland, Frankfurt in Germany, Toronto in Canada, Tokyo in Japan, Amsterdam in the Netherlands, and as we have seen, Sydney in Australia. The evidence also shows that the concentration of financial activity in such leading centres has actually increased over the last decade. Thus in Switzerland, Basle used to be a very important financial centre which has now been completely overshadowed by Zurich (Keil & Ronneberg 1993). And Montreal was certainly the other major centre in Canada two decades ago, and has now been overtaken by Toronto. Similarly in Japan, Osaka was once a far more powerful competitor with Tokyo in the financial markets in Japan than it had become by the late 1980s (Sassen 1991, Chapters 6 & 7).[4]

After a decade of massive growth in the absolute levels of financial activity worldwide, New York, London and Tokyo remain in their position at the top and continue to account for a disproportionate share of all activity. For instance, international bank lending grew from US\$1.89 trillion in 1980 to US\$6.24 trillion in 1991—a fivefold increase in a mere 10 years. These 3 cities accounted for 42% of all such international lending in 1980 and for 41% in 1991, according to data from the Bank of International Settlements. There were compositional changes: Japan's share rose from 6.2 to 15.1% and the UK's fell from 26.2 to 16.3%; the share of the US remained constant. All increased in absolute terms. Beyond these three, Switzerland, France, Germany and Luxembourg bring the total share of the top centres to 64% in 1991, which is just about the same share these countries had in 1980. Similar patterns of concentration are evident in stock market capitalisation and in foreign exchange markets. In the foreign exchange markets, London's daily average at the end of the 1980s was about US\$200

b. New York's $190b and Tokyo's US$110 billion. Next are Zurich with slightly over US$50b, Singapore with slightly less, and Hong Kong with US$50b, Sydney with about US$30b, and Paris with US$25b. Thus, after the top 3 the volume, though extremely large in itself, becomes relatively speaking significantly smaller. Chicago dominates the world's trading in futures, accounting for 60% of worldwide contracts in options and futures in 1991.

We should note again that this unchanged level of concentration is in the context of enormous absolute increases, deregulation and globalisation of the industry worldwide, which means that a growing number of countries have become integrated into the world markets. From Bangkok to Buenos Aires, governments deregulated their stock markets to allow their participation in a global market system. Yet the figures on worldwide capitalisation show immense concentration in leading stock markets. Similar developments took place with other financial markets. The globalisation of the industry has raised the level of complexity of transactions and deregulation has promoted the invention of many new, complex instruments. This has, clearly, raised the importance of the leading centres insofar as only they have the capability to handle such levels of complexity. Among the large financial centres some are more dominated by international business and others more by domestic business. Thus London, with its strong eurodollar markets and foreign exchange markets, is extremely international, while New York and Tokyo with their enormous national economies inevitably are going to have a very large incidence of domestic borrowers, lenders and investors.

If we organise some of the evidence on financial flows, according to the places where the markets and firms are located, we can begin to see distinct patterns of concentration. The evidence on the locational patterns of banks and securities houses points to sharp concentration. For instance, the worldwide distribution of the largest 100 banks and 25 largest securities houses shows Japan, the US and the UK accounted for respectively 39 and 23 of each (see Tables 2.5 and 2.6). This persists in the 1990s, notwithstanding multiple financial crises.

The market value of equities in domestic firms also confirms the leading position of a few cities. In September 1987, before the stock market crisis, this value stood at 2.8 trillion in the US and at 2.89 trillion in Japan. Third ranked was the UK with 728 billion. These values represent extremely high levels. The next largest value is for West Germany, a major economy, but domestic equities only represent 23% of GNP and capitalisation is US$255b, a long distance from the top three. Most of the stock market transactions in the leading countries are concentrated in a few stock markets. The Tokyo exchange accounts for 90% of equities trading in Japan. New York accounts for about two-thirds of equities trading in the US; and London accounts for most trading in the UK. There is, then, a disproportionate concentration of worldwide capitalisation in a few cities. What these levels of capitalisation represent in the top countries is indicated by a comparison with GNP

Table 2.5   US, Japan and UK share of world's 100 largest banks, 1991

	No.	Assets (US$m)	Capital (US$m)
Japan	27	6 572 416	975 192
US	7	913 009	104 726
UK	5	791 652	56 750
Subtotal	39	8 277 077	1 136 668
All other countries	61	7 866 276	1 263 771
Total	100	16 143 353	2 400 439

Source: The *Wall Street Journal*, World Business, Thursday, 24 September 1992, R27.

Table 2.6   US, Japan and UK—50 largest securities firms, 1991

	No.	Assets (US$m)	Capital (US$m)
Japan	10	171 913	61 871
US	11	340 558	52 430
UK	2	44 574	3 039
Subtotal	23	557 045	117 340
All other countries	2	6 578	5 221
Total	25	523 623	122 561

Source: *The Wall Street Journal*, World Business, Thursday, 24 September 1992, R27.

figures: in Japan, stock market capitalisation was the equivalent of 64%; in the US the equivalent of 119%, and in the UK the equivalent of 118% of GNP.

Furthermore, this unchanged level of concentration has happened at a time when financial services are more mobile than ever before: globalisation, deregulation (an essential ingredient for globalisation) and securitisation have been the key to this mobility—in the context of massive advances in telecommunications and electronic networks. One result is growing competition among centres for hypermobile financial activity. In my view there has been an overemphasis on competition in both general and specialised accounts on this subject. As I will argue in the next section, there is also a functional division of labour among various major financial centres; international finance is today a system with multiple locations.

The hypermobility of financial capital brings to the fore the importance of technology. It is now possible to move money from one part of the world to another and make deals without ever leaving the computer terminal. There are disembodied marketplaces thanks to electronics—what we can think of as the cyberspace of international finance. NASDAQ and the foreign exchange markets are examples of disembodied markets, unlike the regular stock market with its trading floor.

Yet, notwithstanding such hypermobility, the trend towards concentration continues unabated. Much of the discussion around the formation

Table 2.7 Select stock exchanges: Market size 1990

	Market value (US$m)		Listed companies (n)		Members
	Stocks	Bonds	Domestic	Foreign	Firms (n)
New York	2 692 123	1 610 175	1 678	96	516
Tokyo	2 821 660	978 895	1 627	125	124
UK (mostly London)	858 165	576 291	1 946	613	410
Frankfurt	341 030	645 382	389	354	214
Paris	304 388	481 073	443	226	44
Zurich	163 416	158 487	182	240	27
Toronto	241 925	–	1 127	66	71
Amsterdam	148 553	166 308	260	238	152
Milan	148 766	588 757	220	–	113
Australia	108 628	46 433	1 085	37	90
Hong Kong	83 279	656	284	15	686
Singapore	34 268	98 698	150	22	26
Taiwan	98 854	6 551	199	–	373
Korea	110 301	71 353	699	–	25

Source: Tokyo Stock Exchange 1992 Fact Book (Tokyo: International Affairs Department, Tokyo Stock Exchange), April 1992.

of a single European market and financial system has raised the possibility, and even need, that the only way for a European financial system to be competitive would be through the centralisation of financial functions and capital in a limited number of cities, rather than maintaining the current structure where each country has a financial centre.[5]

Transnational urban systems

The global integration of financial markets entails the implementation of a variety of linkages among cities that have become so integrated. There is a rapidly growing and highly specialised research literature focused on different types of economic linkages that bind cities across national borders (Castells 1991; Noyelle & Dutka 1988; Daniels 1991; Leyshon *et al.* 1987a; Sassen 1991). Prime examples of such linkages are the multinational networks of affiliates and subsidiaries, typical of major firms in manufacturing and specialised services. There is also a growing number of less direct economic linkages, especially a variety of initiatives launched by urban governments which amount to a type of foreign policy by and for cities.

Some of the most detailed data on transnational linkages binding cities comes from studies on corporate service firms. Such firms have developed vast multinational networks containing special geographic and institutional linkages that make it possible for client firms—transnational firms and banks—to use a growing array of service offerings from the same supplier

(Noyelle & Dutka 1988; Daniels 1991; Leyshon et al. 1987a). There is good evidence that the development of multinational corporate service firms was associated with the needs of transnational firms. The multinational advertising firm can offer global advertising to a specific segment of potential customers worldwide. Further, global integration of affiliates and markets requires making use of advanced information and telecommunications technology which can come to account for a significant share of costs—not just operational costs but also, and perhaps most important, research and development costs for new products or advances on existing products. The need for scale economies on all these fronts contributes to explain the recent increase in mergers and acquisitions, which has consolidated the position of a few very large firms in many of these industries, and further strengthened cross-border linkages among the key locations which concentrate the needed telecommunications facilities. They have emerged as firms that can control a significant share of national and international markets. The rapid increase in direct foreign investment in services is strongly linked with the growing tendency among leading service firms to operate transnationally. Subcontracting by larger firms and the multiplicity of specialised markets has meant that small independent firms can also thrive in major business centres. (Commission of the European Communities 1992a; Sassen 1991; Stanback & Noyelle 1982; see also Lash & Urry 1987). Accounting, advertising and legal services reflect some of these trends (Noyelle & Dutka 1988; Thrift 1987; Leyson *et al.* 1987). In the mid 1980s, the world's 5 largest advertising firms controlled 38% of the western European market, and about 56% each of the Latin American and Pacific Area market (Noyelle & Dutka 1988). The top 10 firms received 27% of worldwide revenues in 1987.

Whether these linkages have engendered transnational urban systems is less clear. It is partly a question of theory and conceptualisation. So much of social science is profoundly rooted in the nation-state as the ultimate unit for analysis, that conceptualising processes and systems as transnational is bound to engender much controversy. Even much of the literature on world or global cities does not necessarily posit the existence of a transnational urban system: in its narrowest form it posits that global cities perform central place functions at a transnational level. But that leaves open the question as to the nature of the articulation among global cities. If one posits that they basically compete with each other for global business, then they do not constitute a transnational system. Studying several global cities then falls under traditional comparative analyses.

If one assumes that besides competing they are also the sites for transnational processes with multiple locations, then one can begin to explore the possibility of a systemic dynamic binding these cities. Elsewhere (Sassen 1991, Chapters 1 & 7) I have argued that in addition to the central place functions performed by these cities at the global level, as suggested by Hall (1966), Friedmann and Wolff (1982) and Sassen (1981), these cities relate to one another in distinct systemic ways. For instance, the interaction among New York, London and Tokyo, particularly in terms of finance and investment, consists partly of a series of processes that can be thought of as the chain of production in finance. Thus, in the mid-1980s, Tokyo was the main exporter of the raw material we call money, while New York was the leading processing centre in the world. It was in New York that many of the new financial instruments were invented, and where money either in its raw form or in the form of debt was transformed into instruments that aimed at maximising the returns on that money. London, on the other hand, was a major market which had the network to centralise and concentrate small amounts of capital available in a large number of smaller financial markets around the world. This was partly a legacy of its older network for the administration of the British empire. This is just one example suggesting that these cities do not simply compete with each other for the same business. There is, it seems to me, an economic system that rests on the 3 distinct types of locations these cities represent.

Further, the possibility of such a transnational urban system raises a question as to the articulation of such cities in their national urban systems. Elsewhere (Sassen 1991) I have argued that the strengthening of cross-national ties among the leading financial and business centres is likely to be accompanied by a weakening of the linkages between each of these cities and their hinterlands and national urban systems. Cities such as Detroit, Liverpool, Manchester, Marseille, the cities of the Ruhr, and now increasingly Nagoya and Osaka, have been affected by the territorial decentralisation of many of their key manufacturing industries at the domestic and international level. This process of decentralisation has contributed to the growth of service industries that produce the specialised inputs to run spatially dispersed production processes and global markets for inputs and outputs. These specialised inputs—international legal and accounting services, management consulting, financial services—are heavily concentrated in business and financial centres rather than in manufacturing cities.

The formation of these types of transnational urban systems results in good part from the growth of global markets for finance and specialised services, the sharp increase in international investment as a key component of the global economy, and the reduced role of the government in the regulation of international economic activity. The formation of such systems contributes to new forms of inter-urban inequality in terms of the geography and characteristics of urban systems (see Sassen 1994, Chapter 3). On the one hand, there is growing articulation at a transnational level among cities that function as international business centres.[6]

This is evident both at a regional transnational level and at the global level; in some cases there are what one could think of as overlapping geographies of articulation or overlapping hierarchies that operate at more than one level. On the other hand, cities and areas outside these hierarchies tend to become peripheralised, or become more so than they had been.

Conclusion

The impacts of economic globalisation on cities discussed in this chapter can be summarised in a more analytic fashion through the following three propositions.

1 *The territorial dispersal of economic activities, of which globalisation is one form, contributes to the growth of centralised functions and operations.* We find here a new logic for agglomeration and key conditions for the renewed centrality of cities in advanced economies. Information technologies, often thought of as neutralising geography, actually contribute to spatial concentration. Yes, they make possible the geographic dispersal and simultaneous integration of many activities. But the particular conditions under which such facilities are available have promoted centralisation of the most advanced users in the most advanced telecommunication centres. We see parallel developments in cities that function as regional nodes: that is, at smaller geographic scales and lower levels of complexity than global cities.

2 *Centralised control and management over a geographically dispersed array of economic operations does not come about inevitably as part of a 'world system'.* It requires the production of a vast range of highly specialised services, telecommunications infrastructure and industrial services. Major cities are centres for the servicing and financing of international trade, investment and headquarter operations. And in this sense they are strategic production sites for today's leading economic sectors. This function is reflected in the ascendance of these activities in their economies. Again, cities that serve as regional centres exhibit similar developments. This is the way in which the spatial effects of the growing service intensity in the organisation of all industries materialises in cities.

3 *Economic globalisation has contributed to a new geography of centrality and marginality.* This new geography assumes many forms and operates in many terrains, from the distribution of telecommunications facilities to the structure of the economy and of employment. Global cities are the site of immense concentrations of economic power, while cities that were once major manufacturing centres suffer inordinate declines.

NOTES

[1] Producer services cover financial, legal and general management matters, innovation, development, design, administration, personnel, production technology, maintenance, transport, communications, wholesale distribution, advertising, cleaning services for firms, security, and storage. Central components of the producer services category are a range of industries with mixed business and consumer markets. They are insurance, banking, financial services, real estate, legal services, accounting, and professional associations. Given the organisation of the pertinent data, it is helpful to group these services under the category of 'mostly producer services', i.e. services produced for firms rather than individuals. It has become customary to refer to them, for convenience, as producer services.

[2] Jobs are far more concentrated in the CBD in New York City compared with other major cities in the US: about 27% of all jobs in the consolidated statistical area were in Manhattan

compared with 9% nationally (Drennan 1989, P. 30). The 90% concentration ratio of finance is, clearly, far above the norm.

3 The data on producer services is creating a certain amount of confusion in the US. For instance, the fact of faster growth at the national level and in medium-sized cities is often interpreted as indicating a loss of share and declining position of leading centres such as New York or Chicago. Thus, one way of reading this data is as decentralisation of producer services: NY and Chicago losing share of all producer services in the US. Another way is to read it as growth everywhere, rather than a zero sum situation where growth in a new location ipso facto is construed as a loss in an older location. In my reading, these patterns point to the growing service intensity in the organisation of the economy nation-wide.

4 Is this tendency towards concentration within each country a new development for financial centres? A broader historical view points to some interesting patterns. Since their earliest beginnings, financial functions were characterised by high levels of concentration. They often operated in the context of empires, such as the British or Dutch empires. Against this pattern of empires, the formation of nation-states represents a condition making possible a multiplicity of financial centres, typically the national capital in each country. Furthermore, the ascendance of mass manufacturing contributed to vast, typically regionally based fortunes and the formation of financial centres in such regions: Chicago and Osaka are but two examples. The overall pattern was one of less concentration in a few centres and in that sense differed both from that of older empires and of the current global economy.

5 These tendencies towards concentration seem to be built into the nature of such financial centres. Centres at the top are characterised by a multiplicity of financial institutions and markets and significant shares of world activity in various markets. They usually have a large number of banks and other institutions that account for a significant share of international lending, foreign exchange trading and fund management. They also have large or significant markets in tradeable securities—whether bonds, stocks or their derivatives.

6 The pronounced orientation to the world markets evident in such cities raises questions about the articulation with their hinterlands and nation-states. It also raises theoretical questions about a key proposition in the literature about urban systems, to wit, that they promote the territorial integration of regional and national economies.

3

Processes of Globalisation, Economic Restructuring and the Emergence of a New Space Economy of Cities and Regions in Australia

Robert J. Stimson

A rapidly changing world

DURING the last couple of decades there have been fundamental changes in the nature of the world economy. 'Economic activity is becoming not only more internationalised but...more significantly, it is becoming increasingly globalised' in a world of 'increasing complexity, interconnectedness and mobility' (Dicken 1992a, p.1). We live in a rapidly changing and increasingly competitive world in which 'global competition is making it essential for regional economies to develop the capacity and capability to make rapid adjustments to changing conditions if they are to remain competitive' (Stimson 1992, p.1).

Production processes and capital flows have globalised in increasingly deregulated environments. Transnational corporations have developed complex global networks. Technological revolutions in transportation and communications have shrunk distance and changed its role as a locational determinant, and revolutions in production and organisation have facilitated the internationalisation and globalisation of production and trade in both goods and services. New waves of migration, an ageing population, and trends towards smaller non-nuclear family households add significant demographic dimensions to the processes of change.

Within this dynamic world, the role of cities and metropolitan regions is changing as they become 'the nexus of the emerging global society' (Knight 1989, p.327). Processes of change are restructuring and reshaping the space economy (i.e. taking into account spatial factors) of nations, regions and cities.

A new international hierarchy of cities is emerging that relates a set of metro-politan areas to one another across international boundaries. These 'metro systems' are the new international trading partners because they both create and cause the flow of commerce. The hierarchy is anchored in the transformations of the international economy as well as adaptive adjustments to internal economic and social environments...Within the hierarchy of cities are [an] incredibly dynamic set of processes that are influencing how functions compete to determine how space is used increating new agglomerations of urban space, new patterns of decentralised activities, and adaptive uses of existing spaces (Blakely 1992, p.3).

This chapter reviews the processes of change that are reshaping and restructuring the Australian space economy as the nation, its regions and its cities adjust to come to terms with the roles they will play in the new global economic system.

Change in Australia: from protection to deregulation and restructuring

During the two to three decades following World War II, the evolution of the Australian economy was characterised by:

- rapid population growth with an increasingly high concentration of its population in the major cities;
- a large-scale assisted immigration programme, focused on the UK and European countries;
- the industrialisation of cities under policies of import substitution and with high tariff protection;
- a reliance on imported technologies;
- the financing of industrial growth from funds earned from the export of the agricultural and mining sectors and from foreign investment (Daly & Stimson 1991).

Australia's development, and particularly that of its major cities, continued to be influenced by forces and linkages related to its colonial origins from Britain. Melbourne, the capital city of Victoria, was the traditional focus for commerce, banking and company headquarters, and it was the place of concentration of 'old capital', even though Sydney, the capital city of New South Wales, was the larger city and the place of less establishment-oriented 'new capital'.

Since the early 1970s Australia has entered a new era, of dramatic restructuring of the economy:

- employment in manufacturing declined, while employment in the services sector increased;
- there was a shift from energy-intensive manufacturing to knowledge-based activities;
- production processes and marketing were increasingly internationalised;
- there was deregulation of financial institutions and integration into global financial markets;
- the nation experienced a number of deep cyclical building and construction 'booms and busts';
- there was a massive increase in foreign direct investment, with a shift of focus from the agricultural, mining and manufacturing sectors to the property and services sector and to new sources of finance from Asia;

- there were a set of fundamental structural demographic changes, affecting both the structure of the population and the composition of households and the patterns of migration;
- and there was an increasing focus of the nation's trading on the Asia-Pacific region (Daly & Stimson 1991).

The post-war era up to the early 1970s was one of increasing prosperity, during which Australia 'rode on the sheep's back' and became a 'quarry for the world'. Its cities grew fat under protectionist policies. Inflation was low, with unemployment never a significant problem. It was the proverbial 'lucky country'. But all that was to change in the post-1970 era. The collapse of the Bretton Woods agreement transformed the world's financial system, but it was not until 1983–84 that Australia reacted to deregulate its financial system and open up its financial markets to world competition. The OPEC oil crisis triggered shock waves, necessitating fundamental changes in production processes. The nation became caught in vicious cycles of high inflation, wage blow-outs, high interest rate regimes, and high levels of unemployment. Reluctantly, but progressively, tariff barriers were lowered, exposing under-capitalised and uncompetitive manufacturing industries to external competition. A downturn in commodity prices meant that they no longer provided the means to balance the trade account against an increasing dependence on imported technology and manufactured goods. Adjustment to the new economic order was sharp and painful, and the impact of change was unevenly distributed, both geographically and socially. The restructuring process is still far from complete. The 'lucky country's' luck had run out.

Towards a services economy

Since 1971 Australia's economic growth has been based increasingly on the activities within the services sector, which averaged 1.7% per annum employment growth. Most of this has been concentrated in the major metropolitan cities and the emerging new urban regions of the eastern seaboard. There was a substantial decline in employment in manufacturing industries, and its share of employment declined from 25% to 15% between 1971 and 1985. In 1959 the manufacturing industries sector had peaked at 29% of GDP, but by 1992 it had slipped to 16% of GDP. In contrast, between 1971 and 1989 employment in the finance, property and business sectors increased by over 90%, by 97% in the community services, by 25% in communications, and by 37% in recreation, personal and other services. Between 1971 and 1986 the employment share of community services increased by 15% to 23% and that of the finance, property and business services increased from 7% to 12% (see Figure 3.1).

Within the mainland State capitals (the major metropolitan cities), the structural pattern of employment showed considerable differences by 1986 (see Table 3.1).

Clearly Sydney's employment structure was dominated more by the finance, business and services sector than was the case for the other cities.

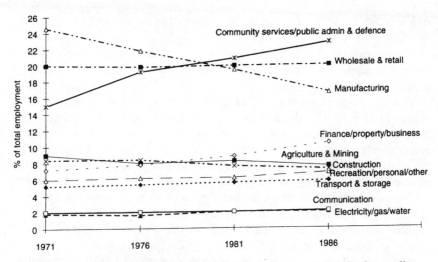

Figure 3.1   Distribution of employment by industry sector in Australia, 1971–86
Source: Australian Bureau of Statistics, Census data.

Table 3.1   Employment share by industry sectors in Australia's metropolitan cities, 1986

Sector	Sydney	Melbourne	Brisbane	Gold Coast	Adelaide	Perth
Agriculture, mining	1.2	1.0	1.6	2.4	1.7	3.1
Manufacturing	16.7	20.6	14.0	10.2	17.1	12.9
Electricity, gas & water	1.9	1.6	1.5	0.7	1.6	1.7
Construction	6.1	6.0	7.7	11.0	6.4	7.3
Retail/wholesale	19.6	19.3	23.1	23.0	20.3	20.8
Transport	6.1	5.1	6.4	3.8	4.4	5.3
Communication	2.3	2.2	2.8	1.9	2.0	2.0
Finance, business services	14.3	11.7	11.9	12.2	10.1	11.9
Public administration	5.5	5.4	7.4	3.7	5.2	5.0
Community services	16.4	17.3	19.6	12.8	21.5	19.6
Recreation, etc.	6.1	5.0	6.0	13.8	6.3	6.6

Source: ABS 1986 Census.

In Melbourne and Adelaide there were higher concentrations of employment in manufacturing industries, as these were the cities where there were the highest concentrations of the post-war protected industrial development. Brisbane and the Gold Coast, which form south-east Queensland 'sun belt' metropolis, clearly had a services-dominated economy with a small manufacturing sector. This was also the case with Perth. But across the metropolitan cities, the financial and business services, the wholesale and retail trades, and the community services sector had dominated employment growth to the extent that by 1986 these sectors employed 48%

of a total labour force of 6.9 million, compared to 38% only 15 years earlier in 1971.

The magnitude of the structural shift towards a services economy is also evident from the pattern of change in industry sector export performances during the 1980s (see Figure 3.2).

The traditional dominant agricultural sector commodities have experienced little growth, and in many cases slumps. In the mining sector it was coal and gold that grew. In contrast, the services sector has displayed considerable export growth in the producer services, and in particular in tourism, which grew rapidly to account for 6% of GDP in 1990–91 to become a leading export sector. There has also been strong export performance in various manufactures, reflecting the positive outcomes of economic restructuring in enhancing export performance in the new capital-intensive and knowledge-based and high value-added new manufacturing activities that began to emerge by the late 1980s.

By the late 1980s it was clear that there had emerged a fundamental shift in the rates of change in employment from the large business to the small and medium size business sectors. The post-1989 recession led to a blow out in unemployment nationally to exceed 11% by late 1992, with highest levels in the older manufacturing States of Victoria and South Australia, and in Tasmania. Job shedding had become widespread, and sizeable groups of long term structurally unemployed and youth unemployed have emerged. It has been big business, which employs about one-quarter of the national labour force, where job cutting has been most pronounced. For example, over the year to February 1992, there was a fall of almost 16% in employment in private sector companies with over 100 staff. However, there was job growth in the small business and self employed sectors to

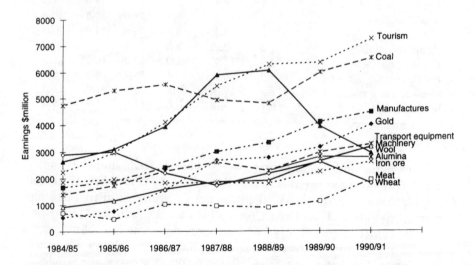

Figure 3.2 Trends in Australian export earnings by industry sector 1984–85 to 1990–91.

partially offset this employment loss, a trend which is similar to that experienced in the US during the 1980s. Shand (1992) suggests that big business is likely to continue this labour shedding trend in order to become internationally competitive as tariff cuts continue and as exporters seek to expand sales.

> Larger businesses, including services, are bench marking productivity levels in domestic plants against their overseas operations, managements are tending to focus on core business and to sell off unrelated activities, and larger companies are pressuring suppliers to cut costs and lift probability (Shand 1992, p.58).

It is likely that only one-quarter of the 1000 biggest enterprises will survive to the year 2000, and that about 700 000 small businesses will have died in Australia (Ruthven 1992). The trend towards employment increases in small business raises a host of policy issues relating to taxation, labour market reform, financial institutional lending policies, and the establishment of enterprise zones.

Patterns of investment

The 1980s was a decade when the pattern of population growth in Australia displayed dramatic shifts in its geographic distribution within a broader framework in which the largest States of New South Wales and Victoria continued to dominate. Between 1981 and 1989 New South Wales and Victoria had growth rates of 10.1% and 9.3%, which were below the Australian average of 12.6%. The resource rich sun-belt States of the north and west had above average rates of growth, as Queensland recorded 20.7% growth, Western Australia 22.4%, and the Northern Territory 27.3%.

Metropolitan development has long been central to the economic growth of Australia. An analysis of the location of commercial building, capital investment and labour force growth by O'Connor (1990) showed that during the 1980s there continued to be a dominance and strong concentration of activity in the two largest metropolitan cities, Sydney and Melbourne. The boom over the last decade in producer services (O'Connor & Edgington 1991) and in media and publishing activity (O'Connor 1991) has also been concentrated heavily in the larger population states and their capital cities.

There were, however, clearly discernible changes in the pattern of distribution of capital expenditure across the states over the decade 1981–1991 (see Figure 3.3).

New South Wales showed a slight increase in its share from 1987–90 for total capital expenditure, manufacturing and non residential construction investment, while those of Victoria tend to be in decline or static, especially for capital investment in manufacturing, financial services and non-residential construction. While the sun-belt States of Queensland and Western Australia have been attracting considerable net gains in population (as discussed later), this has not been matched by their ability to increase a relative share of capital investment on the indices shown in Figure 3.3. In part this was due to the downturn in investment activities in the mining

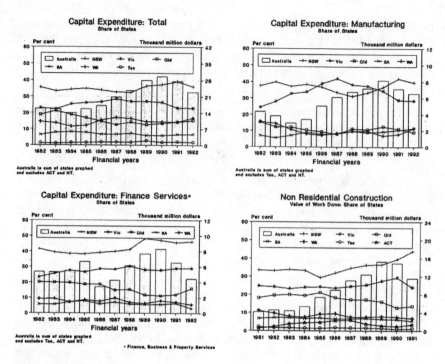

Figure 3.3   Levels and state shares of capital expenditure in Australia, 1981–91: selected indices
Source: Graphs supplied by K. O'Connor, Monash University.

sector. Queensland actually lost share of total capital expenditure from a high point in 1983, when it was level-pegging with Victoria, and it only recovered to third place but with a decreased share by the early 1990s. Its share of non-residential construction actually dropped after 1985, as did its share for finance services until 1990. The dominance of New South Wales was clearly reinforced throughout the 1980s as the major destination of capital expenditure in Australia.

During the 1980s there was a remarkable shift in investment in the metropolitan cities from the manufacturing to the financial, property and services sectors. For example, in Sydney in 1982–83, investment in manufacturing was $A1.15 billion compared to $A1.32 billion in the finance, property and business services sectors. By 1984–85 these levels of investment had changed dramatically to $A0.82 billion in manufacturing and $A1.49 billion in the financial, property and business services sectors. While not matching this scale of investment, similar trends were evident in the other capitals.

The years from 1985–88 were generally boom years for capital city property markets. Sydney emerged as the dominant market in Australia (see Figures 3.4 and 3.5), clearly outstripping other capital cities in both levels of investment and in prime office rents, with capital growth rates averaging

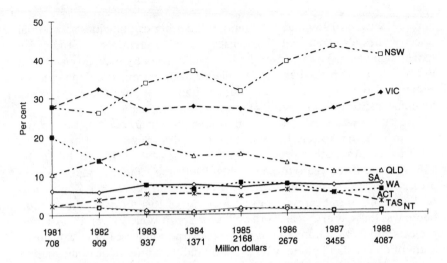

Figure 3.4 Office construction in Australia, 1981–88: state share
Source: ABS Building Activity Survey, Table 2, 1-3-5, Catalogue No. 8752.0, data: value of
work done in each year, summed from Quarterly Publications.

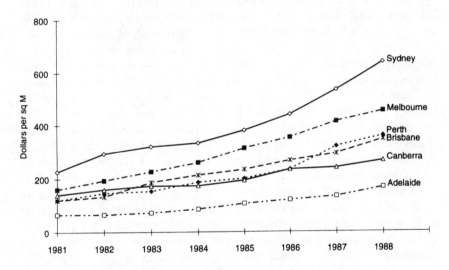

Figure 3.5 Prime office rents in Australian capital cities, 1981–88
Source: Jones Lang Wootton Property Research Pty Limited, Grosvenor Place, 225
George Street, Sydney, 2000.

23% per annum between 1986 and 1988 and total returns reaching 29%
per annum.

But, as with all 'booms', the inevitable 'bust' followed. 'It was a time of
initial response to strong fundamental conditions followed by a state of
euphoria as the fundamentals were swept away by unrealistic expectations'
(Seek & Dickinson 1990). Sydney CBD property prices tumbled by 50%

or more between 1989 and 1992, and in the other capital cities substantial price crashes also occurred. The legacy was an oversupply of CBD office space, high vacancy rates (over 20% in Melbourne, Sydney, Adelaide and Perth, and 14% in Brisbane), the collapse of numerous development and construction companies, and major problems for the banks and other financial institutions in writing off non performing loans.

The 1980s was also an era of enormous growth in construction in the tourism industry sector. In 1982 merely $A400 million was invested in tourist accommodation projects, but this increased to $A1.61 billion by 1989. In the three years to September 1990, the value of major tourism projects under construction or firmly committed more than doubled to $A23 billion. Significantly, the vast majority of this tourism construction boom was located in Queensland and New South Wales. Over 43% of total projects and 37% of hotel rooms under construction were in the 'Sunshine' State of Queensland alone, which clearly had become the leading state in tourism growth in the 1980s.

High tech and R&D concentrations

Work by Andersson & Batten (1988), Tornquist (1983) and Knight (1987) has demonstrated that information and knowledge are underpinning the emergence of advanced manufacturing in the growth of the modern city, and Maillat (1991) has argued that there is a particular 'milieu' that is essential for the development of modern knowledge-intensive industry and services. In an analysis of the geography of R&D activity in Australia, O'Connor (1993, p.16) has demonstrated how there are 'strong spatial biases', given the small base of R&D and the generally low levels of expenditure. The bulk of R&D activity is highly concentrated in New South Wales and Victoria, which retained their dominance throughout the 1980s, even though there were some gains in Queensland and minor losses in South Australia and Western Australia, on some indices (see Table 3.2).

A recent study of 'high tech' activities and 'high tech' office and business industrial park spaces also show the high concentration of new facilities in Sydney and Melbourne, which in 1989 had 20% of new spaces under construction (Jones Lang Wootton 1990). Strong agglomeration in the location of high tech spaces was evident with cities, including the CBD, regional business centres, and a few new specialised business/industry parks, including developments near airports.

Agglomerations of R&D and high tech activities have become magnets for the location of other producer services functions, such as consultancies. The outcome is that 'layer upon layer of new and expanding knowledge-based industries are located in New South Wales and Victoria, and predominantly in Sydney and Melbourne,' and that this 'will go a long way to account for the strengthening role these centres have played in tertiary industry in Australia in recent years' (O'Connor 1993, p.17). Despite the surge of population growth in favour of Queensland, and to a lesser extent Western Australia, this is not underpinned by significant increases in R&D

Table 3.2 R&D activity and performance measures in Australia: State and Territory shares

States and Territories	Location of employment in CSIRO division headquarters in capital city (%) 1990–91	Distribution of Aust. Research Council expenditure (%)		Distribution of national/health and medical research expenditure (%)		Higher degree programme enrolments (%)		R&D expenditure for general government organisation (%)		R&D expenditure for business enterprises (%)		R&D for all manufacturing (%)	
		1984	1991	1984	1991	1984	1991	1986–87	1990–91	1986–87	1990–91	1986–87	1990–91
New South Wales	31.5	35.7	34.3	33.3	24.4	32.6	33.3	42	31	36	37	35	37
Victoria	26.7	23.6	24.0	34.0	42.8	28.4	27.1	46	50	42	40	42	41
Queensland	2.6	11.9	14.2	7.4	10.4	11.5	13.5	5	9	8	5	8	8
South Australia	1.6	14.3	14.2	14.8	12.3	9.3	6.6	4	4	6	9	6	8
Western Australia	3.8	8.8	8.2	8.2	7.1	6.3	11.3	1	2	6	5	5	6
Tasmania	2.6	2.3	3.1	0.8	1.2	2.1	5.5						
Australian Capital Territory	31.2	3.8	3.9	1.5	1.6	9.8	5.5	2	4	2	4	4	3
Northern Territory	–	0.1	0.4	0	0.2	–	1.0						
Total	5126	$22.4m	$23.3m	$23.3m	$85.0m	1008	6965	N/A	N/A	N/A	N/A	N/A	N/A
	Employment	Expenditure		Expenditure		Students							

Source: O'Connor (1992).

activities. However, these activities and public sector functions associated with the Federal Government have been important factors in the recent rapid growth of Canberra, the nation's capital.

Internationalisation and global linkages

Central to the restructuring of the Australian economy in the post-1970 era has been the increasing internationalisation of the economy and the emergence of new sectoral, industry, and metropolitan city and regional global linkages. These are exhibited through trends in foreign investment, international air passenger travel and tourism, and the location of activities and company headquarters which are dependent on global networks.

Foreign investment

Historically, Australia has always been heavily dependent on foreign direct investment to underpin the development of all of its industry sectors, and this was the case with the emergence of the services sectors in the post-1970 era, during which foreign direct investment has been reoriented from the manufacturing, mining and agricultural sectors to the finance, property and services sectors, including tourism.

During the 1980s there was a flood of foreign investment in Australia and a significant change in the source of funds. Between 1983–84 and 1988–89, foreign direct investment into Australia increased on average at 34% per annum, from $A81.9 billion to $A222.9 billion. Even in the manufacturing sector, foreign direct investment had increased by 29% per annum, but it was in the finance, property and business services sectors where the most substantial growth had occurred at 83% per annum, reflecting a high level of interest in investing in Australian property, particularly by Japanese and other Asian investors. Between 1983 and 1989, the share total foreign investment from the US decreased from 24.8 to 18.2%, and that of the UK from 26.5 to 18.8%. Meanwhile Japan's share had increased by 280%, to account for 14.4% of total foreign direct investment by 1989. In the early 1980s, ASEAN and Hong Kong were also significant smaller sources of investment, and since 1990 Singapore, Hong Kong and Taiwan investors became more important sources.

In the second half of the 1980s trading enterprises and banks were the major conduits through which capital entered the country, particularly following deregulation of financial institutions, and Japan was a strong contributor to the transactions of the larger trading enterprises.

The real estate investment boom in the 1980s in Australia was inextricably linked to foreign investment, particularly after 1985 (see Figur 3.6).

In 1985–86 this sector was the destination for 28.1% of total foreign direct investment, but this increased rapidly to 46.1% ($A14.77 billion) by 1988–89, with the Japanese accounting for 33.4% of proposed investments in real estate, followed by the UK with 8.6%, New Zealand 9.4%

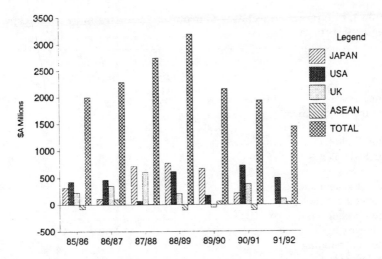

Figure 3.6   Foreign investment in Australia in the finance, property and business services sectors by source of country or region, 1984–85 to 1991–92.

and Hong Kong 7.3%. In 1988–89, when tourism was included as a separate sector in the statistics, it accounted for 15.6% of all foreign direct investment proposals, 70% of which were from Japan. The level of Japanese investments in property and its strong concentration in real estate, and particularly tourism projects, is shown in Table 3.3, p.70.

The geographic distribution of foreign direct investments in Australia in the 1980s showed a distinct and increasing bias towards New South Wales and Queensland. From 1985–86 to 1988–89, New South Wales was the target for 32% of expected total foreign direct investments by value, and Queensland 21%. For real estate proposals, New South Wales accounted for 57%, and Queensland 27%. A significant proportion of New South Wales investment proposals were in commercial real estate projects (44%), while in Queensland 65% were for residential developments. By 1988–89, Queensland attracted 54% and New South Wales 37% of foreign investments by value in tourism, with Japanese investments overwhelming those of other countries (90%). In New South Wales, there were also significant investments in tourism projects from Hong Kong (20%) and ASEAN (18%).

It was the CBDs of the capital cities where the bulk of foreign direct investment in the property and services sectors flowed. In addition, in Queensland the Gold Coast (within the Brisbane–south east Queensland metropolitan region) and Cairns (in far north Queensland) became important locations for Japanese investment in hotels and other tourist facilities.

In the Sydney CBD, between 1975 and 1984, foreign investors had financed about 10% of the total investment in commercial office buildings, and this figure was about 12% in Melbourne. Between 1980 and 1984 these levels dropped to 7% for Sydney and 6% for Melbourne. By 1984, it was

Table 3.3 Japanese purchases of property in Australia by value–type of property, 1986–90

Type	1986 ($m)	1987 ($m)	1988 ($m)	1989 ($m)	1990 ($m)	Total (July) ($m)
Hotel/resort	119.5	16	576.91	593.7	87.3	1393.41
Tourism development site	–	107.024	243.65	118.65	77.7	547.024
Office building	–	55.1	38.175	96.27	111.737	301.282
Retail complex	–	–	150	14.05	–	164.05
Comm./retail development site	–	42.097	–	–	–	42.097
Mixed office/retail complex or retail showroom	–	–	3.8	52.32	41.35	97.47
Industrial building	–	–	–	–	7	7
Residential development site	–	–	35.75	40.155	–	75.905
Residential building	–	–	–	20	–	20
Tourist attraction	–	–	28.8	2.5	–	31.3
Golf course/country club	–	–	13	3.7	–	16.7
Restaurant	–	0.72	–	–	–	0.72
Total	119.5	220.941	1090.085	941.345	325.087	2696.958

estimated that about 15% of CBD office properties in Sydney were foreign owned, 12.5% in Melbourne, 6.1% in Brisbane, 6.7% in Adelaide, and 6% in Perth (Adrian 1984). The boom of the second half of the 1980s resulted in a dramatic upsurge in the flow of foreign funds into major property developments in CBDs around the nation, particularly in Sydney, Melbourne and Brisbane. By 1990, the value of land held by Japanese investors in Sydney's CBD was estimated at $A1.55 billion, most of which had been invested during the mid to late 1980s. In comparison, investments of just over $A300 million were made by UK interests during the whole decade. At the height of the boom in 1988–89, the New South Wales Valuer General placed a value of $A17.4 billion on the land in the City of Sydney (the CBD), thus Japanese interests owned 9% of properties by value. In Melbourne's CBD, where record levels of building activity occurred during the 1980s boom years, foreign investors from 14 nations poured funds into property developments, the Japanese alone investing over $A500 million, with foreign interests adding to make 11% of the CBD property market. In Brisbane foreign investors were also active in the CBD, in which over 40% of the total office floor space was constructed during the seven years to 1990. While UK investors headed the list of foreign owners (30% by floor space), Japanese investors were also active financing 29% by floor space of new buildings.

Since 1988–89, levels of foreign investment have fallen sharply, associated both with the recession and the property bust, and also with the particular internal problems facing Japanese financial institutions in meeting the Bank of International Settlements equity/debt ratio requirements and the downturn in the Tokyo property market.

Australian investment and business involvement overseas

The 1980s was also a decade when there was an increasing involvement of Australian business overseas. While Australian foreign direct investment overseas had increased at a rapid rate of 40% per annum, to increase from $A26.7 billion to $A80.2 billion between 1983–84 and 1988–89, overall there has been generally a restricted level of investment into the Asia-Pacific region, and in particular into the ASEAN, to which Australian exports in manufacturing and services have been growing. Australian foreign direct investment was less than ASEAN foreign direct investment to Australia. Data for 1991 shows that Australia's foreign direct investment continues to be dominated by flows to the US, the UK and New Zealand which accounted for 77% of the total.

During the mid to late 1980s there were a number of studies that were critical of the export performance of Australian business, most of which do not operate on an international basis. While there are notable exceptions, such as BHP and a number of transport and media companies that have developed extensive international operations, in general where Australian businesses have expanded into international markets they have tended to replicate their Australian businesses overseas through acquisition rather than by exporting products from Australia, and they have tended to operate as multi-domestics, rather than as transnational corporations or global firms. But the 1980s was a decade during which there were numerous highly leveraged, and in retrospect disastrous, forays by Australian companies into overseas property markets and manufacturing enterprises.

International air passenger movements and tourism

Indicative of the increasing internationalisation of the Australian economy has been the rapid growth over the period 1983–91 in international air passenger movements (see Figure 3.7, p.72). Sydney has maintained its clear dominance, but Melbourne has been losing share dramatically. The large increase in flows were achieved by Brisbane and the small but rapidly developing tourist city of Cairns, which had particularly rapid growth as the new 'northern gateway' to Australia associated with the boom in international tourism.

International tourism numbers in the 1990s are expected to grow at high rates of at least 7% per annum following average growth rates of almost 11% per annum between 1965 and 1990. Australia's growth in share of Japanese international tourism has been particularly strong. Sydney, Gold Coast/Brisbane and Cairns are increasingly forming an eastern seaboard 'international tourism destination corridor' fed by the rapidly developing Asian market.

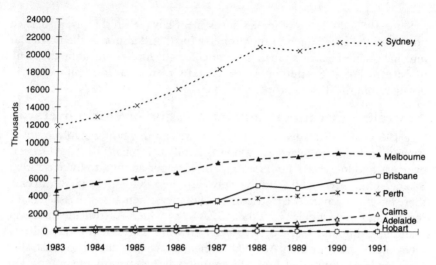

Figure 3.7   International air passenger movements to Australian airports, 1983–91
Source: Department of Transport.

Sydney: international gateway and 'world city'

Sydney stands out as the nation's international gateway and its only 'world city'. Since the 1970s its position of dominance has been reinforced, as illustrated by a range of indices indicating levels of international linkage (see Table 3.4).

Daly & Stimson (1992) have demonstrated that, by the late 1980s, Sydney was the dominant location of activities in the international business and financial sectors. About 150 international institutions were head-quartered in Sydney compared to 43 in Melbourne, and these came from 29 countries, including 48 firms from Japan, 29 from the US and 14 from the UK. Sydney outstripped Melbourne as the headquarter location 10:4 in commercial banking, and 81:6 in merchant banking. Of the 100 largest Australian companies, 60 were headquartered in Sydney in 1989 compared to 45 in 1984, while Melbourne had only 29 company headquarters in 1989 compared to 41 in 1984. In addition, Sydney has a stock exchange that is ranked ninth in the world in terms of capitalisation, it has the eighth largest foreign exchange market in the world measured on the basis of turnover, and it has a futures exchange that is the largest in the Asia region and the eighth largest in the world.

In recent years there has been considerable interest, among Hong Kong firms in particular, in locating regional headquarters in Australia. Federal and State Governments have been active in promoting policies to attract headquarters operations. Overwhelmingly it is Sydney that is the preferred city for such relocations, while the Gold Coast in Brisbane and south-east Queensland is also attracting computer operations, such as that of the Hong Kong Jockey Club to the Bond University Technology Park.

Table 3.4 International activity in Australian metropolitan areas: selected measures

City	Management jobs[a]	Office space[b] ('000 m2)	Hotel rooms[c] ('000)	Hotel occupancy rates[c] (%)	Head offices[d]	International activities[e]	International airport traffic inbound aircraft passenger movements[f] 1985	1985	Population 1986
Sydney	182 471	2 650	14 871	73	50	205	1 395 454	10 093	3 364 858
Melbourne	148 977	1 200	8 815	64	36	156	594 984	5 738	2 832 892
Brisbane	49 115	1 030	6 155	57	4	66	251 372	2 190	1 149 401
Perth	46 686	790	6 160	56	2	79	256 350	1 477	994 472
Adelaide	43 371	460	3 932	69	7	67	55 353	443	977 721
Hobart	7 484	na	na	na	nil	19	7 911	97	175 082

a Group of occupations selected from Australian Standard Classification of Occupations, Census of Australia 1986.
b Jones Lang Wootton (1987) Property 1987, pp. 11–21 for cumulative absorption in each market as per analysis.
c Jones Lang Wootton (1987) Property 1987, p. 63.
d Number of firms in top 100 companies in Australia (Edgington 1983).
e Entries in telephone book labelled 'International'.
f Department of Aviation (1985) International Air Transport.
Source: O'Connor (1993).

Daly & Stimson (1992) suggest that the emergence of Sydney as the nation's 'world city' has been due to its 'ability to absorb the new, its ingenuity, and the opportunities offered by the new system. Melbourne remained conservative and lost ground'. Sydney thus stands out as a 'place where the majority of international contacts can be made' (O'Connor 1990, p. 5). As the Australian economy becomes more internationalised, it can be expected to assume a stronger role as new organisations use it at least as their initial base of operations. In addition, Australian firms dealing with international organisations will find contact with Sydney plays an increasing part of their activity.

Migration trends and the settlement system

Over the past two decades there emerged a new set of demographic forces that have created fundamental changes to the structure of the Australian population and its households, and that have reshaped the magnitude and direction of internal migration flows. Birth rates dropped dramatically to levels around zero population growth. The demographic structure of households changed dramatically, with about half the households becoming one- or two-person households by the mid-1980s. Levels of immigration dropped in the 1970s, and the later upturn in the 1980s was accompanied by a shift away from the traditional UK and European sources to Asian and Middle Eastern countries. When linked with the process of structural economic change, a new structure for the national settlement system has begun to emerge.

Internal migration: the emergence of the sun belt

The immediate post-war decades were characterised by: the movement of populations from country regions to the rapidly developing metropolitan cities; high birth rates; and high levels of immigration, particularly to Melbourne and Sydney. Since the early 1970s, a discernible pattern has emerged in which internal migration flows in Australia have been characterised by: a continuation of a long-established trend of movement within the metropolitan regions from the inner suburbs to the outer suburban areas; a marked shift in the broader regional patterns of net migration gains in favour of the sun-belt States of Queensland and Western Australia, at the expense of the larger States of New South Wales and Victoria and of the small States of South Australia and Tasmania; and movement out of the metropolitan regions predominantly to the emerging urban regions along the eastern, south-eastern and south-western seaboards, in what has been seen as a 'counter urbanisation' trend.

In the 1970s a popular argument was that the changing location of job growth and changing values of metropolitan and non-metropolitan wages were not major factors in the shift in patterns of internal migration. It was suggested that the turnaround in metropolitan migration was population-

led, either by persons not constrained by job location, such as retirees, or
by persons in the labour force prepared to trade to lower wages and lower
material standards of living in environmentally high-quality locations.
However, this view was strongly challenged, notably by Jarvie (1989), who
argued that migration patterns were in response to the changing location of
labour demand. While retirement migration was certainly part of the sun-
belt migration phenomenon, the majority of male net migration gains were
due to labour force factors. There was a significant change in the location
of population growth nationally, and much of this was related to the
structural changes in the Australian economy which began in the 1970s.

Bell's (1992) recent analysis of patterns of internal migration in Australia
between 1981 and 1986 has shown strong migration linkages between the
eastern mainland states, with substantial net gains by Queensland from
New South Wales and Victoria, as well as net gains by Western Australia
and the Australian Capital Territory (see Figure 3.8).

There continued a trend towards counter-urbanisation in the eastern
states, the south-east Queensland region metropolis emerged as potentially
the nation's first truly 'multi-centred growth node metropolitan region', and
high growth was experienced in many of the eastern and south-eastern
seaboard urban centres. Within the major metropolitan regions, the trend

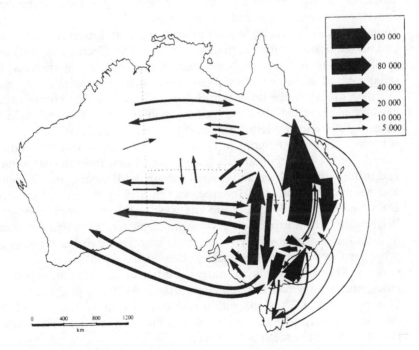

Note: Flows of less than 5000 persons are not shown.

Figure 3.8   Interstate migration flows in Australia, 1981–86
Source: Bell 1992.

towards suburbanisation of the population continued. Bell verified the importance of structural changes in the economy as an explanatory factor in the pattern of internal migration. He concluded that, while there was no simple relationship between migration and regional economic growth, nonetheless there is a close relationship between migration and regional economic performance. The most highly mobile segments of the population and labour force were: the unemployed, wages and salary earners, and the self-employed; and those in the public administration, recreation, professional and para professional, and sales and service industry sectors.

A classification of regions based on internal migration trends

Bell (1992, pp.82–86) also conducted a multivariate analysis of statistical regions in Australia based on characteristics and patterns of internal migration. Eight groupings or classifications of regions were identified to provide a typology of migration regions. The eight groups of regions identified and their characteristics are:

- *Group 1* The inner suburbs of the metropolitan cities of Sydney, Melbourne, Brisbane and Perth, which are areas of net migration loss to other parts of their metropolitan regions, reflecting patterns of outward movement from the centre to the urban periphery, a long-established feature of the development of Australian cities.
- *Groups 2 and 3* These regions occur in all of the state metropolitan cities and are associated with the middle suburbs. The Group 3 regions are similar to those in Group 1, being characterised by net migration loss to other parts of the same metropolitan area through out-migration at rates less than that from the inner suburbs, and they cover many of the older established areas near the central cities, including the inner- and middle-eastern suburbs of Melbourne and the inner- and middle-western suburbs of Sydney. Regions in Group 2, which are mainly the middle suburbs of the metropolitan cities, are more distant from the city centre and include the northern and eastern suburbs of Brisbane, the northern and western suburbs of Melbourne, and in Sydney the belt of suburbs from Liverpool to St George in the west and south from Manly to Warringah in the north. These regions are characterised by medium rates of intra-regional mobility with relatively small net gains and losses to other parts of the city to non-metropolitan areas or intrastate.
- *Group 4* These are the broad outer suburban areas of the metropolitan cities, and are regions of population growth fed by immigration from Groups 1, 2 and 3 regions within that city as well as from elsewhere outside that metropolitan area. There is some net migration loss to other regions outside the metropolitan areas.
- *Group 5* There are 14 coastal regions in a broad and continuous band along the east and south-east coast of Australia from Wide Bay in Queensland to the south-east of South Australia, interrupted by the capital city metropolitan regions, plus the lower western region of

Western Australia and large areas of metropolitan Perth. These are a mixed group of regions and are the principal in-migration places for those leaving the capital cities. They have high rates of intra-regional mobility and they gain also from interstate migration.

- *Group 6* There are seven large statistical regions across four states that make up the interior of Australia. These are the 'outback' areas, characterised by high levels of intra-regional movement and losses through net migration to the rest of Australia.
- *Group 7* The south-east Queensland region was the most rapidly growing region of Australia, comprising the Brisbane and Moreton Statistical Divisions. Apart from those areas of Brisbane included in Groups 1, 2 and 4, this region's large net migration gains are from interstate and the rest of Queensland, and in particular from Sydney and Melbourne.
- *Group 8* Non-metropolitan Queensland forms a distinct set of regions that are characterised by high levels of intra-regional mobility and net gains from interstate migration, particularly directed to the coastal regions including the tourist resort cities, the sugar regions and the manufacturing towns.

A new settlement pattern

Traditionally Australia has been a highly urbanised nation, and during the high growth and boom years following World War II, the primacy of the metropolitan capital cities was strengthened (see Table 3.5). Most states have a poorly developed urban hierarchy with a large gap between the size of the capital metropolis and the lower orders of urban centres.

The nature, direction and magnitude of changes that have occurred in both the demographic and economic structure of the nation since 1971 have given rise to what Paris (1992) sees as a new pattern of national settlement. He has identified significant differentiation with the metropolitan city regions: the 'global city megapolis' covering the Sydney-Newcastle-Illawarra region; the 'Melbourne ring'; and the 'south-east

Table 3.5 Percentage of State population in State capital city statistical divisions, 1971–86

Census counts	1971	1976	1981	1986
New South Wales	63.8	63.3	62.5	62.3
Victoria	71.5	71.4	71.0	70.5
Queensland	47.6	47.0	44.8	44.4
South Australia	71.8	72.3	72.5	72.6
Western Australia	68.2	70.4	70.6	70.7
Tasmania	39.2	40.2	40.2	40.1
Total	63.9	63.8	62.9	62.5

Source: ABS Census data.

Queensland strip'. There are the smaller Perth and Adelaide metropolitan cities. The eastern and south-eastern seaboard regions are developing a linked linear system of rapidly growing urban regions, based on both older coastal resort and service towns and on new low density urban areas, particularly between the Wide Bay region north of Brisbane and the Gippsland region west of Melbourne. There are significant major urban regions developing along the tropical coast of Queensland, particularly around the older and industrial cities of Gladstone, Townsville and the new 'northern gateway' city of Cairns. In the inland regions ('the bush'), the urban system is characterised by a mixture of towns in decline (mining, former industrial and railway towns), and regional centres that are traditional service centres that are strategically located on the road and rail transport network or have specialist features, such as educational facilities.

New spatial structures within cities

Within the major metropolitan cities throughout the world, the transformation of urban economies from a manufacturing production base to a services and information base, along with the suburbanisation trends in settlement patterns, are generating a new urban spatial geography which adapts old spaces to new processes in addition to creating significant new urban spaces. Australian metropolitan cities are exhibiting new spatial structures and forms similar to those emerging elsewhere, particularly in the Pacific Rim Region (Blakely & Stimson 1992).

Services sector and information-based spatial forms

The processes of globalisation and internationalisation, and the increasing dominance of the services sectors in an information economy, have 'unleashed locational forces that have given rise to new urban structural arrangements' (Daniels et al. 1991). Consumer services have become diffused throughout metropolitan space, reflecting the increasing dispersal of residential populations. Producer services have shown strong tendencies towards locational centrality, and the centrality of producer services is not confined to a preference for the CBD, with an accelerated growth in the trend for these to disperse to nucleated suburban locations.

Within Australian metropolitan cities these processes are at work. The CBDs have become more specialised, particularly in Sydney, with its domination of large corporate head offices, multinational enterprises, banking and financial institutions, and other producer services. The CBD is the particular domain of the internationally linked activities. Major suburban business centres, some of which exhibit the 'edge city' phenomenon (Garreau 1991), are becoming major employment and activity nodes that are far more functionally diversified than the regional shopping centres that evolved from the 1960s. The extent of development of these suburban centres in Sydney is illustrated in Table 3.6, which shows that, by the year

2011, it is possible the distribution of employment between the major regional and subregional centres could change dramatically compared to 1981, depending on whether a dispersed or concentrated spatial form evolves.

As current market and technological trends seem to be leading to a dispersed outcome, the result could be a proliferation of freestanding offices, office parks and retail developments, which would account for about 80% of total metropolitan employment, with 21% being concentrated mainly in the CBD and the major nodes of North Sydney and Parramatta, plus a plethora of smaller subregional business centres. Under a concentrated development scenario, the share of the metropolitan region's employment in regional centres increases considerably, as does the

Table 3.6   Potential growth of employment centres in Sydney, 1981–2001

Centre	Employment at 3.25 million (1981)[a]	Employment at 4.5 million (2011)[b] concentrated option	dispersed option
Regional centres			
Sydney CBD	188 919	220 000	150 000
North Sydney	28 750	40 000	40 000
Parramatta	20 360	60 000	35 000
Sub-regional Centres			
Bankstown	9 727	15 000	10 000
Blacktown	10 592	18 000	13 000
Bondi Junction	6 095	10 000	7 500
Burwood	7 355	10 000	7 500
Chatswood	9 363	20 000	20 000
Campbelltown	4 729	30 000	11 000
Gosford	5 233	10 000	6 500
Hornsby	9 637	15 000	10 000
Hurstville	6 978	10 000	7 000
Liverpool	10 904	20 000	12 000
Mt Druitt	1 746	5 000	3 000
Penrith	3 703	20 000	11 000
St Leonards[c]	22 983	20 000	20 000
Sutherland[c]	5 524	10 000	8 000
Wyong-Tuggerah	–	5 000	3 000
Bringelly Sector Centre	–	12 000	7 000
NW Sector Centre	–	8 000	5 000
Total major centres	352 598	558 000	386 500
Share of total employment	23%	30%	21%

[a] Transport Study Group, 1981 Travel Survey. Figures exclude students working part-time.
[b] Department of Environment and Planning, NSW.
[c] Includes industrial and special uses area employment outside centre but in same traffic zones.
Source: Wilmoth 1987.

employment role of the subregional business centres, to account for 30%. Some very significant suburban activity and employment nodes are likely to emerge.

Techno-spaces

Some locations in metropolitan areas are increasingly important as specialised technology-based nodes, or 'techno-spaces' (Blakely 1992, p.9). While some of these are in existing high-activity agglomerations, such as CBDs and emerging 'edge cities', others are at new nodes outside the core city areas, attracted by R&D concentrations, often associated with universities and government laboratories, that are primarily suburban in location. As demonstrated by O'Connor (1993) and Jones Lang Wootton (1990), Sydney and Melbourne in particular have developed new clusters of R&D activities.

Additionally, new manufacturing and commercial complexes are evolving that are associated with international airports (Kasarda 1991), to create new economic landscapes, such as those developing in Sydney associated with the expansion of the International Airport at Mascot. The Brisbane ports area potentially could be a significant component of the development of the south-east Queensland metropolis, as outlined in the 'gateway strategy' for economic development of that region (Stimson 1991). Airport and port activities also formed a significant component of Melbourne's development as a major distribution centre.

Polycentric urban forms and duality within the metropolis

An overall outcome of the changes impacting on cities is the emergence of complex polycentric urban forms in metropolitan regions. These have a new hierarchy of business centres, 'techno-spaces', and industrial-commercial complexes. They reflect the changing nature of information-intensive services and knowledge-based production processes; the development of just-in-time delivery systems; a blurring of what constitutes industrial, commercial and retail functions; and the development of 'smart infrastructure' networks based on new telecommunications and transportation technologies and the necessity for global linkages. The 'global megalopolis' region of Sydney–Newcastle–Illawarra is an example of this form of new city region, as in the evolution of the south east Queensland sun belt metropolis region. Traditional approaches to planning the development of the new form of metropolitan city region are increasingly irrelevant, as they are based on the assumed locational forces that pertained to economic processes of a bygone age (Stimson 1992).

Also, a new form of duality has emerged within cities with different and extreme divisions of social and economic differentiation compared to those that traditionally typified urban areas. Richardson (1989) has referred to:

• a bifurcation of the labour market, the development of 'citadels' and 'ghettos';

- the widening fluctuations in metropolitan economies as exposure to the hypermobility of capital, exchange rates and resource price changes;
- erosion of the metropolitan tax base as a result of competitive industries to attract and retain internationally mobile economic activities;
- a shift from investment in social overhead capital to economic overhead capita; and
- a rise in CBD land and property values.

As a result, highly differentiated housing markets and increasing segregation have developed. Daly & Stimson (1992) have described a duality in metropolitan Sydney that exhibits an 'urban core of high status suburbs around the Sydney Harbour and covering the north shore and inner suburbs, that are linked to the CBD and its growth services sector of 'high flying' activities associated with the international economy and knowledge-based activities. In contrast there are the lower status sprawling suburbs to the west, north west and south west sectors of the metropolis, where the bulk of the population lives and where the painful processes of economic restructuring have impacted most severely, with attendant high levels of structural and youth unemployment'.

A new regional dichotomy?: the 'sun belt' and the 'rust belt'

It would appear that a dichotomy is emerging in the Australian space economy. There are the higher growth sun-belt States of Queensland and Western Australia, and the continuing dominant position of New South Wales with its 'world city', Sydney. In contrast are the rust-belt States of Victoria, South Australia and Tasmania, which have been impacted most severely by the economic restructuring processes of the last two decades and have suffered most from the ravages of the 1990s recession. To an extent the dichotomy reflects difference in the industry sectoral structures between the services-dominated sun-belt states, which also have substantial mining and agricultural sectors, and the legacy of the era of high industry protection which underpinned the manufacturing base to the economies of Victoria and South Australia in the rust-belt. Victoria has been in crisis, fuelled also by the collapse of financial institutions (including the State Bank), escalating State public sector debt, and loss of consumer and investment confidence.

> The State's drama has scared its population and changed self-perceptions, especially in Melbourne. For the first time this century, Melbourne people are seriously confronting the possibility that their property values might not recover (*The Australian*, 30 December 1992).

In contrast, Queensland's economy grows at double the rate of the nation as a whole, migrants continue to flow in at high rates, and the tourism industry continues to boom.

Predicting the continuing impacts of the structural changes occurring in the Australian economy is difficult. Significant new trends have emerged in the labour force, and there are likely to be considerable variations in the potential levels of demand for specific types of labour, particularly between the industry sectors of employment and as well as between regions.

A set of projections prepared by the Commonwealth Government Department of Employment, Education and Training (DEET 1991) show that there may be a somewhat different national pattern of industry output and growth over the remainder of the decade to 2001 compared to the previous two decades, reflecting the increasing reality of the sun-belt and the rust-belt states dichotomy that has been suggested. The study predicts that the rural, mining and manufacturing industries may grow more strongly than overall output, and that there may be a reversal of the trend of decline in the share of manufacturing in output contribution. The recent rapid growth in the services industries, such as banking and finance and community services, is expected to moderate, particularly in their contribution to output, while strong growth is anticipated in the construction, transport (especially air), and recreation and personal services industry sector. Micro-economic reform has the potential to add to overall growth of output, particularly in air transport and communication industries. For industry sector employment, the projections show that increasingly it is the services sector where employment growth is most likely, particularly in the recreation, personal and other services sector. Construction will also have employment growth, as well the wholesale and retail trades, and there would be more modest growth in public administration. Low increases in employment are likely in transport and storage because of micro-economic reform and productivity gains, while in electricity, gas and water, and in agriculture, employment falls are envisaged.

The states are not expected to have equitable shares of future employment growth, with a shift in the labour force from the larger states to the growth states of Queensland and Western Australia, which are projected to have 25% and 17% respectively in the expansion of their working age population. The distribution and expansion of employment across industry sectors within the states over the 1990s will depend on national industry trends, specific State industry mixes and economic and government policy circumstances, and factors such as future mineral discoveries, state transport and infrastructure, and even climatic conditions. The sun-belt State of Queensland is expected to gain above average shares of new jobs in mining, construction, wholesale and retail, and recreation sectors, while it is likely to have below average performance in its share of jobs in manufacturing, finance, public administration and community services. Also, Queensland's rate of decline in the share of new jobs will be greater than average in the agriculture and the electricity, gas and water sectors. A similar pattern is projected for Western Australia.

The pattern of distribution in the share of new jobs and employment share within industry sectors across the States and Territories will not be even, and there will be considerable differences in performance in the 1990s compared to the 1980s. The agriculture, and electricity, gas and water sectors are projected to have declining employment levels, and the States of Queensland, Tasmania, Western Australia and South Australia will have a relatively greater level of fall in employment because of their current higher proportions of labour force concentrations in these sectors. The construction, wholesale and retail, community services, and recreation and personal service sectors are likely to have the largest shares of employment growth, and this will be reflected across all States. The manufacturing sector in New South Wales and Victoria is expected to share in the expansion of employment in this sector, following a period of employment decline in this sector in the 1980–90 decade, during which Queensland and Western Australia were the only States to record employment growth in manufacturing. Over the decade to 2001, the relative share of manu-facturing employment is expected to fall more than it did in the 1980s across all the States. The projected share of expansion of employment accounted for by the construction sector in the smaller States will be far higher than that experienced in the 1980s, with Queensland, South Australia and Western Australia likely to benefit to the greatest relative degree. The finance sector, which accounted for so much of the growth in many States in the 1980s, will provide smaller shares of employment expansion in the 1990s, especially in New South Wales.

The DEET (1992, p.104) study concluded that:

> throughout the 1980s there has been a shift in the working age population and employment from the larger States, New South Wales and Victoria, to Queens-land and Western Australia. The small States of South Australia and Tasmania also experienced falls in their shares of the working age population and employ-ment. These trends are projected to continue during the next 10 years. If the national industry trends are reflected at the State level, there will be a substantial shift of employment within States towards the construction industry and service sector—wholesale and retail, community services and recreation. The signifi-cance of this shift will vary from State to State depending on their current industrial structure. States which already have a relatively large proportion of work forces in these industries will experience a greater shift in their employment structures. The processes on employment growth exerted by this structural change are likely to be different to those experienced by States during the past ten years. Consequently, the growth of each State's employment in the 1990s will be absorbed in the main by a different mix of industries from those over the last ten years.

Challenges ahead

Many challenges face Australia in the 1990s. The impacts of globalisation, internationalisation, and structural economic and demographic change have been profound in the 1970s and 1980s. They have resulted in major changes in the industry sector patterns of economic activity and employ-ment, the pattern and distribution of population and economic growth, the nature of the national settlement system, and the internal structure of metropolitan regions.

Australia entered the 1990s facing a future about which there are considerable uncertainties. The post-deregulation, freewheeling financial world of the 1980s had produced for Australia one of the world's largest international debts, enormous corporate debt, the downgrading of the nation's (and some of the State's) credit rating and that of its largest banks. Historically high real interest rates both accelerated corporate collapses and stood in the way of more production investments. A chronic balance of payments problem was caused by too heavy a reliance on a limited range of commodity exports, and the impact of invisibles in the current account, largely linked to financial relationships (Daly & Stimson 1991). The impacts of change have been distributed unevenly, both in sectoral and geographic terms, resulting in significant changes in the nature of the nation's space economy, and there is considerable uncertainty as to how it will evolve and what it will look like in one or two decades.

International competitiveness

A major issue is the competitiveness and productivity of Australia's industries, and the need for 'world best practice' to be incorporated into management and production processes.

It is instructive to look at the joint IMEDE-World Economic Forum *World Competitiveness Report*, which ranks the 23 OECD nations on a co-point scale based on 326 criteria. In 1989 Australia was ranked only tenth, but had slipped to sixteenth place by 1992, to be well below the average, and to also trail many of the Asian NICs and emerging NICs (see Figure 3.9).

How to address the deteriorating relative international competitive position of the nation is a major challenge. Industrial competitiveness

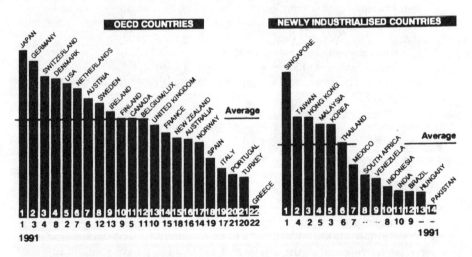

Figure 3.9 1992 world competitiveness report
Source: IMEDE–World Economic Forum.

improvement requires action simultaneously on a whole range of fronts, including continued lowering of industry protection, micro-economic reform (especially in the transportation sector), labour market deregulation and work practice reform, taxation reform, greater levels of commitment to investment in 'smart infrastructure', and higher levels of savings and investment. Fundamental to the competitiveness of Australia's manufacturing and services industries will be the development of strategic alliances and linkages to offshore locations. The Centre for Technology and Social Change (1990) has examined the importance of policies and practices to facilitate the expansion of Australia's highly geographically clustered services industries, most of which operate only nationally or regionally. A challenge for firms in the services industries is to 'find the correct mix of global, regional and national strategies for each component of its operations' (Langdale 1991, pp.10–11).

The debt problem

Australia has a significant level of international debt, which rose spectacularly from a little over $A10 billion in the early 1980s to top $A175 billion in 1992, with predictions that it will pass the $A200 billion mark by late 1994. This represents an increase from 11% to about 45% of GDP within the space of just a decade. Australia's foreign debt is now one of the highest in the world, well ahead of that of Mexico and Argentina, and it is increasing at one of the highest rates. In addition to the level of national debt, there is a chronic current account deficit, which peaked at $A22.3 billion at the end of the 1980s, before settling at about $A15.6 billion in 1990–91. Much of the debt escalation in the 1980s was private debt, which was more than four times the level of public sector debt by 1990–91, and some 42% of it was in the finance, property and business sectors. Most of the private sector debt is short-term, with the majority of the debt being purchased at high interest rates, in a decade when inflation had increased from 9.5% in January 1980 to fluctuate into extended periods of double-digit inflation from 1980–84 and from 1986–90, before being brought back to current levels of about 2%. As a result, capital outflow accounted for a large component of the current account deficit, amounting to 75% in 1990–91. Each year there continues a large gap of about $A16 billion that adds to the accumulated debt, and all up there is a deficit of about $A20 billion each year on services. The current account deficit had risen to 6.1% of GDP.

Thus, Australia has become a dependent nation, based more on capital market relationships than on the traditional forms of dependency based on trade relationships and technological dependence (Daly & Stimson 1994).

It is significant, however, that in the six years to mid-1992 the growth in exports, and in particular in manufactured goods which averaged 15% per annum, occurred as tariff barriers continued to be dismantled, and that some 700 firms had emerged as competitive exporters. Tourism also has

emerged as one of the new opportunities for Australia to improve significantly its capacity to earn foreign exchange, especially as terms of trade of
its export industries (agriculture and mining) have been slowing and with
significant growth prospects in manufacturing exports. Tourism is seen as
a 'competitive, highly assisted, traded goods industry with good prospects
for growth in a growing export market' (EPAC 1991, p.2). Tourism also
has great potential for employment growth in the future (Ruthven 1992,
p.21). The main culprit in the balance of payments deficit appears to be
financial services, and especially interest costs in servicing foreign debt. In
addition, low levels of national savings (5% of GDP compared to 11% in
Japan, 23% in Singapore, and 31% in Taiwan) exacerbate the dependency
on overseas borrowings.

Conclusion

It is not yet clear that the regional dichotomy in Australia's space economy,
as exhibited by the pattern of recent population and economic growth and
as suggested by the projections made in the DEET (1991) study, is
currently being matched fully by patterns of capital investment. The DEET
study of potential levels and patterns of growth in industry sector output
and employment in Australia over the decade to 2001 suggested that much
will depend on increased export performance, through increased productivity and improved competitiveness, on stabilising the level of foreign debt,
and on more rapid progress in micro-economic reforms, in particular in
utilities, transport and communications. The analyses conducted by
O'Connor (1990; 1991; 1992; 1993) suggest the largest population States
of New South Wales and Victoria (traditionally the dominant forces in the
national economy), with the nation's two largest cities, Sydney and
Melbourne, will continue largely to retain their shares of the national
economy. Of particular significance is the high concentration of R&D and
'high-tech' activity and investments in Sydney and Melbourne. Growth
regions such as the south-east Queensland and Perth metropolitan regions
continue to record relatively poor shares of activity and investment, while
Melbourne, despite the devastation of the 1990s recession and the impacts
of economic restructuring, has disproportionately high shares of jobs and
firms in the R&D and 'high-tech' sector. O'Connor suggests that this raises
the question of the important long-term role that 'clusters of research and
development' may have on the pattern of urban growth, particularly when
it is inferred in much of the recent research that 'knowledge-based activity
will be central to the long-term development of nations and cities'.
O'Connor (1993, p.17) concludes that 'in these terms one can expect
Victoria and New South Wales to retain strong shares of the national
economy and national population in the immediate future'.

Alternatively, it might be argued that the Australian regional space
economy is currently undergoing the type of shifts associated with sun-belt

population growth that characterised the earlier emergence of the US sun-belt regions, such as California, Texas and Florida. The south east Queensland metropolitan region in particular is likely to diversify its economic base into higher value-added, knowledge-based activities over the next two decades as its population grows from 1.8 million to a projected 12.8 million, provided the 'smart infrastructure' investments are made, as suggested by Stimson (1991).

Certainly government policies and decisions on where public expenditures occur spatially, will continue to have substantial impacts on the locations where significant new and emerging economic activities might locate. The decision to locate the multi-function polis in Adelaide, a city that is poorly linked globally and which has a struggling regional economy and low levels of population growth, is a case in point, even though initially a site between Brisbane and the Gold Coast had been selected in a region that appears to be more oriented to the 'future' than trapped in the 'past'.

What the future nature and the structure of the space economy of Australia's regions and cities will be is thus somewhat unclear, there being contradictory messages in some trends and potentials. The amenity-oriented population and services industry growth, particularly in sectors such as tourism, seems set to continue in the sun-belt regions, which will gradually diversify beyond their current consumer-oriented industrial sector mix. Many of the new industries and the refinements of old industries will be attracted to the existing agglomerations of R&D, infrastructure, skilled labour markets and international networks of Sydney and Melbourne (O'Connor 1993, p.19). However, environmental deterioration and the physical constraints on growth in the largest cities may become longer term negative factors. The emergence of the settlement patterns identified and proposed by Bell (1992) and Paris (1992) seems highly likely, with the continued dominance of Sydney and Melbourne as major centres of economic activity, and with Sydney being increasingly important as a Australia's 'world city'. Brisbane–Gold Coast and Perth may remain as lower value-added growth metropolises. And it would seem to be fairly certain that a linked linear urban system incorporating new urban developments from Cairns to Gippsland, associated with tourism, retirement, consumer services and local market manufacturing will stretch between the metropolitan spaces along the eastern and south-eastern seaboard. Within the major metropolitan regions, multi-centred spatial structures seem to be the future emerging pattern, incorporating both old and new agglomerations. These include the CBD, 'edge cities', 'techno-spaces', airport-oriented commercial-industrial complexes, and suburban centres. Duality within the cities is likely to intensify because of an increasingly bifurcated labour market, necessitating increasing rates of inter-regional mobility to achieve adjustments between where people live and the availability of employment in industry sectors in growth and decline.

4 Change in the Pattern of Airline Services and City Development

Kevin O'Connor[1]

A case study of Qantas 1970–1990

AIR transport and air travel are important influences on the level and diversity of the current activity within cities, as well as being important to their long-term future. Many activities depend upon the structure and performance of the air transport industry for their vitality. For example, air travel allows face-to-face contact between businesses, provides customers for the tourism industry, and moves a considerable range of freight. As the economy restructures toward knowledge-intensive production, which may require more face-to-face contact, and globalisation of production proceeds, the frequency and connections of air services at a city will be an even more important influence on its vitality.

Air transport is perhaps the most locationally constrained of any transport mode. In most nations, one or two places alone account for the majority of air traffic, so that sectors relying upon air movement of goods or people may have a limited choice of locations. In contrast, sea ports are more dispersed, and rail and road systems seem almost ubiquitous in comparison. For a range of reasons then, the pattern of air transport, especially the number and diversity of air services at the local airport, can be a major influence on activity within cities. The aim of this chapter is to understand the evolution in the pattern of air services between cities, and to use that understanding to contribute to the discussion of the long-term future of urban areas. In very broad terms, the chapter proposes that changes in the pattern of air services to cities depends on changes in technology, markets and regulation; these are analysed and evaluated in turn.

Given the importance of air transport, and its selective geographic impact, it is surprising that there has been scant attention to its dynamics in the mainstream economic geography literature. The impact of transport on urban development is well understood in general terms, though this understanding has been built up from studies of sea, rail and road systems.

There has been little attention to the impact of air transport on city development.

In an overview, Andersson (1986) casts urban history in the context of logistics, and suggests that there could be new locational outcomes as a new logistical problem emerges around the movement of information and people. Air transport is playing a central role in the latter period of Andersson's analysis; this role can be seen in a number of areas. For example, Dunning & Norman (1987) have shown that air services are an important factor in the location of the regional offices of multi-locational firms. Within manufacturing industry, the application of just-in-time production systems and their vertical disintegration across and between continents has been shown to rely heavily upon airfreight (Lieb & Miller 1988). Clark (1992) has suggested that a new form of networked production has emerged in industries where design and customer choice are critical parts of the production process; for these networked production systems, air transport of key personnel, and of some products and services, is central to their method of operation. These studies indicate that for many activities, a regular air service is an important element in deciding on locations for business activity. These examples also underpin some of the findings of Goetz. In one of the very few analyses of links between urban change and air passenger traffic, he found '...increased utilisation of air transport is also a factor in subsequent population and employment growth', although the link was stronger in the 1960s than recently (Goetz 1992, p.235). The overall impact of the air transport industry is underscored by its size and rate of growth. As a guide, the International Civil Aviation Organisation (ICAO 1989) recently estimated that the number of passengers carried by the world's airlines approximated 20% of the world's population, and 25% of the world's trade by value is carried by air. These aspects indicate that there is a large market for air transport; in turn, the air transport industry can be expected to have a big impact on cities.

Some of these local effects are illustrated by the clustering of activity, as Hoare's (1974) pioneering work on Heathrow showed. Many consultant studies show that air traffic has a wide range of direct and indirect effects at and around an airport. In a few cities, the airport provides a major focus in the operation of the metropolitan area, as the activity cluster is so large. The number and diversity of origin of daily flights, as well as the physical scope for airport-related growth, seem to be the main factors that account for the scale of airport-related activities. In this way a city's position in the global air network can be felt in the pattern of its internal structure, extending the influence that air transport has on a place down to the local scale.

The industry has changed rapidly in the immediate past, largely due to changes in technology. Dicken (1986, p.108) has modified a diagram of McHale to show how air transport has 'shrunk the world': in terms of travelling time, the globe in the 1960s was about the same size as Europe in 1850. This shrinkage has changed the locational context for many places and firms. Those that would have been regarded in the past as isolated, can no longer be described as such. For example, O'Connor & Scott (1992) show how Tokyo is now a hub in international, daily, non-stop air services. Only 20 years earlier, it had just local and regional scale connections on a non-stop daily basis. The experience of Tokyo has been repeated in just a few centres within the global airline network, creating a few dominant hubs dispersed around the globe.

Seeing the impact of air transport on cities just as a logistical problem is however a narrow perspective. A very important part of both international and national air transport is the regulatory structure that controls flight frequency and direction. In fact, simple changes in regulatory arrangements could have major consequences in terms of logistics, as has been seen in the US with deregulation of the airline industry. In a sense this does not appear any different from road, rail and sea transport, where regulation of activity has long been a feature of each sector's operation. But the air sector is different, especially in its international dimension, as regulations involve formal international agreements between sovereign nations, often on behalf of nationally operated airlines. This introduces a level of complexity that is not found in other forms of transport. The structure and application of this regulatory regime is a fundamental influence on the operational networks of airlines; the airport activity at a given place will be substantially influenced by this feature of the industry.

This background suggests air transport has been and will remain an important influence on the vitality of cities, and that its effect is felt through the pattern of airline services to and from a city. The core idea in the chapter is that changes in the pattern of airline services to cities reflects the interdependent effects of technology, markets and regulation. This will be explored through an analysis of the pattern of services of one international airline. The objective of the study is to show the contribution that changes in technology, markets and regulation have made to the evolution of the Qantas network between 1965 and 1990, with particular attention to changes in the number of flights to individual places around the globe.

The Qantas network 1965–90

Qantas emerged from domestic service in the isolated parts of Australia to pioneer long-distance flights connecting Australia and England (Gunn 1988; 1990a; 1990b). Details of these early intercontinental flights are provided in the first ABC Air Guide: in 1946, London–Sydney was then a three day journey by plane, and five days by flying boat, which involved accommodation at overnight stops (ABC Air Guide 1946, Tables 11, 14). From the experience of long-distance operation and remote area management involved in flying across southern Asia and the Middle East, Qantas expanded and provided a round-the-world service in the late 1960s, and has become one of the world's major airlines. It was ranked 15th in the world on passengers carried and profit, and 21st on profitability, by *Fortune* magazine in 1992 (*Fortune* 1992).

The Qantas network has changed considerably in the period 1970–90, as it came to terms with jet aircraft technology, new markets and bilateral negotiations in the international marketplace. Figure 4.1 a, b on p.92/93 shows how its network has evolved over the past 25 years.

These maps are based on actual flights listed in the ABC World Airways Guide for the month of June in the years shown. They show that Qantas in 1970 was a worldwide carrier, with a round-the-world service, linking a diverse set of locations across several continents. By 1990, this network had shrunk to focus primarily on the Asia–Pacific segment of the globe, with fewer destinations served across the world. The change took place in a period in which the total number of passengers increased from 790 000 in 1970 to 4 million in 1990.

A more detailed study of these patterns can be carried out by analysing the data in Table 4.1 on p.94, which shows the number of cities served, the total number of flights per week to each city, and the total number of sectors flown.

Outside Australia, Qantas is flying to fewer places, with fewer sectors between cities, but there has been an increase in the total activity, measured by flights per week to all cities. This indicates that Qantas activity has become more concentrated on a smaller number of places; in turn, fewer cities felt the impact of its activity. Within Australia it is serving more locations, so its impact will be spread beyond its home base in Sydney. If the change in the pattern of air transport shown in the Qantas data is typical of all airlines, then changes in air transport between 1970 and 1990 have had an increased impact (due to more passengers and flights) on a smaller number of places around the globe.

More information on the change in the pattern of service can be seen in Table 4.2 on p.94. This distributes the flights per week to all cities across the globe. The table shows that in 1970, 25% of Qantas' flights went to or through an Asian city; in 1990, this share has increased to 42%; in contrast, the Pacific cities have become less frequently served. European cities are still prominent, and account for 30% of the city frequencies of Qantas. North American cities do not figure prominently, and over time the frequency of Qantas there has not changed much. The greatest change is in the role that Indian and Middle Eastern cities play in the Qantas network. In 1970 they were prominent in the table; by 1990 they are relatively unimportant. Attention now turns to explanations of the changes shown in the tables and the maps.

Technology and the Qantas network

The Qantas network has changed in part because of changes in aircraft technology. Increases in the speed, distance flown and carrying capacity of aircraft have made a great difference to the geography of flights. Aircraft

1970

1975

1980

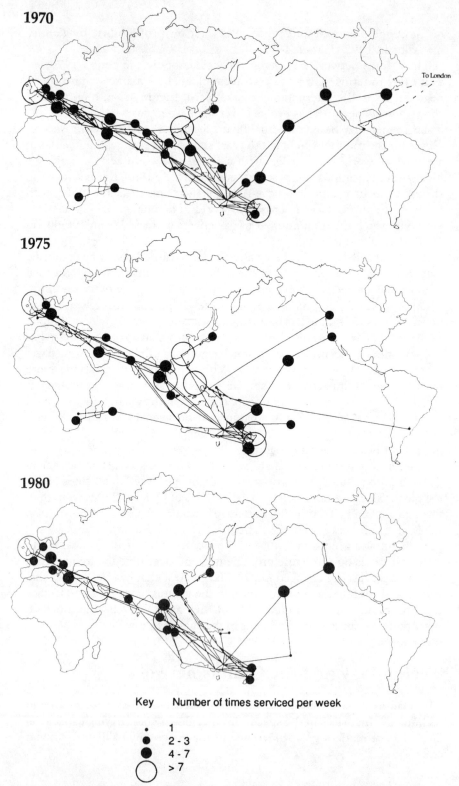

Key   Number of times serviced per week

· 1
● 2 - 3
● 4 - 7
○ > 7

Figure 4.1a) Qantas route network and frequency of services 1970–80
Source: ABC World Airways Guide, flights for month of June in each year.

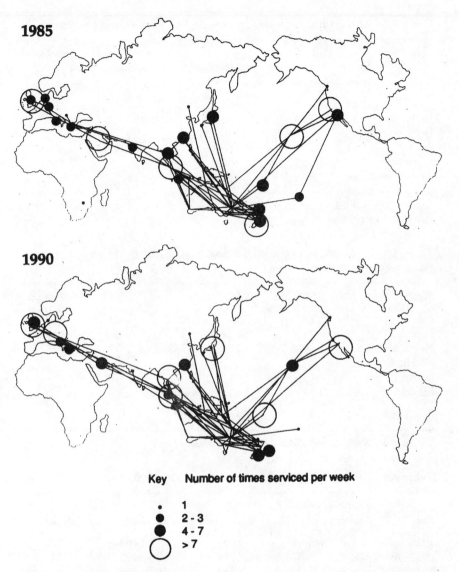

1985

1990

Key   Number of times serviced per week

- 1
- 2 - 3
- 4 - 7
- >7

Figure 4.1 b) Qantas route network and frequency of services 1985–90;
Source: ABC World Airways Guide, flights for month of June in each year.

technology has changed rapidly in the last 30 years, beginning with the refinement of piston engine aircraft, followed by the introduction of jet propulsion. Davies (1992) has reviewed the evolution of aircraft technology from the Ford Trimotor, and his commentary is captured in the data displayed in Table 4.3, p.94.

Nakicenovic (1989) has fitted a transformed logistics curve to the growth of passengers carried over time, and superimposed the different models of aircraft, showing a smooth match in the evolution in plane technology and

Table 4.1   The Qantas network: places, flights, sectors and passengers

	Places in Australia	Served overseas	Total number of flights per week	Total number of sectors between cities	Annual total passengers carried '000
1970	3	33	123	109	790
1975	4	29	137	79	1492
1980	4	25	89	60	1887
1985	6	25	130	63	2567
1990	6	25	160	86	4142

Sources: 'Places, Flights and Sectors', ABC World Airways Guide entry for June in Selected Year Passengers carried: Annual Reports.

Table 4.2   The Qantas network: destinations of flights 1965–90. Share of cities served in major regions

Region	1970	1975	1980	1985	1990
Asia	25	35	33	21	42
Pacific	24	33	13	37	16
North America	9	3	6	12	8
Middle East/ Indian Ocean	18	13	8	12	4
Europe	24	16	30	18	30
Total %	100	100	100	100	100

Source: ABC World Airways Guide. Entry for Month of June.

Table 4.3   Evolution of aircraft technology 1926–1990

Era	Representative aircraft	Speed (mph)	Range	Seats
1926–36	Ford Trimotor	100	400	12
1937–46	DC3	175	700	21
1948–58	DC6	300	1800	60
1959–69	707	550	3600	110
1970–80	747	550	6000	450
1980–90	747	550	7000	450

Source: Boeing Company (1992), p.4.2.

passenger traffic growth. The important feature here is the very much greater gains in range than in speed: flying speed has increased from 300 to 550 mph, but range has increased from 1800 to 7000 miles between the DC6B and the 747. The latter increases reflect the way jet propulsion was able to improve fuel economy, and provide greater thrust for a given level of fuel consumption.

Table 4.4 Sample Qantas flights 1970 and 1990

Sydney–London	1970	Singapore, Bangkok, Bahrein	
		Rome/ or Frankfurt	Flight 733/737
	1990	Bangkok	Flight QF1
Sydney–West Coast US	1970	Nadi, Honolulu	Flight 588/11
	1990	Nil	
Sydney–Tokyo	1970	Hong Kong	Flight 274/21
	1990	Nil	

Source: ABC World Airways Guide.

Qantas has been quick to avail itself of the latest technology, taking delivery of new model aircraft as soon as commercially possible after their release (Gunn 1988). The impact on its network is best seen in the information tabulated in Table 4.4 above.

That shows the stopovers on the major intercontinental flights out of Sydney in 1970 and 1990. The 1970 flight to London had five sectors. There was a range of configurations of the flight, but the common elements were a stop in South East Asia, (Singapore, Bangkok or Kuala Lumpur) the Middle East (Bahrein, Damascus or Tehran), and Europe (Athens, Rome, Frankfurt or Amsterdam) before landing in London. Information on the flights in 1965 shows this flight also included an additional stop in the Indian subcontinent at one of New Delhi, Bombay, Calcutta or Karachi. By 1990 the flight had been reduced to a single stop in south-east Asia, either Bangkok or Singapore. The changes on the Tokyo and US west coast flights illustrate the same effect.

The change in length of sector, or improved range, brought about by the improvements in plane technology accounts for the big fall in the share of flights calling at Middle Eastern and Indian Ocean cities, as shown in Table 4.1. South-east Asian cities remain important on the network as they act as major nodes on the European route, as well as important markets in their own right. Singapore and Bangkok have become relatively more important over time, illustrating that as well as the bypassing of places, the improvements in range have meant the concentration of airline activity in major nodes.

Hence, one important change in the pattern of flights of Qantas can be traced directly to changes in aircraft technology. This has led to a concentration of its activity within one part of the world, and facilitated a retreat from other locations. Non-stop links between major cities now take the place of multi-stop flights. The refinement and adoption of this technology reflected market demands, emerging from the integration of the world economy. Both can be seen in Winchester's (1990) analysis of the development of the 747-400. Recent long-distance aircraft technology has been very important to Qantas, as it has to meet a market demand for links to Australia with distant locations, often with little demand at places en route. This can be seen in the makeup of its present fleet, which consists of 32

747s of various configurations in a total fleet of 47 aircraft (Boeing Company 1990). The remaining aircraft are the long-range but smaller capacity 767s. The technical aspects however cannot be understood without some understanding of the market. That aspect is addressed below.

Markets and the Qantas network

There has been a major change in the market for air travel. Air travel has become the normal means of long-term movement for the majority of the travelling public, and increasingly important in the movement of freight. Boeing Company (1992) data shows that between 1970 and 1990, the span of the present research, the total revenue passenger miles carried by all but the then USSR airlines, increased from 342 million in 1970 to 1144 million in 1990. There were two sources of this growth. One part is business travel, which has grown as the need for closer personal contact has emerged from the complexity of international business. For Qantas, this has been felt in the increased internationalisation of the Australian economy, following waves of British, US and more recently Japanese investment. The investment flowed into manufacturing, mineral and later property and tourism investment. Very recently, too, Australia has had substantial capital inflow as its currency markets have been deregulated, and foreign banks expanded their presence in the country. Both the banking and investment activity would generate business travel into and out of Australia.

It is useful to see the extent to which changes in these business links correspond to some of the changes in the pattern of air services displayed in Tables 4.1 and 4.2 on p.94, and Figure 4.1 pp.92–3. This can be done in part by assembling the geography of Australia's trade in the same framework as that used to analyse the Qantas network in Table 4.2. Critchley (1992) shows that Asia is Australia's major trading partner, with 57% of exports and 40% of imports coming from that region. Europe, in contrast, is less important now than it was in 1970, accounting for only 10% of exports and 5% of imports. This would suggest that some of the network changes of Qantas have been associated with a shift in the business market. European destinations are relatively more important to Qantas than the simple market analysis carried out here would suggest. This will be explored in more detail below.

The second source of growth in the market has been tourism. Rising real incomes, and falling real costs of air travel, together with greater knowledge of distant and different destinations has been felt in the emergence of the international tourism industry (Shaw 1982). The majority of activity in this sector relies on air travel. For Qantas this part of the market has been very important, as the number of international visitors to Australia rose steadily through the 1970 and 1980s. For Qantas, the emergence of the Japanese tourist market has been especially important. Tourism from Japan is encouraged by national policy, and has increased from 4 million people a year in 1979 to 8 million people a year in 1988 (Polunin 1989). The sheer scale of this market has a major impact on air travel worldwide. Asia accounts for 63% of Australia's tourists, up from 51% in 1970 (Critchley 1992). Europe is still an important source of travellers, accounting for 27% of Australia's inbound travellers in 1990. This may help account for the strong part that Europe still plays in the Qantas network.

The tourist and business market segments cannot be easily separated, as most air services carry both business and tourist travellers. It is possible, however, to look at the geography of the total market served by Qantas and explore the shifts in its geography. This involves tabulating the travel to and from the Oceania region by all carriers (Boeing 1990). This is presented in Table 4.5, and shows the growth in total travel from 2.2 million in 1970 to 10.8 million in 1990.

The Asian region has seen the greatest growth in travel from Oceania, and by 1990 accounted for 42% of all travel; it has always been a significant source of travellers, however, with 32% of the total even in 1970. Looking back to the Qantas network, the distribution of flights reflected the pattern of the market in 1975 and 1980, and again in 1990, showing the strength of the market factors in airline operation. Qantas' European activity remains an anomaly in light of the market data: Oceania to Europe accounts for around 10% of all travel from the region over the period 1970–90, yet that same route accounts for 25–30% of Qantas flight frequencies, as shown in Table 4.2. It is possible that the inclusion of South-east Asia–Europe traffic on this route could create a greater market, and account for the stronger Qantas focus.

One particular change in the Qantas air services emanating from market forces was the cessation of the service to London via North America. The withdrawal from the route probably occurred for two reasons. One, that the Australian market could not justify two routes to London, especially when the second was considerably longer, and second the route would involve competition with a large number of carriers across the US and North

Table 4.5 The passenger market: Oceania region 1970–90. Share on major routes

Route	1970	1975	1980	1985	1990
Oceania–Asia	32	34	34	39	43
Intra Oceania	36	34	29	26	23
Oceania–North America	20	15	19	21	21
Oceania–Middle East/Indian Ocean	6	4	4	4	3
Oceania–Europe	6	13	14	10	10
	100%	100%	100%	100%	100%
Total passengers (million)	2.210	4.812	6.470	7.370	10.896

Source: Boeing Company (1992).

Atlantic. This has usually been seen as one of the most competitive markets; without a strong home demand, it would be difficult for a carrier like Qantas to stay in this market. The cessation of this service actually occurred as the technology of long-distance flights had begun to improve, illustrating the impact that market forces can have on the utilisation of technology.

Some of the information discussed above can be related to the services at individual cities. The dropping of the Indian and Middle Eastern cities makes sense when their small role in the market is acknowledged. There are few market reasons for Qantas to call, and each city's air service really related to its technical role in an earlier era. At the same time, the continued use of cities like Bangkok and Singapore reflects Australian linkages with these destinations, either as business or tourist destinations.

In overview, shifts in the market exert an influence on the pattern of airline services. Superficially, it seems the latter could change in response to demand, so long as landing capacity is available. A change like that has brought Cairns onto the international air network as a new gateway to Australia, as it became recognised as a tourist destination. If airlines do shift flight activity from one city to another in response to demand, some cities could face a volatile future. The change in Qantas services shows there has been a major reorientation in flight patterns in a 20-year period. Change, however, is constrained by regulation through bilateral agreement, and an airline can only change the number of services subject to its bilateral agreement with the destination country. Attention now turns to that aspect.

Regulation and the development of the Qantas network

Networks of international airlines are determined in large part by bilateral negotiation between countries. These negotiations result in legal country-to-country agreements, and involve substantial details concerning flight frequency, aircraft size and in some cases specify cities to be served. The agreements draw upon a set of international rules ('the freedoms of the air') established in a series of meetings in the immediate post-war period. The regulations were established by countries eager to protect their own air travel market, and to benefit from the development of international aviation. This philosophy was consistent with the 1948 context as the airline industry was just beginning, and many countries wanted to establish or foster government-owned airlines. O'Connor (1971) has shown international regulation of the air industry requires an understanding of aspects like nationalism, national pride, market protection, infant industries and tourist development. Mascarenhas (1992, p.5) has outlined these aspects in a spirited defence of the current bilateral system:

> ...national interest has led all countries—however small—to have their own airlines, be it to promote trade and tourism, provide national security or simply show the flag.

The general approach, and the institutions built to administer the system, share a common spirit with other international economic institutions like the IMF and GATT, where a focus is on world-scale development subject to individual country freedom to maximise individual rights and benefits. Hence the regulation of international air traffic is an extension of general national development policy, and the deal achieved for an airline depends in large part on the bargaining strength and commitment of its national government (Thornton 1970). In many situations, governments own airlines, so the government has a direct interest in the outcome of the bilateral negotiations.

International regulation in effect determines the pattern of air services, as no carrier can fly without an agreement in place, and this agreement will have emerged from negotiations carried out under the broad structure set in place by the Chicago Convention in 1944. In this context the outcome for individual airlines is dependent in part on what the country has to offer in bilateral negotiations. An airline from a nation with a big market will attract offers from other airlines wanting access to their market, or to use a hub or gateway. A convenient hub or gateway will mean that there will be opportunities for airlines to compete for traffic to and beyond the nation. Alternatively, airlines from small markets, or from locations that do not attract any through traffic, will have less to offer in negotiations, and will be hard-pressed to win commercially attractive deals. Governments can adopt protective attitudes to their own airlines, restricting access of other carriers to their own market. Their success in this approach depends upon the importance of their market, and their determination to succeed at negotiations.

In this set of circumstances Qantas is not well placed. It represents a small nation, in a location that can offer little in fly-past opportunities. Its negotiating position is further weakened as in most cases there are more people from foreign countries wanting to visit Australia than there are Australians wanting to travel in return. In addition there has been criticism of the Australian government's efforts to represent Qantas in bilateral negotiations (Findlay 1985). All these aspects can be seen in the outcome of negotiations with the Japanese. Two Japanese carriers have rights into Australia, but Qantas has difficulty obtaining additional rights into Japan. The traffic is of course heavily biased in favour of Japan, as the number of Australians travelling to that country is very much less than the outbound tourist market from Japan to Australia. It is clear that for some airlines, the lack of a home market has not been a disadvantage. In the case of Singapore Airlines, the hub position and technical-stop advantages of Singapore, together with the strength of the south-east Asian market, have provided that airline with something to trade. It has negotiated a wide range of bilaterals, and used aggressive marketing and price tactics to achieve patronage and growth from those bilateral agreements. Over time it has developed from a regional to a world carrier. Other small national airlines, say for example Sabena, have not had the same success. Perhaps Qantas sits between these two extremes, developing a market subject to bilateral deals with much larger European, Asian and US carriers.

Although it is difficult to relate regulation directly to changes in patterns of air services, it is important to recognise the influence of regulation, and understand that any change—like more or less frequent flights between countries or cities—requires the approval of two governments. Hence the services shown in Figure 4.1 are a reflection of these bilateral negotiations, just as much as they are a reflection of the technology and market situation. In many cases the bilateral negotiations specify particular cities to be served, and could be directly responsible for the service available at a given place. Negotiations over Qantas' access to Japan, for example, involves debate over rights to cities other than Tokyo, and Qantas has begun with charter services from cities like Sapporo, which it has then negotiated into formal landing rights. Often the bilateral merely specifies the number of cities to be served, without naming them, so that the airlines will built their pattern of services depending on the local market. Any city attempting to improve its air service needs to come to terms with these forces.

Other forms of regulation are important here. These are the technical regulation of airline operation, and safety, which have had an impact on the development of networks. For example, a regulation banning the use of twin-engine aircraft on long-distance ocean crossings has recently been amended, which has changed the operational scope of airlines with the Boeing 767. This aircraft, especially in its extended range form, can be used in North Atlantic and Pacific crossings to satisfy smaller market demand than that currently met by the 747. Some airlines may use the 767 to provide links between smaller cities. This illustrates that air services can be a creature of the regulatory environment.

Regulation over the use of airspace at airports, too, has played an important role in the evolution of air services, limiting the flights into some places, and making other centres more attractive due to lack of curfew. Here regulations are generally adopted to achieve local objectives to reduce traffic noise, but are also used by home-based carriers to control competition by limiting the number of available landing slots. A decision to impose (or remove) a curfew effectively adjusts the capacity of an airport by 25–30%, and will require adjustment in flight frequency and timing to fit take-offs and landings around times available elsewhere in the system. In this way as well, airline services reflect regulations, and it is a mistake to see them simply as an outcome of market or technical factors.

Overview: influences on the pattern of airline services

The aim of the research was to isolate the relative importance of technology, markets and regulation on changes in the pattern of airline services, and so attempt to account for airline activity at particular cities. Though treated as separate forces, all three interact and together shape the pattern of services. For example, improvements in technology interacted with the market and produced lower fares; the emergence of new markets like in Japan produced demand for longer distance, larger planes. All this took place under an established regulatory regime that allowed individual countries to negotiate market shares for airlines.

In assessing the role of each factor, it seems technology was the dominant force for change in the period from 1970 to about 1975. In this period the refinement and the broad-scale adoption of the Boeing 707 and similar aircraft, followed by the Boeing 747, provided greater carrying capacities, higher cruising speeds and longer sectors. As seen in the Qantas network, the pattern of air services was concentrated on a few key destinations, with fewer stops along the way. Cities were bypassed in this era, and other places became more important as hubs or gateways.

Airline services will continue to evolve in response to technological change. There seem to be two prospects. The first is faster, greater carrying capacity mid-range aircraft, providing airlines with larger capacity to move people and goods non-stop in the 4500–5000 mile range. This will provide more capacity, and frequency, on non-stop links between Asian cities, across North America and between Europe and North America. That future is represented by the Boeing 777 and the Airbus A330. A second future could involve an evolution of the 747, providing a high speed, high passenger load capacity aircraft, to halve the time of big long-haul flights like Los Angeles–London, or Los Angeles–Tokyo (Boeing Company 1992). That future seems consistent with past trends, which have seen flight times fall as passenger loads and length of sector have increased. However that scenario may be constrained by the lack of a market (Davies 1992). In addition, flexibility may be a more important factor for airlines than size and capacity. The improved speed and carrying capacity of the mid-range aircraft may be the dominant wave in technological change. Under this scenario, it is possible that more cities will attract a greater number of flights, as airlines have smaller, more flexible aircraft, suited to smaller markets, at their disposal. However, aircraft technology alone is unlikely to be the conditioning factor here, and the long-term outcome for cities will probably be determined by the interplay of other forces discussed below.

The market was perhaps the dominant force in the 1975–85 period, when large-scale international tourism, especially out of Asia, became a factor for the first time. At the same time, the consolidation of the role of the Asian tigers in the world economy focused air traffic into places like Hong Kong, Taipei, Bangkok and Singapore to a greater degree than in the past. Markets will naturally remain an important influence, though there will be a shift in the factors involved. In the past, pressures for changes in networks came from the demand side, in the form of new destinations, or the demand for more travel on existing routes. In the immediate future, aspects of demand associated with computer reservation systems, and frequent flier programmes and interlining could be more important, as they are used to channel demand on particular routes by particular airlines. In this way, the supply side of the market may be the most important dimension in the change in airline services in the immediate future.

The 1985–90 period can perhaps be seen as the period when matters involving regulation were the most important influences on the pattern of airline services. The full effects of the US deregulation of the airline industry began to be felt by then, and some other countries and regions began to follow the US lead. It is unlikely that the national experience of places like the US or Australia will be transferred to the international context, but there could be some changes in the bilateral system in the immediate future. Tentative steps in this direction in Europe (Wassenbergh 1992), in the Andes region, and debate within ICAO provide some pointers to the potential for change (ICAO 1992).

Other factors are now beginning to emerge as important influences on the pattern of airline services. The first is *corporate restructuring* within the airline industry. Corporate restructuring has long been understood as an important influence on change in the geography of activity (Taylor & Thrift 1986). An initial impact of this process can be seen in the demise of Braniff International (Nance 1984). The subsequent impact has been more dramatic, largely due to the pressure of change emanating from the US deregulation. Dempsey (1992) has shown how the 'mega carriers' emerged from a spate of takeovers, mergers and bankruptcies that have changed the character of the US airline industry in the space of the last 10 years.

Privatisation of some former government airlines, and *cross-ownership* between different airlines, have been two other parts of the corporate restructuring process. At the time of writing, the Australian government has just sold a 25% stake in Qantas to British Airways, and merged Qantas with a domestic operator, Australian Airlines. These changes could reinforce Qantas' role as an Asian Pacific carrier, linking with British Airways flights at places like Singapore, Hong Kong, Tokyo and Los Angeles, and also strengthen its role within Australia. Other cross-ownership moves have produced a bewilderingly complex pattern of corporate connections, tying together 55 separate airlines with at least a 5% stake in one another (Woerth 1992, p.166). There are also a set of *interline* and *code sharing* agreements that allow passengers and freight to move across the globe on two or more airlines on single documentation. A common thread in all these corporate arrangements is the concentration of customers on one or two airlines' flights. In the US, recent experience of the new corporate structures, especially as orchestrated through computer reservation systems and frequent-flyer programmes, has concentrated traffic in a smaller number of hubs, and has constrained the spread of airline services to smaller centres (Dempsey 1992). If the future of the international airline industry is built around similar approaches to market power and influence, there seems less chance that airline services will disperse away from their present concentrated patterns.

A second new factor that could change the pattern of airline services is airport and airspace congestion. This problem is a creature of the processes discussed above, together with the pressures exerted by the growth in traffic. Once again it seems that there is potential for the pattern of airline services to change, to favour new, less congested locations. Present approaches to solving congestion will probably deny that outcome, and maintain the current pattern of concentration. One approach, involving charges for landing rights, and airport slots, is already in use in some congested US airports. The long-distance intercontinental operators, who will be spreading the cost of landing across a much larger number of passengers than the small regional flights, will benefit from this approach. Hence this approach will probably have less impact on the pattern of long-distance services than on smaller local services.

A second approach to the problem of congestion is to substitute fast rail services for aircraft in the 200–300 mile range around large cities. This could mean that some passengers could reach hubs by rail, with integrated booking services making the connection a simple one. There has been experience with this approach in Germany (Davies 1992). If rail services do develop in this fashion, they could both deliver passengers into airports and reduce congestion from local traffic; the hub airports will be even more attractive for the intercontinental and transcontinental services in this situation. The role of rail will be especially important in Europe, where greater population density, and an established rail network, will strengthen the likelihood of transfer of customers from air to rail in intercity markets. The approach is also being considered in parts of the US. If congestion is addressed in this way, the scope for change in the pattern of airline services, and for new services for other cities, will be limited.

The problem of congestion is also being addressed by constructing new airports. A large number of new airports, or additions to existing facilities, are being planned or under construction. In almost all cases, these are within currently large and well serviced metropolitan areas, as that is where the congestion is felt. As these projects come to fruition over the next decade, it is likely the pattern of airline flights will continue to focus on its current arrangement of big hubs and gateway cities. In addition, improvements in air traffic control could reduce some of the congestion, especially in Europe. Once again the established centres will be reinforced, reducing the scope for change in the network.

Conclusions

The analysis carried out on the pattern of air services of Qantas has shown considerable change in the geography of flights in the last of 20 years. Qantas flies more frequently to fewer places than it did 20 years ago. If that experience is reflected in the activity of other airlines, it seems international aviation has become more concentrated over the past two decades. These changes reflect changes in Australia's business and tourist markets, as well as changes in aircraft technology, and the outcome of bilateral negotiations. Looking to the future, technology and markets will continue to interact and shape the pattern of airline flights. Their joint impact will be sharpened by the outcomes of corporate restructuring, which will bring pressure to bear on the current regulatory framework. Preliminary thinking on these aspects suggests they will reinforce the present concentrated pattern of services. This will reduce the opportunity for cities to attract new air services, unless they have major new market attractions. In turn, firms that need good airline connections will probably be slow to move away from the big hub cities of today.

NOTES

1 This chapter is based on initial data analysis carried out by Janet Critchley. Ron Speirs and Sue Park supplied the data from old editions of the ABC World Airways Guide. In Melbourne, John King of Avmark provided access to Boeing Company statistics, and two recent conference proceedings. Ann Scott's comments on an earlier draft helped focus the research. The author gratefully acknowledges the assistance these people have made to the project.

PART 2 The Changing Urban Hierarchy

HISTORY IS A CONTINUAL reminder of the inevitability of change in urban systems—of growth and decline, of winners and losers. Cities have always been in competition with one another, but the pace and scale of *territorial competition* in the late 20th century is unprecedented. Why this is so is the focus for chapters in this section of the book.

The forces of *globalisation* are addressed by several authors in Part 2 as one of several key factors that expose national urban hierarchies to competition. Globalisation exposes regional markets to a wider spectrum of competition (threats) as well as providing possibilities (opportunities) for urban centres to shift from the more restrictive central place/hinterland orientation of economic interaction to one which is based on international *global-local networks* (Goddard, p.135). 'Winner cities' are those that are global in orientation (Wegener, p.148).

The shift to a *new post-industrial base for urban growth* also explains variations in the economic performance and well-being of cities within national urban hierarchies (Cheshire and Gordon, Chapter 5). This new industrial base incorporates the information-intensive industries and producer services industries as well as corporations which have restructured in order to emerge as more flexible and less hierarchical organisations with an enhanced ability of being able to functionally and spatially separate their activities. Organisations in this category tend to operate under a different territorial logic to that of, say, traditional (Fordist) industries (Goddard, p. 131).

New technologies and their associated infrastructures are not uniformly accessible to all regions and the industries they accommodate. Yet the combined effect of *communications* and *air transportation*—and *high speed rail* (HSR) in selected countries—is defining new opportunities for industry: significantly reducing the time and cost of moving goods, information and people; generating one-day business trips to an increasing variety of destinations; expanding market areas and labour force regions; enhancing methods for either control or collaborative operations within and between firms, increasing the ease with which 'traditional' locations can be abandoned in favour of alternatives. *Connectivity* is critical. Japanese experience suggests that centres where several network nodes (air, HSR and communications—e.g. teleport) are integrated tend to generate additional synergy for economic growth and development (Taniguchi *et al.*, Chapter 9). At the same time there are regions with similar infrastructure endowments which display divergent economic performance outcomes. In confronting this apparent paradox, several contributors (Newton, Chapter 8, Kunzmann, Chapter 14, Willoughby, Chapter 13) highlight the significance of *human capital* and an *innovative milieu* to the *utilisation* of new technology, especially that related to communications and information technology (CIT). New technologies clearly provide necessary but not sufficient conditions for regional economic development.

The attractiveness of urban centres as places to live and work acts as a magnet for population as well as industry. The larger cities in Europe, North America and Australia continue to be the favoured destinations for *overseas migrants* (Wegener, p.150), sustaining their growth in the face of low rates of natural increase and *inter-regional migration* which is commonly negative. The latter class of migration, driven by production and consumption motivations, generates contrasting population streams with differing impacts on cities and regions. Production-based migration tends to sustain the human capital requirements of city-based industry; consumption-based motivations are decanting populations on fixed incomes (retirees, social security beneficiaries, those not in the labour force etc.) from the cities to non-metropolitan regions—intensifying differences in economic well-being between city and country (Newton, Chapter 8).

None of these demographic, economic and technological forces operate in vacuo. Change in global *politics* is paralleled by change in regional alliances such as the Single European Market (SEM), North American Free Trade Agreement (NAFTA) and Asia Pacific Economic Community (APEC)—all of which seem likely to engender change in national urban hierarchies within the respective trading blocs (Cheshire and Gordon, Chapter 5, Wegener, Chapter 7, Goddard, Chapter 6). Within countries, local governments as well as city and regional governments engage in territorial competition via a range of 'capacity building' initiatives designed to capture investment capital. At all spatial scales, *political fragmentation impairs the ability to compete* (Cheshire and Gordon Chapter 5).

The dynamics of change of national urban systems, if viewed as an end in itself, can mask the considerable change that is occurring within cities. The city–suburb nexus is undergoing transformation and is the subject of analysis by several contributors to this part of the book.

Stanback (Chapter 11) assembles data to demonstrate that *central cities have been transforming their economies* over the past two decades, specialising in the higher value added information-intensive producer services, which are tending not to decentralise. These industries therefore face higher levels of competition from their counterparts in the central cities of other major cities than they do from their own suburbs. This extends to a range of other industries, including tourism and conventions, educational services, medical services etc. This places pressure, as Daniels (Chapter 12) highlights for London, on the infrastructure providers and governments of large cities to create a built environment and *office milieu* that is attractive to skilled information and knowledge workers and managers. The negative economic transformation of inner cities is also creating opportunities for *inner city revitalisation* of residential precincts via gentrification of original housing stock as well as new construction and conversion of surplus (and obsolescent) office floorspace to residential use.

Meanwhile the suburbs are increasing in maturity, but are mostly attractive to services that are routine in nature or do not require close informational linkages with other firms, as well as retailing, manufacturing and personal services. The emergence of *'edge cities"* on the margins of some of the larger metropolises represents a new element of metropolitan growth heavily dependent on highway systems to access markets and labour.

The *spatial deconcentration* of population and industry common to all cities has negative as well as positive consequences. The negative environmental effects such as traffic congestion and environmental pollution are the focus for chapters in Parts 4 and 5. The adverse social consequences of urban and technological change are identified in chapters (7 and 8) by Wegener and Newton who point to a *polarisation of the populations and industries* that co-exist in the major metropolitan areas. There is little evidence that such polarisation can be halted or reversed. As Goddard observes (Chapter 6):

> ...the uneven diffusion of information and communication technologies through the urban hierarchy, social hierarchy and firm size hierarchy favours the largest cities, the most advantaged social groups and the largest firms.

This also tends to be borne out in Chapter 9 by Taniguchi *et al.*, where high speed rail (HSR) was seen as a mechanism for redistributing urban functions out of Tokyo to realise a more balanced pattern of national growth and development. However, HSR increasingly serves Tokyo's metropolitan commuter market, dramatically increasing its economic sphere of influence. Cheshire and Gordon, in Chapter 5, highlight the horns of the dilemma currently facing urban strategists and policy makers: territorial competition (designed to capture a larger share of investment and business activity for an area) is often in conflict with regional policy (designed to reduce spatial disparities).

5 European Integration: The Logic of Territorial Competition and Europe's Urban System[1]

Paul C. Cheshire and Ian R. Gordon

Introduction

IN the United States the concept of an urban growth machine is a familiar one, with a history going back to the competitive urbanisation of the Frontier during the 19th century (Molotch 1976; Logan & Molotch 1987). In this process, small coalitions of landowners, developers, and railroad operators with a direct stake in the economic development of a particular town were readily able to perceive their shared interest and could be joined in this by localised finance capital. In Europe, with its more established pattern of urban settlements, coalitions of this kind were more likely to be assembled around competitive suburbanisation at this time, although a number of new resort developments were exceptions to this rule. As with the mercantile cities of a previous era, the success of these resorts clearly depended upon place-specific investments and reputations to an extent which was not true of contemporary industrial cities.

The urban political machines, which legitimised the capture of urban government by dominant economic interests in many American cities, were also alien to the experience of most European countries. Rather than securing the political recruitment of newly urbanised voters through the provision of particularistic favours, such as offers of housing or work, partisan attachments in Europe came to depend more on ideological, religious or class identification, particularly where urbanisation preceded enfranchisement. Anglo-American comparisons in particular suggest that territorial politics and class politics are more or less mutually exclusive bases for the organisation of a political system. Where the scale of regional inequality (or 'territorial injustice') threatens a cross-cutting cleavage, the natural response of the main parties in a class (or ideologically) based system is to seek ways of redistributing resources between areas on the basis of national criteria—as in the case of *regional policy*—in order to reassert the integrity, even-handedness and paramount importance of the national polity (Gordon 1990).

So the emergence across Europe over the past decade of activities characteristic of 'urban growth machines' engaged in processes that have been characterised as territorial competition (Gordon & Jayet 1991) is a remarkable development. This is not simply another case of trans-Atlantic copycatting, as with a series of previous innovations in urban policy, but reflects both the shift to a post-industrial basis for urbanisation and the impact of international economic *integration*, accelerated by conscious policy and symbolised by the completion of the Single European Market.

European economic integration is expected both to increase the exposure to competition of the leading European cities and to widen existing regional inequalities across the EC (Cheshire *et al.* 1991). Up to now, with substantial non-tariff barriers to trade in services, and with chauvinistic public procurement policies, the major cities in Europe have enjoyed a degree of monopoly within their nations not available to their American counterparts. In those countries with more hierarchical urban systems, national capitals enjoy this monopoly. Fifty per cent of Europe's 300 largest indigenous companies are headquartered in London (28%) and Paris (22%) (Pumain & Rozenblat 1991). In Spain none are headquartered outside Madrid and only Germany exhibits a more evenly distributed pattern, although even there three cities account for two-thirds of the headquarters. Consequently London[2], for instance, in the past escaped the violent swings in economic fortune experienced by New York, which in other respects seems its close counterpart (Fainstein *et al.* 1992b). The Nomura Survey of international cities (as cited in Rimmer 1991) identifies 12 first-rank cities in western Europe, spread across 9 countries, with only the federal states of Switzerland and Germany having more than one city in this category. By contrast, there were 8 such cities within the US and 3 in Canada. Completion of the Single European Market will be a substantial step towards an integrated European urban system, even if a United States of Europe is some way off. Within this system, with so many first-rank cities (almost all) compressed into its core area, competition for status is likely to become quite as intense as in North America.

Empirical evidence presented by Cheshire (1992) for 120 of the largest urban regions in Europe shows that since the late 1970s integration has been associated with a strengthening of the position of those regions clustered around the economic centre of gravity of the Community. Polarisation has been further enhanced by a tendency for the largest urban regions to improve their position at least relative to medium-sized urban regions (the smallest urban regions were excluded from the analysis). This marked a sharp reversal of trends from the early 1970s, when there had been some regional convergence. Taking the last two decades together, the gap between better-off and poorer regions within the EC has perceptibly widened. At the same time within the nations of Europe, commitment to regional policies designed to reduce these disparities has demonstrably weakened. Increasingly it has been left to cities or regions to mitigate the effects of recession and industrial restructuring on their territories. Limits on their capacity to achieve this are indicated by evidence that change in an urban region's economic position was primarily determined by forces outside the region or the control of policy makers. However, when these external factors were controlled for, there were signs that local policy could have some influence, both for good and for bad.

This chapter seeks to build on this analysis and examine the implications of these processes for the formation of local policy in Europe; in particular, the implications of European integration and of other forces acting on the development of Europe's urban regions for the policy process—both locally and at the national and Community levels—and for the future form of Europe's urban system.

The logic of territorial competition

By territorial competition we mean a process through which groups acting on behalf of a regional or sub-regional economy seek to promote it as a location for economic activity in competition with other areas. In principle it involves *both* active local economic development measures of various kinds *and* a self-conscious strategy to guide policy development and implementation, with regard both to the future economic role of the area concerned and with respect to its principal competitors. Part of this competitive activity is inevitably addressed to the attraction of investment, with some discrimination between more and less desirable functions. But part is also concerned with enhancing the share (and social profitability) of existing local businesses in the product markets they serve.

This function thus combines the concerns of

- traditional property-oriented growth machines (Harding 1991),
- the newer city marketeers oriented both to image manipulation and re-creation of the 'place product' (van den Berg *et al.* 1990; Ashworth & Voogd 1990),
- New Left urban authorities seeking industrial restructuring in the social interest (GLC 1985; Benington 1986), and
- French (or Japanese) planners of regional technopoles (Cohen & Simmie 1991).

Common to all these concerns is the existence of externalities of various kinds which mean that policies pursued by individual agents will not be socially optimal from the viewpoint of the citizen of the territory as a whole; and the existence of quasi-public goods which pose a problem for provision by independent 'clubs' acting on behalf of businesses if prospective members think that the club will either be ineffective or else they can 'free-ride' (cf. Olsen 1965). The potential geographic scope of territorially competitive policies is also likely to exceed the scope of traditional units of regional or sub-regional government. Even assuming that such policies can be effective, the capacity of an existing local governmental area to engage effectively in them cannot, therefore, be taken for granted.

Changes in the economic environment mean, however, that city-regions will increasingly need to pursue effective competitive strategies if they are to prosper. There are two types of reasons why this is so.

First, it follows first from shifts in the balance of economic activities (notably the processes of economic transformation and of producer service growth) which have proceeded most rapidly in the biggest cities. The new producer service bases, encompassing a range of activities from finance and corporate governance to scientific research and software development, make more specialised demands on an urban environment than did their industrial predecessors.

Secondly, for the core activities in particular the issue is not simply access to markets, inputs or a fixed supply of labour; but such factors as access to key decision makers, up-to-the-minute information, a range of other specialised services, and an accommodating regulatory environment, underpinned by a quality of life that facilitates the recruitment of skilled and highly mobile professionals.

This combination of environmental and agglomeration factors seems much more open to manipulation by intelligent planners than were many of the traditional industrial location factors. The tourist industry, in all its varied forms, is another increasingly important activity for which this is true.

The imperative for territorial competition in Europe arises from a general process of international economic integration, which has been given a deliberate and powerful political boost by the creation of the EC. The removal first of tariff and then of non-tariff barriers to trade, together with a free movement of capital and labour, is creating a European economy in which activity becomes more 'footloose'. Companies are increasingly restructuring themselves to serve the European market as a whole rather than a set of national markets. They eliminate national headquarters and have just a European headquarters; they have European-wide marketing strategies; they streamline their product range and concentrate their production.

One recent case illustrates this process: Tambrands, the US-based multinational making Tampax, used to have four nationally based companies in Europe. These were located in Britain, France, Belgium and Spain. They were each entirely independent and between them had 220 different packages. There is now one European company based in the London metropolitan region, just two packages, one with six languages and the other with seven, and one European-wide advertising campaign. In the two years since this reorganisation in 1989, sales increased in the European market by 48% and sales per employee increased by 21% (*Independent*, 1 Oct. 1992).

The early phases of European integration principally affected trade in goods, where the effects of integration may have now largely worked themselves through. The changes initiated by the Single European Act in relation to removal of non-tariff barriers (including open tendering procedures for government contracts) and the mutual recognition of professional qualifications, have much more substantial implications for service functions, however. These changes are likely to have a particular impact on advanced metropolitan regions, where information-intensive and decision-making activities tend to be concentrated. Activities such as research and development, high level business and financial services, media, or marketing will be exposed to quite new competition from outside national borders.

Restructuring of European companies will involve the relocation and rationalisation of existing headquarters designed to serve a company's old national markets from a base in the national capital or another leading urban region. Such relocations of corporate control functions will also reflect the comparative advantages of rival centres, and will have knock-on effects for producer service activity. In addition, industrial economists have argued (Itaki & Waterson 1990) that production will tend to become concentrated in few, or more capital-intensive plants in more central but larger regions.

These expected changes add up to a prospective restructuring of the European urban system, from a set of distinct national systems into something like a single integrated urban system. The implications of this process, in terms of the likelihood of significant growth or decline, will be greatest among those cities towards the peak of the old national systems. Those cities will be losing their effective monopoly of high order service functions, but some of them will be gaining European-wide functions. An integrated European urban system can support a very large number of cities the size of Salisbury, Burgos or Limoges; even of Newcastle, Firenze or Hanover. The position of the largest cities—such as London, Paris, Barcelona, Madrid, Bruxelles, Milano, Roma, Amsterdam, Hamburg, Frankfurt, Berlin, Düsseldorf or München—is, however, much less certain. Since the growth of secondary centres, such as Reading (London), Terrassa (Barcelona), Monza (Milano), Leuven (Bruxelles) or Augsburg (München), depends in part on the success of their associated leading city-region, in the longer term there will be repercussions for selected smaller cities, too.

Thus the effects of integration produce a situation in which there is a greater incentive for urban regions to compete for economic activity. Not only is more activity becoming even more (potentially) mobile, but the system is becoming more volatile. And with other cities competing, a city which does not compete is more likely to lose.

The organisational base for territorial competition

In contrast to traditional regional policy, territorial competition is essentially locally based, involving activity by groups within an identifiable area seeking to promote the economic development of that area. These groups may be of local governments (e.g. a regional government coordinating the activities of a set of municipal governments); or they may be private sector agencies such as developers or chambers of commerce; or, perhaps increasingly, they may be partnerships of both public and private sector groups or agencies. Although their wider aim is to promote the economic development of their regions they may pursue this aim in different ways. They may seek simply to sell or promote their region; they may attempt to modify their region's economic and physical environment in specific ways (e.g. tax abatements, making sites available or providing supporting infrastructure) that will attract mobile investment; or they may engage in supply-side policies and 'capacity building' to modify the local social, economic or physical environment in a more general and fundamental way. Funding may be partly or wholly drawn from national governments or EC funds, but the distinctive feature of territorial competition is that it is local in both origin and perspective.

The organisational bases for competitive activity in a specific territory are thus likely to be crucial influences on the character of that activity—in terms of its scale and range; the biases towards particular policy instruments, concerns, and potential beneficiaries; its degree of parochialism; and the adequacy of its understanding of the area's competitive situation. These organisational bases may be analysed in terms of: the set of local interests motivated to support the activity (i.e. 'join the club'); the sources of *leadership* able to translate support into activity; and the *constraints* on permissible competitive activities imposed by external regulators, such as the relevant nation state or the EC.

The 'product' of territorial competition can be viewed as a quasi-public good, from whose benefits it may be hard to exclude those who have not contributed to the activity. If an area's economy develops, the beneficiaries are not just those agents who engaged in or resourced the successful policy, but all earners of economic rents or quasi-rents in the area. Where the area is a locality within a larger functional economic region, the effects of successful policy will also leak out to benefit rent earners in the rest of the region. There is thus potentially a major problem of motivating groups to invest in the activity. They are more likely to do so when substantial gains are anticipated for relatively few participants within an economically self-contained area.

The groups likely to benefit most will generally be the owners of land sites and commercial developments and other rent or quasi-rent earners (who may in some cases include public authorities, cf. Harding 1991)—that is to say those with non-existing or imperfectly transferable assets in fixed supply whose returns are essentially dependent on the local (or specific, in the context of skills) level of demand for those assets. Hence the most likely form of competitive activity to emerge from a more-or-less voluntary club is the type of property-oriented promotion associated with traditional urban growth coalitions. Such activity may proceed even when it has negative impacts for many non-property owners on fixed incomes, who face rising rents and a deterioration in their environment as a likely concomitant of success, but who lack the incentive or resources to establish a rival anti-growth 'club'.

Employees with skills in relatively less elastic supply and would-be employees, whether unemployed or inactive, are other potential beneficiaries of competitive success. Given their number, transactions costs of such beneficiaries are likely to be high relative to individual gains and so their participation in collective action cannot be counted on unless they are effectively represented by a major economic or political actor, such as a traditionally dominant industry, a political party or a trades union grouping. The extent of shared economic interests in a particular community and the recognition of the need to do something about advancing or sustaining them is thus a crucial determinant both of the support which can be mobilised for competitive activity and of the form which it takes. A relatively specialised urban economy, with a high degree of integration among long-established businesses, may be the most promising economic base for the organisation of competitive activity, but that activity will tend to reflect the perceptions and interests of those particular businesses more than a strategic view of current competitive prospects and so may be less effective. It will work well when it represents a sector in which the area still has a comparative advantage (such as finance and banking in Frankfurt or London). It may prove a handicap where it attempts to prop up historically dominant sectors which have lost comparative advantage.

The weakness of the purely voluntary 'club' as a framework within which to organise the provision of quasi-public goods means that some leadership is likely to be required from a governmental or quasi-governmental body in order to initiate, manage and fund promotional activities (cf. Bennett & Krebs 1991, p.22). This leadership may be provided by a single regional authority, a 'club' of local authorities of the same administrative tier (e.g. the London Planning Advisory Committee—a consortium of London Boroughs) or a regional authority and its component municipalities (e.g. Ile de France and Paris or Lothian and Edinburgh under the auspices of Scottish Enterprise). Increasingly common, however, is a partnership of both public authorities, public agencies and representatives of the local private sector. There are examples of such arrangements in Birmingham and Sheffield as well as in continental Europe.

Whatever the specific arrangements, a number of general points may be made about the role of public agencies. The first is that public sector involvement provides a vehicle for pursuing a particular set of local economic interests, probably as a response to political pressure; it is not a substitute for those interests. The second is that public agencies have or develop their own interests (whether these are electoral/fiscal/bureaucratic survival or the pursuit of ideological interests by a dominant group) and that these also influence both the adoption of a leadership role and the form of the resulting activity (Gordon & Jayet 1991). In the UK, for instance, the search for a new role within town planning departments was one factor in the emergence of physically oriented local economic development initiatives in many urban areas. The third observation is that voluntary cooperation may not be much easier to organise among public agencies than it is among private interests, and that concerted action is more likely to emerge when fewer independent agencies need to be involved, when the agencies are regionally based, when agencies are united by some ideological communality (e.g. under the control of the same political party) or when one agency holds the purse strings.

The final set of factors likely to influence the form and extent of territorial competition involves explicit or implicit constraints on the actions of relevant agencies. One example of an implicit constraint has already been given: the administrative boundaries of a region. A region such as London which has no tier of government representing its functional territory, will be constrained in terms of the actions it can take and also in terms of the types of action which secure the highest pay-off. So, too, will a city such as Dublin or Athens where the resources available to it and the powers of local government are severely constrained. These latter are extreme cases of local governments which have no fiscal resources of their own at all, but British local governments have fewer discretionary resources of revenue available to them than regional or local governments in France, Germany, Spain or the Benelux countries. Similarly in the private sector there are significant differences in the resources and powers of local agencies. Chambers of commerce in Italy or France have a more active role and more extensive activities than their equivalents in Britain. In Germany, membership of chambers of commerce is compulsory and they have both a legal status and legally defined responsibilities which make them far stronger agents for territorial competition. For example, the chambers of commerce of many of Germany's larger cities are directly represented in Brussels. Yet a further implicit constraint may be the constitutional powers of local or regional governments which may be so defined in some countries as to preclude certain sorts of territorially competitive policies.

Explicit constraints may be imposed by national governments or by the EC. Particular types of expenditure may be prevented or controlled and certain sorts of development may be impeded (for example by central controls of industrial development). National constraints on local initiatives may be influenced by party political strategies or attempts to maintain a national hegemony in 'economic' policy making (as the dual state theorists would contend). In particular, they may form part of a regional policy intended to reduce territorial inequities. But they also clearly reflect attempts to avoid wasteful and inflationary competitive bidding between areas for a larger share of the national economic cake. The translation of urban competition to an international level substantially alters this calculus as far as national governments are concerned, encouraging a more positive attitude to local economic initiatives. At the European level competition policy presently constrains only certain types of activity such as direct subsidies on operating costs but, in the future, more direct controls may be imposed in order to regulate the process of territorial competition itself.

Specific propositions

From these general considerations certain propositions can be derived as to conditions which will tend to produce variations in the development, form and likely effectiveness of territorial competition policies. Territorial competition is more likely to be engaged in and more energetically pursued where:

1 there is a smaller number of public agencies representing the functional economic region, with the boundaries of the highest tier authority approximating to those of the region;

2 there is a strong sense of cultural (or political) identity within the region, since this makes it easier for agencies to come together and cooperate;

3 there is a stronger representation in local decision making of 'rent earners' than non-rent earners. Since rent earners are the chief beneficiaries of local economic development they have a stronger motivation to engage in territorial competition than other groups, some of whom may actually lose[3];

4 the local economic structure contains a strong representation of firms who are sensitive to environmental conditions and/or agglomeration economies, and well integrated into the local economy;

5 there is the potential for significant positive *or* negative change in the regional economy. Without this, and the perception that it is possible to influence outcomes, there would be little interest in territorial competition. Since European integration is expected to produce most change (and potential gains) in strong regions rather than lagging ones, the former may paradoxically be more active in pursuing territorial competition, even if it seems less necessary there;

6 there is a smaller number of leading local firms and other actors, since this will tend to reduce the transaction costs involved in forming a 'club' to undertake, or pressure for, competitive policies;

7 leading local actors have a major stake in the local economy. Long-established local firms, especially if they own land and premises or rely on local businesses, are more likely to join such a club, because their potential gains are greater, and because the 'exit' option would be more costly for them. Hence a territorially competitive policy that emphasises inward investment may, in the long run, become weaker, since incomers may have less incentive to participate in the club and gains may increasingly leak, therefore, to non-participants;

8 there are potential lead agencies free to allocate adequate resources to competition policies without severe constraints from superior tiers of administration;

9 the interests of local agencies and actors are relatively homogeneous and/or complementary.

Many of these hypotheses also have implications for the mix of policies likely to be emphasised in particular areas. For example, we should expect that the mix of policies will vary so as to maximise the potential to exclude non-participants from benefits given the conditions under which the particular agency operates. For example, agencies which are smaller relative to their functional economic region will tend to pursue a different policy mix than agencies which are larger.

Strategic (as opposed to ad hoc) policies seem more likely to emerge where agencies are strongly motivated to engage in territorial competition but the local economy is more diversified. Factors enhancing commitment to competitive policies will also generally support a more strategic approach. This is especially true if there is an administrative unit representing an approximation of the functional economic region; or there is a cultural sense of identity; or there are fewer leading firms with stronger stakes in the area but in a more diversified economic structure[4]; and with there being a stronger probability of change occurring.

The evidence

The anecdotal evidence for increasing territorial competition is Europe-wide. Conferences organised by EURICUR on city-marketing have been held on a regular basis since 1986. After the early successes of the distinctive local policies initiated by Bologna, Lyon and Glasgow, more and more cities have been engaging in similar developments. Rotterdam, Birmingham and Barcelona have recently been actively engaged. Paris and the Ile de France region have been trying to promote themselves as Europe's capital city region. The city of Lille is heavily engaged in the process, having secured the main junction of the TGV network in competition with Amiens. Now London and Edinburgh have launched projects to study and promote their competitiveness (LPAC 1990; Coopers & Lybrand Deloitte 1991; LBS 1992a; 1992b; LEEL 1992). The process extends to urban regions outside the EC as the effects of integration and enlargement are anticipated. The Stockholm region has recently launched a strategic plan for the development of the Stockholm–Mälar region (Linzie & Boman 1991).

Unfortunately, systematic as distinct from anecdotal evidence on the growth of territorial competition is not available for the EC as a whole. Evidence for England and Wales on growth of local authority resources devoted to local economic development, presented in Table 5.1 (p.118, provides some indication of the pace of change, however.

Economic development activity in the local public sector is not the only manifestation of territorial competition but it is important, particularly in the UK which lacks powerful local private sector organisations such as chambers of commerce. Activity, measured in terms of staff employed, increased sharply in all types of local authority between 1980 and 1985 but most sharply in the more strategically orientated county authorities. This trend contrasts strongly with the run-down in regional policy in Britain (and

Table 5.1 Change in staff working on local economic development, 1980–85, by authority type

Greater London Council	+30.6%
London boroughs	+5.9%
Other metro counties	+52.5%
Metro districts	+53.8%
Non-metro counties	+71.9%
Non-metro districts	+38.9%

Source: Bennett and Krebs (1991).

in most of Europe) over the past 20 years. Wren (1990), for example, calculates that the real value of expenditure on regional policy in the UK approximately halved between 1980 and 1989 from £983m to £477m. There is thus not only evidence of an increasing level of territorially competitive activity but also evidence supporting proposition 1 above.

More detailed information on the form and scope of local economic development policies can be used to throw light on a number of the propositions advanced in the previous section. Local authorities are not the only agencies involved in economic development activities in Britain, nor is local economic development policy the only manifestation of territorial competition, but local authorities are the most significant actors in the majority of areas, and the only ones who are present in every area and required (since 1990) to publish an economic development plan if they are involved in this field of activity. All available plans were collected and the policies in which they engaged to promote local economic development were categorised into types. The results of this analysis are shown in Table 5.2, where the percentage of each class of local authority engaging in each type of policy is presented.

The geographic range of this enquiry was the English 'greater south-east', one of the leading European regions now exposed to international competition. As defined by Hall (1989b), this territory encompasses parts of four standard regions, stretching 100–175 km out from the core of London. In important respects, however, this area now represents the effective London region within which businesses have access to key agglomeration economies of the metropolitan area. Arguably, therefore, this is the scale at which territorial competition with the other leading European regions will be conducted. In fact, however, its highest tier authorities are 33 London boroughs (including the City of London) and 18 'shire' counties (each including an array of districts with their own economic development roles). A limited coordinating function is carried out by a Standing Conference of planning authorities in the south-east region (SERPLAN), but this body has limited resources and is more preoccupied with reconciling the needs for regional housing with the opposition of local NIMBYist[5] lobbies than with region-wide economic initiatives.

Table 5.2

Type of authority	per cent of authority type engaged by type of policy												
	1	2	3	4	5	6	7	8	9	10	11	12	13
County council N = 11	9	18	64	18	18	18	100	36	0	9	9	18	..
District N = 36	3	17	50	17	17	11	75	61	8	3	0	0	19
London borough N = 23	13	37	59	26	52	20	48	22	13	0	13	..	..

Key to types of policy:
1 Childcare as a means of helping females re-enter the labour force
2 Vocational training beyond statutory requirements especially for disadvantaged groups
3 Free business advice
4 Grants
5 Advice on premises/location
6 Loans
7 Marketing and promotion of area
8 Development of premises or provision of subsidised premises
9 Information technology training/advice
10 Local labour market information
11 Support to voluntary sector to promote local economy
12 Assistance for local business networking
13 Developing Science Park.
Source: Local authority local economic development plans.

Some clear patterns emerge. Much the most common type of policy, for example, was promotion and marketing, but while all county councils engaged in this, less than half of London boroughs did. London boroughs were considerably more likely to provide advice on premises and location but not to subsidise or develop premises. This was most common for district councils and is consistent with the proposition that agencies with boundaries more closely reflecting functional economic areas will be more active in pursuing territorial competition. It is also supportive of the proposition that the type of policy engaged in is influenced by the circumstances and functions of the agencies pursuing the policy.

London boroughs represent only a small fraction of the functional economic region of London and are far from self-sufficient local economies. They need to promote themselves in ways which tie the activity they seek to their territory. If the London borough of Ealing, for example, advertises the merits of locating in Ealing any mobile investor whose notice is attracted is, when he or she looks around, just as likely to choose a neighbouring borough. On the other hand, if Ealing provides advice on premises it can tie the investor into its local economy. Indeed in the survey a number of boroughs identified not cities other than London, but neighbouring London boroughs, as their main economic competitors.

It is the authorities representing the most self-sufficient local economies, the counties (Level 3 regions in the NUTS system of European regions) which engaged most in marketing and promotion. The probability is that little attracted activity will end up in other administrative units. By providing advice on premises and location or by subsidising and developing premises (the balance as between London boroughs and districts between these activities probably being explained by funding and the definition of functions), an authority ties the activity to its locality. When its locality is only one element in a larger functional region this is obviously a significant subsidiary aim of policy.

The evidence available from Table 5.2 and other sources is obviously fragmentary and too limited to evaluate all of the propositions advanced in section 4 (p.116). Nevertheless there is support for a number of the hypotheses.

First, it does appear that having a smaller number of authorities representing a functional economic region is helpful to the pursuit of competitive policies. Not only does the difference in behaviour between counties and London boroughs reported in Table 5.2 support it, but it is significant in the contrast between London and Paris, for example, and also in that between Birmingham and Manchester. Bennett & Krebs (1991, p.174) also note that all three of their paradigms of effective local economic partnership, Sheffield, Duisburg and Hamburg, are :

> large, relatively free-standing cities [with] a relatively close correspondence of labour markets with government boundaries and all have coordinated governmental functions

Secondly there is some evidence that a sense of cultural and/or political identity matters. Barcelona, for example, consists of 28 separate municipalities and is under socialist political control. The surrounding region, Catalonia, is under conservative Catalan nationalist control. But the unifying cultural identity Catalans have, at least during the 1980s, provided a strong enough unifying force for a powerful territorially competitive policy to develop. A similar unifying force can be provided by political control. It was not until the city of Lille and the surrounding Nord-Pas de Calais region both came under socialist control that Lille developed strong, territorially competitive policies.

The London case shows that heterogeneity and inconsistency in local interests, or those of potentially key actors, hinders the pursuit of territorial competition; the interests of public agencies conflict and the interests of the important financial sector do not coincide with those of manufacturing industry. This is also a problem in the case of many English cities where important agencies are run from Whitehall rather than locally. Such agencies may be more motivated by the aims of their agencies vis à vis other agencies—the Department of the Environment retaining control of urban policy against competition from the Trade and Industry or Employment departments, for example, or the Department of Education and Science retaining control of vocational education—than with furthering local economic development. Here and elsewhere the imposition by national government of an Urban Development Corporation (UDC) may generate a conflict between the UDC and the local authorities (see Cheshire *et al.* 1992).

The degree of representation of 'rent earners' in local decision making seems a likely component of the explanation of why London boroughs are less engaged in territorial competition and the home counties are most engaged. In general there is a correlation between political control by left or right and the influence rent earners have in decision making. Not only are London boroughs less engaged in the most obvious forms of territorial competition (marketing and promoting their areas and developing or subsidising premises) but, as shown in Table 5.2, they are more likely to identify policies which benefit non-rent earners as being for the purposes of local economic development. Thus 13% identified providing childcare for female workers so they could return to the labour force, 37% identified providing vocational training beyond statutory requirements, and 13% providing information technology training; all policies designed more to benefit the disadvantaged, non-rent earners than rent earners. In each case the highest probability they will be provided is in London boroughs.

Bennett & Krebs (1991) provide some interesting evidence for Germany on the question of whether strong or weak regions are more likely to be active and effective. This is summarised in Table 5.3.

Table 5.3 shows that on every measure chambers of commerce were more engaged in territorial competition in economically stronger regions. Only in terms of numbers of partners involved was there more involvement in weaker regions. But, as argued earlier, this represents a weak organisational base for competitive action.

There is evidence also that cities with a combination of favourable factors can (unlike the general run of local authorities surveyed in greater south-east England) formulate a strategic approach through consideration of their strengths and weaknesses. Thus Frankfurt perceives its strength to be in financial services and communications and tries to secure the proposed European Central Bank and strengthen its airport; Stockholm builds on information-intensive activities; Birmingham attempts to offset its image as a cultureless city by building up its symphony orchestra, building a new

Table 5.3 **Involvement of chambers of commerce in local economic development projects, 1989 (FRG) by type of region (means for groups shown)**

Type of region	Zero projects	Number of projects	Staff per project	Capital '000 DM	Share of business	Number of partners
Strong local economy	10%	72	26.1	5361	47	13.2
Weak local economy	26%	46	12.2	2243	14	14.0

Source: Adapted from Bennett & Krebs (1991) Table 7.8.

world-class concert hall and attracting the Sadlers Wells company from London. In London itself we note that by far the best resourced research study oriented to questions of competitive strategy is being carried out for the City of London—a tightly knit body representing a small functional economic area with rather homogeneous economic interests, a shared culture, and considerable financial independence (LBS 1992a, 1992b).

The implications for Europe's urban system

The evidence presented is not conclusive but certainly supports the interpretation of territorial competition offered in the earlier sections of this chapter. The logic which has given rise to the development of territorially competitive policies is essentially a by-product of European integration. But when such policies are viewed from the perspective of the provision of public goods, it can be seen that the capacity of different areas to effectively engage in the process will vary systematically. So far as evidence exists against which to test the propositions advanced in earlier sections, those propositions perform well. Since there is also evidence that effective territorial policies can positively influence the economic success of urban regions, just as ineffective policies or no policies can have a negative impact (Cheshire 1990; 1992), it is also to be expected that territorial competition will have an impact on the structure of Europe's urban system.

That territorial competition can have an impact does not mean that it is desirable from some broader perspective of economic and social welfare. There appear to be three not mutually exclusive possibilities.

1 It may be pure waste in the sense that the resources expended have no impact at all on local economic development.
2 It may be pure zero sum (i.e. no growth) from an EC perspective although non-zero sum from a regional or even national perspective. Successful regions may gain at the expense of unsuccessful ones.
3 Territorial competition may act successfully on the supply side and on those characteristics of regions from which their inhabitants derive welfare; it is capacity building and makes those regions which successfully engage in it better places in which to live, adding to output, real income and welfare across Europe as a whole.

Although there is, as yet, inadequate systematic evidence for Europe, it does seem likely that much territorial competition is pure waste. It is misdirected in terms of the real demands of firms and is essentially fulfilling bureaucratic goals. If a territory establishes an agency to promote itself and encourage local economic development, that agency has to justify itself. Advertising and marketing may be highly visible ways of spending budgets and justifying an agency's existence, but they are not necessarily very effective in terms of attracting mobile investment, still less in terms of local capacity building. One might provisionally classify policies as follows:

Zero sum	Growth enhancing
Pure promotion	Training
Capturing mobile investment	Fostering entrepreneurship
Investment subsidies	Helping new firms
Subsidised premises	Business advice
	Uncertainty reduction
	Coordination
	Infrastructure investment

One problem is that in reality no clear-cut distinction can be made. Few policies are likely to be purely zero sum or purely growth enhancing. In addition, not only should a policy be growth enhancing but it should offer a good use of resources. Infrastructure investment, for example, seldom has a zero pay-off but the fact that it has some positive rate of return does not necessarily make it worthwhile. The rate of return has to be equal to that available elsewhere in the local economy for it to be worthwhile, and it is always possible that resources spent by the local private sector would offer a better rate of return than infrastructure investment by the local public sector.

There is yet a further consideration. It was argued above that one reason why there may have been a reduction in the efforts of national governments to regulate territorial competition is that regional policy concerns have become weakened. At the European level, however, there has been a progressive strengthening of regional policy. This has now well-established priorities (see CEC 1987; 1992) and those objectives will frequently conflict with those of territorial competition. As already discussed, the incentives to engage in territorial competition tend to be strongest for the advanced metropolitan regions nearest the centre of gravity of Europe's economy, i.e. precisely those regions which have most to gain from European integration. In contrast, the priorities of the Community's structural funds are to reduce spatial disparities and promote the development of peripheral and lagging regions.

The growth of any urban area is a complex multivariate process (see Glaeser et al. 1992 for a recent analysis for the US). The effectiveness with which the region engages in territorial competition is only one factor. The implications of the analysis presented here, however, are that its impact will be greatest in city-regions in the core of Europe; they have most to play for. It will have most impact in altering the position of cities nearer the top of the urban hierarchy; they, too, have most to play for but in addition, because their geographical extent is greater, there is likely to be a greater variance in the extent to which the boundaries of public agencies coincide with the boundaries of functional urban regions. Thus Paris with the near coincidence of the Ile de France administrative region with the functional reality of Paris, is likely to gain. For similar reasons so too is Berlin.

Bruxelles and London are at a disadvantage, although in the case of Bruxelles this disadvantage of administrative fragmentation may be partly offset by a sense of regional identity—compared to the Flemish-speaking north and Wallonian south. And almost regardless of what territorially competitive efforts are made in the case of Brussels, it gains from being Europe's capital.

The advantages of Berlin and Paris in comparison with London are repeated for the majority of German and (to a lesser extent) French cities. Germany has powerful private agencies as well as a system of urban and regional government which makes discretionary resources available at a local level and favours local autonomy and identity. It also has an institutional and political system which helps unify labour and industrial interests and reduces transactions costs in the establishment of territorially competitive policies.

Elsewhere conditions favour some Spanish cities more than those of Italy and Portugal. Spain has not only devolved significant powers to a regional level but has a strongly developed sense of regional identity. During the 1980s and early 1990s Barcelona seemed to set the pace. Although it continues to enjoy advantages not directly related to local policy, objective circumstances appear, in the long run, to favour Madrid in the territorial competition game. Barcelona is fragmented into at least 28 municipalities and many more if the functional region is considered. There is a regional tier to coordinate action but this is under different political control, and is likely to remain so. In contrast, the region of Madrid is almost coincident with the functional reality of Madrid, and there is only one central municipality. While a distinctive sense of being Catalan may have provided a binding agent in the recent past for Barcelona (as did the exceptional event of the Olympics), as regional identity becomes a commonplace, political differences may be re-engaged; and the sense of Castillian distinctiveness is likely to grow in Madrid as other regions become more independent.

Italian cities are still hamstrung in the development of territorial competition by an inefficient and corrupt public sector. The new regional basis of politics and the attempt to clean up Italian politics[5] may strengthen the position of at least northern cities in the future; so too may the introduction of a tier of metropolitan regional government announced for 1993. But these are highly uncertain and the potential for the existing regions to paralyse the creation of the proposed metropolitan regions is considerable. Relative to other Italian cities, Milan has more of those characteristics identified as favouring the development of territorially competitive policies than others. It has a more diversified economy, whereas Torino is dominated by the car industry which itself is likely to be in crisis; and the functional area of Milano coincides quite closely with that of Lombardy.

The importance of this last factor is illustrated yet again by the cases of Rotterdam, Birmingham, Glasgow and Edinburgh. Rotterdam, like other

Dutch cities, is not well placed in administrative terms. The city boundaries are not so narrowly drawn as those of Bruxelles but its administrative area is far smaller than its functional region. Recognising this, the city was attempting during 1992 to form a new regional tier to which both the city and the surrounding municipalities would give up substantial powers (a proposition opposed, as would be expected, by many of the smaller municipalities on the grounds that such a regional authority would be dominated by Rotterdam city). Birmingham, as noted, has been perhaps the most successful developer of territorially competitive policies in England. Its administrative boundaries are more generously drawn than those of any other major English city. Glasgow and Edinburgh have even stronger senses of regional identity and, in strong contrast to English cities, have a regional authority which has helped initiate strategic policies and coordinated local public and private agencies.

Thus the logic of European integration is to intensify the process of territorial competition. Such policies contribute only marginally to the growth or decline of individual cities, and even when they contribute positively, do not necessarily represent a net gain or even an economic use of resources. But unless the process is regulated it is increasingly going to be in the perceived interests of city-regions to compete: if others are competing they will not be able not to compete successfully. In the absence of supranational regulation this in turn will increase the competitive process.

The general effects of this competition will be to strengthen the position of core cities as a group compared to peripheral cities and stronger cities compared to weaker ones because peripheral and weaker cities will not perceive a significant pay-off. Within the core cities there are likely to be cumulative losses for city regions in which conditions do not favour effective policy formulation and cumulative gains where they do. The greatest variations in the short and medium run are likely to be in the city-regions nearest the top of the national (and EC) urban hierarchies, but in the slightly longer term the success or failure of the largest metropolitan regions will filter through to their associated satellite cities.

But even within this logic of divergence there is a self-correcting logic too. The process of European integration has created the conditions giving rise to the development of territorial competition and reduced the incentive for national governments to regulate it. But at the same time is has created a logic for the EC to regulate the process and, as territorial competition grows and more frequently and visibly generates net resource costs, so the incentive for the EC to regulate the process will increase.

NOTES

[1] This paper was prepared as part of a wider ranging project within the ESRC's Single European Market initiative. The authors gratefully acknowledge the support of the ESRC under grant No. W113.251003. They would also like to acknowledge the contribution of Ruth Winter and Christopher Evans, researchers on this project.

[2] Which in 1971 had 68% of the headquarters of Britain's largest 500 industrial companies compared to 33% of the US largest 500 headquartered in New York (Evans 1973)

[3] That is, they may lose even if the policy is successful in increasing the level of local economic activity and net incomes relative to other places. Since policies consume resources if they are unsuccessful, there is a net loss to the area as a whole.

[4] If the local economy is dominated by a single industry, a restricted range of industries or only a few large firms, the competitive strategy adopted is more likely to favour those interests regardless of the prospects for success than it is to take an objective, strategic view.

[5] NIMBY is the acronym for 'not in my back yard'.

[6] This was originally written early in 1993 when it was deemed too early to judge the impact of these efforts. This was still the case in mid 1994 when the book went to press.

6 Information and Communications Technologies, Corporate Hierarchies and Urban Hierarchies in the New Europe

J. B. Goddard

Introduction

T HIS chapter explores the relationship between communications, new organisational dynamics and territorial development within the evolving European space. It examines the spatial implications of the emerging Single European Market, focusing on the effect on European enterprises. The key question is whether new corporate forms, underpinned by the use of advanced information and communications technologies (ICTs), will reproduce, on a European scale, the spatial division of labour that characterised a Europe of nation states dominated by national companies.

This question is addressed by highlighting the contrast between fordist and neo-fordist forms of industrial organisation in terms of design and development, manufacturer–supplier relationships and the roles of ICTs in R&D, lean production, and supply chain management. The chapter concludes by discussing the implications of new corporate forms for both spatial policy and communications policy, arguing for a greater integration of two domains which have hitherto been separate in both academic and policy debates.

The single European market: the macro-economic effects

Any discussion of spatial change in Europe must begin by considering the impact of the Single European Market (SEM) which, unlike European Monetary Union, is a reality. The likelihood that the SEM would presage rising regional disparities was recognised in the legislation which endorsed the 1992 project *and* embraced a major reform of the so-called structural funds to compensate for the uneven effects of removing the barriers to trade between countries—hence the Single Act embracing the SEM project and the reform of the Structural Funds; as a result, expenditure under these funds will have risen from 18% of the Community budget in 1988 to 30% this year.

Mackay (1992) has analysed regional disparities in unemployment and shown that growing economic integration has been associated with rising regional disparities within the Community. After a period of convergence in the 1970s, divergence became the norm in the 1980s. The fact that the ten most prosperous regions of the Community are to be found in Germany is particularly significant in the context of the SEM, since it in part reflects the fact that Germany as a whole has a trade surplus in manufacturing goods with ten of the other twelve member states. Situated at the heart of the EEC, many of the German Lande, with their particularly strong institutional structures, have been able to gain the greatest trade advantages from European integration.

Will the final completion of the SEM heighten these disparities? Any analysis must take account of the one-off effects of the lifting of non-tariff barriers to trade and the dynamic effects on corporate strategy. Regarding the one-off effects, Emerson (1988) has suggested that, with a few minor exceptions, countries with the most pronounced regional problems do not have excessive concentrations of sectors linked to government or regulated markets.

Notwithstanding these specific exceptions, it is generally acknowledged that the dynamic effects of the SEM are likely to have far more significant regional consequences, particularly in the way in which enterprises located in different regions are able to exploit the comparative advantage of local factor endowments, resulting in distinct patterns of industrial specialisation and inter-sectoral trade. In these terms, the most obvious comparative advantage of Europe's less favoured regions is relatively low labour costs which would make these attractive to labour-intensive industries. However, these advantages are not of great significance if compared to locations outside the EEC, locations which do not have the burdens of the additional employment costs likely to be realised with a full enforcement of the social charter.

More significant than the factor endowment considerations are those associated with economies of scale and externalities—these chiefly impinge on patterns of inter-industry trade in which firms in particular regions specialise in individual products within an industry. In terms of patterns of trade, it is clear that the main growth of the Community has been within, rather than between, sectors. The chief gainers would appear to have been firms producing the most technologically sophisticated products with a high R&D and skilled labour content. Although it is not possible to demonstrate these patterns through trade statistics directly, the Emerson analysis of the 40 key sectors most affected by the SEM legislation and the pattern of trade associated with these, shows that the core countries have a strong comparative advantage in R&D and capital-intensive activities.

If it were possible to disaggregate the national analyses to the regional level, it would be apparent that the older industrial regions assisted by the European Regional Development Fund remain specialised in those sectors which, although subject to significant restructuring during the 1980s, have now reached the limit of technical improvement: these remain the industries of the third and fourth Kondratiev, with the areas concerned being unlikely to gain from the growth of information industries (electronics and producer services) which are heavily concentrated outside the problem regions.

The single European market: micro-economic effects

There clearly are limits to a sectoral analysis of the impact of the SEM. Sectoral changes conceal what is happening at the micro-economic level in terms of the decisions of the key players in the economic arena, namely the largest national and multinational corporations. In the context of large multi-site organisations, economies of scale do not apply to single plants but to the corporation as a whole, and in addition to production, embrace pre-production (R&D), distribution and marketing activities. But the changing internal division of labour within these largest companies could be more important for the regions.

There is strong evidence to support the view that completion of the SEM has acted as a spur to increasing concentration of enterprise ownership (Amin *et al.* 1992). The number of mergers and acquisitions made by Europe's thousand leading companies increased from 303 in 1986–87 to 622 in 1989–90. In the last year, the number of mergers involving companies in two countries exceeded domestic mergers for the first time. If the motives for these mergers are considered, it is clear that in a quarter of the cases recorded in 1987–88 strengthening market position was the primary reason—this compared with only 7% giving this as a primary reason in 1983–84. Mergers are also getting larger. In 1986–87 there were 88 deals which created new companies with a turnover of more than five billion ECUs. In 1989–90 there were 257. Finally, there has been a rapid rise in joint ventures involving companies in several countries, the principal motive being the sharing of R&D costs.

The full locational implications of this increasing concentration of ownership are not clear cut. Within the UK, concentration has continued to be associated with centralisation of corporate control, with 74% of the total turnover amongst the 500 largest UK enterprises being controlled from London, a figure which has hardly changed since 1971. Within Europe the importance of a few leading cities as corporate control centres is recognised. European policy makers and the media have referred to this phenomenon as the 'hot banana', an area containing the headquarters and R&D centres of leading European companies (Masser *et al.* 1992). This area includes the long-established triangle based on London, Paris and Amsterdam, but then extends through southern Germany and northern Italy before spreading west into southern France and to Barcelona in Spain. Just as national companies shifted their headquarters to capital cities to be close to centres of national government influence, companies wishing to have a European presence are shifting headquarters into the 'banana. For example, Pilkingtons have recently announced the move of the head-quarters of their all-important plate and safety glass division from St Helens in Lancashire to Bruxelles. According to Sir Anthony Pilkington,

> the move to Bruxelles is intended to make people think in a different way about how the business should develop: they will have to have a more European orientation.

The reasons for such locational shifts are not hard to find. The cities of the banana are the hubs of the European transport and communication network as reflected in the high-speed train and airline systems. Other cities are attempting to compete by promoting innovations like teleports as a means of remaining in the race to attract these high-level command and control functions. Historically, a counterpart of centralised administration was the growth of external control in less favoured regions, as independent firms were incorporated into national enterprises and their internalised production systems. At the same time, a branch plant phenomenon with few local industrial and service linkages and an emphasis on blue-collar occupations came to typify many problem regions.

In addition to corporate control, accelerating rates of technological change associated with the 5th Kondratiev point to highly concentrated patterns of innovation. Drawing on R&D data in three key sectors (artificial intelligence, biotechnology and aerospace), on the concentration of public R&D and on evidence on public and private networks of research collaboration the CEC Monitor/Fast group have identified ten major *islands* of innovation in the European Community. This particular terminology highlights the isolation of these localities from the rest of Europe. Whilst these islands can be depicted as major European *regions*, in fact they focus on individual cities, notably London, Paris, Amsterdam, Dortmund, Frankfurt, Stuttgart, Lyon, Grenoble, Milano and Torino. Community and business programmes have tended to reinforce these islands of innovation. Thus data for awards under the BRITE programme which supports innovation in engineering industries show the south east receiving 73 out of the 197 awards or 37% of the total, compared with a 25% share of UK manufacturing employment.

New organisational dynamics

Will the completion of the SEM contribute towards an even greater centralisation? There are several reasons to suggest why this may not be the case. The most important reason relates to changes in the structure of organisations that have emerged in the 1980s. These changes that have been facilitated by the spread of information and communications technology. The classic corporate model which neatly mapped onto geographical space was the multi-layered hierarchy. Because of the limitations of communications—the telephone was not an effective substitute for face-to-face contact—many layers were necessary in the hierarchy as information was processed and refined to support strategic decision making at the centre. In manufacturing, the production of standardised products was dispersed to peripheral regions of the European nations to take advantage of relatively cheap and unskilled labour—this partly explains the convergence of regional disparities observed in the 1970s.

However, by the end of the 1970s a host of factors challenged the inflexible hierarchically organised and vertically integrated national and multinational corporation. These factors included increasing competition on a global basis, more rapid rates of technological change, especially the incorporation of information technology into a wide range of products and services, and the availability of information and communications technologies to manage and control more flexible organisational forms, including relationship between customers and suppliers.

As with the fordist era of mass production, the car industry is the harbinger of new forms of industrial organisation, in this instance spearheaded by the Japanese and their concept of 'lean' production (Morgan 1992). In order to understand the spatial implications of this concept, it is useful to contrast it with fordist forms of design, development, production and purchasing. These in essence relied on a long product life cycle in which new products emerged as the outcome of distinct stages, beginning with current development and followed by feasibility testing, product design and development, pilot production and final production. Product development was in essence a 'relay race'. Each of the functions involved in these stages were specialised and separated, often geographically. A major weakness of the approach was that products were seldom designed with manufacturing in mind, because the manufacturing teams were not involved at the conceptual stage; this weakness was exacerbated by shortening product life cycles during the 1980s. There were related weaknesses in terms of relationships with suppliers. These were characteristically based on multiple sourcing, switching to keep suppliers on their toes, and price-based purchasing. As producers sought to achieve economies of scale in production by concentrating production in fewer and fewer sites, long supply chains developed with capital locked up in goods in transit.

Lean production has been developed as a means of resolving these problems. It is a form of organisation whose locational implications are far from determined (Charles & Feng 1993). The aim is to strip out all forms of waste from that production and to increase flexibility and responsiveness to changes in market demands. In terms of the development process, 'value engineering' is used to break down the production process into a large number of steps and then to continually drive down the costs of each during the lifetime of the product. This in turn requires 'simultaneous engineering', involving a continuing interaction between interdisciplinary teams drawn from all stages of the production process. The risks of product development are then shared with a limited number of suppliers (co-makers), with the ultimate aim of single sourcing of particular components.

In this way a large part of the production process is out-sourced. The management of the supply chain thus required makes transport and communication a strategic function for the company. It requires coordination of routine information transaction *and* good movement using electronic data interchange (EDI) and advanced logistic systems. The supply chain becomes an extension of the assembler's production process, with the minimum of buffer stocks and coordination of production processes within two firms (i.e. more than just-in-time delivery). Frequent deliveries in small volumes may take place, but as part of a long-term contract which provides economies of scale in purchasing.

Taking these considerations together, it is clear that high levels of information exchange within and between firms is essential. Diagrammatically this can be expressed in terms of an increasing integration of the previously separate spheres of design, manufacturing and coordination, an integration that has been facilitated by ICTs. It also implies tighter linkages between customers and suppliers. The critical question is whether this all points to new locational patterns.

In the case of manufacturing it is clear that lean production does not mean small-scale production—on the contrary, it has made production economies of scale a more feasible proposition in a range of sectors. Individual locations may gain or lose production sites within the SEM and this may be influenced by a range of local considerations such as the age of the premises, flexibility of the workforce, communication facilities, etc.

In the case of R&D, pressures for decentralisation and reintegration with production sites may be reinforced by the tight labour market for scientists and technologists increasingly found in the major R&D concentrations. At the same time there are arguments for a continued centralisation and geographical separation, embracing considerations such as the internal economies of scale in R&D, the fixed cost of existing laboratories, the ease of face-to-face contact with other scientists, and the reluctance of R&D staff to move. One way in which these conflicting pressures can be resolved is to use ICTs *and* frequent travel to build R&D teams which operate on a multi-site basis, thus overcoming the need for major spatial reorganisations.

Finally, in the case of suppliers, the shift to single sourcing implies a search for the most appropriate supplier, wherever it is located. ICTs can then be used to manage long supply chains to ensure predictability of supply regardless of the length of the chain. This may involve the exploitation of existing agglomerations rather than the creation of new ones—although there are also strong pressures on closely integrated suppliers to co-locate in order to drive down inventory to minutes rather than hours or days.

Mobility, fixity and ICTs

The preceding discussion suggests that it will be increasingly difficult to read off implications for space and place arising from the new corporate dynamics. If anything, the corporate map of the new Europe may look more like a bowl of fruit salad than a banana! At a more theoretical level it is clear that the improvement of communications flowing from the use of ICTs is leading to a reassessment of the power of geography or geographical differentiation. As Harvey (1989) has pointed out:

> global corporations are now paying much closer attention to relative locational advantages, precisely because diminishing spatial barriers gives capitalists the power to explore minute spatial differentiations to good effect. Small differences in what the space contains in the way of labour supplies, resources, infrastructures and the like become of increased significance...This approaches the central paradox—the less important the spatial barriers the greater the sensitivity of capital to the variations of place within space, and the greater the incentive for places to be differentiated in a way that is attractive to capital.

Building on this perspective, Gillespie & Robins (1989) have discussed a number of implications concerning the impact of ICT on location.

1 They suggest that however much the use of ICT is associated with organisational and locational flexibility, this by no means signifies the final transcendence of spatial barriers. Rather, they can bring about 'new and more complex articulations of the dynamics of mobility and fixity' within which uneven development remains the norm.
2 The locational changes associated with ICT are less absolute than many analyses suggest; inherited and place-bound social, industrial and institutional structures will shape and constrain the changes associated with ICTs in a way which serves to preserve elements of continuity.
3 The nature of the locational changes under way in the present period cannot be determined by ICTs—for example, a universal tendency towards either centralisation or decentralisation of functions. Rather, ICTs increase the spatial/organisational repertoires that may be utilised to ensure that the opportunities of geographical differentiation can be maximised whilst at the same time maintaining overall control of production/distribution processes.

Given that ICTs have the potential to underpin greater mobility, it is appropriate to examine the changing pattern of globally mobile investment into the SEM. On the basis of interviews with 30 multinational companies and 20 trade and business associations within the UK, the Netherlands, Germany and the USA, Bachtler and Clement (1990) suggest that the globalisation strategies of Japanese companies involves the:

> establishment of a major presence in every world market and 'insideration', renegotiating the operating autonomy in these markets to increase the quality of investment.

For some Japanese firms, the SEM implies 'a greater integration of headquarters and management control functions through the setting up of European regional offices to deal with manufacturing, sales, marketing, logistics and distribution and a greater integration of European subsidiaries, including the diminution or elimination of country specific autonomy'. Most German and Dutch companies have already established a European-wide network of plants and offices with a high level of interdependency between them and the centralisation of functions such as marketing and the decentralisation of administration.

US companies report a greater concentration of production and other business in existing locations and more specialisation in areas such as R&D, manufacturing and distribution, based on the locational advantages of countries/sites—in other words a realignment of functions. Such changes may be associated with the creation of 'platform' products which are tailored to specific national markets.

A further element of mobility is associated with the introduction of 0800 services into the UK. Richardson (1993) demonstrates how functions have been removed from high-street front offices of financial and other service firms, centralised and then dispersed to major cities which have a sufficient supply of skilled labour. For example, although 90% of British Airways Tele-sales originated in the south-east of England, this demand is met from centres in London, Manchester, Glasgow and Leeds. Leeds alone has attracted 2000 jobs in telemediated service functions in the past two years, including the headquarters of the largest telebanking operation, First Direct, a subsidiary of a major retail bank, the Midland. Such large cities are gaining in this process at the expense of smaller cities. At the same time international 0800 numbers provide the basis for attacking the European market opened up by the liberalisation of information markets. Individual cities may win or lose in this process.

European regional policy

How do these locational disparities relate to European regional policy? The answer to this question must be seen in the context of the objectives of the European *industrial* policy, which has been to facilitate the development of European enterprises able to compete in global markets with US- and Japanese-based firms. The evolving rules of the SEM support this objective by offering firms economies of scale via trade liberalisation/standardisation. In addition, European science and technology policy supports the industrial policy objective by assisting precompetitive research, again principally in the largest companies.

Set against this implicit support for large enterprises through industrial policy, the main plank of European regional policy has been to promote indigenous development via small and medium sized enterprises. This would seem to be predicated on the belief that any trend towards centralisation of industry associated with ownership concentration can be mitigated by improving the competitiveness of industry indigenous to the LFRs. Following the bad experience of national governments with assisting large firms to establish branches in lagging regions in the 1960s and 70s, virtually no attention has been paid in European policy to the changing division of labour within larger enterprises, particularly the regional opportunities, as well as threats arising from processes of globalisation and localisation.

Under the regulations of the Community Support Framework Programmes for regions, the main emphasis of policy is now placed on development from below. Policies have been designed to build up a critical mass of local factor inputs, a strong institutional base and a dense network of linkages between local firms. The objective is to sustain a spiral of cumulative growth resulting in local comparative advantage in external trade. Specific measures include: support for inter-firm cooperation; easier access to local capital; vocational training in high level skills; support for communications; cooperation between local interest groups such as banks, chambers of commerce, trade unions, and central and local government; support for the supply of advanced business services; and improvement to the quality of local, social, cultural and environmental amenities.

This solution to the regional problem has immediate appeal, as it suggests the possibility of a Europe composed of independent regional economies converging in their growth rates through product specialisation and local effort.

But how realistic is this solution for lagging regions? There is much evidence to suggest that declining industrial regions provide an adverse environment for small-firm-based entrepreneurs. Moreover, the SEM and improved communications will eliminate many of the barriers behind which local small firms have traditionally sheltered. In addition, many of the conditions of success for small-firm-based local development are simply not transferable from areas like Prato in Italy, or Baden-Württenburg in Germany, to the Community's LFRs. Finally, and more importantly, it is clear that the networked small firm regional economy may have just been an interim stage of development between the hierarchical patterns of the 1960s and 70s, and the global/local corporate nexus of the 1990s; it certainly cannot be claimed to be a hegemonic model of organisation on which to base policies for all types of regions for the current decade.

Telecommunications policy

The evolving corporate geography has profound implications for telecommunications policy. Outside the largest metropolitan areas business demand for telecommunication is likely to be increasingly for point-to-point services to underpin intra-corporate links; there may be limited demand for broad-band switched services, because there are few local agglomerations of strongly linked enterprises. From the telecommunication supplier's perspective, a large share of the revenues and profits will be derived from multi-site global enterprises, whilst the highest cost will accrue amongst small users, particularly in areas where there are few large users to provide network externalities. Suppliers will also have to face volatility in demand arising from the ability of large users to readily reconfigure the distribution of information-intensive functions between sites.

In response to these changes in demand and in the context of increasingly deregulated telecommunications markets, we are seeing the replacement of a narrow range of services, basically telephony and data transmissions provided to all locations under principles of universal service, by a wide range of specialised services provided to those locations generating sufficient demand. In the UK, competition in local telecommunications is focusing on the larger cities where cable companies are 'cream skimming' the most profitable business; peripheral regions in contrast are confronted by a monopoly supplier.

In the light of such considerations, telecommunications is now becoming part of the armoury of urban and regional policy. Of the 93 largest urban district authorities in the UK, 60% claim to be developing local tele-communications policies of some description and 20% to have well-established policies in this area (Graham 1992). In the US, it has been asserted that in developing federal economic strategies it is 'not possible to resist the current electronic blitzkrieg with an unyielding defence' (Jacobsen 1987). Similar emphases are emerging in France (Vedel 1987) and Japan (Newstead 1989).

While there are a mixture of motives behind these local interventions, Graham (1992) has suggested that they are driven by concerns about the uneven diffusion of ICTs through the urban hierarchy, the social hierarchy and the firm-size hierarchy. More specifically the changes induced by ICTs are favouring the largest cities and enterprises and the most advantaged social groups. These interventions are however being made in the absence of any statutory municipal role in this field and notwithstanding the fact that local authorities are often among the largest users of ICTs in a particular area.

Graham suggests five local policy roles in local electronic infrastructure development.

1 Brokerage and risk subsidy between telecommunications suppliers and consumers—for example, through promoting the development of a teleport as is happening in Edinburgh;
2 quasi-public provision of ICT-based goods and services—for example, by providing a community-based information service, as is the case with Manchester City's HOST computer network;
3 by developing the economic and social networks which can exploit the opportunities created by the spread of ICTs—for example, Sheffield's Information 2000 Strategy;

4 by promoting particular technologies and related services like broad-band cable television services and associated community programming;

5 by using electronic infrastructure as a catalyst in urban regeneration based on physical development, as in Trafford Park Manchester where the Urban Development Corporation is promoting a 'communications park'.

From the UK perspective it is clear that electronic infrastructure policies are being used to address a wide variety of problems. In peripheral cities, ICTs are seen as tools for addressing geographical marginalisation and strengthening local–global linkages. In some large industrial cities the emphasis is on local community development strategies with a desire to ensure that the social benefits of ICTs are widely diffused. In the congested metropolis there is a greater emphasis on urban functioning, including traffic management, decentralisation and remote working.

However, these initiatives are all as yet very tentative and the modes of intervention poorly developed. Whilst there are clearly strong interactions between physical transport and communications—wires and wheels are complementary and not competing technologies—communication has yet to assume the role in economic development and land-use planning ascribed to transport.

More generally, there is a failure to link telecommunications and spatial policy, a situation which parallels the gap between various parts of the academic community concerned with the impact of ICTs on economic and social development. For example, the telecommunications policy research community is a very specialised group with few links for those concerned with urban and regional development.

Conclusion

The body of the chapter has concluded by reference to telecommunications policy not because telecommunications should be regarded as the answer to problems of uneven development. Rather it has been used to suggest that telecommunications is an important adjunct to processes of globalisation and localisation of large enterprise, processes which are undermining long-established spatial hierarchies within Europe. The threats and opportunities to cities and regions that are arising clearly cannot be addressed only by policies to support small and medium sized enterprises. Territorial policy must focus on the leading enterprises currently located in an area or which might be attracted to it, and their future requirements in terms of broadly defined infrastructure.

This infrastructure is likely to include strong regional technology transfer mechanisms; policies to create more local embeddedness of leading regional enterprises, for example through support for key suppliers; external promotion to attract reinvestment from global parent companies; targeting of new investors with the potential of complementing the supply capacities in the region; promoting the provision of appropriate regional communications infrastructure; identifying and developing strategic sites for new investment; and last but not least, support for specialised training to enhance the technical, scientific and managerial skills needed to cope with rapid technological and organisational change. Such a key enterprise strategy is not to deny the continuing importance for policies designed to foster indigenous development based around small and medium sized enterprises; rather it is to suggest the need to link such policies more explicitly to an agenda of regional competitiveness in a global economy.

7 The Changing Urban Hierarchy in Europe

Michael Wegener

The urban transition

AFTER the decline of the cities of the Mediterranean in the wake of the fall of the Roman Empire, the urban system of Europe re-emerged in the 10th century. From then until modern times it remained relatively stable. Growth of cities was slow and, apart from devastations by wars, epidemic diseases or natural disasters, urban decline was rare (e.g. when trade routes changed such as in the case of Venezia, the port cities of Flanders or the Hanseatic League).

However, starting in the second half of the 18th century, an unprecedented wave of urban growth swept over the continent. This period marked the beginning of industrialisation in England. Basic inventions such as the steam engine and the railway made large-scale production of goods possible in mechanised factories. The new industries located in the cities close to their markets and developed a large demand for labour. At the same time mechanisation of agriculture made rural labour redundant and led to the first wave of rural-to-urban migration, which resulted in the growth of industrial cities at the expense of the countryside. In this period, urbanisation was the consequence of the first phase of the *economic transition*, the transition from agricultural to manufacturing employment.

This primary phase of urbanisation first took place in the industrial cities of the British north-west in the second half of the 18th century, and during the following 100 years spread to the continent, first to the countries of north west Europe, Belgium, the Netherlands, north-west France and Germany. It took well into this century before massive industrialisation occurred in northern Italy and even until after World War II before it occurred in southern Germany and southern France. Large regions in the Mediterranean countries are only now passing through this first phase of the economic transition.

Growing affluence and advances in medicine and hygiene in the early 19th century reduced mortality, in particular infant mortality, with the effect that population growth accelerated and more people moved to the cities. Now urban growth became even faster; many cities multiplied in size in a few decades. Only much later fertility also started to decline when social security systems made a large number of children for old-age support unnecessary, so population growth slowed down or even turned into decline. The sequence of declining mortality and subsequent declining fertility, the *demographic transition*, ended the period of urban growth; and where there was no international immigration, cities started to decline in population. The demographic transition occurred first in those countries which also first went through the economic transition, and has only recently arrived in the countries of southern Europe which still have much higher birth rates than their northern neighbours.

The wave-like diffusion of the economic and demographic transitions from the north-west of Europe to its south-west, south and south-east helps to explain the different phases of urbanisation coexisting in Europe at one particular point in time. In the north-west, where both the economic and demographic transitions have almost been completed, deindustrialisation and deurbanisation is found except where through extraordinary efforts the next phase of the economic transition, the shift from manufacturing to services, has been achieved. In the regions of the second wave of industrialisation, the south-east of England, the south of Germany, the north of Italy and southern France, the second phase of the economic transition is most advanced, here the post-industrial city is emerging. At the same time in parts of Spain, Portugal, southern Italy and Greece, some cities are today experiencing the growth period of early industrialisation and urbanisation (Hall & Hay 1980; Cheshire & Hay 1989). In an analogy to the terms economic and demographic transition, the shift from the industrial city to the service and post-industrial city can be called the *urban transition* (cf. Friedrichs 1985).

Counter-urbanisation

In the mid-1980s, 90 cities in the European Community had a population of more than 250 000 (see Figures 7.1 and 7.2).

However, eight out of ten Europeans lived in smaller communities. Champion (1989) presents evidence that in a wide range of countries in north-western and central Europe there was a reversal of traditional migration flows between centre and periphery, and that population growth was strongest in remote rural districts.

It may be speculated that at the heart of the counter-urbanisation trend is a tendency to equalise the differences in density between city and countryside. If this hypothesis is true, both extremes, the rural village and the metropolis, would lose importance as a form of human settlement. Small and medium-sized cities with good accessibility to the traditional centres would be the winners because they represent the settlement form of the future, the continuum between city and countryside, towards which both city and countryside asymptotically develop.

Size of city

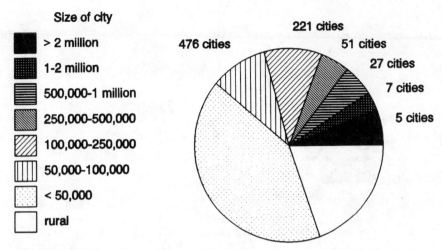

> 2 million

1-2 million

500,000-1 million

250,000-500,000

100,000-250,000

50,000-100,000

< 50,000

rural

Figure 7.1 Urban population in the EC by city size in the 1980s
Source: Kunzmann & Wegener 1991.

More recent data seems to demonstrate another trend reversal towards a new phase of urbanisation. Figure 7.3 (p.143) shows that in the second half of the 1980s urban decline was confined to Italy, Switzerland, Belgium, the UK and parts of France (Commission of the European Communities 1991).

In eastern Europe urban growth continues to be dominant. The speed of urban growth in southern and eastern Europe has declined, while in central Europe there is again urban growth due to international migration. In particular the recent growth of cities in west Germany has been based on massive immigration from east Germany and eastern Europe.

Urban hierarchies

The Oxford Dictionary defines a hierarchy as an 'organisation with grades or classes ranked one above another'. The notion that there is a 'hierarchical' rank-order among cities dates back to the time when cities were seats of local and provincial governments or of bishops and archbishops and represented the hierarchical order of the State or the Church. Christaller and Lösch reinterpreted this hierarchical structure in functional terms by demonstrating that hierarchical 'central-place systems' are efficient in terms of service or market areas and hence should either develop naturally or be promoted by policy.

More recently, it has become fashionable to talk of urban networks instead of urban hierarchies. Pumain (1992) has pointed out that in geography, a hierarchy is a special kind of network, a tree. While it is true that the term network seems to place more emphasis on the links than on the nodes, its use does not clarify the structure of an urban system any better than the term hierarchy. The question of interest here is whether the

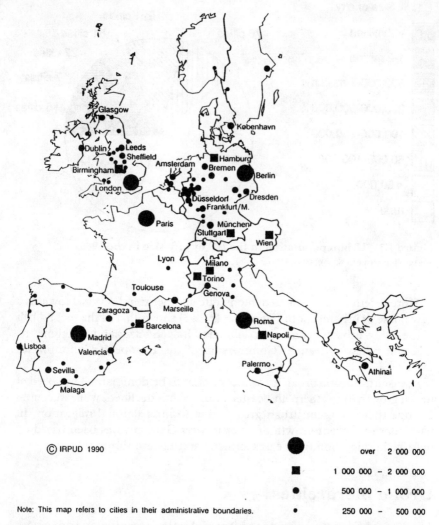

Figure 7.2 Cities in western Europe with more than 250 000 population
Source: Kunzmann & Wegener 1991.

urban system is organised in clearly distinguishable levels of cities with significantly different functions, i.e. whether it is a central-place system.

However, a look at Figure 7.2 does not suggest a system of central places in Europe. There have been numerous attempts to discover a hierarchical order among the cities in Europe. Figures 7.4–7.7 show recent examples of such attempts.

Törnqvist, as early as 1970, developed the notion of contact networks, hypothesising that the number of interactions with other cities would be a good indicator of the position of a city in the urban hierarchy. Figure 7.4 illustrates the results of a recent application of this method to cities in

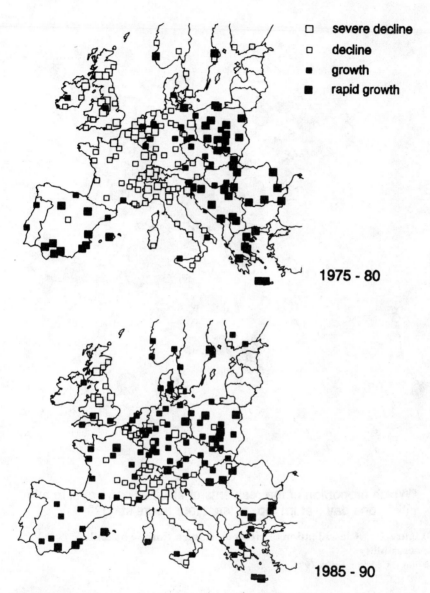

Figure 7.3 **Urban growth and urban decline in Europe 1975–80 and 1985–90**
Source: adapted from Commission of the European Communities 1991.

Europe (Cederlund *et al.* 1991). Keeble *et al.* (1988) analysed the centrality of economic centres in Europe using a potential indicator; the resulting centrality contours are shown in Figure 7.5 (p.145). Brunet (1989) rank-ordered 165 agglomerations with a population of more than 200 000 using the 16 indicators listed at the bottom of Figure 7.6 (p.146). The study was

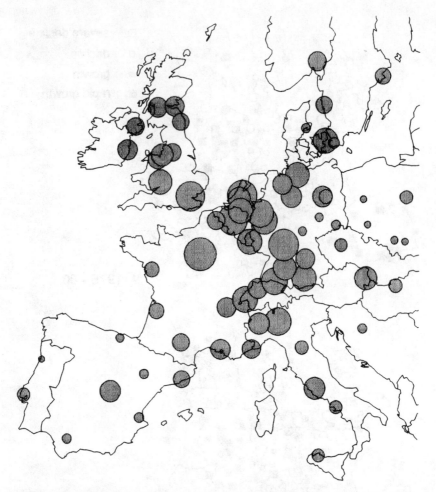

"Which proportion of the residents of all cities can travel to this city in one day (return trip, at least four hours stay)?"

Figure 7.4   Classification of cities in western Europe by daily accessibility
Source: Cederlund *et al.* 1991.

designed to support the hypothesis that the largest concentration of conomic activity in Europe is in the 'backbone' between south east England and northern Italy (the 'Blue Banana')—and not in France. Bruinsma & Rietveld (1992) used a similar potential indicator as Keeble to establish a rank-order of accessibility of population centres (Figure 7.7).

The common result of the four exercises is not surprising: there is no clear urban hierarchy in Europe. No matter whether one chooses functional indicators (as Brunet) or accessibility measures (as the other authors), one finds all gradations of centrality irregularly scattered across the continent.

Figure 7.5   Classification of cities in western Europe by accessibility of economic centres
Source: Keeble *et al.* 1988.

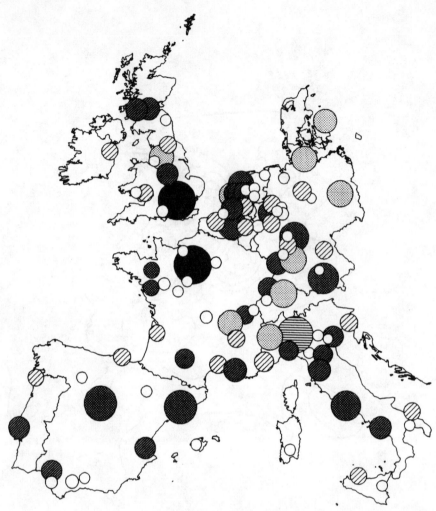

Sixteen indicators:

1 Population	9 Airports
2 Population change	10 Ports
3 Multinationals	11 Culture
4 Infrastructure	12 Trade fairs
5 Engineers	13 Congresses
6 R&D	14 Press and Publishing
7 Universities	15 Telecommunications
8 Financial Services	16 Special Features

Figure 7.6   Classification of cities in western Europe using 16 indicators
Source: Brunet 1989.

$$A_i = \sum_j \frac{Q_j}{c_{ij}^{\alpha}}$$

Accessibility of city i (Population potential)

Population in city j

Travel time between i and j

Figure 7.7   Classification of cities in western Europe by accessibility of population

Source: Bruinsma & Rietveld 1992.

However, certain spatial clusters of higher intensity or accessibility show up in all four maps: the London region, the Paris Basin, the Randstad, the Rhein-Ruhr and Rhein-Main agglomerations and, to a lesser degree, the industrial cities of northern England.

Based on similar considerations, Kunzmann & Wegener (1991) proposed the following four-level 'hierarchy' of cities of European importance (Figure 7.8):

- At the top of the hierarchy are Paris and London, the only two global cities of western Europe.
- They are followed by conurbations such as Rhein-Main (Frankfurt), København/Malmö, Manchester/Leeds/Liverpool, the Randstad (Amsterdam/Rotterdam), the Ruhr (Dortmund/Essen/Duisburg) and the Rhein basin (Bonn/Köln/Düsseldorf). Of similar importance on the European scale are a number of larger European cities ('Euro-metropoles') such as Athínai, Bruxelles, Birmingham, Wien, Lyon, Milano, Roma, Madrid, Barcelona, Hamburg, München and Zürich. These cities perform essential economic, financial or political and cultural functions for Europe. After the reunification of Germany, Berlin too may become again a city of major European importance and in the long run even become a candidate for a global city.
- A third category comprises national capitals and other cities of European importance such as Dublin, Glasgow, Lisbon, Strasbourg, Stuttgart, Palermo, Torino and Napoli. These cities are completing the network of cities of European importance, although their function is mainly a national one.

Below this level, and depending on national definitions of central places which exist in a few European states (e.g. Denmark, the Netherlands, Austria or Germany), various levels of lower urban hierarchies follow. The growing integration of Europe will gradually replace these national urban hierarchies by one integrated urban hierarchy in Europe.

Spatial mega-trends

In the late 1980s and early 1990s the pattern of urbanisation in Europe has become more complex. The changes occurring can be linked to four major developments, two of them technological, the other two political:

- The first development is the acceleration of transport and the fact that transport is vastly underpriced in relation to its social and environmental costs. Forty years of massive road construction, mass-motorisation and low fuel prices have made every corner of the continent accessible for car travel and truck transport. Domestic and intra-European air travel has grown dramatically and have made one-day business trips between major European cities a normal thing.

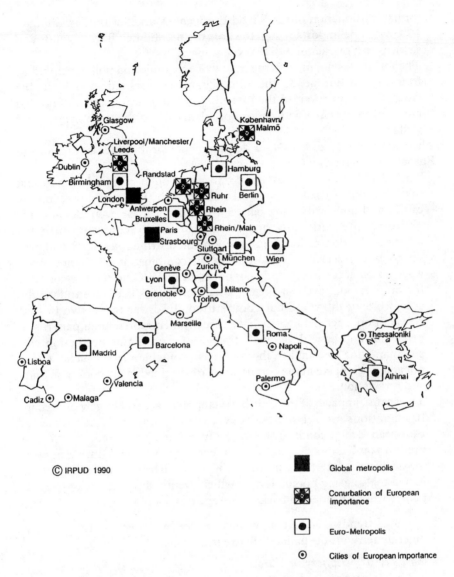

Figure 7.8 in legend:
- Global metropolis
- Conurbation of European importance
- Euro-Metropolis
- Cities of European importance

© IRPUD 1990

Figure 7.8 A hierarchy of cities of European importance
Source: Kunzmann & Wegener 1991.

- The second development is the explosive growth of telecommunications and the diffusion of computerisation and automation in more and more industries with the effect of universal flexibility in terms of products, materials, labour, logistics ('just-in-time'), and internal and external organisation of firms.
- The third development is the unification of Europe. Despite many delays and frictions, the process of European integration is moving ahead. The

formal introduction of the Single European Market at the beginning of 1993 was one important step towards the uninhibited flow of goods and services in a market of 345 000 consumers.
- The fourth development is the inflow of economic and political migrants from eastern Europe. As these immigrants are first attracted to the big cities, they have reversed the prevailing counter-urbanisation trend and are confronting cities with serious problems of housing, employment and welfare.

The *spatial* impacts of these four developments for the cities and regions in Europe are far-reaching:
- The first three developments combine to open regional or national markets, which used to be protected by high transport costs, import duties and non-tariff trade barriers, to European or even global competition. Internationally operating firms have acquired a hitherto unknown freedom to locate production and distribution facilities all over the continent, seeking the most profitable combination of labour supply, wage costs, accessibility and economies of scale. Cities and regions, on the other hand, have lost their spatial monopolies and have to face the cold wind of international competition. In order to survive in that competition, successful cities have turned to entrepreneurial patterns of behaviour and have developed strategies for economic promotion and city marketing to establish themselves a profitable market niche. Cities not able to compete in the competition are destined to fall back in income and population.
- The spatial impacts of the fourth development are still largely uncertain. Immigrations are spatially selective, and the immigration policies of European Union countries are far from being coordinated, therefore at present some countries are more affected than others. Immigration is mostly seen as a burden, however this may change in countries with an increasingly ageing population, in which eventually an inflow of younger people will be needed to balance the age distribution.

The spatial trends sketched above have worked together to produce three distinct dimensions of spatial disparity in Europe:

1 Core versus periphery in Europe
- The internationalisation of regional and national economies has tended to favour cities in the European core (Belgium, Germany, the Netherlands, central and northern France and south east England). Cities in the core of Europe have comparative advantages. Their accessibility by air and train is vastly superior, they offer more and better high-level services (banking, insurances, consultancy services, etc.), and their trade fairs, convention facilities and cultural events have higher international reputation. This is why they attract more international corporations and subsidiaries of foreign enterprises. The highly urbanised triangle between London, Paris and Bruxelles, and the densely populated urban corridor from Amsterdam to Basel are

winners in the polarisation of the urban system in Europe. The losers are cities in the European periphery (Scotland, Ireland, Greece and Portugal). Although transport costs overall are low, in the tightening competition small differences count, so not being connected is a disadvantage.

2 North versus south in Europe

• Just as on the global scale, in Europe there is also a divide between 'north' and 'south'. Population growth in the cities and urban regions of the north—which also includes parts of central Europe (Germany, Austria and Switzerland)—has virtually come to a halt, whereas the cities in the south continue to grow as they are still in an earlier phase of industrialisation and urbanisation. The difference in phase also means a difference in income, and this gap is widening despite all equalisation policies of the EU. In recent years significant economic progress has been made in some regions of Spain, however the 'Spanish miracle' has yet to prove that it is permanent. On the other hand, birth rates in Italy have fallen in recent years, and urban growth rates in Italy may start to stagnate or even decline in the coming decade.

3 West versus east in Europe

• The unexpected opening of the once impenetrable Iron Curtain to eastern Europe has recalled an old spatial divide, the east–west divide. It may well supersede the north–south divide and become the dominant political issue of the next decades. Some German cities (Berlin, Hamburg, Hanover) may benefit from the new geopolitical situation, but also cities in Denmark, Austria and northern Italy. The future development of Berlin and its role in the urban system in Europe are still very uncertain.

In summary, the urban system in Europe is in a process of polarisation between 'winners' and 'losers', between centre and periphery, north and south, and east and west. There is little evidence that this polarisation can be reversed or halted. Policy efforts to narrow the gap between centre and periphery and between north and south have only been partially successful, although, by creating jobs, providing public services and improving living conditions in assisted areas, they have contributed to slowing down the widening of regional disparities.

Urban deconcentration

The common experience of winner and loser cities is spatial deconcentration. The evolution of transport systems made the expansion of cities over a wider and wider area possible. In particular the diffusion of the private automobile brought low-density suburban living into the reach of not only the rich. Suburbanisation was not caused by the car but is a consequence of the same changes in socio-economic context and lifestyles that were also responsible for the growth of car ownership: increases in income, in the number of working women, smaller households, more leisure time, and a related change in housing preferences. Yet the car certainly contributed to pushing people out of city centres through congestion, lack of parking space, and noise and pollution, as did housing shortage and high land prices.

Offices and light industry and retail started to decentralise later, following either their employees or their markets or both, or taking advantage of attractive suburban locations with good transport access, ample parking and lower land prices; in particular greenfield shopping centres have become a threat to inner-area retailers. Still more recently, manufacturing industries, taking account of new tendencies in plant layout, organisation of production and just-in-time logistics, have started to prefer low-density, environmentally attractive suburban locations with good road access, ample space for expansion and still moderate land prices.

The results of the deconcentration process are both positive and negative: suburban living represents the preferences of large parts of the population. However, the consequences of urban dispersal are less desirable: longer work and shopping trips, high energy consumption, pollution and accidents, excessive land consumption and problems of public transport provision in low-density areas. This makes access to car travel a pre-requisite for taking advantage of employment and service opportunities and thus contributes to social segregation. The counterpart of suburbanisation is inner-city decline.

All over Europe therefore cities have undertaken efforts to revitalise their inner cities through restoration programmes, pedestrianisation schemes or new public transport systems. In some cases these efforts have been remarkably successful. Besides cities in the Netherlands (e.g. Delft), Germany (e.g. Celle) and Scandinavia (e.g. Roskilde), Italian cities such as Bologna and Firenze are examples of this trend.

Recent figures indicate that the exodus from the inner city may have passed its peak and that there may be a 're-urbanisation' phase. However, if the term re-urbanisation is understood in quantitative terms as a *reversal* of the decentralisation of population and employment, no instances of true re-urbanisation are likely to be found. Actually there is a superposition of two counteracting trends, the continuing outward movement of traditional suburbanisation and the inward movement of a relatively small number of very mobile households whose location preferences may change quickly. In addition, the new back-to-city movement at least in part consists of people with relatively high incomes who tend to occupy large flats and so consume more housing space than the former poorer tenants. The displacement of lower-income households by more affluent ones ('gentrification') is one of the problematic aspects of the otherwise desirable re-urbanisation. Because of the displacement effect, a permanent upward turn of inner-city population and employment can only be expected where substantial increases of floor space in the city centre have occurred. Without that the re-urbanisation in European cities is a *qualitative* phenomenon.

The renaissance of inner-city living and shopping demonstrates the vitality of the European city with its history and cultural heritage as well as the increasing diversity of urban lifestyles in Europe. Never before have the urban centres of Europe attracted so many visitors to their historical monuments, museums and theatres, never before have there been so many new shopping arcades, boutiques and restaurants, new office buildings, hotels and convention facilities. Historical buildings and traditional residential neighbourhoods are carefully restored everywhere. A journey through urban Europe today is a fascinating series of discoveries.

However, this is only the bright side of a two-sided coin. Just as in the competition between cities and regions, so intra-regional spatial change involves equalising and polarising tendencies. On the one hand the exodus of people and jobs from the centre reduces the difference in density between core and periphery. On the other hand, under market conditions, both suburbanisation and re-urbanisation tend to aggravate existing social differences within urban regions.

Suburbanisation had already a significant impact in terms of social segregation. Households leaving the inner city for the suburbs tend to be the younger and economically more active; the older and less mobile remain in the old inner-city housing areas. Disinvestment and neglect produces the 'rent gap', which makes the rehabilitation and upgrading of run-down housing areas profitable. However, this means the replacement of the original low-income residents by more affluent tenants who are able to afford the higher rents, and the destruction of the remaining cheap housing, with the effect of the spatial marginalisation of those who are already socially marginalised.

On the top level of the urban hierarchy, in London and Paris, but also in cities like Bruxelles, Frankfurt, München and Milano, this has led to massive real estate speculation and exorbitant increases of real estate prices and building rents, which threaten to make the central areas of these cities unaffordable as places to live for the majority of the population.

If these trends continue uncontrolled, the 'successful' metropolis is likely to be divided into three different 'cities' (see Siebel 1984):

- The most visible city is the 'international' city with airport, hotels, banks, office buildings and luxury flats and its prospering downtown shopping zone, but also high-class residential areas usually in the western parts of the city.
- Hidden behind the international city is the 'normal' city for the native middle class in the low-density suburbs and high-rise housing areas at the urban periphery.
- In the shadow remains the 'marginalised' city for the old, the poor and the unemployed and the migrant workers in the run-down inner-city

housing areas, in most cities east of the traditional centre, and in devalued under-utilised transition zones at the urban fringe.

Re-urbanisation and gentrification are likely to accelerate this partition. Where gentrification expands the 'international' city, the poor are further pushed out into the worst segments of the housing market which in the future tend to be the low-quality, high rise tenement blocks of the 1950s and 1960s.

Future trends

All four mega-trends identified above—advances in high speed transport, telecommunications and computerisation, European integration, and large-scale immigration—are likely to stay in effect. The most relevant urban issues resulting from these continuing trends are the following (Kunzmann & Wegener 1991):

1 *Dominance of global cities*
  - The dominance of the only two global cities in Europe, London and Paris, will become even stronger with further advances in tele-communications and the continuing integration of global markets. The future role of Berlin in the urban system in Europe is yet unclear. The city can only return to its pre-war importance if the economies of eastern Europe are able to overcome their present stagnation, which is beyond forecasting. However, during the next decade Berlin is unlikely to join Paris and London as a global metropolis.

2 *Polarisation through high speed transport infrastructure*
  - The emerging European high speed rail network will bring another contraction of distances in Europe (Figures 7.9 and 7.10) and so will reinforce the position of those cities which are connected to the new network.
  - Cities in the 'grey' zones between the high speed transport corridors are disadvantaged. Access to international airports will continue to be a key factor for urban development. Cities without airport access will have difficulties in attracting high-tech jobs and skilled labour.

3 *No borders, new hierarchies?*
  - Cities at inner-European borders may benefit from the Single European Market (e.g. Aachen, Strasbourg). The unification of Germany has brought new impetus to some cities, which in the past three decades were peripheral (e.g. Hanover, Braunschweig). The opening of eastern Europe may improve the position of cities which before the war had traditional links to east European markets (e.g. Hamburg, København) or of cities bordering east Europe (e.g. Thessaloniki, Frankfurt an der Oder). Cities in east European countries such as Praha or Budapest may regain their pre-war position in the urban hierarchy of Europe.

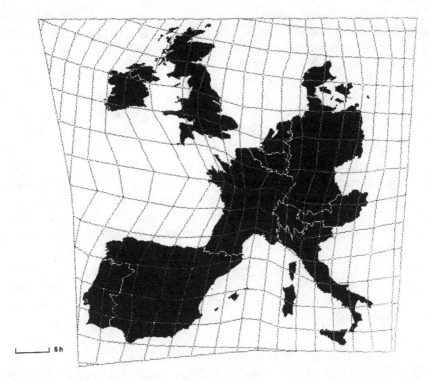

Figure 7.9   Time-space map of Europe based on rail travel times of 1991
Source: Spiekermann & Wegener 1993.

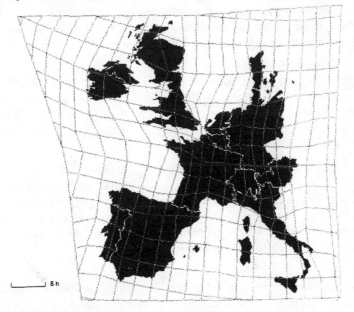

**Figure 7.10   Time-space map of Europe showing the contraction of
space by the future high-speed rail network**
Source: Spiekermann & Wegener 1993.

4 Pressure on European gateway cities
- Cities bordering eastern Europe (e.g. Thessaloniki, Frankfurt an der Oder), port cities on the Mediterranean (e.g. Marseille or Trieste) and cities with large international airports (e.g. Paris, Amsterdam, Frankfurt) will experience increasing pressure by immigration flows from eastern Europe, Africa and the Middle East. They will not be able to absorb the masses of arrivals, and their economy will be burdened by the large number of unskilled workers.

5 East-west or south-north?
- The greatest challenge of the next decade will be to overcome the gap in economic prosperity between the cities in western and eastern Europe. The infrastructure of east European cities is obsolete, their environmental conditions are desolate, their housing stock is far below west European standards, and local governments are almost incapable of managing their own development.

6 Cities at the European periphery: forgotten?
- With increasing importance of the accessibility to the cities in the core of Europe, cities at the periphery will increasingly become 'forgotten' as locations for international information exchange and communication.

7 Further decline of industrial cities?
- Unemployment will continue to be a major problem in declining industrial cities. Despite the success of some cities in restructuring their local economies, many others will still have to struggle. The combination of economic decline, a poor environment and a bad image makes it difficult for such cities to escape from the vicious circle of disinvestment and physical degradation.

8 Port cities under pressure?
- Port cities that have failed to modernise and specialise their infrastructure will be in danger of further decline. They will be affected by the growing competition of the large European ports and their attractive services and efficient transport links to the continental hinterland.

9 High-tech and garrison cities: victims of disarmament?
- Due to the new political situation, cities which used to be locations of defence industries and defence-related R&D or large military installations may stagnate. The success of a conversion of their economic base will depend on the capability of local industries to diversify and survive in much more competitive markets.

10 Just-in-time urban regions?
- Just-in-time production complexes will affect the spatial structure in regions where car production is concentrated. Such regions are increasingly being dominated by the infrastructural requirements of

the automobile industry and their forward and backward linkages. Although these regions flourish at times of economic prosperity, they become heavily affected in times of recession.

11 Rurban belts: the ubiquitous city

- 'Rurban' belts along important transport corridors will grow further. They will become a favourite location for formerly urban industries and population attracted by lower land prices and a better natural environment, and will be prime locations for goods handling and distribution centres, but will also bring development pressure on green areas.

12 Unguided growth: large cities in the south

- Unguided urban development will continue to be characteristic for growing large cities in southern Europe. Local governments in these cities will not be able to cope with managing rapid urban growth, so squatting and strip development will become the rule.

13 The future of urban form

- The two global cities, London and Paris, will continue 'mega-projects' such as the Docklands and *les grands travaux*. Cities like Bruxelles and Frankfurt, and possibly Berlin, will make efforts to live up to their growing European importance by creating glamorous convention and cultural facilities and expanding their networks of urban motorways and metros. Other 'Euro-metropoles' such as Lyon, Milano or Barcelona will try to follow. Cities not having the energy and resources for such change are likely to fall further back.

14 Declining urban infrastructure and services

- Whereas affluent cities will be able to generously improve their infrastructure and expand their services, the less affluent cities in Europe will be faced with growing problems of ageing infrastructure (roads, bridges, public transport networks, water supply and sewage systems, schools, hospitals).

15 Urban poverty

- People unable to learn the new skills required by the high-tech industries, migrants from peripheral areas of Europe and developing countries, and the tendency to reduce government involvement in social security, public housing, social services and health care, will increase the number of urban poor. In large cities with rising land prices poverty will often turn into homelessness.

16 Urban land markets: a time bomb

- While a functioning land market is a vital ingredient of a prosperous city, inflated land prices that are no longer related to the value that can be generated on land are a serious threat to a balanced urban development. They make large parts of the inner city unaffordable as a place to live for local people with low incomes. First signs of this harmful process can be observed in London, Paris and Madrid, but also in München and an increasing number of other European cities.

17 Urban transport: the reappearing problem

- Urban transport is reappearing as another fundamental urban question, in particularly in prosperous, growing cities where the 'final gridlock' is becoming predictable. In the short run it is necessary to apply a complex mix of 'synergetic' policies encompassing traffic management and regulation, taxation and pricing, street design and pedestrianisation. In the long run, however, only a reversal of the trend to ever more urban mobility is required.

18 Urban environmental problems

- In particular in prospering, successful cities, growing traffic volumes, uncontrolled land use development and negligence of ecological concerns may endanger the quality of the urban environment. In the fast growing cities of southern Europe, lack of public finances is a prime bottleneck for improving sewerage, waste disposal and energy generation systems.

Conclusions

It can be concluded that the urban hierarchy in Europe will change in many aspects, but that its basic structure will remain the same. The most dominant trend will be further polarisation. Under the proviso that the future will not be overshadowed by major military conflicts or economic turbulences, London and Paris and the other cities in the top levels of the hierarchy, including the small and medium-sized cities in their hinterland, can look forward to a bright prospect of affluence fuelled by unprecedented levels of exchange of people and goods. They will be able to upgrade their economy to the most advanced technologies and services, polish their physical appearance and transport infrastructure and attract the most creative and innovative talents in politics, business, science and the arts. Given the wealth and opportunities lying ahead of them, the next decade may become a great period for these cities.

On the other hand, there are serious risks. The greatest danger is that the success of these winner cities might go at the expense of the much larger number of potential losers. The most likely losers are cities that are not linked to the new high-speed transport infrastructure, cities at the European periphery or cities that do not succeed in liberating themselves from their industrial past and finding their own particular niche in the wider European market. This is the negative side of polarisation and it is in direct conflict with the stated equity goals of the regional policy of the European Union.

And there are the negative side effects of growth itself. Even the apparent winner cities may in some respect become losers if they do not manage to cope with the undesirable consequences of economic success such as

The European grape

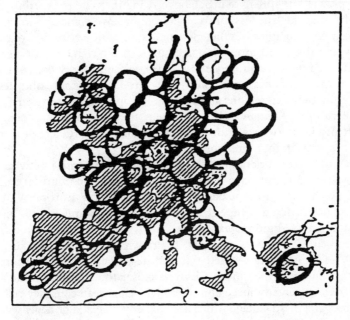

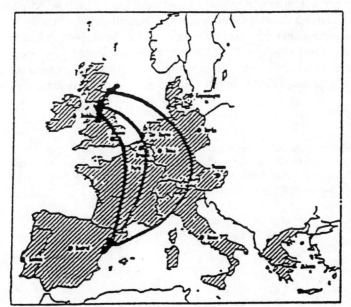

not the (blue) banana

Figure 7.11   The 'European Grape' as a Leitbild for the polycentric urban structure in Europe
Source: Kunzmann & Wegener 1991.

exploding land prices, traffic congestion, environmental degradation and urban sprawl. The spread of urban poverty in otherwise prosperous cities is a warning that the 'success' of some cities may have come about by relying too much on principles of efficiency and competition without concern for the less able that need protection and support. If a reduction of disparities between the regions, and hence also cities, is the primary goal of regional policy, it may be necessary to study how the indispensable and desirable competition between regions and cities in Europe can be complemented by an element of *cooperation* and mutual help among regions and cities. The support of city networks and of collaborations of border regions by the Union are steps in that direction.

A regional policy oriented at the reduction of disparities between cities in Europe might also have a second look at other community policies guided by the principle of competition. The Single European Market, aiming at unrestricted competition in the whole Union territory, may give a boost to inter-regional exchange and trade, but may make life for peripheral cities, which used to have their own undisturbed markets, more difficult. The new high-speed rail and advanced telecommunications networks will bring the cities of Europe closer to each other, but by first linking the large already successful cities in the European core they will put all those cities at a disadvantage which are not yet connected or never will be.

These considerations may also suggest a different and more 'cooperative' *Leitbild* for urban development in Europe than the 'Blue Banana' which is the pure expression of the competition between the regions in Europe. The 'European Grape' (Figure 7.11), may be more suited to represent the polycentric structure of the urban system in Europe and the fundamental *similarity in diversity* of the interests and concerns of its member cities.

NOTE

I am grateful to Klaus R. Kunzmann for the permission to use parts of our joint work on urbanisation in western Europe, in particular his drawing of the 'European Grape', and to Klaus Spiekermann for the permission to reproduce the two maps in Figures 7.9 and 7.10 from our common research on time-space maps.

8

Changing Places? Households, Firms and Urban Hierarchies in the Information Age

Peter W. Newton

T HE future prospects of cities and regions in advanced industrial societies depends upon whether they can continue to compete, both locally and globally, as places where it is attractive and affordable to live, as places where it is attractive and profitable to work, and as places where it is attractive and accessible to visit. Issues of consumption and production represent the dual motivations of households and firms for changing places. The sum total of such revealed preferences, at any point in time, is reflected in a nation's settlement hierarchy (measured typically by concentrations of population or industry or both). Over time, such hierarchies can be observed to change—in response not only to the 'faster' local and international processes associated with business, profit and housing cycles, but also in response to 'slowly changing' long-wave infrastructure factors. It is the latter class of factors which, some argue, control and direct the faster processes of change (Andersson 1986). Five key global infrastructure networks (and their associated secondary networks) can be identified which have led to long periods of economic expansion and urban development. These are (after Richardson 1991): canals (turnpikes), railways (streets), seaports (water, gas, telegraph), highways (electricity, radio, telephone) and airports (telecommunications, computers). The latter era is envisaged to span a period from the 1980s through to the mid-21st century. Each generation of infrastructure has provided new opportunities for restructuring space, generally by increasing the logistical capability of networks developed for the exchange of people, materials, goods, capital, information and knowledge.

The urban structures established during prior technological eras provide the platforms from which new trajectories of development can take place. So it is with Australia's urban system. Formative port functions remain as part of the contemporary economic base of two-thirds of Australia's current top 20 cities. Evolution of a national railway network tended to reinforce the positions of the State capitals and major urban centres, given the generally radial pattern of their fixed networks. Albury–Wodonga on the

NSW–Victoria border represents one of the few large urban centres in Australia to come to prominence via rail, primarily as a result of a lack of agreement between States for a national standard gauge track: twin towns on opposing sides of a State border emerged to handle 'break-in-bulk' transfers (Table 8.1).

Greater flexibility was inherent in the evolving national highway system, providing opportunities for lower order centres to bypass the larger cities and establish networks at all levels of the urban hierarchy. Evidence presented in Table 8.1 provides a measure of support for the cumulative causation models of Myrdal (1957) and Pred (1975) which propose that 'large cities change their population rank infrequently as population and economic functions concentrate around previous growth foci' (discussed more fully in Burnley 1980a, p.50). However, given sufficient time, together with the emergence of revolutionary space-transforming technologies (embodied in aerospace, computers and telecommunications) and changes in the relative attractiveness of regional economies as places to work and live, opportunities emerge for certain locations to advance while others retreat.

Table 8.1   Australia's largest 20 cities in 1991 with comparisons in 1976 and 1891

City	1991 population ('000)	Rank 1991	Rank 1976	Rank 1891
Sydney	3 698	1	1	2
Melbourne	3 153	2	2	1
Brisbane	1 327	3	3	4
Perth	1 197	4	5	11
Adelaide	1 062	5	4	3
Newcastle	432	6	6	12
Canberra–Queanbeyan	315	7	7	–
Gold Coast–Tweed Heads	274	8	11	–
Wollongong	239	9	8	15
Hobart	185	10	9	7
Geelong	152	11	10	8
Townsville	115	12	12	–
Sunshine Coast	115	13	25	–
Launceston	94	14	13	10
Albury–Wodonga	90	15	17	–
Cairns	84	16	22	15
Toowoomba	83	17	15	14
Ballarat	82	18	14	5
Burnie–Devonport	78	19	16	–
Darwin	77	20	23	–

Source: Coopers and Lybrand (1992) and ABS 1991 Census.

New engines for growth—new patterns of development

The latest logistical revolution we are now beginning to experience embodies growth in information processing and communication capacity and their convergence; an expansion of the knowledge base; and further improvements to the air transport system. These technologies are fundamental to the continued growth of all sectors of the economy, particularly the information sector. There is now little debate concerning the existence of the information sector in urban economies, although the size and importance of the sector will vary between countries according to their level of development. It has continued to exhibit growth over a long period, as information has become more important in the economy; most notably since the 1940s (Beniger 1986, p.24). In Australia, the size of the labour force involved in what could be classed as information-related activity is now approaching 40% (see Table 8.2); other studies have placed it slightly higher (Lamberton 1987; 1988). By the turn of the century the information sector will be the major contributor to GDP as well as employment.

The increased importance of information to urban and regional economies is evident from the continued growth of the information sector in the economic base of all major metropolitan areas in Australia. Their transformation has been from centres of production and distribution of material goods, to centres of information exchange, service production, consumption and manufacturing. The evolution to a service- and information-oriented society should continue to stimulate demands for telecommunications and high speed transport infrastructures, given that information is a key factor in production across all sectors of the economy.

Indeed, Goddard (1989, cited in Hepworth 1989, p. xvi) argues that it is more appropriate to our understanding of the evolution of industrial activity to recognise that manufacturing and service sectors are becoming increasingly dependent on effective information management. The information sector grows as it services the information production, processing and distribution needs of other sectors as well as generating growth in its own right via commercialisation of information and communication technologies and their infrastructures. As the Office of Technology Assessment study (1990, p.112) also notes: '...as productive processes become increasingly complex in advanced industrial societies, the largest reserve of economic opportunities will be in organising and coordinating productive activity through the process of information handling'.

With few exceptions, the trend appears to be that urban centres with a higher than (national) average involvement in the information economy also enjoy higher than average economic health, as measured by the employment to population ratio, recognised as one of the best available measures of economic well-being of a metropolitan area (Drennan 1989). This is partly due to the more flexible economic base characterising centres

Table 8.2 Employment trends by major industry sector 1954–86: major urban centres

| | Industry sector (% persons employed) | | | | | | Employ-ment to population ratio |
| Urban centre | Manufacturing | | Services | | Information | | |
	1954	1991	1954	1991	1954	1991	
Sydney	37	14	41	39	20	39	0.44
Melbourne	40	18	39	37	19	37	0.44
Brisbane	28	13	46	40	23	39	0.43
Adelaide	36	15	43	36	18	39	0.43
Perth	25	11	50	40	22	39	0.42
Newcastle	41	15	39	40	12	33	0.39
Canberra	6	3	34	29	55	60	0.50
Wollongong	45	20	34	36	9	33	0.39
Gold Coast	13	9	59	50	17	31	0.38
Hobart	26	11	46	37	25	43	0.41
Geelong	46	22	39	38	12	32	0.39
Townsville	23	8	54	40	20	44	0.44
Darwin	6	5	48	39	40	46	0.46
Toowoomba	25	12	32	41	38	38	0.39
Launceston	30	14	49	40	18	34	0.40
Ballarat	37	14	41	38	18	40	0.36
Cairns	21	8	57	49	16	34	0.43
Rockhampton	30	11	50	46	16	34	0.41
Australia	28	13	39	38	17	36	0.42

ASIC classification: Information sector (communications, finance, property, business etc., public administration, community services).
Source: ABS Census.

which have shifted furthest from dependence on a relatively specialised industrial base to diversity among an increasing concentration of inform- ation-intensive companies; companies which exhibit greater capacity for moving in new directions as opportunities for growth or change emerge.

Infrastructure availability and utilisation

The ability of both the private and public sector to install new transport and communications technologies and use them in innovative ways is viewed by many academics and bureaucrats to enhance the competitive advantage of regions, attract development and promote a centre's standing within the national settlement hierarchy (Edwards et al. 1989).

Telecommunications

Recent studies which have sought to examine the relationship between the economic growth performance of urban regions and their telecommunications infrastructure have engendered opposing positions. On the one hand there are those who argue that with the growth of IT-dependent business functions, communications will become a key determinant of whether cities and regions attract or lose business (Keen 1991; Edwards *et al.* 1989). On the other hand there is the thesis which, in its most basic form, suggests that telecommunications is a necessary but not sufficient condition for regional development; that communications issues tend to rate relatively low in surveys of industrial relocation decision making; that communications represents a relatively small (albeit rising) proportion of the operating budget of organisations; and that communications is a relatively ubiquitous infrastructure, access to which is generally not a problem for business (Loveland 1990). In other words, little significant difference exists between competing cities with regard to telecommunications infrastructure—certainly none that could not be relatively quickly overcome, should the need arise. Similar arguments have been advanced by Bar (1989, cited in Urey 1991) who argues that telecommunications technologies are useful indices of a region's economic activity, but are not magnets for economic development:

> [I]t is also difficult to envision cases where the telecom network would determine a firm's [choice] between two neighboring cities because telecom networks act quite differently from railroads: the telecom network reaches everywhere, and can easily be extended—much more easily and economically than a railroad could. Furthermore, at least within the United States, if a company is large enough substantially to affect economic activity in a region or locale, it usually can get virtually all of the communications facilities and services it needs for its business regardless of where it chooses to locate. While it was hard for firms to build their own railroad, they have found it very easy to install satellite dishes or microwave equipment when the network in place did not meet their needs (Bar 1989, p.10).

Is such a thesis valid? Are there no spatial implications of telecommunications technology and telecommunications infrastructure, save for the potential that telecommunications has to collapse distance completely? To explore the nexus between telecommunications, geography and economic development it is useful to consider how telecommunications confers productivity benefits to organisations and to regions. At least two of the key dimensions are outlined in Figure 8.1 (p.166): on one axis we have the necessary conditions of physical capital and its availability; on the other the sufficient condition, human capital and its utilisation of technology and physical capital.

In the sections which follow, we briefly explore, for a range of telematic network services, patterns of telecommunications availability and utilisation from a spatial perspective.

PSTN (public switched telephone network)

The PSTN hosts, in addition to voice services, a variety of other value-added services such as 008, fax, automated subscriber trunk dialling (STD) and international direct dial (IDD) services. Their availability is virtually Australia-wide.

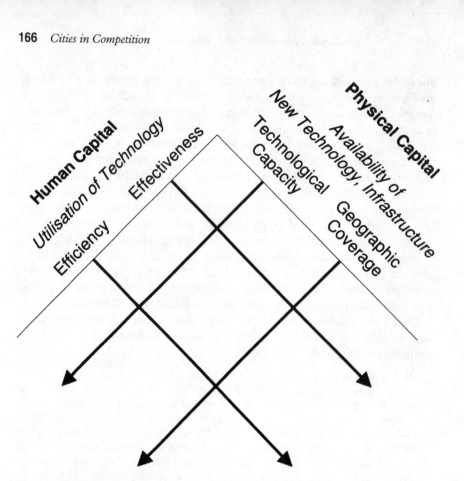

Figure 8.1   Dimensions of productivity of transport and telecomunications systems

Introduced in 1979, 008 services permit long-distance telephone calls at local rates and commonly on a 24-hour basis, and have the potential for extending the market catchment for a particular private or public sector enterprise to national boundaries and beyond. The industry and settlement implications are difficult to anticipate. In theory, 008 services provide opportunities for the location of business establishments in either urban or non-urban environments. In practice, they appear to be centralising business in the most populous states (Newton 1992). The take-up of G3 facsimile transmission on the PSTN within Australia has been dramatic and the statistics on penetration by urban centre reveal a clear division between the five largest cities where telematic services (as measured by fax) are concentrating at levels appreciably higher than city size alone would suggest (Newton 1991). By way of contrast, Newcastle and Wollongong constitute Australia's major centres of heavy industry (iron and steel); centres which appear, on this evidence, to be in the process of being bypassed as foci for the more information-intensive type of activities common to most of the capital cities.

STD (Subscriber Trunk Dialling)

Penetration of STD is now universal within Australia. Until national coverage was attained, however, the more remote country regions were disadvantaged by lack of 24-hour automatic switching services and the access that provides to the rest of the nation and the world. Furthermore, virtually all non-metropolitan centres have a higher proportion of long-distance versus local calls, reflecting the strong metropolitan orientation of telephone traffic within all Australian States (Holmes 1983; Newton 1992). Such ratios also point to an added disadvantage faced by country centres in attracting information-intensive industry—the impost that STD rates impose during business hours. Extension or elimination of STD zones may encourage companies to move out of cities, but would deprive tele-communications organisations of revenue.

CMTS (cellular mobile telephone service)

CMTS is now operating in all State capital cities and major regional centres, covering over 75% of the Australian population. Highway coverage is also being expanded, with almost continuous coverage currently available between Melbourne and Albury, Canberra and Sydney, and Sydney and Brisbane along the coast road. From the limited data available in this highly competitive telecommunications area, it would appear that take-up has been strongest initially in the States of Victoria and NSW (Newton 1992).

ISDN (Integrated Services Digital Network)

Computer networking began commercially in Australia in 1969 with the introduction by Telecom of a modem service that would operate on the PSTN to provide digital-analog conversion necessary for computer-to-computer communication. The technological limitations of this service have been progressively overcome as other networks were introduced during the 1980s, with the introduction of the most recent service, ISDN, in 1989. Lazak (1987) comments from a German perspective that as ISDN is now available, advanced users of network systems are trying to install it as soon as possible, yet it is estimated that it will take from 30 to 40 years to switch over from old to new networks throughout that particular country. The spatial implications for the transition period are a 'landscape of ISDN islands, both nationally and internationally'. At the international level, Lazak (1987, p.229) envisages advanced networks with access and distribution nodes, (teleports) providing user access to the most advanced communication systems worldwide, irrespective of the status of networks throughout particular countries. At national levels, Zamanillo (1985) describes what is typical of phased introduction of ISDN over the PSTN: trunks first, then exchanges, then subscriber lines—initially in areas considered most likely to exhibit high levels of demand.

Against such forecasts of significant time lags before certain regions are connected to the latest telematic networks, what has the experience been to date with ISDN in Australia? Two years after its introduction, 36% of exchanges in the State of NSW and 29% in the State of Victoria had ISDN capability (Crawford *et al.* 1991). Clearly, at national/regional levels there are 'ISDN islands'. In Victoria by mid-1991, 83% of exchanges with ISDN capability were in Melbourne. Within the major cities there has also been rapid penetration: in Melbourne, over 85% of business and commercial centres have ISDN access; in Sydney the figure is in excess of 90%. In terms of utilisation of ISDN, NSW and Victoria again exhibit levels slightly above that which their share of business establishments would suggest (see Newton 1992).

IDD (international direct dial) and globalisation

Many large, multi-functional corporations were operating on an international basis prior to the 20th century, but their growth and proliferation have been most pronounced since World War II. Reasons for this trend have been examined elsewhere (Taylor & Thrift 1986) and need not concern us here, save for the integrating role that advanced information and communications technologies have played in the process. To our way of thinking, globalisation from a telecommunications perspective requires, as a minimum, 24-hour point-to-point real time connectivity for voice and text services between communities located anywhere in the world. This has increasingly become a reality for urban communities in advanced industrial societies over the past decade or so. Real time transmission of large data and image files must await the introduction of broadband services in the mid-1990s, however. This will then open up the prospects for the internationalisation of an even wider array of key business activities, including operations, service, technology development, design, human resource management, procurement, marketing, sales and dispatch. Under these circumstances, the firms which seek comparative advantage by allocating their activities among a number of countries to gain the optimum advantage, will increasingly be able to integrate their distributed information systems and undertake operations as if in a single locality (see Newton *et al.* (1993b)). From the time of its introduction in 1976, IDD has sustained average rates of growth of between 20 and 30% per year, as local exchanges have been progressively connected to the service (penetration now stands at greater than 97%) and as overseas countries become interconnected (13 in 1976, over 200 in 1991). Availability of an automatic IDD facility is not, however, reflected in utilisation patterns—as we have found with other telematic services.

Table 8.3 reveals a number of important contrasts in the pattern of international telecommunications traffic which is outgoing from each of Australia's major urban centres.

Sydney generates by far the greatest volume of international communications, with almost 50% share of business traffic among the top 20 cities. Melbourne and the Gold Coast are the only other cities which have a share equal to or above that which their location in the urban hierarchy would suggest.

Table 8.3   Share (%) of international telecommunications activity: Australia's major urban centres

Urban centre	Population share (1991)	Outgoing business IDD share (1992)
Sydney	28.3	46.3
Melbourne	25.2	25.9
Brisbane	10.4	7.0
Perth	9.3	8.4
Adelaide	8.7	4.1
Canberra	2.5	1.9
Newcastle	2.4	0.5
Gold Coast	2.1	2.2
Wollongong	1.9	0.3
Central Coast (NSW)	1.8	0.3
Hobart	1.2	0.4
Geelong	1.2	0.5
Townsville–Thuringowa	0.9	0.3
Toowoomba	0.7	0.1
Darwin	0.6	0.5
Launceston	0.6	0.1
Ballarat	0.6	0.1
Cairns	0.6	1.0
Bendigo	0.5	0.1
Rockhampton	0.5	0.1
Total	100.0	100.0

Source: OTC, ABS.

The bulk of international business traffic linking Australia's major urban centres to offshore destinations is focused on 15 countries which account for an 80% share of all outgoing business traffic (Figure 8.2).

While several of the cities, especially Sydney, Brisbane and Melbourne, have similar traffic profiles, other cities appear to have forged strategic business alliances which diverge to some extent from the trio of eastern seaboard capitals. For example, on the Gold Coast, Japan accounts for over 25% of all business traffic; Darwin has a higher relative share with Indonesia and Singapore than other capitals; and Perth has relatively stronger links with the UK and Singapore. When city-level associations occur between communications and other sectors of transport such as air and sea, an infrastructure platform is created capable of supporting international trade into the 21st century. It is clearly in the interests of Australia's cities to attempt to diversify their linkages—establishing alliances which are complementary and synergistic from among what is now a wide array of global options (trading blocks notwithstanding). This will require increased capital expenditure, however, on nodal rather than link infrastructure.

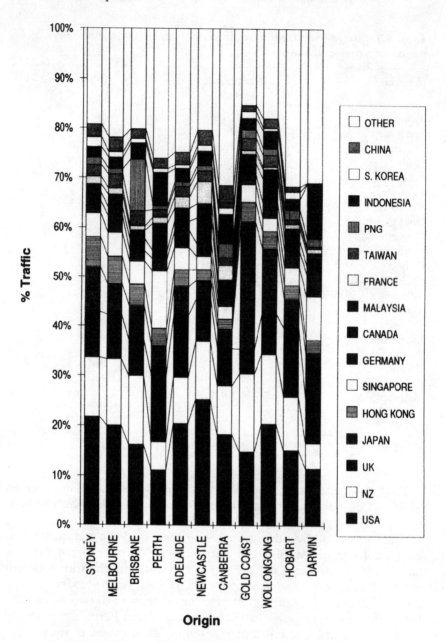

Figure 8.2 Dispersal of outgoing business traffic from major Australian cities, 1992

Air transportation

For the role it plays in the movement of business personnel for face-to-face meetings, for tourism and for the shipment of high value-to-weight goods for use in manufacturing and retailing, air transport continues to expand its role in advanced industrial economies.

Australia's cities, however, vary widely in their ability to capture airport infrastructure and to subsequently provide greater opportunities to local business to establish network-based industries that do not rely to any degree on physical propinquity.[1] As Batten (1992, p.235) remarks:

> Contiguity of places and even of regions is becoming less significant. The old tree-like transportation network structure, with its sparse, scale-oriented links, is gradually being supplanted by a multi-layered grid of road, air and telecommunications networks possessing completely different characteristics...Together these new networks possess tremendous potentials for synergy...at practically any location within the...developed world.

Sydney is Australia's key gateway to international passengers, receiving 50% of all in-bound movement in 1990, with Melbourne having a 20% share (Table 8.4).

Both cities have lost a small percentage of their share as other state capitals have attempted to secure a greater proportion of international flights. Congestion and curfews currently inhibit Sydney's capacity (prior to its securing additional capacity via a third runway and second airport)

Table 8.4   International and domestic air passenger traffic to major urban centres, 1982 and 1990

Urban centre	Inbound passengers					
	International			Domestic		
	Number 1982	('000) 1990	% Change	Number 1982	('000) 1990	% Change
Sydney	1 226	2 112	72.3	3 010	3 838	27.5
Melbourne	506	878	73.6	2 452	3 100	26.4
Brisbane	223	562	155.7	1 456	1 885	29.4
Adelaide	5	93	1 750.2	931	937	11.7
Perth	225	439	95.2	547	740	35.3
Newcastle	–	–	–	111	78	-29.9
Canberra	–	–	–	416	499	19.7
Wollongong	–	–	–	–	–	–
Gold Coast	–	–	–	295	410	39.1
Hobart	7	9	38.1	233	268	15.2
Geelong	–	–	–	–	–	–
Townsville	14	9	-34.5	231	189	-18.0
Darwin	18	46	154.4	152	193	27.0
Toowoomba	–	–	–	–	–	–
Launceston	–	–	–	197	189	- 4.1
Ballarat	–	–	–	–	–	–
Cairns	8	145	1 703.0	221	399	80.7
Rockhampton	–	–	–	103	85	-17.4
Bendigo	–	–	–	–	–	–
Total	2 232	4 293	92.3	10 355	12 810	23.7

Source: Department of Transport and Communications.

and the Federal Government limits the opportunities of cities such as Melbourne to directly negotiate agreements with international airlines to headquarter their operations out of cities other than Sydney.

The rate of growth of domestic air travel has been significantly less than that for international, and while Sydney retains a higher level of traffic than Melbourne (30% versus 24%) neither have lost overall share during the past decade. The gains by Perth, the Gold Coast, Brisbane and Cairns have been mostly at the expense of the lower order cities and mirror trends in inter-regional population redistribution identified in Flood *et al.* (1991) and Wulff *et al.* (1993) in relation to sun-belt migration.

Towards a conceptualisation of telecommunications and spatial development

The information and communication sector has grown primarily in response to the increasing complexity of the economic system (Robinson 1986; Englebrecht 1986; Karunaratne 1986) and the increasing demands this places on information handling—given that each economic transaction is necessarily associated with a flow of information. As the flow of economic transactions intensified and extended beyond local economies to national economies and ultimately to a world economic system, it precipitated what Beniger (1986) has termed a crisis of *control*, which demanded new means of communication. The communications infrastructure we have today is an accumulation of networks developed over many decades, designed to direct and mediate, in the most efficient possible manner, the flow of information, people and goods between areas and sectors of the economy.

Most studies that attempt to assess the manner in which telecommunications affects the geography of economic activity (Kellerman 1984; Johnson 1991) end up with conclusions that are essentially equivocal. They are that the computer–telecommunications environment of the 1990s (PCs, workstations, intelligent terminals, distributed database technology—LANs, MANs and WANs) permit organisations to choose centralisation, or decentralisation, for a wide range of business functions. These could include management decision making, inventory management, operations, service, technology development, human resource management, infrastructure management, procurement, marketing and sales.

Keen (1986) argues, on the basis of an assessment of organisational pros and cons, that telecommunications allows *decentralisation with centralisation*. While computer-communication technology has now reached a sufficient state of maturity to permit integration of geographically distributed information systems without centralisation (Orlowska *et al.* 1992), there appears to be a common tendency among the larger organisations for telecommunications to '...increase central control through ownership of data and the ability to monitor decentralised units' (Keen 1986, p.68).

Telecommunications clearly is making its greatest impact with regard to what Lion and Van De Mark (1990) term *structural decentralisation*—the ability of businesses to separate their management and production functions and facilities—although geographic decentralisation is also occurring. Noyelle & Peace (1988) point to an increasingly wide range of jobs, including higher level jobs, decentralising (for technological, labour and capital reasons). Yet these forces appear insufficient to generate relocation outside cities in the US for a variety of market, technology, infrastructure and organisational reasons.

We now proceed to explore some of these issues in the context of contemporary trends in the locational behaviour of business and population within Australia.

Corporate headquarters

Large organisations view telecommunications as a strategic business tool. They have the capacity to generate innovative applications for new telecommunications technology. They are sufficiently large to absorb the cost of new technology. They typically have the skill base to fully exploit new opportunities provided by the technology (indeed many have established their own private networks, bypassing the public network). They typically locate their key headquarters operations in environments rich in infrastructure and human capital. The settlement implications are clear: Sydney and Melbourne dominate the corporate hierarchy to an even greater extent than they do the urban hierarchy (see Table 8.5), bringing with them the need for the sophisticated systems of control which are embodied in modern telecommunications.

It is therefore not unexpected that, of Australia's top 100 communications companies, over 90% are also headquartered either in Sydney or Melbourne (*Australian Communications Networks*, December 1988). Key producer services firms (e.g. McKinsey, Arthur Anderson, Ernst & Young) also tend to be located close to their corporate clients. The nature of the symbiotic relationship between major corporations and producer services is well illustrated in Steven's (1992) report on the Adelaide-based Sagasco SA Gas Company. In attempting to marshal support against a recent takeover, Sagasco has put pressure on the State government and the city's business and investment communities

> ...by proclaiming loudly that the State can ill afford the loss of another Corporate Headquarters...Sagasco argues that the loss of the company would be a blow to South Australia's fight to maintain itself as a viable home for big companies...It is not just the loss of the companies, it is the loss of the support services...senior accountants, senior lawyers and those other vital services. They are leaving this State...and the very good ones will not be back.

The current pattern of demand for higher speed networks (see Table 8.6) is also centred most strongly in Sydney and Melbourne, and the central business districts of the other States.

Table 8.5   Location of Australia's largest corporations, 1982–90

| | Distribution of headquarters (per cent) | | | |
| | Top 500 companies | | Top 100 companies | |
	1982	*1990*	*1982*	*1990*
Sydney	46.0	45.6	52.0	54.0
Melbourne	31.8	28.6	40.0	33.0
Brisbane	5.4	6.8	4.0	3.0
Adelaide	6.0	6.4	2.0	4.0
Perth	5.0	6.6	1.0	4.0
Newcastle	0.2	0.2		
Canberra	0.6	0.6	1.0	
Wollongong		0.2		
Gold Coast		0.4		
Hobart	0.6	0.6		
Geelong				
Townsville	0.2			
Darwin		0.4		
Toowoomba		0.2		
Launceston	0.2	0.2		
Ballarat				
Cairns				
Rockhampton				
Bendigo	0.2			
Other urban centres	1.0	1.6		
Overseas		0.6		2.0
Location not identified	2.8	1.0		
Total Australia	100.0	100.0	100.0	100.0

Source: *Australian Business*; Riddell's *The Business Who's Who of Australia*.

The higher concentrations found outside the central cores in Sydney and Melbourne reflect the existence of suburban concentrations of information-intensive industry not evident to the same degree in the other State capitals (discussed in a subsequent section). The CBD concentrations in all States reflect the telecommunications intensity demanded of head-office control functions associated with the locations of Australia's top corporate headquarters and government departments. The series of case studies reported later in the chapter provide a clearer indication of the functional and spatial separation of activities within major business organisations.

Information-intensive industries

In attempting to gain a broader perspective on the distribution of information-intensive industry, we embrace the conceptual frameworks of earlier workers such as Porat (1977) and Hepworth (1989) who conceive of key groupings related to information production, information processing and information distribution; while abandoning their route to measurement.

Table 8.6 Spatial concentration of high speed data lines*, 1990

Locality	Number of services/ lines	Per cent of state total	High speed lines (% of total)	Share of business estab-lishments (%)
NSW			46.1	35.5
Sydney CBD	1 819	46		
Remainder Sydney	1 989	50		
Remainder NSW	169	4		
Total	3 977			
ACT	531			
Victoria			33.5	25.8
Melbourne CBD	1 943	59		
Remainder Melbourne	1 288	39		
Remainder Victoria	48	2		
Total	3 279			
Queensland			6.3	18.2
Brisbane CBD	455	74		
Remainder Brisbane	125	20		
Remainder Queensland	37	6		
Total	617			
WA			7.0	8.8
Perth CBD	548	80		
Remainder Perth	91	13		
Remainder WA	44	7		
Total	683			
SA			6.8	8.9
Adelaide CBD	525	78		
Remainder Adelaide	130	19		
Remainder SA	15	3		
Total	670			
Tasmania			0.3	2.8
Hobart CBD	17	50		
Remainder Hobart	2	6		
Remainder Tasmania	15	44		
Total	34			
Australia Total	9 791			

* Lines operating at 48 Kbps and 2 Mbps.
Source: Telecom Australia.

Use of census-based industrial or occupational taxonomies which reflect now outdated industrial groupings or divisions of labour, are replaced by the currently superior detailed business group divisions of the Telecom Yellow Pages Business Directory.

Maps depicting the locational patterns of primary information industries (a combination of information production, processing and distribution industries) in Australia's two largest cities, Sydney and Melbourne, are found in Figures 8.3–8.6, pp.177–180. (These maps are based on data extracted from Telecom's Yellow Pages CD-ROM Directory according to the taxonomy listed in Table 8.7 and aggregated to districts according to the telephone prefix in the listing.)

They reveal that there is no spatial ubiquity with primary information industries at the intra-metropolitan level; they reveal distinct clustering or agglomeration tendencies which result from the benefits which they gain from locating near complementary primary information industries (producers, processors, distributors) as well as client industries which 'subcontract' their services. The CBD exhibits absolute concentrations of primary information industries several times above the level found in the middle-ranking centres in the middle-ring suburbs—although when the data is standardised according to the total number of business establishments in each area, the inner city loses its dominance (especially in Melbourne) to be replaced by suburbs where information industries constitute a high percentage of the overall stock of business organisations. The outer suburbs generate relatively few primary information industries at all. The attraction of the CBD to such industries is at least twofold: proximity to the offices of major business corporations and those of both federal and state government and the necessary infrastructure support, including telecommunications located at the core. The attraction of particular inner and middle-ring suburbs to information industries relates primarily to the labour market and housing market characteristics of those areas.

With Melbourne's primary information economy we essentially have a tale of two cities. Information workers and information industries occupy similar geographic spaces within the metropolitan area (see Newton 1992). Being environmentally benign, information industries do not warrant or

Table 8.7   Taxonomy of primary information sector industries

Sector of information industry	*Constituent industry groups*
Information production	Knowledge production (e.g. R&D), consultant services (e.g., engineering, economic), market search, information collection and coordination.
Information processing	Information transformation and storage (e.g. data processing), brokerage industries, advertising and marketing, non-market coordination industries, insurance and finance industries, public information and regulatory services.
Information distribution	Education (primary, secondary, tertiary), communication (all media).

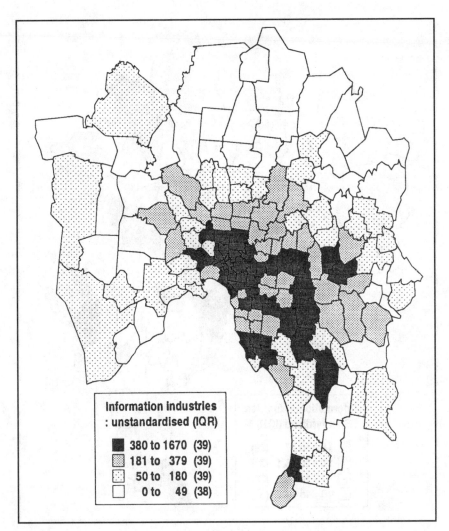

Figure 8.3   Melbourne's primary information industries—
unstandardised

attract the exclusionary zoning meted out to light industry. Furthermore, perhaps more than any other sector of industry, the key resource base of the information-intensive sector is skilled labour and knowledge workers. In such cases, industry tends to follow workers—subject to suitable real estate—to the professional-managerial heartland of the major cities, maintaining and in some cases accentuating the patterns of social class segregation established in earlier periods.

Telework

In addition to concentrations of information industries and information workers, we can add yet another layer to the innovative milieu that characterises particular regions within the major cities—that of telework. A

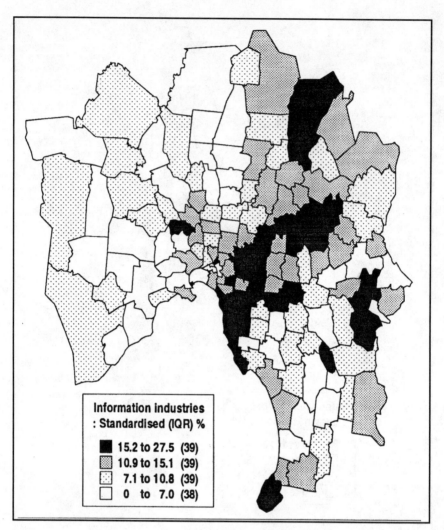

Figure 8.4   Melbourne's primary information industries—standardised

number of writers (e.g. Keen 1991; Mulgan 1991) have observed that while firms have historically brought people to work and relied heavily on organisational structure as the basis for operations and strategy, it is now becoming possible for firms to consider bringing work to the employees— to consider a redesign of the organisation with the lessening of constraints on time and place. Possibilities exist for both knowledge-based work as well as routine information collection, processing or distribution tasks to be relocated back to the home, re-establishing the link between work and residence that existed prior to the industrial revolution. The technological underpinnings for such a shift are the emergence of flexible, high speed communications networks and increasingly cheaper computing power. Communications and information technologies can effectively 'place'

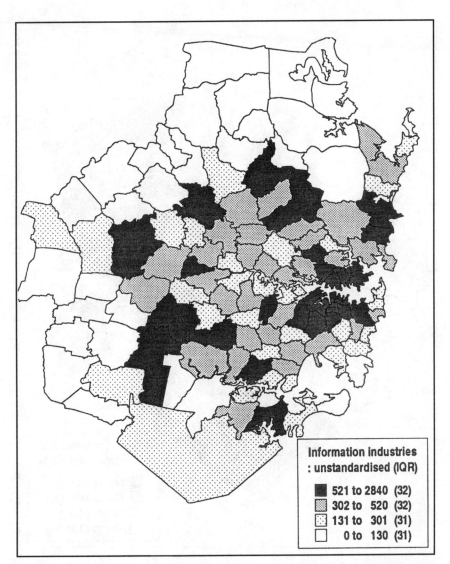

Figure 8.5 Sydney's primary information industries—unstandardised

hardware and software, databases, knowledge bases and multimedia information which 'permanently' reside in capital city business establishments into homes in the suburbs or non-metropolitan areas. Voice services can also effectively provide an organisational umbrella for home-based operatives of telemarketing services.

There is considerable scepticism, however, as to the likely future significance of telework (see Forester 1992). It is also a particularly difficult phenomenon to measure. Data identifying those suburbs within Melbourne where houses have on average more than one line to the local telephone exchange (indicative of some level of home-based telework activity, e.g. fax,

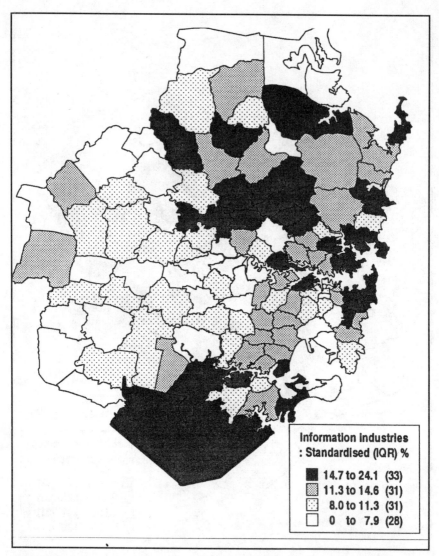

Figure 8.6 Sydney's primary information industries—standardised

modem; Telecom Australia, Development Forecasting Group, Melbourne, pers. comm. 1993) suggests that a 'seedbed' capacity and associated infrastructure is being developed in these localities to accommodate the increasing variety of telematic activities capable of being undertaken within the home. For many it could be providing an extension of the daytime office into the residence for that class of employee who 'takes work home' or has a more flexible work schedule. Again, these suburbs are mostly east and south of the city centre—areas where education levels are higher than average for the city as a whole, where incomes are above average and where the highest concentrations of workers in Melbourne's information-intensive industries tend to reside.

Network firms

The previous analyses paint an aggregate picture of the geography of the information economy in Australia's largest cities, providing several insights into the revealed locational preferences of information-intensive firms, but lacking sufficient detail to provide the basis for examining such hypotheses as the tendency of organisations to *decentralise with centralisation.*. We conclude by documenting case studies of large, multilocational firms—a class of organisation which has been pioneering in their use of computer-based communication networks.[2] This class of organisation must be seen in context, however (Figure 8.7).

Information and communications networks are a ubiquitous infrastructure in advanced societies, supporting a range of interactions from individual through to corporate levels.

Intra-organisational networks are used primarily to *control* the flow of goods, ideas and capital within the organisation, incorporating capability for feedback, command, strategy development and surveillance. As the size and complexity of organisations have increased over past decades (viz. through mergers and internationalisation), the need for better control technologies has also grown. The technological underpinnings for such developments have been networked computing and distributed database technology,

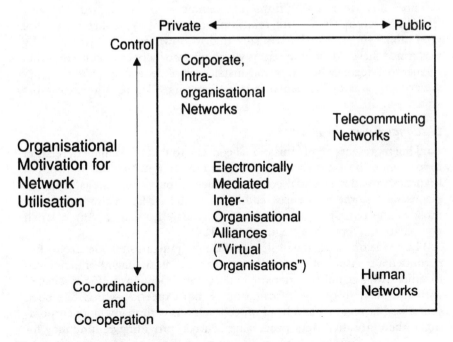

Type of Host Network Infrastructure

Private ← → Public

Control ↑ — Corporate, Intra-organisational Networks

Telecommuting Networks

Organisational Motivation for Network Utilisation

Electronically Mediated Inter-Organisational Alliances ("Virtual Organisations")

↓ Co-ordination and Co-operation

Human Networks

Figure 8.7   Organisational logic of communication networks

personal computing systems linked to LANs and WANs and low-cost transmission.

Inter-organisational alliances are also intensifying as the large national and international organisations are experiencing difficulty in sustaining growth and profits in the face of global competition and the rate of change in business and product cycles. Many large firms are being broken into smaller enterprises or businesses, coordinated in part by electronic networks. Decisions are being made whether to manufacture components internally or purchase them from a contractor. Overall there seems to be a trend away from vertical integration towards the proliferation of smaller, more flexible firms which are capable of responding more quickly to demand impulses. In many enterprises these changes are precipitating shifts in organisation and network philosophies—from those centred on control to those based on *cooperation and coordination*.

Coles Myer Ltd

The geography of Australia's major retailing corporation reveals a national head office which relocated in the late 1980s from the CBD to the fringe of the inner ring of suburbs in Melbourne; most of the state and regional control centres are also in the middle ring suburbs—all within a labour market area capable of supplying the full complement of personnel for this strata of corporate activity (see Figure 8.8, which overlays the distribution of information workers on the nodes of Coles Myer establishments), and readily accessible to those housing submarkets preferred by professional and managerial employees.

Those headquarter functions that remain in the CBD are relatively limited by comparison and relate most directly to department store operations—operations historically tied to downtown and still usefully performed there. Most of the firms which offer key services to the corporate retailer to a degree which warrants installation of dedicated networks (e.g., advertising, customs, specialist ICT, banking) also agglomerate most strongly in the inner city.

National Australia Bank

Banking represents an organisational contrast to that of retailing in relation to location of business establishments and use of networks. Its network of branches provides a spatial coverage which favours those areas of higher disposable income to the east and south of the CBD. In this sector also most of the bank's regional offices are located, each serving a small constellation of branches (Figure 8.9, p.184).

The creation of this new level in the organisational hierarchy has permitted a measure of labour down-sizing and de-skilling at branch level, together with a significant increase in telecommunications traffic associated with the need to query information or tap expertise which has been centralised at regional and head office levels. As with many large corporations, there are dual data processing centres, providing redundancy for

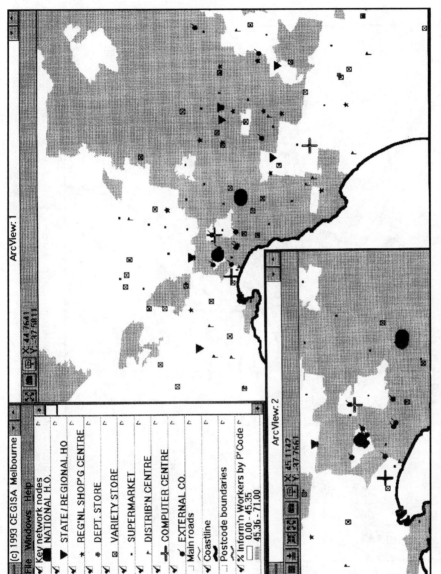

Figure 8.8   Coles Myer establishments, Melbourne, 1992

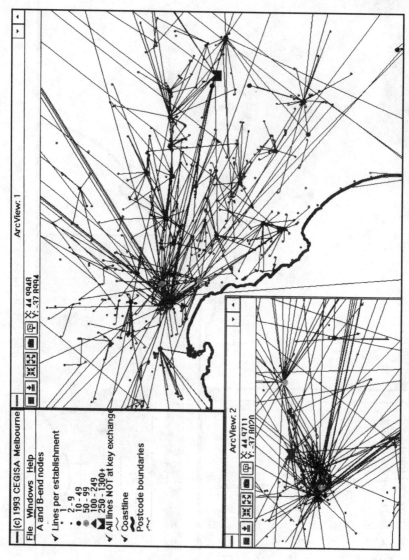

Figure 8.9 National Australia Bank communications network, Melbourne, 1992

emergency situations where backup is required. Other 'generic' corporate features include a main data processing centre located in Melbourne's middle-outer suburbs and a collection of head office functions located in the CBD.

Australian Airlines

Australian Airlines (recently amalgamated with Qantas) is one of Australia's two domestic carriers. The national headquarters of the airline are based on the north-west perimeter of the CBD, optimising freeway access to Melbourne's international airport (Figure 8.10 p.186).

Key elements of the airline's network, apart from airport linkages and its suburban computer centre, are the dense network of links to ticket sales offices and travel agencies, each provided with the capacity to tap the centralised airline reservation systems.

Inter-regional migration: the movement of human capital

Recent research on population redistribution in Australia has suggested that longer distance moves have increasingly been motivated by consumption-related factors rather than economic or productive factors. Flood *et al.* (1991, p.19) define consumption moves as those that take place when 'people move to where they want to live rather than where they have to live'. These moves are almost entirely location driven either in terms of physical features of areas, goods and services available, or the social characteristics of the locale. In the US, Frey & Speare (1988) note that the importance of employment as the major factor in long-distance moves has been steadily diminishing with time, and the strong relationship between net migration and job-related indicators identified in earlier studies has diminished. The same trend has been documented in Canada as well by Shaw (1985) who has analysed intermetropolitan migration patterns from the 1950s through the 1970s. Shaw explains this shift from employment-related moves (prompted by traditional market forces such as wages, cost of living or unemployment) to consumption moves as a result of increasing affluence in society. As societies have become wealthier over time, their members tend to be less motivated by monetary considerations alone and, further-more, wealthier societies tend to install more 'safety nets' to protect citizens from some of the consequences of economic stress (e.g. unemployment benefits). Consequently there is less pressure to move to find work. This explanation seems to fit the pattern of long distance movements in Australia. Inter-regional wage differentials are very low (due to the system of centralised wage-fixing) and therefore have been poor indicators of migration behaviour. In addition, the widespread availability of pensions and benefits, which do not change with location as they do in many other

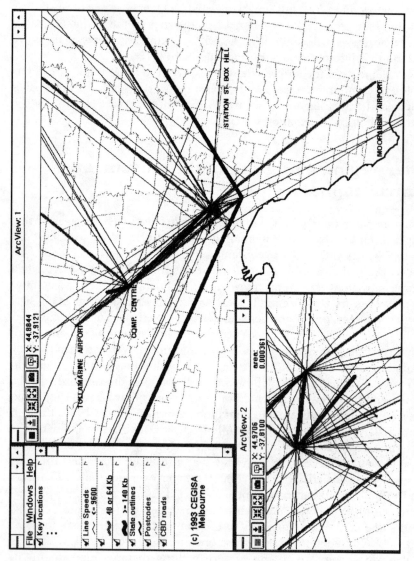

Figure 8.10   Australian Airlines communications network, Melbourne, 1992

countries, tends to favour moves which seek to lower the cost of living or to improve the quality of life (Hugo 1989).

Studies such as Hugo & Smailes (1985), Burnley (1988) and Flood *et al.* (1991) have shown that in Australia 'employed' and 'not employed' groups are often moving in opposite directions, with the not employed persons (not in the labour force, unemployed or children) moving out of the city towards coastal and high-amenity areas, apparently in search of cheaper accommodation and better living conditions. The employed, on the other hand, continue to move into the cities. These trends are shown in Figure 8.11 for 30 Australian regions.

Those not in the labour force (NILF) are a heterogeneous group of about 45% of the adult population, and appear to be the key group contributing to net inter-regional migration in Australia. The group includes spouses in single income families, pensioners and beneficiaries, retirees, and people on 'fixed incomes' such as owners of shares and residential properties. It would appear that the NILF group (along with the smaller unemployed group) are moving out of the cities, towards coastal regions in New South Wales, Victoria, and especially Queensland. The employed group have been moving to the cities, where jobs and a range of consumption opportunities are available (Flood 1992).

Conclusions

Cities are assuming increasing significance as the focus for economic growth in post-industrial societies. The ability of cities to maintain or increase their standing in national or global hierarchies will depend upon their ability to attract the necessary human capital, physical infrastructure capital, as well as investment capital linked to particular sectors of industry. Cities are in competition on all three counts, nationally and internationally.

In the present chapter, focus was primarily upon new urban infrastructure—especially telecommunications—and the role that the utilisation of such infrastructure can have on the pattern of urban development. Research reported in this and earlier papers (Newton 1991; 1992) indicates that it takes time for new communications networks and telematic products to spread throughout a nation. Response times required to install much of the new telecommunications infrastructure are relatively short, however, by comparison with other network infrastructures and need not represent a barrier to economic development. The clear core-periphery trend in relation to diffusion of new telecommunications products and services reflects dominant market-pull effects, accentuated by a recent change in business philosophy of Australia's major telecommunications company following the emergence of competition in the communications sector (i.e. shift from engineer-driven to customer-driven).

For telecommunications products and services which could be classed as 'mature' and where there is, generally speaking, universal access, patterns

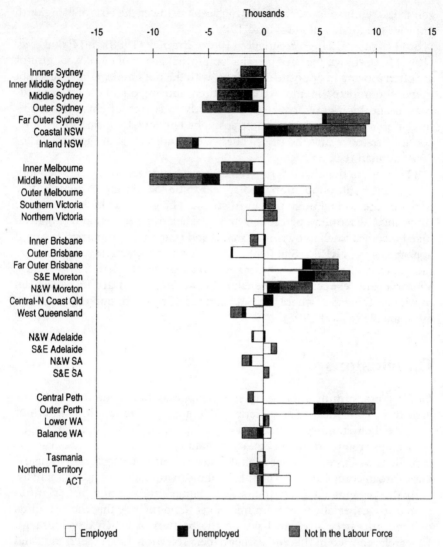

Figure 8.11   Net internal migration by workforce status, Australia, 1985–86, statistical regions, population 15 years and older
Source: Flood et al. 1991.

of availability do not equate directly with patterns of utilisation. Several of Australia's urban centres have levels of utilisation well above that which might be expected on the basis of their size of population or business base alone. This reflects a concentration of information-intensive industries and information workers in particular cities and underpins a proposition advanced by Moss (1988) that ICT is creating a new urban hierarchy which reflects the different concentrations of information-intensive industries nationally and internationally. There appears to be some evidence in the USA that ICTs are changing the face of national settlement systems, yet in the UK they appear to be reinforcing the old order (Goddard 1990). In Australia, ICTs do not appear to be reordering the existing urban hierarchy to any great extent as yet. However, there are clear signs that Sydney is assuming world city status—harnessing aerospace, information and computer technologies to comparative advantage. Melbourne is struggling to keep pace with Sydney, yet is clearly in second place and is maintaining its ascendency over the remaining capital cities in relation to investment in office and manufacturing infrastructure that will support basic, export-oriented industry (see O'Connor & Stimson 1993). Brisbane–Gold Coast and, to a lesser extent, Perth are staking their claims as rising stars—based in large part on the sun-belt industries of tourism and retirement migration (see Stimson, chapter 3 this volume). The major cities continue to be attractive to the younger and more highly skilled individuals and households (from within Australia and overseas) as locations where their business, career and consumption interests can be satisfied and advanced. Those groups which are older and no longer in the labour force or who are unemployed appear to be detaching themselves from the larger, more expensive, cities and shifting in increasing numbers to warmer and smaller centres. The social justice implications of polarisation tendencies at inter- and intra-metropolitan levels has been little studied (Wulff *et al.* 1993).

Within the major cities, specific areas have emerged where concentrations of primary information industries are highest. A centralisation of upper management functions is taking place in key urban locations in key cities; CBDs are retaining their roles as control centres for medium- and large-sized organisations. Meanwhile, decentralisation of a wide range of information production, processing and distribution activities is occurring to particular sectors within cities (there is a general failure of smaller urban centres and country regions to attract information industries in any significant number). A measure of functional and geographical 'centralisation with decentralisation' is occurring for the larger corporations—but apparently not outside the cities—as our case studies clearly indicate. There are strong moves by industry to locations favoured by information workers as places of residence. In this sector it is clear that jobs follow workers, generating an innovative milieu within those cities where new work practices (e.g. telework, inter-organisational alliances) are being employed and where new information industries will be spawned. These areas represent the economic engines of our 21st century cities.

NOTES

[1] The national government, through its Federal Airports Corporation and Civil Aviation Authority, regulates this sector of industry.
[2] Geographic information systems (GIS) can be used to powerful effect in developing visualisation tools for examining how organisations are currently structuring their operations (who does what, where) and what communications networks are being used to support the various business functions of the organisation. A persistent problem confronted by research of this nature in the spatial sciences has been, firstly, the difficulty of adequately representing

the structure of major organisations and their networks and, secondly, locating those operations within the wider social space framework which provides the geographical context of the organisation and its activities. GIS now provides us with the means of doing both. GIS possesses the functionality for configuring the communication network of a particular organisation, given data on each node and link in the network. Each node can be further characterised by the principal type(s) of economic activity undertaken at that location and each link can be differentiated in several ways according to the service type (e.g. ISDN, packet switched, PSTN/modem), network speed and volume of traffic. Clearly, an organisation's communications network should not be viewed in vacuo. Additional layers of information relating to such features as workforce location patterns (as indicative of labour market access); spatial pattern of personal taxable income (as indicative of disposable income); and metropolitan network infrastructure, can assist in understanding why organisations locate their operations in a particular fashion. GIS permits an overlay of an organisation's communications network onto a variety of *landscapes*: transport and communications networks (for distribution and access); suburb income (purchasing power); area occupational structure (labour market); industry profile (agglomeration potential); etc.

GIS is a methodology offering an organisation-by-organisation approach for studies of the geography of business operations; as well as opportunities for aggregation and synthesis across organisations or comparative evaluation between firms within the same business sector; and the monitoring of change over time in the evolution of organisations and their communications networks (see Newton et al. (1993a) for more details).

[3] The author is grateful to Marina Cavill and John Crawford for their involvement in ongoing research on the geography of telecommunications. Acknowledgment is also given to *Prometheus* for permission to reproduce material from 'Australia's Information Landscapes'.

9 The Changing Urban Hierarchy of Japan: The Impact of the High Speed Rail

Mamoru Taniguchi, Dai Nakagawa and Tsunekazu Toda

Introduction

WE can call the coming decade the new railway age. Many countries have already decided to introduce new high speed rail (HSR) systems or to improve their existing HSR technologies for intercity passenger trips. European countries are in the process of constructing an extensive HSR network with the integration of the EC and construction of the Channel Tunnel. The United States also has executed feasibility studies for HSR in many states. In Asia, South Korea and Taiwan have researched all avenues in order to introduce HSR, and Japan has begun to construct a new test line for Maglev.

These HSR technologies will play an important role in achieving the realisation of productive and sustainable cities. Hall (1991a) pointed out that HSR technologies are now competitive with air transport on journeys of less than 500 km. A downtown HSR station provides higher productivity than a suburban airport for existing business services in the city. In addition, automobile and air transport cannot provide as timely and stable services as HSR. Moreover, compared with other transport modes, HSR has fewer negative impacts on the environment. Based on these points, HSR must be considered as the most satisfactory technology for future intercity passenger services.

Even though HSR will bring us these many benefits, there are other implications of the technology for us to consider. The construction of a HSR network sometimes brings unexpected changes in the urban hierarchy, such as competition between cities, uneven distribution of urban functions and dramatic expansion of metropolitan areas. It is very important to forecast these changes. We can only achieve a productive and sustainable environment by controlling these urban systems. With this point in mind, Brotchie (1991) investigated the major regional impacts caused by the Shinkansen in Japan. This chapter focuses on examining the influence of HSR on the urban hierarchy by using updated Japanese data.

The case study: Japanese Shinkansen

It is not easy to estimate the impacts caused by HSR on the urban hierarchy. Socio-economic changes such as migration or industrial shift take a long time to be fully manifested. Furthermore, it is very difficult to identify the effects caused by the HSR itself. The Japanese Shinkansen is probably the best example to evaluate HSR's impacts accurately, based on the following reasons:

1 Among the world's HSR systems, the Shinkansen has transported the largest amount of people. Japan's first HSR, the Tokaido Shinkansen, was opened in 1964. Annually, about 300 million people use the Shinkansen.
2 The Shinkansen is the most important mode for intercity travel in Japan. For example, the Shinkansen and other railways have a 71.3% share of total inter-city passenger trips with a distance of between 500 and 750 km. Also, they retain a 40% share of trips between 750 and 1000 km. Air transport has a 14.3% share between 500 and 750 km, and a 46.2% share between 750 km and 999 km. The share of automobiles is only about 10% between 500 and 1000 km.
3 Japan is an island without internal borders. The location of ports for international communication are limited to the large metropolitan areas such as Tokyo, Osaka and Nagoya. In this regard, regional conditions outside these large metropolitan areas are very similar.

Competition among cities

HSR provides opportunities for cities to activate their local economies by improving their accessibility dramatically. The focus of this section is the observation of the influence of the Shinkansen on the competition among cities.

Figure 9.1 (p.194) shows the process of Shinkansen construction in Japan.

The first Shinkansen line, the Tokaido line, was opened in 1964 and the Sanyo line in 1975. Presently, each line carries roughly 134 million and 68 million passengers respectively, per annum. In 1982 the Tohoku and Joetsu lines were opened. These two lines were extended directly into Tokyo in 1985. We can categorise the Shinkansen history into three different stages. The first stage was pre-1975, the period of the construction of the primary lines. The second stage was from 1976 to 1982, the period of completion of basic trunk lines. Finally, the third stage was from 1983 until now, i.e. the period of upgrading lines.

The growth rate of each city, which is the seat of the prefectural government, between 1960 and 1990 is shown in Table 9.1.

Figure 9.1 also shows the location of these cities. Thirty years is a sufficient period to examine the long-range impacts of Shinkansen investment. Though there was only a limited aviation service in 1960 in Japan, a

Table 9.1   Population increase of seats of the prefectural governments

No.	Name of city	Population increase 1960–90 (%)	Shinkansen station XX: 1960–75 X: 1976–82	Jet airport
1	Sapporo	171		XX
2	Sendai	116	X	XX
3	Ooita	98		XX
4	Hiroshima*	95	XX	X
5	Fukuoka*	81	XX	X
6	Utsunomiya	79	X	
7	Miyazaki	73		XX
8	Matsuyama	69		XX
9	Mito	69		
10	Kumamoto	64		XX
11	Kagoshima	60		XX
12	Kouchi	58		
13	Okayama*	53	XX	
14	Morioka	50	X	
15	Niigata	50	X	XX
16	Akita	48		X
17	Maebashi	44		
18	Kanazawa	42		XX
19	Wakayama*	39		
20	Naha	37		XX
21	Fukushima	36	X	
22	Aomori	36		
23	Takamatsu	35		
24	Nagano	35		
25	Matsue	35		
26	Shizuoka*	34	XX	
27	Tsu*	34		
28	Toyama	33		
29	Tottori	33		
30	Yamagata	32		X
31	Gifu*	31		
32	Saga	31		
33	Fukui	30		
34	Tokushima	30		
35	Yamaguchi*	26		
36	Koufu	25		
37	Nagasaki	15		X

* Located in Pacific Belt.

complete network of conventional railway already existed. In other words, 1960 was the eve of the high speed transportation era. Japan consists of 47 prefectures, and national government assigns the same level of adminis-

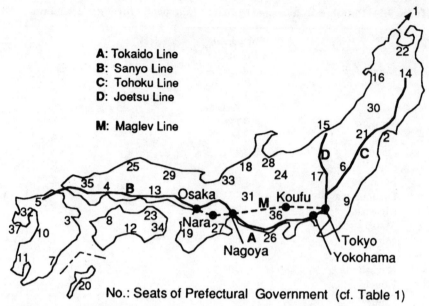

A: Tokaido Line
B: Sanyo Line
C: Tohoku Line
D: Joetsu Line

M: Maglev Line

No.: Seats of Prefectural Government (cf. Table 1)

Figure 9.1   The network of Shinkansen and proposed Maglev in Japan

trative functions to these cities. In most cases (42 out of the 47 prefectures), the most populous city in the prefecture is the capital of the prefecture. Accordingly, if a Shinkansen line comes to the prefecture, the station tends to be constructed at the seat of the prefectural government. Because of the difficulties of comparison, the three largest metropolitan areas, Tokyo, Osaka and Nagoya, and seven other cities which are located inside these three metropolitan areas, are excluded from this table.

The table shows a variety of population increase within these 37 cities, and we can infer many interesting findings about the Shinkansen's influence on the competition among cities:

1 Cities that have Shinkansen stations have seen relatively higher growth rates than cities without such stations. However, this factor alone cannot provide a full explanation about differences of population increase rates among cities.

2 Most of the high growth cities without Shinkansen stations have jet airports. Among the top 15 cities where the population increased by more than 50%, only Mito (ninth) and Kouchi (twelfth) have no Shinkansen station, and had no jet airport until 1982. The reasons why these cities saw some of the fastest rates of population increase were that the city of Mito is located at the fringe of the growing Tokyo metropolitan area, and the city of Kouchi, though having no jet airport, has a local airport with a very frequent service.

3 In Japan, the so-called 'Pacific Belt,' the corridor area between Tokyo and Fukuoka is the largest megalopolis, comparable to the north-east

corridor in the US. Though the cities located in this area used to enjoy the agglomeration effects on each other, the growth rates of these cities have not been so high in the past 30 years. Obviously, the advantages of high speed transportation are much larger than conventional locational advantage.

4 In this table, two cities with Shinkansen stations, Fukushima (21st) and Shizuoka (26th) show relatively low growth rates. The common point with these two cities is that they are located about 150 km from Tokyo. In Japan, the main objective of constructing local airports is to connect local cities to Tokyo directly. As these two cities are too near to Tokyo, they have had no chance to construct an airport until now.

We can conclude from this analysis that either the Shinkansen or a jet airport are key factors for the growth of the centre city in each region. Moreover, it is more important for the city to possess both a Shinkansen station and a jet airport than just one of them.

Concentration of urban functions to Tokyo

We can find obvious differences between the competitiveness of local cities whether or not they enjoy high speed transportation services. The next problem is the difference in competitiveness between Tokyo and local cities. The important point has been that most of the investment for high speed transportation has been spent for improving access between local cities and Tokyo, and not between local cities.

Table 9.2 shows the concentration of urban functions towards Tokyo over the past two decades.

Although Tokyo saw a decline in the share of gross manufacturing output, it has gained share in all other fields. However, the share of the total number of employed workers has not increased as fast as the rise in wholesale trade and the number of workers employed in the business

Table 9.2 Concentration of urban functions in Tokyo metropolitan region

	Share of Tokyo met. region (%)	Fiscal year		
Manufacturing output	35.1	1970	33.7	1989
Total number of employees	34.2	1972	35.9	1986
Wholesale trade	39.5	1974	42.5	1988
Business service workers	55.1	1972	60.1	1986
The amount of bill exchange	53.8	1970	84.7	1989
Sales of stocks	57.4	1970	74.4	1989

Source: Kokudocho 1991.

service sector. Moreover, Tokyo has almost achieved a monopoly in the fields of bill and stock sales. Monocentric high speed transportation investment contributes towards an uneven distribution of information and business opportunities, and this has brought such a concentration of urban functions to Tokyo.

Commuting from outer suburban areas: changing urban hierarchy?

Obviously, the aim of the Shinkansen was the provision of an intercity passenger service, not a commuter service. However, Shinkansen services have begun to change metropolitan commuting patterns. Figure 9.2 indicates the drastic increase in the number of Shinkansen commuters. They are certainly changing the idea of conventional commuting and eroding the common concept of metropolitan areas.

Most of the Shinkansen commuters have their offices in downtown Tokyo, and live 70–120 km from Tokyo station. Although these areas had not been previously considered suitable for commuting to Tokyo, the following social-economic changes have encouraged this trend:

1 there are no more suitable places to develop new housing within an easy commuting distance of Tokyo;

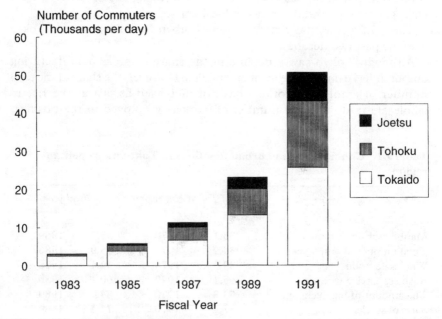

Figure 9.2 Number of Shinkansen commuters

2 the price of land and housing have rocketed. It is very difficult to find comfortable dwellings with a reasonable price in the Tokyo metropolitan area;
3 the attitude of people has gradually changed. More people tend to prefer outer suburban life with a good environment to metropolitan life;
4 a one-hour Shinkansen commute with a seat is more comfortable than standing on an overcrowded train for the same time; and
5 some companies support their employees by providing subsidised commuter tickets.

The total number of daily Shinkansen commuters has now reached 50 000. This trend will bring important changes to cities that are located within the Shinkansen commuting zone. Some cities that were independent from Tokyo will become the bedtowns of Tokyo. In 1992, the 'Nozomi'-type super express debuted along the Tokaido line. This super express connects Tokyo and Osaka (550 km) within 2.5 hours. Theoretically, it is possible to commute from Osaka to Tokyo every day now using the railway system.

What will come next? The effect of the Maglev line

The Maglev linear motor system has already overcome basic technological problems, and a decision on construction is imminent. As Amano *et al.* (1991) pointed out, the first route will connect Tokyo and Osaka, in one hour, to relieve the saturated Tokaido Shinkansen. A new test line of Maglev is now under construction in the suburb of Koufu city, and this will form part of the Maglev mainline in the future.

If this Maglev line is constructed, what kind of changes will occur in the Japanese urban hierarchy? This HSR will bring about great changes not only in cities along the Maglev line but also in neighbouring cities. Though estimating the impact on each city is not so easy, the following accessibility index helps in forecasting the impacts:

$$ACS_i = \sum_j \frac{P_j}{\exp(axT_{ij})}$$

i = seat of the prefectural government, $i = 1, 2 \ldots, 47$
j = prefecture, $j = 1, 2 \ldots, 47$
P = population in 1990
T = time distance calculated from the following equation:

$$T_{ij} \frac{(18 - S_{ij})}{Z}$$

S = maximum possible duration of stay calculated from train and airport timetable of Japan.

The idea of using the maximum possible duration of stay to specify time distance is a novel one. This duration is calculated on the effects of real service frequency and schedule. Under the conditions of same speed, and different frequencies of train service, conventional time distances of these two conditions are the same. But this new 'time distance' that considers duration can show differences in convenience between these two services. Accordingly, a more realistic accessibility index can be calculated, rather than using simple trip time. The conditions of the calculation are as follows:

1 supposed Maglev line and stations are as shown in Figure 9.1;
2 comparison is done between 1990 accessibility and accessibility with new Maglev system;
3 Shinkansen, air transport and conventional rail services are considered for base 1990 calculation; and
4 time savings created by the new Maglev service is the only additional condition to calculate accessibility with the Maglev.

Figure 9.3 shows how the 47 cities' share of accessibility will change after the construction of Maglev.

The key findings are as follows:

1 Maglev stations will provide a drastic increase in share of accessibility not only for the Maglev station area, but also the neighbouring prefectures of the stations such as Yokohama city;

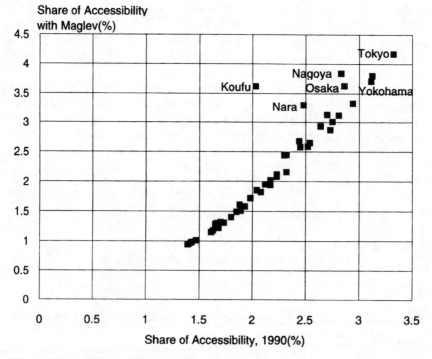

Figure 9.3   Change of accessibility share by Maglev

2 cities such as Koufu and Nara that do not have Shinkansen stations will enjoy the great increase of accessibility; and

3 the differences of accessibility share between prefectures become larger. It is necessary to consider the redistribution of urban functions from Tokyo to realise balanced growth.

Conclusion

We can conclude that the effects of the Shinkansen on the urban hierarchy have been extensive in Japan. The information presented underlines the fact that either a Shinkansen station or a jet airport is indispensable to realise a productive and competitive local city. Cities that received earlier investment for high speed transportation have enjoyed its advantages much more than other cities.

It was also evident that the benefits of the Shinkansen were mainly concentrated in Tokyo, the central node of the high speed transportation network. This trend will be accelerated by the construction of Maglev. Thus, to realise balanced and sustainable growth, it is necessary to prevent an extremely uneven distribution of urban functions. Expansion of metropolitan areas should be controlled more carefully, too. Finally, the decentralisation of government functions, as proposed by Amano *et al.* (1991), is highly desirable for the coming very high speed age.

10 Innovation and Urban Development

John W. Dickey

THE question of how innovation affects urban development and form is a very complicated one. First, as will be discussed below, innovation is more than 'visible' technology. It can include 'invisible' techniques and procedures as well as human capabilities to employ all these beneficially. Second, major innovations still seem to take a long time to develop and spread, despite recent efforts by companies to reduce the 'cycle time.' Third, urbanisation (at least in the United States) seems to be slowing and becoming more even in its growth. The question then is how innovation is contributing to these trends.

The answers are far from obvious and certainly not fully addressed in this chapter. The purpose of this discussion, therefore, is really to shed light on some relevant issues in the hope that they will eventually lead to a greater understanding of the innovation/urbanisation phenomenon.

The nature of innovation

Innovation, in our rudimentary definition, is simply the creation of a new idea and its implementation. Despite this simplicity, there are several complex ramifications that relate to the impact of innovation on urban development and form:

1 Innovation is not just creativity, but involves implementation. This is the 99% perspiration that Edison said accompanies the 1% inspiration. Yet it is not just perspiration in implementation either. Since new ideas ultimately bring on change, the innovator (or, more likely, the innovating organisation) has to deal with the management of change. This in turn brings about competitive clashes for resources, positions, and economic and political gain—untidy and often uncomfortable situations.
2 Innovation can involve both small and large items. Most people tend to think of the 'big' innovations like television, the computer, the airplane and the fax. Certainly these have had momentous impacts, but so also

have small ones. Yasuda (1991), for example, points to the 20 million ideas (where an 'idea' is defined as an implemented suggestion) that have been generated in Toyota's Suggestion System. These have been estimated to have an economic value of about $100 each, for a total of $2 billion.

3 Innovations can be 'invisible.' Most of the 20 million ideas mentioned above probably went into process improvements rather than visible technologies. They were 'invisible' innovations. As another example, Sam Walton built Wal-Mart stores and became the richest man in the US in large measure due to innovations like his principles of management (*Business Week* 1992):

   • The customer is boss.
   • Employees and suppliers are partners.
   • Managers are servants.
   • Costs must be driven down to keep prices low.

The last principle has had particular ramifications for land use, in that most Wal-Mart stores are built outside municipal boundaries, in part to save on property taxes.

Interestingly, Dr Yoshiro Nakamats, apparently the world record patent holder with 2300 patents to his name (including the audio CD), claims that 'invisible' inventions are more powerful and far-reaching than visible ones (Thompson 1992).

4 Innovation can also bring into consideration human capabilities (imagination, skills, education, and the like) to utilise both the 'visible' and 'invisible' inventions.

While these statements all may be true, our concern here is for visible innovation (which we take to mean 'technology'), and so the remaining discussion will be constrained to that aspect.

The pace of innovation

The speed of creativity and implementation has become much faster lately. The number of patent applications filed, for example, went from 113 000 in 1980 to 177 000 in 1990. Trademark applications almost tripled in the same time period—from 46 837 to 127 346. (US Bureau of the Census 1992, p.535)

Small and even medium-sized innovations can happen fairly quickly. The 20 million ideas mentioned above probably took about a month each to put in place. Apple Computer and Hewlett-Packard are now working on a six month cycle time for some of their new products. Intel is bringing out a new chip about every three years (*Business Week*, 22 Feb. 1993a).

Major new innovations, which lead to new industries, take much more time, however—perhaps even centuries. As an example, patents for the fax machine date back to 1843. The Italian priest, Giovanni Caselli, was the inventor. A later version, called the 'pantelegraph,' was used for five years (1865–70) to send reproductions of pictures from Paris to Lyons (Flatow 1992, pp.62–63).

There are several claims to the first invention of television. One is by Philo T. Farnsworth. In 1922 he devised a system while in high school, then moved to San Francisco to perfect it. This was accomplished and patented in 1927. It was 12 years, however, before he was able to license it (to RCA). The company first thought it was no good, then tried to harass him out of it (Flatow 1992, pp.94–97). Current work on HDTV also is taking a long time, in part because of the fierce technical and political competition involved in standard setting (*Business Week*, 22 Feb. 1993b).

As another example, Paul Saffo (1990) from The Institute for the Future, points out that both the computer 'mouse' and 'windows' were suggested and tried before 1965 by Doug Engelbart (a resident of Silicon Valley). Yet they were not adopted until about 20 years later.

Once a major invention catches on, however, its progress in aggregate can be quite smooth. Moravec (1988, p.64) has demonstrated this in terms of the development of the computer since 1900 (see Figure. 10.1).

The plot is of (the log of) computational power per unit cost (constant dollar) versus time. The regularity of the relationship seems to argue for the inevitability of growth in this technology, regardless of the individuals and organisations involved.

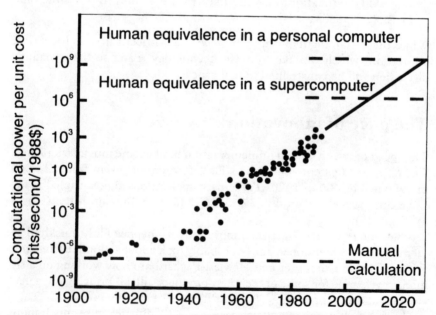

Figure 10.1 Computational power per unit cost vs year
Source: Moravec 1988, p.64.

Innovation and productivity

Moravec's chart brings up a highly intriguing question about the impacts of technology. If computer power per unit cost has risen a trillion-fold since 1900, why has not productivity done likewise? As an illustration, the index of output per hour in the US business sector rose from 65.6 in 1960 to 110.0 in 1991 (US Bureau of the Census 1992, p.409), while computer power per constant dollar jumped about 10 000 times in the same period.

Several explanations are possible here. One possibility is that the previously mentioned human capabilities to use the technology might be lagging far behind. Another is that productivity indices simply do not capture the 'real' impact of technology. The overall productivity index, for instance, would treat a $1200 PC in 1980 as equivalent to a $1200 one now, but it is obvious there are manifold differences.

It also is intriguing to note that, while semi-conductor manufacturers themselves were first among selected industries in the average annual per cent change in output per employee hour (12.0% from 1975–90), the next ones in line were iron, copper and coal mining (6.3%, 7.8%, and 5.9% respectively) as well as railroad transport (7.6%) (US Bureau of the Census 1992, p.408). Could it be that innovations in these basic industries really are the main contributors to change?

Urban development in the United States

There are many aspects of urban development and form that potentially could be affected by innovation. We obviously cannot discuss all of these but will take the more basic ones like rural/urban division, metropolitan population growth, and income.

In 1950, 64% of the US population lived in urban areas. That figure jumped to 69.4% in 1960, then 73.6% in 1970. From there it has remained almost stable, to 75.2% in 1990 (US Bureau of the Census 1992, p.16). A speculation here is that innovation played only a partial role, yet a very broad and subtle one, in leading to these rural/urban divisions. Race relations, desires to live in a warmer climate, and cheap labour all played their parts. Yet migration would not have been as easy without better roads and communications, as well as the technologies that brought economic growth and job opportunities to the cities.

Still, the rural/urban mix now is fairly stable at the time when innovation by almost any measure is increasing. Can it be that technology, perhaps in conjunction with social 'safety nets,' is now acting to help keep people in place? Is there more 'electronic' rather than physical migration of labour?

Another view of urban development comes from looking at the fastest and slowest growing (or declining) metropolitan areas. As seen in Table 10.1, of the 10 fastest growing areas, 6 were in Florida.

With the possible exception of Orlando and the Melbourne area, none of these is particularly known for technology. A warm winter sun and a haven

Table 10.1   Metropolitan areas (PMSAs or MSAs) with population growth rates over 40% in 1980–90

Fort Pierce, FL	66.1%
Riverside–San Bernadino, CA	66.1%
Fort Myers–Coral Gables, FL	63.3%
Las Vegas, NV	60.1%
Orlando, FL	53.3%
West Palm Beach, FL	49.7%
Melbourne–Titusville–..., FL	46.2%
Austin, TX	45.6%
Daytona Beach, FL	43.3%
Phoenix, AZ	40.6%

Source: US Bureau of the Census 1992, pp.30–32.

for the elderly with adequate social security cheques might be better explanations. Moreover, in Orlando, the main technology is 'entertainment' and in the Melbourne area it is 'space' (the Kennedy Space Centre at Cape Canaveral). Health care technology also is of presumed importance. None of these technologies normally is associated with urban development.

Still, transportation and communications can be playing a subtle role. The elderly and vacationers need to get back and forth to Florida easily (by auto and air). They also need to communicate easily (by phone). Even electronic banking pays a subtle role through such measures as direct deposit of social security cheques in any selected financial institution.

The remaining metropolitan areas on the list are equally interesting. Austin, TX, might be the only one that plays the role of the prototypical 'high tech' locality (note that the Silicon Valley and Research Triangle areas are not in the top ten). Riverside–San Bernadino might be the only prototypical, expressway-bound suburban area. Phoenix, AZ, probably owes much of its existence to water- and air-conditioning technologies. Las Vegas, meanwhile, is known for a different set of innovations.

Table 10.2 presents the other side of the ledger.

The biggest losers in population appear to be strongly related to technologies, but in this case to 'declining' ones like agriculture, steel, construction, oil production and mining. Pittsburgh is a particularly curious inclusion on the list. It obviously has lost in the steel industry, but also is known for its advanced computer work. Moreover, it was cited just a short time ago as one of the best places to live in the nation. This dichotomy in impacts has been brought out by Newton (1993) in his study of Melbourne, and shows that technology can have a split personality, even in one place.

It is also curious that two metropolitan areas in West Virginia, a state considered almost synonymous with coal, are in the steepest decline. Productivity increases (and presumably innovations) in coal mining and rail transport (of coal), as brought out above, have been among the greatest. Can technology be a despoiler as well as growth generator?

Another perspective on urban population growth can be developed from Table 10.3.

Table 10.2 Primary statistical metropolitan areas with population loss rates over 5% in 1980–90

Davenport, IA	− 8.8%
Pittsburgh–Beaver Valley, PA	− 7.4%
Peoria, IL	− 7.3%
Youngstown–Warren, OH	− 7.3%
Huntington–Ashland, WV–KY	− 7.1%
Charleston, WV	− 7.1%
Gary–Hammond, IN	− 5.9%
Saginaw Bay–Midland, MI	− 5.3%

Source: US Bureau of the Census 1992, pp.30–32.

About three-quarters of the metropolitan areas that had a population growth rate greater than the national average (11.4%) in the 1970–80 decade had a lower rate in the 1980–90 decade (national average = 9.8%). On the other hand, 62% with lower than average growth rates in the earlier period *increased* them in the later one. In addition, of those metropolitan areas with a negative growth rate in 1970–80, 82% were more positive in 1980–90 (not shown in table).

The trend, it appears, is toward a more equitable distribution of growth rates. This contrasts, by the way, with the polarity hypothesis of Breheny (1993) for the UK. It is difficult to imagine that this trend can occur despite the great strides in technology. So innovation must be leading to more equalisation.

Population growth, of course, is far from the only indicator of urban development. Another one of interest is income. Table 10.4 gives the 8 central (incorporated) cities, out of the top 50 by population size, that had per capita money income increases of greater than 70% over the period of 1979–87.

While it may be somewhat speculative, we can identify some of the main technologies that may be at work in these cities.

Boston, of course, is well known for computer technology and Atlanta now for communications, especially television. New York and Charlotte, both with diverse economic bases, might be best known for finance and banking. Nashville is a surprise, but is linked with 'entertainment' technology.

Table 10.3   Comparative population growth rates of metropolitan areas (PMSAs and MSAs) in two decades

Number of metro areas with growth rates in 1970–80 decade:		
	1980–90	*< 1980–90*
1970–80 > 11.4% & 1980–90 > 9.8%	53	26
1970–80 > 11.4% & 1980–90 < 9.8%	21	0
1970–80 < 11.4% & 1980–90 > 9.8%	0	10
1970–80 < 11.4% & 1980–90 < 9.8%	32	43

Source: US Bureau of the Census 1992, pp.30–32.

Table 10.4   Top 50 cities* with over 70% increase in per capita money income (1979–87)

Boston, MA	98.1%
Atlanta, GA	78.8%
New York, NY	77.8%
Charlotte, NC	75.4%
Nashville–Davidson, TN	72.9%
Virginia Beach, VA	70.6%
Jacksonville, FL	70.1%
Baltimore, MD	70.0%

* A 'city' here is the central incorporated area in a MSA or PMSA.
Source: US Bureau of the Census 1992, p.455.

Virginia Beach and Jacksonville, while both offering recreational opportunities, also were influenced in this period by military technology (namely, naval bases). So we see a broad range of possible technologies that can come into play.

Some possible issues

The preceding discussion—based on some examples, readily available data and deductions—leads to some possible issues on the nature and directions of the relationship between innovation and urban development. Most of these need a great deal more analysis, but at least point the way to possible fruitful studies:

1 Innovation, which includes 'invisible' inventions, should be the focus, not the more limited variable of 'technology'. This will require defining and delineating those processes that are 'innovative' (as opposed to 'standard') changes.
2 Equal attention should be given to the mass of small innovations as well as the major 'breakthrough' ones. The net impact of the small innovations may turn out to be of equal if not greater magnitude than the major breakthroughs.
3 Neither technologies nor techniques can be very productive if not accompanied by corresponding innovations in human capacity to develop and employ them.
4 The pace of innovation is significant. Many companies are pushing hard to reduce product cycle times, but do major breakthroughs still take as long as before?
5 More needs to be known about how the vast increases in the power of some innovations (like the computer) get translated into productivity. What happens to all that power? Does most of it end up being diffused and dissipated?
6 Innovation, along with many other factors, is causing urban development to stabilise in terms of both the rural/urban mix and population

growth differentials between metropolitan areas (see Table 10.3, for example).

7 Some innovations—like those in intercity transportation and communications—appear to be playing very subtle roles in the stabilising urban populations. The Interstate and electronic banking systems, for instance, make almost all places in the country accessible, so that there is less need to relocate for, say, business purposes. Is this due to the fact that these innovations are relatively ubiquitous?

8 The types of innovation that affect urban form can be fairly diverse. They include, for example, mining, water supply, entertainment, banking, and defence, as well as the 'traditional' ones of transportation and communication.

9 Some innovations can have a deleterious effect on urban growth. Major increases in mining productivity, for instance, may have led to depopulation. So too might have an over-reliance on 'declining' technologies like steel, agriculture and oil production.

10 In general, innovations might be divided into two categories—those that are site-specific (like water supply from the Colorado River), and those that are more ubiquitous (like communication). The latter appear to be becoming more powerful in their impacts on urban development.

A future glance

The past is only partial prologue to the future. The trend in computer power per unit cost in Figure 10.1 seems to indicate a very predictable future. Yet the type of outcome that such power allows may be quite different from what might be expected. Just as the computer chip made more of an impact on games than as expected with desktop systems (Saffo 1990), so too might some unpredictable directions emerge in the future.

The immediate future brings the interactive multiplayer, which can be hooked to a TV; is 3-D; allows for user interaction; can be employed with a computer, audio CD, CD-ROM, and VCR; and can be networked via cable or telephone with other homes or businesses (*Time* 1993). The apparent effect of this is to keep even more people at home and thus to further stabilise urban development. Or will, perchance, this technology be built into cars and planes and thus lead to more travel?

As seen in Figure 10.1 and forecast by Moravec (1998, p.64), by the year 2010 there should be a supercomputer with the equivalence of the human brain. By the year 2030 that equivalence will be available in personal computers, which (who?) will be our 'friends'. Shortly thereafter, they will no longer need our advice (Moravec 1988, pp.116–24). Will we then sit at home, playing games? Will robots go to work and travel elsewhere in our stead?

The impacts on urban development and form could go in many different directions.

11 Putting City-Suburb Competition in Perspective

Thomas M. Stanback, Jr

THROUGHOUT the post-war period the urban system of the US has experienced dramatic changes. Some central cities have shown high rates of growth while others have grown little or have even declined. The outlying areas of the metropolitan areas (the suburbs), however, have tended to grow more rapidly than central cities—typically *far* more rapidly, both in population and employment.

During the early stages of their growth suburbs gained employment largely through establishment of manufacturing plants and from increases in retailing and other local services. Over the years, however, they have not only increasingly taken on functions associated with the provision of basic services for their residents but also become hosts to a variety of other services and institutions, including headquarters and divisional sales offices and back-office activities of corporations ranging widely in size. A fairly recent development has been the appearance outside large central cities of sizeable agglomerations of economic activity—sometimes called 'edge cities'—tied closely to markets and labour supply through ready access to major highways and beltways.

In recent decades the central cities have experienced marked changes as well. Central business districts have been expanded and refurbished through new office complexes, hotels and convention centres; and historic areas have been renewed. Yet many of these cities have increasingly been beset by a variety of economic and social problems that have brought increased burdens to taxpayers and increased operating costs to firms located there. Continued outmigration of manufacturing has frequently eroded employment and the tax base, while outmigration of middle class families and a continued inflow of minorities and immigrants have brought new problems of providing for the educational, health and housing needs of very low income residents.

These developments—the rise of social problems, costs and diseconomies in the central cities and the increasing maturity of suburban economies—have raised a critical question regarding the future roles of central cities and suburbs in America's urban system. Are the erstwhile strategic advantages of central cities being eroded by competition from their neighbouring areas or is the process at work simply one in which functions are being rearranged with city and suburb becoming increasingly inter-dependent and the metropolitan area as a whole strengthened? In short, is the relationship between central city and suburb one of competition or symbiosis?

The purpose of this chapter is to shed light on this question through examining the findings of three recent studies. Section one examines findings relating to industrial composition, growth, and developmental characteristics of 14 major metropolitan areas based on the writer's recent work *The New Suburbanization, Challenge to the Central City* (Stanback 1991). Section two highlights the results of a study by Breandán Ó hUallacháin and Neil Reid (1992) relating to the intra-metropolitan location of those services covered by the *Census of Services 1987*. Section three looks at evidence from a study by Alex Schwartz (1992) relating to the location of providers of selected corporate services purchased by corporate headquarters.

The 14 metropolitan area study

In highlighting evidence from the 14 metropolitan area study, we look first at industrial composition of employment and at earnings levels in these central cities and suburbs toward the close of the 1980s (1987), and then at patterns of growth and development over the two preceding decades. Following this, evidence is presented relating to the nature and extent of development of large centres of business activity in a number of major metropolitan area suburbs.

Industrial composition, 1987

Although metropolitan economies differ in size and industrial structure, there are, nevertheless, certain characteristic differences between central cities and their suburbs in most metropolitan areas. One major difference is readily observed in Table 11.1: city employment is more heavily concentrated in a variety of service type activities (excluding retail); suburban employment in construction, retail and manufacturing.

Shares in construction and retail were larger in all or virtually all suburbs. The larger shares of employment in retailing indicates the greater importance of these local sector services which must provide not only for the require-ments of residents who work in the suburb but for a host of residents who daily commute to the central city and are an integral part of the city's economy. The larger shares of construction employment in the suburbs simply reflects the greater importance of building in the suburbs where growth rates have been higher than in the central cities. Finally, the larger shares accounted for by manufacturing in the suburbs show that these areas, with their lower densities and land cost, coupled with lower taxes and

Table 11.1   Distribution (%) of US employment among industry groups and summary of location quotients, 14 central cities and suburbs, 1987

	Percentage US	Location quotients[a] Average[b]	C>S	LQ>1.0
City				
Construction	5.50	0.69	0	0
Manufacturing	15.51	0.85	4	5
TCU	4.85	1.16	11	9
Wholesale	4.97	1.25	10	12
Retail	16.83	0.86	1	2
FIRE	7.95	1.24	11	11
Other services	26.74	1.15	12	12
Federal, civilian government	2.49	1.17	12	5
State, local government	11.14	0.84	10	3
Other	4.01c			
	100.00			
			S>C	LQ>1.0
Suburbs				
Construction	5.50	1.15	14	13
Manufacturing	15.51	1.08	10	9
TCU	4.85	0.93	3	5
Wholesale	4.97	1.03	4	7
Retail	16.83	1.09	13	13
FIRE	7.95	0.93	3	5
Other services	26.74	1.04	2	8
Federal, civilian government	2.49	0.60	2	3
State, local government	11.14	0.85	4	3
Other	4.01c			
	100.00			

a Location quotient is the ratio of the share of employment in a given industry group to the corresponding share in US employment.
b Modified average: highest and lowest values dropped.
c Includes primary industries and military.
Note: Other services include business-related services (SICs 73, 81, 89); repair services (SICs 75, 76); social services (SICs 80, 82, 84, 86); consumer services (SICs 70, 72, 78, 79).
Source: Data supplied by the US Bureau of Economic Analysis.

costs of congestion have been generally favoured over cities as locations of manufacturing.

Earnings levels

Earnings measures provide valuable additional insights in differences between city and suburban economies. A higher level of average earnings per worker in a particular industry classification in the central city compared with its suburbs reflects the joint effects of some combination of higher skill levels, greater sheltering of workers through unions and credentialling, greater use of full-time employees, or the employment of fewer young workers, women or minorities.[3]

Table 11.2 summarises the results of an analysis of earnings levels in the 14 central cities and their suburbs.

In each city and its suburbs, average earnings per worker in each of the various industrial categories have been expressed as a multiple of *average earnings for all industries combined in the central city*. Modified averages of these measures are shown along with tallies of the number of instances in which central cities earnings are greater than, equal to, or less than comparable suburban earnings. Except in manufacturing, earnings are higher in central cities than in suburbs in virtually all comparisons. The relationship between city and suburban indexes varies widely among industry groups however (e.g. the FIRE average index for the cities is more than 70% higher than the comparable index for the suburbs, whereas for retail it is only 12% higher). There is evidence here that these large cities have tended to become much more specialised in certain sophisticated higher value-added services (especially in the FIRE type services) than in the suburbs.

Table 11.2   Comparison of indexes of city and suburban worker earnings levels, by industry group, 14 metropolitan areas, 1987

| | Average indexes[a] | | Tally[b] | | |
	city	suburbs	C>S	=	S>C
Construction	1.21	1.05	12	—	2
Manufacturing	1.29	1.21	8	—	6
TCUc	1.37	1.16	12	—	—
Wholesale trade	1.24	1.14	12	—	2
Retail trade	0.55	0.49	11	2	1
FIRE	1.01	0.59	14	—	—
Other services	0.90	0.74	13	—	1
Federal, civilian government	1.16	1.04	11	—	3
State, local government	0.95	0.85	12	1	1
Total	1.00	0.83	14	—	—

[a] Averages of indexes are modified: lowest and highest values are excluded.
[b] Where differences are 0.01, city and suburb are scored as equal.
[c] TCU data not available for two suburbs.
Note: In each city or suburb indexes were computed in the form of ratios of the average earnings per worker in each industry to average per worker earnings for the *city's total workforce*. Other services include business-related services (SICs 73, 81, 89); repair services (SICs 75, 76); social services (SICs 80, 82, 84, 86); consumer services (SICs 70, 72, 78, 79). Source: Data supplied by the US Bureau of Economic Analysis.

Employment growth and transformation

An initial observation from Table 11.3 is that, with minor exceptions, suburbs grew more rapidly than central cities during both the 1970s and 1980s.

Table 11.3 Annualised rates of JIs, JDs, and net change, 14 central cities and suburbs, 1969–79, 1979–87

	1969–79			1979–87		
	JI	*JD*	*Net change*	*JI*	*JD*	*Net change*
Cities						
New York	0.2	1.4	−1.2	1.9	0.7	1.2
Chicago	1.0	0.6	0.4	1.2	0.9	0.3
Philadelphia	*	1.9	−1.9	0.9	1.1	−0.2
Los Angeles	2.4	*	2.4	2.0	*	2.0
Atlanta	2.3	0.2	2.1	2.2	0.1	2.1
Boston	1.1	1.4	−1.3	2.0	0.5	1.6
Cincinnati	1.3	0.2	1.1	1.7	0.6	1.1
Columbus	3.0	0.4	2.6	2.9	0.2	2.7
Dallas	3.7	0.2	3.6	3.7	−	3.7
Detroit	0.3	0.9	−0.6	0.3	1.7	−1.4
Minneapolis	2.7	*	2.6	2.4	*	2.4
Pittsburgh	0.9	0.5	0.5	1.3	1.6	−0.2
St Louis	0.4	1.3	−0.8	*	2.5	−2.5
Washington	1.0	0.6	0.4	1.4	0.6	0.9
Suburbs						
New York	2.3	0.1	2.2	3.0	*	3.0
Chicago	3.9	0.3	3.6	3.7	*	3.7
Philadelphia	2.6	0.5	2.2	3.0	0.3	2.7
Los Angeles	7.1	*	7.1	4.3	*	4.3
Atlanta	5.3	*	5.3	7.3	−	7.3
Boston	2.2	0.1	2.1	3.1	0.1	3.0
Cincinnati	4.1	*	4.1	3.3	−	3.3
Columbus	2.5	0.5	2.1	1.7	0.6	1.2
Dallas	6.0	−	6.0	7.9	−	7.9
Detroit	3.8	0.5	3.4	2.7	*	2.6
Minneapolis	3.4	*	3.4	2.4	*	2.4
Pittsburgh	1.8	0.2	1.6	1.0	1.6	−0.6
St Louis	2.7	*	2.6	3.4	*	3.4
Washington	4.3	0.4	4.0	4.9	−	4.9

Note: For explanation of JI and JD see text. Asterisk indicates rate of JI or JD is less than 0.1%.
Source: Data supplied by the US Bureau of Economic Analysis.

Examination of the job increase (JI) and job decrease (JD) rates adds an additional finding: in many of the central cities there was difficult transformation in one or both periods, with relatively high levels of JDs in some industrial groups partially or totally offsetting JIs in the remaining industry groups.[4] In the suburbs, however, JIs dominated. In only three suburbs during the 1970s and two suburbs during the 1980s were annual rates of JD as high as 0.5%.

Table 11.4 presents a summary of an analysis in which job increases and job decreases in each city and suburb are distributed separately across industry groups in each period. Among central cities we observe that the principal source of job declines was manufacturing (JD in 11 central cities during the 1970s; in 12 during the 1980s; with median shares of 49 and 69% of JD, respectively). Over the two periods, the principal sources of JI were growth in Other services and FIRE.

In the suburbs where employment increases typically occurred in virtually all industry groups, the largest shares of JI are found, typically, in other services, FIRE and Retail. In the relatively few instances of JD, employment declines occurred largely in Manufacturing. Even in these instances, however, the tally shows declines in manufacturing employment in only 2 of the 14 suburbs during the 1970s and 5 during the 1980s.

Changes in earnings levels

The differences between central city and suburban earnings levels have tended to increase, not decrease, during the 1970s and 1980s (Table 11.5). Among central cities, ratios of earnings to comparable average US earnings increased in 10 or more cases in Construction, Manufacturing, TCU, wholesale and FIRE, whereas among suburbs ratios increased in 10 or more cases only in Manufacturing, State, local government. In short, as central cities transformed and their economies became more specialised in certain activities, they upgraded these activities, employing relatively larger numbers of better trained, better paid employees (this was especially true in the FIRE categories with 13 of the 14 central cities showing gains in ratios). In contrast, the suburbs upgraded to a lesser extent or not at all. In the FIRE categories, ratios declined in 11 of the 14 comparisons, suggesting a strong tendency to take on lower paying back-office type functions.

Still another finding is that as central cities have become more specialised and have upgraded as service centres, they have become more dependent on suburban commuters. Evidence is found in Table 11.6 (p.216), which presents for 1969 and 1987 residence adjustment estimates expressed as a percentage of total earnings at place of work (central city or suburb).

Residence adjustment estimates are estimates of the net commuter earnings that must be subtracted from or added to total earnings at place of work to estimate city or suburban resident earnings. Accordingly, a negative sign preceding a percentage indicates that the residence adjustment is negative and that there are *net incommuter* earnings; a positive sign preceding a percentage indicates that there are *net outcommuter* earnings.[6] We observe not only that the residence adjustments are negative for all central cities in both 1969 and 1987, but that in 13 of 14 cities they became increasingly negative over the years (in St Louis they remained unchanged), indicating that incommuters were receiving increasing shares of total central city workforce earnings.

Table 11.4 Summary of analysis of distribution of JI and of JD among industry groups, 14 cities and suburbs, 1969–79, 1980–1987

	Median of shares JI or JD[1]				Tally; no. of cities or suburbs with:			
	1969–79		1979–87		1969–79		1979–87	
	JI	JD	JI	JD	JI	JD	JI	JD
Cities								
Construction	1.4	9.5	4.0	2.3	6	8	10	4
Manufacturing	7.7	49.1	1.8	69.0	3	11	2	12
TCU	3.4	7.4	3.4	7.4	5	9	6	8
Wholesale	7.4	11.9	4.2	9.2	8	6	5	9
Retail	16.4	16.5	8.7	4.4	8	6	11	3
FIRE	12.4	4.6	14.0	2.1	10	4	13	1
Other services	44.5	–	59.2	20.5	14	0	13	1
Federal, civilian government	3.8	7.1	0.4	1.2	5	9	7	7
State, local government	13.2	–	4.7	7.2	14	0	9	5
Suburbs								
Construction	5.8	31.6	7.4	–	12	2	14	0
Manufacturing	7.7	41.6	4.8	80.6	12	2	9[2]	5
TCU	3.6	–	3.4	43.4	14	0	12	2
Wholesale	8.0	–	6.8	0.3	14	0	13	1
Retail	20.4	–	16.9	–	14	0	14	0
FIRE	10.8	–	10.6	–	14	0	14	0
Other services	31.8	–	48.2	–	14	0	14	0
Federal, civilian government	0.4	13.6	0.9	–	8	6	14	0
State, local government	11.2	–	4.6	8.8	14	0	8	6

1 Shares are expressed in percentages. Where there were an even number of observations the median is the average of the middle two.
2 Includes one no change.
Source: Data supplied by the US Bureau of Economic Analysis.

Table 11.5 Number of cities and suburbs in which ratios of average earnings to US average earnings increased (+), decreased (–), or showed no change (NC), 1969–87

	Central city			Suburbs		
	+	NC	–	+	NC	–
Construction	10	1	3	8	1	5
Manufacturing	10	2	2	11	2	1
TCU[1]	11	0	3	4	0	8
Wholesale	11	2	1	10	1	3
Retail	7	2	5	6	0	8
FIRE	13	0	1	3	0	11
Other services	8	0	6	5	2	7
Federal government	6	2	6	6	3	5
State, local government	4	3	7	10	1	3
Total	9	2	3	4	3	7

1 Data not available for TCU in two suburbs.
Notes: Other services include business-related services (SICs 73, 81, 89); repair services (SICs 75, 76); social services (SICs 80, 82, 84, 86); consumer services (SICs 70, 72, 78, 79).
Source: Data supplied by the US Bureau of Economic Analysis.

In part, the increasing shares of central city workforce earnings accounted for by net incommuter earnings was due to an increase in the number of incommuters as a share of the cities' workforces, but it was also due in large part to the fact that these incommuters from the suburbs are, on average, better skilled and better paid workers. In 1980 average earnings of incommuters were well above average earnings of residents working in the 14 central cities—44% higher, on average (Stanback 1991, p.50).

Paradoxically, the residence adjustment measure also declined in 10 of the 14 suburbs (Table 11.6), indicating a declining importance of commuter earnings as a source of income within these economies. The paradox is explained by the much more rapid growth of employment in the suburbs and the fact that they are increasingly receiving incommuters from outlying areas. In spite of the larger number of commuters moving into the city daily, their net earnings amount to a smaller percentage of the total earnings of persons at work in these suburbs.

Suburban centres

As noted above, in recent years economic growth of the suburbs of larger metropolitan areas has increasingly focused on a number of very large centres. These centres—variously designated as 'suburban downtowns', 'suburban centres' and 'edge cities'—have been identified and discussed by several authors (Hartshorn & Muller 1986; Cervero 1989; Garreau 1991).

For firms in these 'suburban downtown' areas there are important urbanisation economies and other locational attractions. Each firm profits

Table 11.6   Residence adjustment as a percentage of total earnings at place of work in 1969, 1987

	Central city			Suburbs		
	1969	*1987*	*Change*	*1969*	*1987*	*Change*
New York	−23.2	−28.5	−	47.1	30.7	−
Chicago	−9.7	−15.0	−	48.6	43.9	−
Philadelphia	−29.0	−33.2	−	31.6	22.8	−
Los Angeles	−7.4	−11.4	−	32.9	15.4	−
Atlanta	−39.3	−48.3	−	51.9	34.2	−
Boston	−50.2	−52.6	−	16.6	6.7	−
Cincinnati	−19.4	−25.9	−	104.0	70.2	−
Columbus	−6.9	−12.8	−	18.8	39.9	+
Dallas	−10.5	−23.9	−	69.4	118.3	+
Detroit	−16.2	−19.2	−	32.9	20.9	−
Minneapolis	−11.5	−21.7	−	11.5	24.4	+
Pittsburgh	−8.6	−12.3	−	24.9	27.6	+
St Louis	−52.8	−52.8	=	38.2	15.4	−
Washington	−54.0	−61.5	−	46.5	19.6	−

Note: Residence adjustment is computed by subtracting in-commuter from out-commuter earnings. Negative sign (−) indicates net in-commuter earnings.
Source: Data supplied by the US Bureau of Economic Analysis.

by the presence of the other. Corporate offices can more readily receive visiting executives, salespeople and customers because of attractive hotel and restaurant facilities; retail customers gain opportunities to shop at a variety of stores; individual stores, in turn, gain from the heavier traffic that large retail agglomerations make possible; and workers of all types are provided opportunities to shop during lunch hour and after work or to lunch with friends away from the company dining hall. Finally, the location of these centres adjacent to major highways or beltways makes it possible to draw upon workers and customers located at relatively distant points.

The 14 metropolitan areas study sought to identify major suburban agglomerations and to shed light on which among the business and financial services have located there. Because only county data was available much detail was lost. Yet it was possible to identify 17 counties that appear to be centres of important concentrations of economic activity. These counties were designated 'magnet counties'.

Two types of evidence were examined in each suburban county. The first is the ratio of employment to population. Here I looked for high ratios indicating substantial agglomeration and also for significant increases in the ratio, especially since 1979, indicating those counties where there have been large employment increases relative to any population gains.

Also examined in each suburban county were levels and changes in levels of the residence adjustment expressed as a percentage of total earnings of the county's workforce. A drop in the residence adjustment percentage indicates that net outcommuter earnings have declined relative to all wages and salaries earned in the county workplace, or where the sign has switched from positive to negative, that net outcommuter earnings have given way to net incommuter earnings.[7]

On the basis of these two analyses, the 17 magnet counties were identified. Magnet county–central city comparisons of employment shares and of earnings per worker were then made for each of the major industrial categories and for subclassifications within the other services category (Stanback 1991, pp.65–77). In general, the earnings evidence complemented the evidence gleaned from the employment data. Central cities demonstrate a strong comparative advantage in the combined FIRE activities and in legal services: central city employment shares in these activities were larger and earnings higher. It is here that the special agglomeration economies of the central business district are most pronounced, although a few types of FIRE activities, such as credit agencies and insurance carriers, may operate successfully as a part of the export base of the suburban economy. Moreover, the relatively low earnings levels in the FIRE in the suburban magnet counties indicate the predominance of routine consumer banking and real estate services and suggest the presence of back-office activities, whereas lower earnings levels in legal services indicate that most suburban law firms are relatively unspecialised with few highly paid personnel.

Among the business-related services within the Other services category, these well developed suburban counties are often quite successful in attracting firms that pay, on average, wages and salaries much closer to central city levels than is true for most other industrial classifications. In some activities, such as data processing and R&D, firms may find favourable conditions for locating. Professional business-related services also often do well in these counties, although the data suggests that engineering and architectural firms are more frequently found in the suburbs than are accounting and auditing firms.

Relatively low suburban earnings levels were observed in consumer services, reflecting to a considerable extent a difference in mix between central city and county, the central city typically showing relatively more employment in hotels and amusements; the suburbs, relatively more employment in personal services. Earnings in the social services were for the most part higher in the central city, but the differences were typically not great.

The Ó hUallacháin–Reid study

This study examines central city–suburban locational characteristics of those services classified within the 'other services' group.[8] Although other industry classifications are not treated, the study is highly relevant here since the 22 service classifications that are included account for an important share of total employment in both central cities and suburbs (see Table 11.1, p.210).

As part of their analysis, the authors compute for each metropolitan area the ratio of the central city's share of metropolitan area employment in each service category, along with the central city's share of total employment. Table 11.7 shows that the mean central city share for the combined other services category (50%) is higher than the mean central city share of total metropolitan employment (35%) and that, among the 22 individual services, the central city shares range from 36–74%. In short, the tendency for all of these services to be centralised (i.e. to be located disproportionately in the central city) is much stronger for some services than for others.

A second finding based on regression analysis is that the various services fall roughly into two groups. Services in the first group tend to become less centralised as the size of the metropolitan economy becomes larger (i.e. the central city's share of metropolitan employment is negatively associated with metropolitan population size). The explanation would appear to be

Table 11.7 Central city shares of metropolitan area employment, for total employment, other services (combined) and 22 services: 74 metropolitan areas, 1987

Classification	Mean share (%)	Group 1	Group 2
Total employment	35		
Other services	50		
Legal services	74		X
Advertising	65		X
Accounting, auditing	59		X
Mailing, reproduction	58		X
Misc. business services	57	X	
Personnel supply services	57	X	
Educational services	56		X
Services to buildings	55	X	
Credit agencies	55	X	
Engineering, architectural	52	X	
Hotels	52		X
Computer programming	48	X	
Automobile repair	47	X	
Management, public relations	47		X
Personal services	46	X	
Misc. repair services	44	X	
Health services	44	X	
Research, development	43	X	
Services, NEC	43		X
Equipment rental	40	X	
Amusement, recreation	37		X
Social services	36	X	

Note: For definitions of group one and group two see text.
Source: Ó hUallacháin & Reid (1992), Table 1 and pp.340–42, 346–50.

that in larger metropolitan areas where suburbs have over the years become more extensive and more mature in terms of economic development, greater opportunities exist in terms of market size for successful operation of these service activities in the suburbs than in smaller metropolitan areas where suburbs are less developed and the central city is more dominant.

Services in the second group do not tend to decentralise as metropolitan size increases (i.e. central city share of metropolitan employment in these services is not significantly associated with metropolitan size). A separate analysis demonstrates, however, that most of these services play a larger role (i.e. constitute a larger share of employment) in the combined 'other services' sector of the central city in larger metropolitan areas.

Table 11.7 identifies group one and group two services. Group one includes services that do not appear to require the close informational linkages with other firms that characterise large city central business districts. Many of these activities are drawn increasingly to the suburbs as the outlying areas become larger and more densely developed (e.g. auto repair services, health services, social services, equipment rental services). Still other services within group one, such as computer programming, engineering/architectural, research development labs and credit agencies, may well rely heavily on central city demand but, nevertheless, find that they can serve customers adequately under the lower cost operating conditions of the suburbs.

Group two services are more clearly bound to the central city. Some, like advertising and accounting/auditing, are heavily dependent on the informational linkages that are possible only in or close to the central business district of the city where related firms are located in close proximity. Others, like hotels and amusement/recreation services, are service classifications containing large numbers of firms and institutions that play important roles in supporting the central city as a centre for arts and recreation, conventions and tourists. They also undergird the city's economy by supporting the flow of persons involved in the private and public sector affairs of the metropolis.

The Schwartz study

The study by Alex Schwartz is quite different from the preceding two in that it deals only with the location of providers of 13 non-routine, high value-added financial and professional services purchased by corporate head-quarters (see Table 11.8 for services included).[9]

Locations of both corporate clients and their service providers were identified according to metropolitan area (metropolitan statistical area or consolidated metropolitan statistical area, as applicable) and, within metropolitan area, according to location in the central city, suburbs or satellite city. The 13 services clearly vary widely in terms of their importance. Taken together, however, they provide a picture of the linkages between corporate headquarters and their high-level service providers. Analysis leads to several findings:

Table 11.8   Service providers for central city and suburban corporate headquarters (per cent distributions), 1991

	Client companies in central city Shares of provider linkages in:			Client Companies in Suburb Shares of provider linkages in:		
	Central city (%)	Suburbs/ satellite cities (%)	Other metros (%)	Central city (%)	Suburbs/ satellite cities (%)	Other metros (%)
Services						
Thirteen services, combined	46.8	2.0	51.2	45.9	12.9	41.1
Actuarial consulting	48.3	3.3	48.4	51.0	15.6	33.5
Auditing	76.3	1.8	21.9	61.8	23.2	14.9
Business insurance brokers	59.0	3.9	37.1	51.0	20.1	28.9
Commercial insurance carriers	20.4	2.4	77.2	23.5	11.4	65.2
Foreign bank relation	21.0	0.0	79.0	24.3	2.3	73.4
Investment banking	21.9	0.3	77.8	36.1	4.0	59.8
Legal counsel	70.4	1.2	28.4	65.0	13.5	21.6
Major bank relation	52.5	1.1	46.3	48.6	14.0	37.6
Master trustee	46.7	0.4	53.0	53.1	2.9	44.1
Medical insurance carrier	30.6	3.3	66.1	29.7	12.4	58.0
Pension consulting	38.7	4.6	56.8	43.4	17.1	39.5
Pension manager	24.6	1.8	73.6	28.2	4.9	66.9
Transfer agent	36.4	3.2	60.3	44.9	8.8	46.3

Source: Compiled from Tables 4 and 5, Schwartz (1992), pp.12–13, 16–17.

1 For both corporate headquarters located in central cities and in suburbs only a small share of client–provider linkages were with service provider firms located in the suburbs or in satellite cities (Table 11.8): for all 13 services combined, only 2% of linkages of city corporate headquarters and 12.9% of linkages of suburban corporate headquarters. For suburban corporate headquarters, the shares of client-provider linkages accounted for by suburb/satellite city service providers ranged widely among the 13 categories of services (from 2.3% for foreign bank relations to 23.2% for auditing), although for city corporate headquarters the range was much narrower.

2 Patterns of linkages differed sharply between corporate headquarters located in very large metropolitan areas (New York, Los Angeles, Chicago and San Francisco combined) and corporate headquarters located in all other metropolitan areas (combined) (Schwartz 1992, pp.10–17). In the four very large metropolitan areas, corporate headquarters depend less on providers located in other metropolitan areas and more on their own central city and suburb/satellite city providers. Among corporate headquarters located in the suburbs of these very large metropolitan areas, linkages with suburbs/satellite providers are significantly larger than for suburban corporate headquarters in the

remaining metropolitan areas—for all services combined, 20.1% in the four largest metropolitan areas versus 7.2% in the remainder.

3 Among corporate headquarters located outside the central city, those with the lowest annual sales volume were significantly more dependent on suburban/satellite city providers than were those with the higher annual sales volume (Schwartz 1992, p.20).

4 Although corporate headquarters relied less on service providers outside their metropolitan areas in some service categories than in others, these outside linkages were more important than linkages with suburban providers in virtually every service category (Table 11.8). Accordingly, for city service providers, competition from other metropolitan areas was greater than from the suburbs.

Table 11.9, which presents the extent of outside linkages of central city corporate headquarters in each service category for five large metropolitan areas, provides additional information.

We observe that in each service classification the extent to which corporate clients are dependent on outside providers varies among the five cities. Indeed, in the case of New York, the dominant centre for investment banking, outside relationships account for only 8.4% of linkages in this service category compared with 78.6–100% of linkages in the remaining four cities, and in Boston there are no outside linkages for master trustee services, although in the other cities there is heavy dependence on outside providers.

Table 11.9   Shares (%) of central city corporate headquarters linkages with providers in other metropolitan areas, 5 cities, 1991

	New York	Boston	Cincinnati	St Louis	Atlanta
Actuarial consulting	20.8	23.1	64.7	28.6	17.4
Auditing	14.4	10.1	10.3	2.4	8.9
Business insurance brokers	19.1	25.0	31.3	21.1	21.7
Commercial insurance carriers	53.1	44.4	77.8	86.7	50.0
Foreign banking relationship	25.9	100.0	50.0	88.9	94.4
Investment banking	8.4	100.0	78.6	95.0	80.0
Legal counsel	13.0	25.0	16.1	4.0	17.1
Major banking relationship	18.1	33.3	46.2	49.1	27.8
Master trustee	40.3	0.0	75.0	60.1	50.0
Medical insurance carrier	51.1	41.9	65.2	52.9	87.5
Pension consultant	29.1	42.9	64.3	62.5	50.0
Pension manager	44.8	45.8	78.8	85.1	67.6
Transfer agent	20.7	38.9	81.3	47.8	56.5
Total, 13 services	26.3	32.3	55.1	50.7	42.0

Source: Unpublished data supplied by Alex Schwartz.

The principal finding, however, is that in virtually all services in all 5 cities there are linkages *both* with local service firms and with firms in other metropolitan areas. It is clear that outside providers and local service providers compete in all 5 cities in all or most services.

5 A final major finding from the Schwartz analysis is that providers of these high-level corporate services are located principally in a relatively small number of cities. Table 11.10 shows that the cities ranked among the top 10 (in number of corporate clients served by each city's service providers) accounted for from 50–91% of all client linkages in the 13 service categories; cities ranked among the top 20, for 69–97%.

Given the fact that there were 189 cities in the US with a population of 100 000 or more in 1985 (64 cities with a population of over 250 000), it is clear that there are a large number of smaller central cities in which negligible or no high-level services are produced. For most if not for all of these smaller cities, attraction of high level corporate service providers is probably not an option, although provision of business services for smaller businesses may well be.

Implications

The preceding discussion goes some distance in putting into perspective the significance of suburban–central city competition in recent decades.

First, it makes clear that the suburbs with their more rapid growth have continuously taken on additional activities as they have become increasingly

Table 11.10   Share (%) of corporate headquarters served by service providers in top 10 and top 20 ranked cities, 13 services

Service categories	Top 10	Top 20
Actuarial consulting	65.0	84.0
Auditors	50.3	69.1
Business insurance brokers	58.2	74.9
Commercial insurance carriers	76.5	95.2
Foreign banking relationships	90.8	96.6
Investment banker	90.2	95.6
Legal counsel	58.6	75.4
Major banking relationships	67.4	80.4
Master trustee	72.2	88.5
Medical insurance carrier	65.1	78.8
Pension consultant	65.3	87.3
Pension manager	74.5	87.6
Transfer agent	79.5	95.0

Note: Rankings are based on number of corporate clients served by service providers located in a given city.
Source: Based on unpublished data supplied by Alex Schwartz.

more attractive locations due to larger market size, improved transportation and an expanding labour supply. Moreover, there is evidence that in recent decades sizeable agglomerations of economic activity have developed in the suburbs of a number of larger metropolitan areas, resulting in still further locational attractions for a variety of activities, including some business services.

Second, many large central cities have undergone significant transformation both in their physical infrastructure (e.g. office skyscrapers, convention halls, hotels, medical centre hospital structures, and arterial linkages to the interstate highway system) and in the concentration of for-profit and not-for-profit service activities in which their economies have tended to move toward greater specialisation in higher value-added activities.

A third observation, a corollary to the above, is that whereas some services have adapted to suburban locations, especially in larger metropolitan economies, others have tended to remain highly resistant to decentralisation. Among the latter group a number of sophisticated, relatively high value-added corporate services play an important role in the economy of many central cities. For these service providers the greatest competition is often with firms in other metropolitan areas, although some corporate headquarters located in the suburbs, particularly the smaller corporations, are linked to suburban providers.

So it seems clear that central cities have experienced competition from their suburbs and that such competition will doubtlessly increase. Whether or not such competition damages the city's economy depends on whether or not that economy is able to strengthen or renew those activities in which it enjoys a fundamental comparative advantage.

But today's central cities may be threatened by a number of other developments, the most important of which may be the economic decline in the prosperity of the regional hinterland served or of the industrial complex of the immediate area. Examples abound: the plight of the 'rust belt cities' during the 1980s; current problems in Seattle (aerospace) and Los Angeles (defence and aerospace).

Accordingly, the significance of suburban competition must be viewed in a broader context. Where the regional hinterland or the industrial complex served by the metropolitan economy is in vigorous growth, the loss of activities, income and employment to the suburbs will be readily replaced by the appearance of new firms and the expansion and upgrading of old. Suburban growth and central city prosperity will be seen as complementary. Similarly, changes in national or international markets may bring growth and development of the economic base of certain cities which will offset erosion through suburban competition. On the other hand, central cities may be adversely affected by economic decline in hinterland areas or industries. If suburban encroachment is added to this, the city's troubles will clearly be exacerbated.

Moreover, to a significant extent these cities compete with one another. There are a number of services in which inter-metropolitan competition may overshadow competition from the suburbs. Cities may compete not only in corporate services but in a variety of other services including convention trade, tourism or even medical services or universities. Accordingly, there is likely to be considerable potential for gain that is dependent on the innovativeness of the city's providers of export-type services in challenging those located elsewhere and on the attractiveness of the city's environment and business climate for new entrepreneurs.

Finally, a strong case can be made that policy measures to strengthen the central city's export base through increasing its competitiveness vis-à-vis other metropolitan economies will also be useful in preventing move-outs of firms to the suburbs. Improving the transportation system (providing quicker access either to metropolitan airports or to suburban commuter residences), making the city safer, cleaner and more attractive, encouraging the development of an improved housing stock, improving public education—these and a variety of other measures which are largely dependent on public sector initiative can act to foster the growth of firms and institutions most suitable to the city's economy and to reduce incentives for relocation of firms that might otherwise elect to depart.

Notes

[1] The data analysed is county data. The central cities of New York, Philadelphia, St Louis and Washington, DC are conterminous with a county or counties, and use of county data poses no difficulty. In the case of the remaining cities—including two major cities, Chicago and Los Angeles—this procedure is open to criticism because the central city substantially overbounds the municipal limits of the central city. Yet all of these central cities dominate their counties and county data should provide an acceptable approximation of the central city itself.

[2] Table 11.1 presents, for 1987, average location quotients of shares of employment, indicating relative size of each industry group's share of total city or suburban employment, on average, along with tallies indicating the number of cases in which city share exceeds suburban share (or suburban share exceeds city share), and the number of cases in which the location quotients are greater than one. The location quotient for a given industry group is simply a measure of the size of employment share (percentage of employment) in a central city or suburb expressed as a multiple of the corresponding share for the entire US (shown in Table 11.1).

[3] There is also an overall tendency for wages to be somewhat higher in central cities than in their suburbs, reflecting higher costs of living due to higher rents, costs inherent in congestion, and higher taxes due to higher public sector expenditures. But overall city–suburb wage differentials do not explain the quite wide variations in differentials among industries observed in Table 11.2.

[4] Job decreases are the total *net* decreases in employment in those industry groups in which employment declined; job increases are the total *net* increases in employment in those industry groups in which employment grew. Both rates of JI and JD are based on percentages of *total* employment at the beginning of the period. (NB: The rate of JI minus the rate of JD equals the rate of net change.)

[5] Table 11.5 is based on an analysis of central city and suburban average worker earnings in the several industrial classifications in which such earnings were computed as ratios of US average earnings in comparable classifications. Increases or decreases of ratios from 1967 to 1987 were then tallied.

[6] The rather wide range of percentages of residence adjustment to total earnings of workers employed in the cities—from 11% to 60%—should not be regarded as necessarily indicative of differences in the importance of commuting. At least in part they are accounted for by

variations in the extent to which central city counties overbound the cities themselves (for evidence, see Stanback 1991, p.11).

[7] It should be noted, however, that the *value* of the residence adjustment (i.e. net incommuter earnings) could increase in dollar terms and yet decline as a percentage of workforce earnings (if the latter are increasing more rapidly).

[8] The authors analyse data relating to the 22 service categories within the 'services' classification (redesignated 'other services' in this paper) for the 74 metropolitan statistical areas and consolidated metropolitan statistical areas with a population of at least 500 000 in 1987 based on data from the *1987 Census of Service Industries*.

[9] The primary data source was the computer tape for the *Corporate Finance Bluebook* (1991), which provides a variety of information on approximately 5000 of the nation's major public and privately owned companies covered by the *Fortune* list of 1000 largest industrial and service corporations, *Forbes* list of 400 largest private companies, and the *Inc.* lists of 500 fastest growing private companies and 100 fastest growing small public companies. The unit of analysis for the study was the linkage between a company headquarters and its service providers. For each corporate client, a linkage was defined as a relationship between that corporate headquarters and one supplier. Company response rate varied among the 13 services. The highest response rates were for auditors (85%), banking (74%) and legal services (67%); the lowest rates were for foreign banking relationships (11%), pension consultants (19%), and commercial insurance carriers (20%) (Schwartz 1992, pp.7–9).

12

Office Development and Information Technology: Sustaining the Competitiveness of the City of London?

P. W. Daniels

Introduction

IT is now 30 years since the foundation of the eurobond market (by S.G. Warburg & Co.) which was the platform upon which the City of London developed the most diverse collection of financial services in the world. In 1992 the volume of eurobond new issues ($US270 billion) represented a 400-fold increase on 1963 (after adjusting for inflation). The concentration of financial business activities in London (securities, currency dealing, insurance, reinsurance, commodity broking, futures, investment banking, and many others) has been increasing markedly during the 1980s. One frequently cited index is the number of foreign banks and foreign securities houses operating there (Table 12.1); most have located their offices in the City and 'there was a 26% increase in foreign bank employment in London in 1986 alone' (Harris & Thrift 1987, p.65).

Table 12.1 Change in the number of foreign banks and foreign securities houses located in London, 1980–89

Year	Foreign banks	Foreign securities houses
1980	383	104
1981	399	111
1983	445	127
1984	459	130
1985	454	134
1986	462	143
1987	464	155
1988	478	156
1989	482	158
Absolute change 1980–89	+99	+54
% change	+25.8	+51.9

Source: *Financial Times*, 29 November 1990, pp.36, 38.

During the mid-1980s demand for office space was actually outstripping supply, with annual demand reaching a level of 3.5 million square feet in 1986. This growth has had two broad implications for the Central London office market. First, it has generated a significant requirement for high quality office accommodation. Second, during the initial phase of a foreign company's presence in London, there is a marked preference for locations within the traditional 'core area', i.e. the City. This behaviour is most clearly expressed by Japanese firms that have acquired offices in London or by overseas investors from the oil-rich states of the Middle East. Following an initial period of settling into the London market, firms have expanded and become more locationally 'footloose', usually requiring larger premises at more economic rents. In the case of US institutions, for example, this second phase location choice has led them to move quite considerable distances (by City standards) to office buildings located on the South Bank of the Thames or in the West End.

Retaining its competitive advantage for the location of advanced services is vital for the future of the London (as well as the UK) economy. Recent competition from New York and Tokyo or from the newly emerging international financial centres such as Singapore or Hong Kong has led to much self-examination; can the City of London of the 1990s maintain its position at the apex of the Golden Triangle? (see for example, King 1990; London Planning Advisory Committee (LPAC) 1991; Budd & Whimster 1992; Leyshon *et al.* 1987b). In this chapter it is suggested that retaining a productive and sustainable financial services centre does not just depend upon the City's (and London's) prestige, tradition, relatively liberal regulatory environment for international business, or excellent international transport and telecommunications infrastructure (its 'traditional' virtues). Also required is a built environment that is able to meet the rapidly changing and constantly evolving requirements of increasingly sophisticated national and international service (as well as manufacturing) multinationals at the leading edge of the globalisation of the economy (see for example Dicken 1992b; Daniels 1993; Enderwick 1989; UNCTC 1990). For the City this, of course, largely means suitable office space and supporting infrastructure. The chapter explores the importance of IT in the design and construction of City office buildings, the extent to which it is actually utilised by firms that have recently leased space, and its significance for their choice of building and location.

Some 75% of all employment in the City is office-based and offices occupy over 71% (66.7 million sq ft/6.2 million m2) of the total floor space (Corporation of London 1989). But 75% of the office stock was constructed before 1980 and is often ill-suited to the requirements of the modern office user. This, together with the pressure of demand, led in the mid- to late-1980s to the largest office construction programme central London has ever seen, even by comparison with the reconstruction after the Great Fire and the Blitz (see for example King 1990; Duffy & Henney

1989). The key features of this restructuring have been quality and flexibility. The quality of office space is determined by the planning system, the development industry, the investing institutions and, to an increasing degree, the precise requirements of users (see for example, Harris 1991b). Quality, for the purpose of this chapter, is defined not just in terms of the aesthetic appearance (external and internal) of office buildings but also includes 'functionality'. This is especially important in relation to the way in which information technology (IT) services is incorporated in office buildings.[1]

Information technology: influence on office building design and demand

The importance of IT in business practice has risen markedly over the last 20 years. The position per se of information is that it is 'no longer an instrument for producing economic merchandise, but has itself become the chief merchandise' (McLuhan, cited in Eco 1987, p.135). Given that information has become pivotal, it is not surprising that IT has assumed such importance in the workplace. A truly precise definition of IT is problematic but 'the creation, manipulation, storage, duplication and transfer of different types of information in an electronic format' is probably appropriate here (CALUS 1983 15; see also Hepworth 1989).

The IT equipment that is used in the late 20th century office requires a substantial supporting infrastructure. A clean and reliable power supply is required; such is the value of the records, data, images, etc. held by office information systems that some organisations must also contemplate providing (if they have not already done so) their own 'back-up' power sources or they must depend upon small-scale 'local' power stations to insure against the information losses triggered by a major breakdown in power supplied by the utilities. In most circumstances, networked links to other items of hardware within the same building or at other locations is required. This is manifest in the need for large quantities of cabling, which for safety and convenience must be located 'out of harm's way' while remaining easily accessible. Thus, in the IT-friendly office the raised floor is seen as an essential element. Technology-intensive offices also generate 'wild heat' which is given off by IT equipment while it is in operation, and this must be removed and/or controlled by air-conditioning. To be functional and successful, the modern office building must be capable of meeting all of these challenges, while simultaneously incorporating a potential for expansion or improvement that is dictated by IT and the changing space planning needs of existing and future tenants.

The design of office buildings has therefore moved from the purely aesthetic towards a more comprehensive functional approach. This also embraces the requirements of the work force as well as of IT equipment.

Thus

tenants in all sectors now seek to upgrade offices to accommodate new operating practices, to raise the corporate profile and to attract high quality staff—an increasingly scarce resource (Peach & Jones 1988, p.86; see also Henderson 1988).

Whilst building design considerations aimed at facilitating the use of IT have been reasonably well documented, the 'humanisation' of the office remains an essentially abstract concept. Over the past decade companies have become more conscious of the asset value of employees, especially during an era when demographic trends are reducing the size of the labour pool. The changes in office design related to changing working practices do not just relate to ergonomics or preventing conditions such as repetitive strain injury (RSI). The 'social dimension' of the work experience is now being addressed: this involves designing buildings in ways that allow for the creation of 'social spaces' where employees may interact with each other rather than just interacting with a VDU around which the principle tasks of many office workers now revolve. Other aids to maintaining staff satisfaction and morale within a given working environment include environmental control of heat, light, noise or office 'landscaping'. Current trends therefore point towards more varied working environments with a greater emphasis on the ability of individuals to control aspects of their own part of the office milieu. Many of the new variables incorporated in the design of office space have emerged only during the last 10–15 years; many only in the last 5 years. They have revolutionised the life expectancy of existing buildings and have modified the expectations for the most recently completed office buildings.

In common with other commodities, the costs of office construction have risen markedly over recent years. It is estimated that during the 1960s the costs of construction were approximately £425 per m^2. By the mid-1980s this had risen to £650 per m^2. Not only has the absolute level of expenditure risen but the relative amounts spent on the different components of an office building have changed. Most significantly, the cost of the building shell relative to the cost of building services has declined (Table 12.2).

Table 12.2   Relative costs associated with UK office construction—1960s to 1980s

| | Office construction costs (%) | |
Factor	1965	1985
Building shell	70	40
Building services	20	40
Scenery	10	20

Source: Duffy 1986, p.61.

The building services cost component has increased significantly as the importance of new technology has risen, i.e. information technology places a whole new set of demands upon buildings in the form of a higher standard of building services. Loe (1987) has shown how office automation has exerted an increasing influence on the cost of office buildings since 1970, while the life expectancy of various parts of buildings has been shortened so that within 7 years both office automation and the fittings need to be changed. This has in turn had an impact on the design of office buildings.

It has already been noted that the changes in the appearance of the office building during the 1980s have not just been confined to aesthetic considerations. The 'functional' design of the structures has also changed markedly. Whilst the external appearance of a building remains important

> there is a growing realisation within the development industry that office buildings must be designed from the inside out, thereby taking full account of how a potential occupier or range of occupiers is likely to use the building (Pentecost & Love 1985, p.887).

In 1984 it was

> no longer sufficient to provide a basic building, whether new or refurbished, and expect it to let—it is vital to offer a well-conceived product where a high degree of consideration has been given to the design for modern occupational needs. New buildings must be built to satisfy tenant demand and not institutional structure (Peach 1985, p.34).

Thus, the Lloyd's building (designed by Richard Rogers) is aesthetically appealing yet still very functional. By placing the services on the outside of the building a large amount of internal space is saved, possibly as much as 30%. Not only does this release more useable/leasable space, it also allows the building services to be upgraded or changed more easily because they are readily accessible and the work involved will minimise disruption on activities taking place within the building. It is evident that as 'the electronic office grows servicing is now the most critical element' (Duffy 1986, p.60).

Another factor which has an impact on office design is the increase in the aggregate level of space per employee. The floor space/worker ratio has increased from 173 sq ft (16.1 m^2) in 1961 to 255 sq ft (23.7 m^2)in 1981 (Egerton-Smith et al. 1985, p.31). This 47% increase in space provision has not, however, been generated by improvements in space standards for individual employees; rather it is derived from the increases in floor space allocated to IT equipment and related facilities, conference and meeting rooms, and plant and machinery.

The ability of the development industry to respond to changes in office design has been assisted by major changes in the techniques of office construction. Generally, UK office buildings have relied upon reinforced concrete technology, whereas the 'fast track' method involves construction with steel frames and metal deck floors. The adoption of 'fast track' construction methods has compressed the time scale for construction although not necessarily the total time from initiation of a development proposal to final completion: 4 to 5 years remains the usual time for the complete process. Nevertheless, the time taken to construct a building has, on average, been reduced (with the added benefit of easing the developers' debt burden). The versatility of the finished product, which can be altered structurally at a later date to accommodate new ductwork or new building services, is enhanced. Recent examples of the use of this method are the Broadgate development (adjacent to Liverpool Street station) and the main tower in the Canary Wharf project in Docklands. Such buildings will be easier to demolish and to replace at a reasonable cost in the future.

Following its introduction into the UK, 'shell and core' completion of a building was considered very radical. The concept requires that

> the developer completes the building but leaves the internal office areas in a shell condition. The developer can then either complete the office areas to a building standard or give a cash equivalent to the tenant to finish them in accordance with his specific requirements (Pentecost & Love 1985, p.887).

This saves both the tenant and developer the considerable expense and inconvenience of refitting a building which is not to the tenant's liking or convenience. One of the first major projects within the UK to adopt the practice of fitting out to shell and core specifications was the Broadgate development. The number of developers actually adopting the shell and core approach in full remains uncertain however. Research completed by Mathews and Goodman shows that the

> cautious approach is for a developer to take a building through to the finished article, and our recent research shows that in most cases this is now happening. (cited in Webster 1989, p.92).

However, new office buildings are not the only ones on the market. Older properties with relatively lower standards of design, construction and materials are still standing and need to be utilised unless they have reached the end of their economic life. Obviously there are problems in the letting of such buildings, as the contemporary office tenant is far more sophisticated and discerning than their counterpart in the 1950s and 1960s when these buildings were first occupied. Whilst many of the buildings completed during the two decades after World War II do come in for extensive criticism on a number of grounds (including poor construction and lack of versatility), there are those who take a more lenient view so that where

> the location is good and the building sound, the standard 1960s office building has proved more flexible than we are today prepared to credit (Elwood 1988, p.92).

Whilst the market for buildings in 'original' condition is limited, the level of demand for office space and low levels of supply (especially during the late 1980s) meant that refurbishment was a practicable and profitable option for the owner in many cases. A good example is the City Tower in the heart of the City which has attracted peak level rents, despite its age and design. However, the high rental levels may simply reflect a locational premium rather than the standard of the accommodation.

The cost associated with major refurbishment can in some cases prove very restrictive. Since

the economics of adaption show that costs range from £250 to £1110 per m², and the average construction cost of new buildings is approximately £330 per m², one can see the clear possibility of certain buildings being rendered useless as a secure investment for the future (Bateman 1985, p.146).

A full-scale refurbishment of a 1920s office block in Mayfair cost the Prudential Assurance approximately £1000 per m² when it was completed in 1984. The refurbishment, which includes raised floors, suspended ceilings and new cabling shows that Prudential were prepared to go to a great deal of trouble to equip the building in a way that would see it through to the end of the century. The cost of achieving this (if indeed it is possible) may well mean that in many instances the only option for maximising income from a site is to demolish the building and start again. However, in the present depressed property market the refurbishment option may be seen as the ideal way to maintain an office building for the next few years.

The context for office location decision making has also been revised as a result of the changes in the level of technology utilised in contemporary office buildings. Indeed

many of today's expanding tenants put the need for such items as high-velocity air-conditioning and large floor areas with multi-access raised floors, higher on their list of priorities than location, which has, for many years, been the dominant feature affecting the relocation decision (Egerton-Smith et al. 1985, p.31).

Office users have, over the last decade, become more sophisticated in their demands, and in many respects more aware of their precise requirements (Harris 1991b). This improvement in demand specifications has, when combined with the scarcity of good quality office accommodation in the City, opened the way for the development of schemes on its 'periphery' (such as Broadgate), and the consequent movement of some advanced service firms away from traditional City 'core' locations.

This is not to say that the factor of location does not enjoy some importance in the context of successful development but

occupiers are now evaluating and comparing alternative office buildings in much greater depth as regards accommodating their organisational structures and servicing the modern office technology (Pentecost & Love 1985, p.886).

The notion that the new demands generated by advanced business technology have in some respects overridden or reduced the importance of a 'central' location for businesses is not, of course, universally accepted. Some observers of the central London office market have argued that it is only in certain limited circumstances that location is replaced as the prime factor. Where

tenants have a choice, the continued preference is for a central location or a lower rated borough. Where no choice exists, the keen demand for buildings of the right design and specification has overcome much of the reluctance that previously existed on the part of tenants to locate within the higher rated boroughs (Egerton-Smith *et al.* 1985, p.32).

It should be noted, however, that in the long term the influence of rateable values has been removed following the introduction of the Uniform Business Rate in 1990.

The location of offices has undoubtedly been affected by IT, but in subtle ways. The idea that IT can be a facilitator, allowing companies more locational freedom for some functions, is now widely accepted. As tele-communications facilities have been improved by IT (such as telex, fax, data transfer), traditional locations can be vacated in favour of more economic sites with IT facilities, cheaper accommodation, a more readily available work force and lower labour costs. Since the deregulation of the Stock Exchange (or 'Big Bang') in 1986, traders have been less tied to locations near the Bank of England. There has been a discernible willing-ness to contemplate locations which would never have been considered during the mid-1980s. Advances in IT have enabled a move away from floor-based trading to on-screen trading, making the market more efficient but also allowing securities dealers to seek out the benefits of lower costs or more suitable offices at locations some distance from the financial core of the City. If this process gained momentum, as appeared possible in the late 1980s, it would 'pull' dependent services in similar directions, thus loosening the tightly organised functional and physical fabric of the City.

It is of course important to be sure that the actual office premises requirements of firms that consider the City (or are already located there) is actually related to the level and types of technology used by companies. This is likely to vary according to the business sector in which they operate. An excellent summary of the premises requirements of office firms by sector is provided by Duffy & Henney (1989). Comparisons of requirements are made across a wide range of sectors, from large investment banks to accountants and services with markets strongly oriented to the City. Emphasis is given to the variation between sectors rather than differences that can related to the size of organisations. Although each individual firm will of course be unique in its requirements and organisation, some broad generalisations emerge.

Starting with the outside of the building, Duffy and Henney note that the 'image and style of the building' is very important for investment banks, foreign commercial banks and headquarters of major UK clearing banks. Solicitors and accountants, on the other hand, tend to prefer more 'modest' premises. The size of the premises and the size and shape of individual floors are ranked highly by investment banks, larger insurance brokers and UK clearing banks; they all generally require large rectangular floor areas. By comparison, solicitors tend to require smaller amounts of floor space (apart from the very large companies) which can be subdivided into individual offices.

The other key building requirement is the provision made for IT facilities. Although it has been suggested that 'the precise facilities related to information technology in any one office are impossible to predict' (Bateman 1985, p.137), the demand by companies for certain IT support facilities can be taken as an indication of the extent to which their efficient operation relies upon IT. Investment banks, clearing banks, foreign commercial banks and securities companies all cite this as an important provision; it is only the solicitors and accountants who feel that this is unimportant. The level of building services required is closely correlated with the demand for generous slab-to-slab heights; firms requiring maximum slab-to-slab heights also require a high level of building services.

Assumptions vs reality: survey of firms occupying office space since January 1985

This short summary of existing research does suggest that IT is significant in the office building choice decision of firms in and around the City. There is more ambivalence about the part played by building location in the decision-making process. It is therefore useful to examine the experiences of advanced service firms that have recently relocated to office space that has been occupied from new or as a refurbished building since January 1985. To what extent has their choice of building been governed by IT requirements or have other factors such as cost or location been more, or less, significant in the final decision?

The survey universe was compiled from an extensive, and regularly updated, database of property deals within London made available by Applied Property Research (APR).[2] It was decided at an early stage that some parameters would have to be set to narrow the search for relocating firms by time, size of relocation and sub-sector. Records were initially selected from the APR database by year of move into present office premises: only moves which occurred between January 1985 and March 1991 were considered. This coincides with the period of intensive activity within the City in anticipation of, as well as following, 'Big Bang', when an unprecedented level of new office construction and rehabilitation was under way. Secondly, with reference to the size of relocations, an upper limit (in terms of floor space occupied) was not set but a minimum level of 10 000 sq ft (930 m^2) was used in order to exclude relocations by very small companies. Finally, a sectoral or functional filter was applied to the data set to eliminate retail and manufacturing companies.

APR recorded 537 office moves into office accommodation between January 1985 and March 1991.[3] These consisted of moves between origins and destinations within the City and West End as well as from outside; some companies had moved offices more than once. Different divisions of the same company, each with slightly different requirements, were often involved. Each mover was subsequently approached by mail and telephone.[4] The structure of the responses by sub-sector very closely matches that of the original data set (Table 12.3).[5]

Table 12.3 **Comparison of survey responses with original data set: by sub-sector**

Sub-sector	% of APR data	% of responses
Banking and insurance	48.3	43.9
Professional services	25.8	28.0
Financial services	5.0	7.3
Business services	6.6	6.1
Industrials	14.2	14.6

Source: APR and Postal Survey, 1991.

Reasons for relocation of offices

The majority (26.6%) of the respondents moved into their new premises in 1989. There was a steady increase in movement activity between 1985 and 1989, followed by a decline (Figure 12.1).

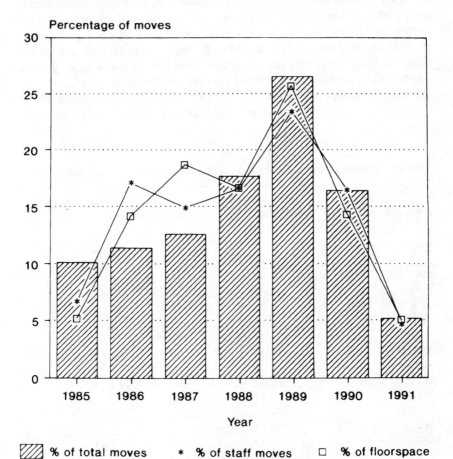

Figure 12.1 **Office moves: number, staff, floor space**

There are clear indications that relocation was accompanied by some deconcentration; some 60% of the relocated firms originally occupied offices in just three postal districts (EC2, EC3 and W1) (Table 12.4), but after they had completed their moves they occupied space in four postal districts (EC2, EC3, EC4 and W1).

There were significant increases in buildings located in the postal districts contiguous to the City such as E14 (0–4.9%) or WC2 (4.6 to 8.5%). The proportion of movers opting for the traditional prime locations in the City (EC1 and EC2) decreased from 10.3% to 3.7% and from 31.0 to 18.3% respectively.

Respondents were asked to rate the importance of several factors relevant to their decision to move on a scale from one to five (1 = not important, 5 = very important) and 92% provided useable information (Figure 12.2).[6]

By far the most dominant reason for considering relocation was expansion and the perceived need for more space, followed by quality of accommodation and consolidation. The notion that the search for new office accommodation within central London is at least initiated by a desire for a higher quality of accommodation does seem to conform, at least in part, with the evidence; London does require a stock of good quality office space to satisfy demand from tenants who require more overall utility from their buildings than in the past. But, as Cowan (1968) had already observed much earlier, consolidation of office functions within central London is also an active influence on relocation. For example, several organisations had moved from two or three separate buildings in different London boroughs.

Table 12.4   Relocations within central London by postal code areas

Postal code area	% of total respondents prior to move	% of total respondents after move
E1	1.15	2.44
E14	–	4.88
EC1	10.34	3.66
EC2	31.03	18.29
EC3	14.94	15.85
EC4	9.20	14.63
N1	1.15	2.44
NW1	–	1.22
SE1	2.30	4.88
SW1	4.60	7.32
W1	14.94	13.41
WC1	2.30	–
WC2	4.60	8.54
W2	1.15	–
Other	2.30	2.44

Source: Postal Survey, 1991.

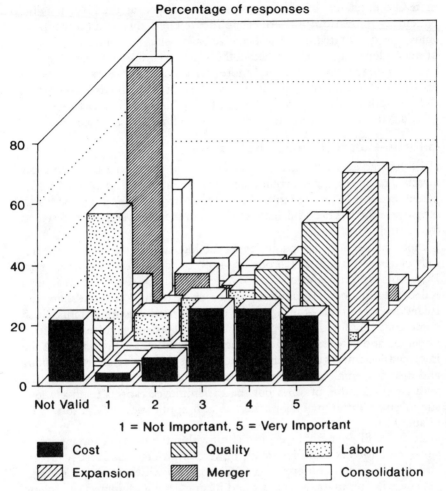

Percentage of responses

1 = Not Important, 5 = Very Important

- ■ Cost
- ◩ Quality
- ▦ Labour
- ▨ Expansion
- ▨ Merger
- ☐ Consolidation

Figure 12.2 Factors initiating office moves

While there is no doubting the hypothesis that a stock of readily available office space of adequate proportion, specification and flexibility is of great importance, on the questions of enhanced IT provision and better staff facilities, respondents were evenly split as to whether their new building offered such benefits. This is even though companies actually ranked IT and associated services at the top end (scoring 4 to 5 on the scale) of their list of priorities for new premises. It is possible either that occupiers are not satisfied with what developers describe as state-of-the-art buildings, or that their building requirements change so quickly that initially attractive office specifications are soon out of date. If either, or both, of these is correct then the consequences for London's competitive advantage in terms of its office building stock could be considerable. Reflecting the impact of changes in corporate structure and organisation (see for example Enderwick 1989), increased efficiency of office organisation emerged as the most tangible benefit of relocation; 82% of respondents cited this as an advantage. Enhanced corporate image was perceived as a benefit of relocation by 69% of respondents. It is unclear whether this is related more to the new location in a fashionable area or indeed more site-specific concerns such as the appearance of a building and its internal layout. Cost of accommodation did not score highly (only 28% cited this as an advantage of their new location) in the list of priorities relating to the final decision to move.

Information technology utilisation

If IT figured less prominently than expected in the list of advantages attached to occupying new buildings, how far can that be explained by the actual levels of IT utilisation by City office firms? An IT index and a PC/employee ratio has been calculated for each of the major sectors represented in the study (Table 12.5) .[7]

Financial services companies enjoy the highest IT utilisation level, not only of a broad range of IT products (4.08) but also in terms of a very high PC/employee ratio approaching one to one (0.93). These high figures reflect the functional dynamic of a sector where both money and shares are traded in large volumes using highly computerised dealing systems. Such companies rely for their very existence and competitive advantage upon obtaining and processing large volumes of data to enable them to make the informed judgments necessary to trade in their chosen markets. Banking and insurance rank in second place with respect to their utilisation of IT with an index value of 3.31, but the PC/employee ratio is relatively low, suggesting a different bias in the use and application utilisation of IT (Table 12.6).

Access to IT (per employee) is highest within the banking and insurance sub-sector for fax facilities, mainframe computers, data feeds, private circuits and video-conferencing facilities. The widespread use of large networked computers in banking and insurance would appear to account for the relatively low PC/employee ratio. Some types of IT still border on the insignificant: video-conferencing (although it seems likely that such facilities will become increasingly accessible as costs of hardware decrease, and costs of sending employees abroad, for meetings increases), wide area

Table 12.5   Sectoral IT index and PC/employee ratios

Sector	IT index	PC ratio
Banking and insurance	3.31	0.58
Professional services	2.44	0.43
Financial services	4.08	0.93
Business services	2.83	0.70
Industrials	2.46	0.76

Source: Postal Survey, 1991.

Table 12.6 IT Hardware utilisation per employee by sub-sector and item

IT items	Banking & insurance	Professional services	Financial services	Business services	Industrial
BT phones	1.331	1.234	1.455	0.906	1.193
Mercury phones	0.657	0.835	1.455	0.906	0.420
Fax machines	0.215	0.044	0.044	0.041	0.130
Mainframe computer	0.030	0.006	0.015	0.006	0.004
Training rooms	0.006	0.002	–	0.111	0.008
LANs	0.015	0.007	0.007	0.023	0.013
WANs	0.003	–	0.004	0.006	0.004
Data feeds	0.200	0.008	0.073	–	0.013
Private circuits	1.274	0.006	0.109	0.012	0.013
Computer rooms	0.003	0.002	0.004	0.006	0.008
Video conferencing	0.001	0.001	–	–	–

Source: Postal Survey, 1991.

networks (WANs), dedicated computer rooms (mainly of importance to those companies with mainframe computers), or training and presentation rooms with IT facilities (a relatively new concept which may well grow in importance over time) all have low scores. Professional services consistently reveal a comparatively low level of IT use which is probably a function of the difficulty of computerising the many non-routine aspects of their work. Increasing the level of office automation within this sector will be extremely difficult as solicitors and accountants both rely more on human value judgments than raw information.

Both telephones and fax machines enjoy the most universal usage by office staff even though the latter only really came onto the market during the 1980s (Figure 12.3).

The data clearly demonstrates the increased and continuing importance of telecommunications to contemporary business. The effect of deregulation in the telecommunications sector is demonstrated by the inroads made by Mercury; over 40% of companies possessed a link into the network provided by British Telecom's competitor in 1991. It offers particularly advantageous rates for long-distance national and international calls. Other items of IT hardware, especially those associated with communications, vary widely in their use. Local area networks (LANs) were mentioned by more than 70% of respondents, but wide area networks (WANs) were used by less than 30%. By comparison, the use of computing equipment within offices has become almost ubiquitous during the last 10 years; over 95% of respondents had PCs within the workplace and over 60% had direct access to a mainframe or mini-computer (all companies that responded to the survey had a computer of one form or another). The range of IT facilities available and their application seem closely related to the functional dynamic of the organisation that requires City office space.

Type of hardware

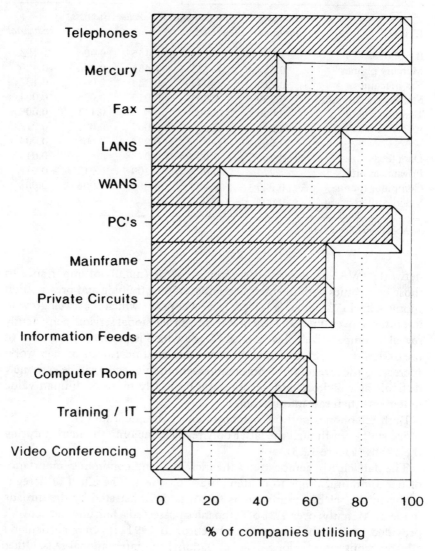

% of companies utilising

Figure 12.3 Levels of utilisation of IT hardware

The highest rated items for the performance of office tasks are telephones, fax machines, computers and data or information feeds (Figures 12.4a and 12.4b).

All these place quite specific and 'weighty' demands upon the design of a building. Not only is the IT hardware itself important, but as a consequence of its widespread use, facilities to allow the full exploitation of new technology are of the utmost importance to users. This means that office occupiers are now more careful in their choice of new premises, examining

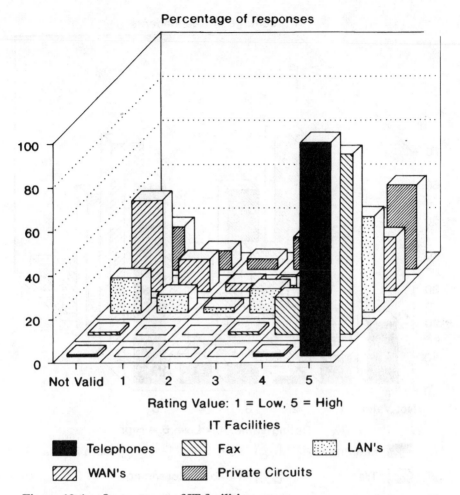

Figure 12.4a Importance of IT facilities

not only cost and appearance, but also air-conditioning, cabling facilities and power generation.

Information technology support services

In terms of the facilities available within premises that are ancillary but essential to the IT operational needs of occupiers, the full range was found to be present in most of the office buildings occupied by the companies in the study. Raised floors, suspended ceilings and clean power supplies are available in over 60% of offices examined (Figure 12.5)

All three are necessary for the efficient operation of banks of IT hardware, especially computers, as they allow, respectively, access for cabling, removal of wild heat and prevention of disruption to equipment. Stand-by power is also available in the majority of buildings; indeed this is becoming widely accepted as a necessary part of the standard specification for new office buildings. High-density telecommunications were found in less than 50%

Percentage of Respondents

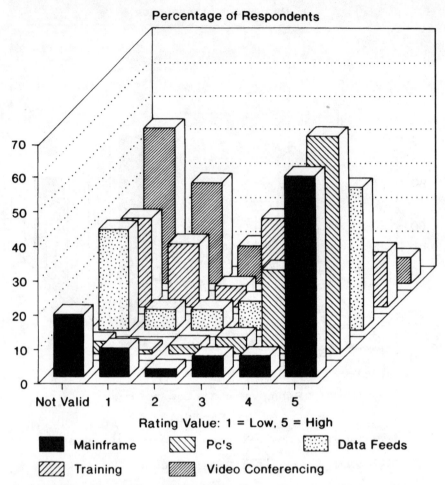

Figure 12.4b   Importance of IT facilities

of premises, reflecting the sectoral differences within the sample financial institutions tending to require this facility much more than administrative head office functions, for example. There are considerable variations in the provision and quality of air-conditioning; variable air volume (VAV) air-conditioning is more responsive to internal office conditions and thus more highly suited to IT-intensive organisations for which 'standard' air-conditioning equipment is inadequate for coping with rapid shifts in the microclimate of the internal office environment.

Respondents were asked to categorise the servicing provision within their buildings into one of three categories: high, medium and basic.[8] Almost 9 out of 10 office occupiers responded to this question with 61.1% indicating that their premises had a high level of service provision. Almost 17% described service provision as medium and 22% as basic. Since the definition of each category provided to assist respondents was quite precise, it is assumed that these responses are representative. This provides further

Support Facility

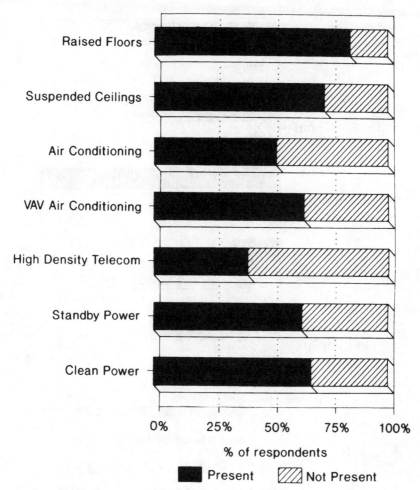

Figure 12.5   IT support facilities

corroboration for the initial assumption that high-quality office servicing facilities are widespread and have become the accepted 'norm' in building specification rather than, as in the past, an optional extra. There are clear differences between sub-sectors in the rating of building service quality (Figure 12.6) with 100% of the financial services firms citing high quality, compared with less than 20% of business services and some 60% of banking, insurance and professional services firms.

Business services are more likely to occupy buildings with servicing of intermediate quality, perhaps because they do not need to make such large-scale investments into state-of-the-art IT in order to provide the service which clients require. Their 'dependent' role in the economy of the City as distinct from the 'lead' or central role of financial, banking and professional

Services Sub-Sector

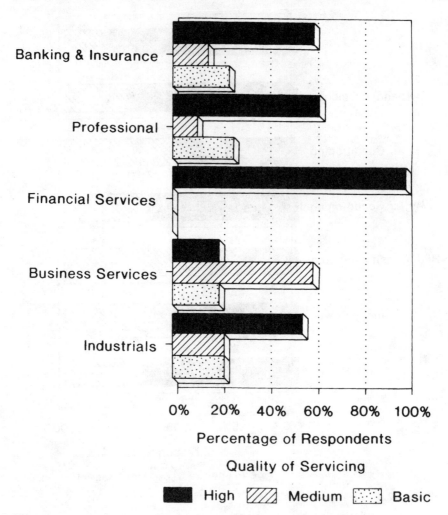

Figure 12.6 **Servicing facilities, by sub-sector**

services, also ensures that they are unable, or cannot afford, to compete for the highest quality or best-located office space.

For the most part, occupiers are satisfied with their premises, with the largest proportion rating a number of attributes as good or very good (Figures 12.7a and 12.7b).

There are, however, two notable discrepancies: operating costs and overall satisfaction with the building. The issue of operating costs, which embrace a variety of elements including energy (gas/electricity), business rates, service charges and perhaps most costly of all, rents and service charges, is a particularly notable problem. High rental costs in central

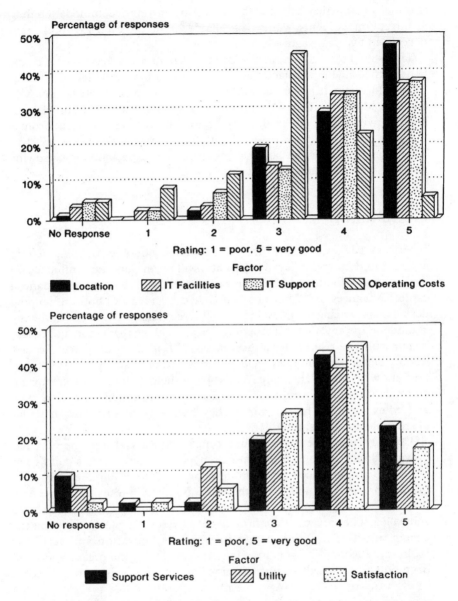

Figure 12.7a,b Occupiers' assessment of premises

London have long been a source of complaint by tenants, and especially those signing leases during the late 1980s when rental levels were either at or approaching their peak. During the period 1989–92, which includes the survey reported here, it became increasingly apparent that occupiers were highly disgruntled about the lease structure for offices in the City, with 25 year leases and upwards-only rent reviews the norm. Such is the depth of the current property recession and the glut of available (new) office space that many companies now find themselves occupying accommodation that, at current market rents, would be available for 50% or even 33% of the rent negotiated two years ago.

The overall satisfaction rates of occupiers tend towards good or average, belying their higher satisfaction rates for certain individual factors such as IT and location. In large part, the lower overall level of satisfaction with office buildings can be traced to operating costs and a continuing discrepancy between what office developers assume that clients want from a building, and what the occupiers themselves want (see also Harris 1991b). That being said, the gap between provision and expectation is narrowing due to the work of companies such as DEGW.

Conclusion

It has been suggested that structure and servicing of office buildings within central London, and especially the City, will be an important influence on the future development of London as a national and international financial centre. Advances in IT have been matched by advances both in building design and servicing facilities; indeed, all three elements have been evolving in tandem. Improvements in building design, making them more utilitarian, began with the rise of the importance of IT in the workplace. Several leading developers commissioned research during the early 1980s to try to find out what occupiers really wanted from their buildings and then went on to act upon it. IT suppliers have also been examining their client base and what servicing requirements they might need for their particular operational configuration.

The survey of recently relocated offices within and into the City has provided some support for the working hypothesis but it seems that IT has not necessarily replaced other, more conventional, requirements such as prestige or the right address in the office building choices made by firms that are operating in international markets and are often foreign-owned and controlled. The survey has confirmed that IT does influence in extremis the design of office buildings, their internal configuration and servicing facilities, reflecting the pervasive influence of IT on the organisation and operation of all the sub-sectors examined in this study; IT has now become as essential to many firms as members of staff. While the levels of IT utilisation is shown to vary between sub-sectors, there are certain minimum requirements that buildings must meet in order to satisfy tenant demand, failure to provide such facilities make a building virtually unlettable.

Meanwhile, London competes not just with its partners in the 'Golden Triangle' and with other secondary centres around the globe but also with locations on mainland Europe. Paris and Frankfurt are the two centres most often cited as the main competitors to London. Both have been marketing themselves aggressively and have been undertaking massive building programmes both of office space and infrastructure to hone their competitive edge over London. In both cities, rental costs and overall occupation costs are lower than London, although the contrast is now not quite so marked since, as a by-product of the worldwide recession, there is now a large volume of good quality vacant office space (over 5 million sq ft/464 000 m^2 in April 1993), consequent large reductions in rental levels and a more flexible attitude towards leasing arrangements in the latter. London's other crucial advantage is its comparatively loose regulatory environment which is generally regarded as more favourable for attracting and retaining international service and manufacturing office functions, but the introduction of the Single European Market (SEM) in 1993 is modifying the playing field.

In terms of tone, comparisons between London and its potential competitors tends towards examining its flaws rather than highlighting the limited 'absolute' advantages of other centres. London has been consistently criticised by both the media and researchers for its poor infrastructure and high costs (LPAC 1991). These sources of criticism would seem to be enough to encourage those public bodies with responsibility for such matters into some form of action. However, most official responses lack vision and offer very few cast-iron commitments. The poor and deteriorating transport infrastructure could well contribute as much to the demise of the City as the quality of its office space. The key may not be to allow the current recession to lull developers and policy makers into a false sense of security about the City's ability to compete on the basis of its current attributes when, as it surely will, the demand to locate there expands again. Virtually without exception (and usually unprompted) office occupiers and interviewees who participated in the present survey were highly critical of transport costs and reliability, as well as the costs of operating a business in London. Perhaps the time is right (at a time of declining public transport ridership and of office space demand) to embark on the infrastructural investments that will position the City in the most advantageous way possible for the late 1990s.

NOTES

[1] Since the demise of the Greater London Council in 1986, the London planning system has been in disarray. London is one of the few global cities without a strategic planning authority (LPAC 1991) and, although various unitary development plans (UDPs) are in preparation, strategic planning guidance has only been provided by the Department of the Environment (DOE 1989). There is plenty of scope for confusion of roles, responsibilities and relationships with London First; the London Transport Board; London Buses Ltd; London Forum; London Underground Ltd; the traffic director for London; the commissioner of police for the metropolis; the parking director for London; 32 London borough councils; the City Corporation; the London planning advisory committee; serplan, nine training and enterprise councils; four DOE task forces; the Docklands Development Corporation; parts of four regional health authorities; 50 district health authorities; health service trusts amongst just some of the bodies with operations covering all or part of the metropolitan area. Major land-use issues for London are resolved at the local level, there is no integrated transport strategy;

there is no mechanism for securing a world-beating financial services sector in the City (or in London as a whole'.

[2] This information is usually supplied to other organisations on an on-line basis at some substantial cost. Since it was to be used for academic research, the information required was supplied free of charge on the understanding that it would under no circumstances be disseminated to third parties.

[3] Subsequent quality checks showed that there were some entries that were unusable and this reduced the working data set to 368 records.

[4] The overall survey response rate was 35%; 65% of those mailed failed to respond in any way. Only 7.9% of those mailed refused to complete the questionnaire (citing confidentiality or pressure of work as reasons), 22% returned usable forms, and the remainder (5.1%) returned forms that, for various reasons, could not be used. A total of 82 fully completed questionnaires was received.

[5] Classification of enterprises for the survey:
Banking and insurance:
8140, Foreign bank; 8140, UK clearing bank; 8150–1, UK merchant bank; 8200, Insurance, Insurance company, Assurance company; 8200, Reinsurance market; 8320, Insurance broker, Actuary.
Professional services:
8350, Solicitors; 8360, Accountants; 8370–1, Architects , Quantity surveyors, Consultant engineers; 8370–1, Design consultants; 8380, Advertising agency.
Financial services:
8150–2, Investment management; 8150–2, Futures and options; 8150–2, Broker/dealer; 8310, Stockbroker; 8310, Money broker; 8310, Financial services
Business services:
8394, Computer hardware; 8394, Computer services, Computer software; 8395–1, Management consultants; 8395–2, Public relations; 8395–4, Information services; 8395–4, Telecoms
Industrials:
8396*, Headquarters of manufacturing, public sector concerns.
* This categorisation encompasses the office functions of companies involved in primary, secondary and tertiary sectors not classified elsewhere.

[6] The decision to relocate office premises is explored via questions that have been so structured as to provide information on three aspects of the relocation decision: the initiation of a decision to relocate; the final decision to move to a particular location; and an analysis (by the respondent) of the perceived benefits of the new location for that particular company

[7] Respondents were asked to indicate the numbers of items of different categories of IT equipment within their organisations, and also to provide a PC/VDU ratio, i.e. the number of PCs/VDUs per person employed. From the number of IT items available within a given company, and knowing the number of employees, a ratio (per person) was calculated for each item of IT by dividing the number of items by the number of employees.

[8] The servicing facilities are categorised as follows:
- **High volume VAV** air-conditioning with zoning control down to 200 sq ft (18.6 m^2), clear raised floor of 6 inches (150 mm), more than 50% of net floor area can be highly serviced.
- **Medium capacity VAV** with zoning control down to 400 sq ft (37 m^2), clear raised floors of 6 inches, 25% of net floor area can be highly serviced.
- **Basic capacity VAV** controlled floor to floor, three compartment trunking, less than 25% of net floor area can be highly serviced.

[9] A grant from the ESRC (No. R00023 2040) to support the research reported in this chapter is gratefully acknowledged. Particular thanks to Mark Bobe, Service Industries Research Centre, University of Portsmouth, for his invaluable assistance in conducting the project.

PART *3* The Creation of Technopoles

W HAT COMES FIRST, technology firms or technology parks? Urban economists have been engaged in a two-decade long debate over whether the place or the people or some combination of them can explain the *agglomeration* of high technology firms. At the most simplistic level, *high technology* should be like any industrial form. That is, there are a certain set of physically defining *locational endowments* that are prerequisite to the success of the firm. This concept was modified as industrial technology became ubiquitous and manufacturing firms were able to relocate to take advantage of *labour force endowments* rather than merely *physical assets*. Economic geographers and regional scientists constructed sophisticated models to demonstrate the *regional life cycle* of manufacturing firms. These models formed the backdrop of modern economic development theory. It is clear that the theoretical constructs devised to support a domestically oriented manufacturing economy fall very far short in explaining the current new/high technology regional development patterns. Nonetheless, regional scientists remain wedded to models that do not fit the contemporary conditions nor explain economic development prescriptions.

The evidence with regard to high technology locational patterns is not disputed. High tech firms are congregated in a few metropolitan environments that are rich in *intellectual resources*. Saying this is not saying very much. What conditions are present in these localities that are not present in other places? Why do some regions grow within the same state and other areas lag? Why do some areas spawn or grow new technology firms and others do not?

It is this last question that policy makers and economic development professionals focus on in their debates on development options. This question is at the heart of the Japanese technopoles, Sophia Antipolis, the Australian Multi-function Polis (MFP), and a myriad of similar new settlements all over the world. These new settlements are intended, according to their proponents, to create *entrepreneurial and innovative environments*. In essence, these new settlements are intended to recreate artificially what Silicon Valley and Route 128 did spontaneously. Obviously, this macro-planning approach to simulating new enterprise formation has its critics. However, even with a short track record it is apparent that some of these new technopoles are enjoying moderate to quite spectacular success. Again, the question is, are the successful technopoles in successful places or places where the same thing could have and would have occurred without any external intervention? While the debate continues on this topic it is important to distinguish between place and firm success. Nevertheless, it is becoming increasingly clear that place or locality matters a great deal in high technology firm success. It is this issue of *place versus firm* that forms the centre of a very hot debate among regional planners.

The notion of creating a place started with the tech park. 'Tech park' has now been expanded to tech city and tech region. This was a natural evolution. The reason for the movement from park to whole community relates to the debate among regionalists over whether the place attracts or induces the firms. This concept was carried to its logical conclusion when the Japanese decided to alter their entire urban fabric to stimulate techno-logy development. The Japanese reasoned that the *'soft structure'*, not the *hard infrastructure,* determined high technology economic development success. The soft structure is composed of all of the social and community institutions that give life and meaning to a community. The only way to develop the correct 'soft structure' is to control the total community environment. Thus, the Japanese technopoles are an outgrowth of regional development theory. The chapters in this section are based on the concept of 'soft or regional institutional structures'. Technopoles are, therefore, a state of mind as well as identifiable places.

The state of mind issue is central to Kunzmann's analysis of Ruhrgebiet, Germany (chapter 14). He shows that the social structure of this predom-inantly iron and coal area had been very adaptive until international restructuring occurred. Now the region is in a deep crisis because its old institutional base of government and large private firms cannot respond to a new innovative technology-based economic order. He describes the need to unleash the region's creative potential by redesigning its social insti-tutional base.

The Australian MFP presented by Hamnett (chapter 16) represents the epitome of the technology soft structure concept. Hamnett shows how wishful thinking, as opposed to creative problem solving discussed by Kunzmann, is not nearly enough. The MFP is not likely to cross all of the hurdles necessary to bring it into existence, for two reasons. First, Hamnett shows that the Australian government placed too much emphasis on the location of the MFP and too little on the institutional and organisation's supports for technology development that form the base for incubating new technology firms. Second, Australians are suspicious of the foreign enclaval potential of the project. On the other hand, the MFP (a new suburb) represents an exciting venture for any city in the world. Yet, Hamnett believes that the project was doomed almost from the start by high levels of cynicism and suspicion because it was inspired externally by the Japanese and had few local roots. However, in spite of the flaws in the project, few projects in the world today have gained as much name recognition and generated as much interest as the MFP. One can only speculate as to the reason for this. Could it be that many other MFPs are in their planning stages around the world?

If Taiwan is used as an sample, then the answer is clearly yes. The chapter (15) by Blakely discusses and describes the design and development of a new high technology city in Taiwan. This city, Hsinchu, is meant to create a new international 'soft structure' for Taiwan. The idea is that Taiwanese science is handicapped by the enormous brain drain to the United States. These brains will only return to Taiwan if the physical and social environment reflects the best of a high technology community. The scale of this high technology development is as large as the Australian MFP. It is meant to be an international community too. The Taiwan government is struggling with the physical form and internal organisation of the Hsinchu Science City. However, unlike the Australians, the Taiwanese want internationalisation. Ironically, while the Hsinchu Science City project is a success on many dimensions, it has yet to develop the kind of soft structure that is attractive to the internationalists it seeks.

Willoughby's chapter (13) provides the answers for both the MFP and the Hsinchu Science City project. He develops a clear, theoretical base that moves the concept of soft structure to a more empirically and sophisticated conceptual base with his '*innovative milieu*'. The innovative milieu, he says, requires, '...a self-sustaining economic capacity [of] local networks of dedicated and...complementary organisations...' This network, he demonstrates through his work on competitive biotechnology regions, has shape and form through a dense system of supporting institutions. Willoughby argues that it is impossible to induce the development of high technology firms through the usual physical development of tech parks or direct government actions such as tax breaks. Rather, he suggests that an innovative milieu must be designed as the underpinning for a form of high tech development. As he says,

> ...there is no magical 'key' to turn which will guarantee success. It is not possible to just create the desired development (high tech firms), it is only possible to create the conditions which may indirectly lead to the desired end product.

The desired end product for most of the world is a high technology community that produces wealth and very little social or physical by-products. As both Boston and San Francisco can attest, neither one of these conditions can exist for long unless the institutional network is strong enough to fashion both a physical and industrial future.

13 The Local Milieux of Knowledge-Based Industries:
What Can We Learn from a Regional Analysis of Commercial Biotechnology?

Kelvin W. Willoughby

Introduction

A lively debate has recently emerged in the academic literature over the relationship between technological change and regional form. This debate has been accompanied by a corresponding array of practical experiments in the policy arena. The rise of various 'high technology' regions, particularly in the United States (e.g. Silicon Valley), has stimulated efforts by national, regional and local governments throughout the world to emulate the formation of similar regions, or high technology industrial nodes, within their own territory. Instruments devised for this purpose include technology parks, innovation centres, training programmes, taxation supports, targeted research funding, regulatory streamlining, direct subsidies to firms, or special financial schemes to aid small start-up high technology firms (Carter 1981; Blakely & Shapira 1984; Whittington 1985; Schmandt & Wilson 1987).

Interest in the relationship between technological change and regional form has become even more pronounced with the emergence of bio-technology. Alongside popular concern about the social impacts of biotechnology (see Perpich 1986) interest in its economic dimensions has grown (e.g. Hacking 1986), with some research addressing the local economic development impacts of this emerging field of technology (Feldman 1983; Hall et al. 1988). Some scholars (e.g. Kenney 1986) have observed that the biotechnology industry differs from other industries in critical ways, such as the pivotal role played by university professors in commercial ventures. Biotechnology is widely perceived to be less dependent than other forms of new technology upon a pre-existing industrial base, and therefore as having the prospects for ubiquitous development. Many communities which failed to become substantial players in the rise of micro-electronics look to biotechnology as a source of industrial hope, believing that they may possess the requisite ingredients to become successful 'biotechnology regions' (Blakely & Nishikawa 1989). Using biotechnology as a case (and one prominent international example—

the New York biotechnology industry), this chapter will explore the links between regional form, local milieux and the emergence of industry based upon new technology.

The chapter begins by discussing the convergence of regional studies and technological innovation studies, and by reviewing the literature on high technology and regional form. After briefly introducing biotechnology as an industry, and documenting the basic geography of the industry in New York, we will examine the locational behaviour of biotechnology firms, and the importance of regional context for the incubation of a local biotechnology industry will be explored. This analysis is used to develop some simple principles, based on the concept of the local biotechnology milieu, to guide regional economic development planning related to biotechnology.

The convergence of regional studies and technological innovation studies

A sizeable body of scholarly literature was published during the 1980s around the subject of technological change and regional economic development (Malecki 1981; Thwaites & Oakey 1985; Amin & Goddard 1986), with much of the research concerned primarily with the spatial aspects of high technology. The reasons for the emergence of the debate over technological change and regional form, together with the concomitant policy experiments, were varied.

1 Technological innovation was recognised as a determinant of the economic performance of industrial firms and sectors (Hill & Utterback 1979; Pavitt 1980; Dosi 1984; Rothwell & Zegveld 1985; Freeman 1987; Teece 1987). Changing market conditions, increasing costs of industrial inputs, greater emphasis on information flow as part of the economic process, more complex trading patterns, complex regulatory requirements, and sophisticated product-standard environments—to name some of the key pressures facing contemporary businesses—were seen to place a premium on an organisation's competence in adopting and managing new technology. Technology, furthermore, came to be seen less as something which emerged miraculously out of the 'black box' of science and engineering, exogenous to the processes of the economy, but rather linked with the economic and managerial context (Rosenberg 1982).

2 Growing international competition and interdependency in trade gave technological innovation an even higher profile in economic policy, and firms came under increasing pressure to innovate in order to remain in business (Granstrand 1982; Rothwell & Zegveld 1983; Zysman & Tyson 1983; Krugman 1986; Furino 1988). As a consequence, the strategic management of technological innovation became an important component of corporate management, and most national and provincial governments established some kind of ministry concerned with technology policy.

3 During the 1980s the phenomenon of 'uneven development' received considerable scholarly and political attention. The uneven distribution of wealth has been debated widely since the classical work of Adam Smith and Karl Marx, but some recent manifestations of this scholarly tradition, fed by contributions from a number of disciplines—including political economy, geography, city and regional planning, and sociology (e.g. Massey & Allen 1988; Marshall 1987)—has exhibited a strong spatial emphasis, stressing the theme that the economic disparity between regions and within regions exhibits structural features, changing in consonance with national and international macro-economic forces.

4 Uneven participation in state-of-the-art technology development and application came to be seen as an explanation of uneven economic development between and within regions (Armington et al. 1979; Maillat 1982; Oakey 1984; Office of Technology Assessment 1984; Chapman & Humphrys 1987; Hamilton 1987; Sharp & Shearman 1987; Oakey *et al.* 1988; White *et al.* 1988; Willoughby 1990). Some of the research dealing with this theme was based on particular technology-based industry sectors in particular places, such as microelectronics in Britain (Morgan & Sayer 1988), but a body of literature also emerged aimed at producing concepts or policy principles which transcend particular geographical regions, fields of technology and industry sectors (Sweeney 1987; Storper & Walker 1989).

5 As a consequence of the above themes emerging within scholarly debate, cities, or urban regions, became recognised by some scholars as the locus for leading-edge technological development, with a number of prominent 'international' cities or regions receiving the greatest attention: for example, the San Francisco Bay Area, the greater Los Angeles region, Cambridge in Massachusetts, Tokyo, or Cambridge and the M4 Corridor in Britain (Saxenian 1983; Hall & Markusen 1985; Segal Quince Wicksteed 1985; Boddy et al. 1986; Tatsuno 1986; Hall *et al.* 1987; Scott 1988a).

6 Given the prominence of a relatively small number of 'international' high technology regions, and their apparent interdependence, scholars sought to understand both the ways in which advanced technology industries affected regional form, and the ways in which regional form affected the prospects and form of local advanced technology industry complexes (Brotchie et al. 1985; Brotchie *et al.* 1987; Aydalot & Keeble 1988a; Tarr & Dupuy 1988). No generally accepted theory has yet been distilled from these efforts, but a consensus does appear to have emerged that a shift from an industrial style of economy (with its emphasis on the flow of resources and goods, and the accumulation of tangible assets) to an advanced industrial style of economy (with its emphasis on the flow of information and the accumulation of knowledge) will be accompanied by a shift away from the '19th century agro-industrial' city form (with its simple centre–periphery land use patterns) to something more complex and probably more decentralised.

The convergence during the 1980s of two fields of scholarly endeavour—technological innovation studies and regional studies—has been mirrored in the national policy arena, with the emergence of deliberate efforts to create modern cities in which high technology and its associated social forms may flourish. Examples include the 'technopolis' regions in Japan, and the 'multifunction polis' idea in Australia (Glasmeier 1988a; Mandeville 1988; Masser 1989).

In the wake of these developments, some analysts have sought to produce general theory to describe and explain the confluence of regional and technological change. Accordingly, the notion of a 'high technology regional form' has appeared in the literature. Something of the spirit of this notion is reflected in the following extracts from a recent futuristic paper in this field, with a focus on North American cities (Gappert 1987):

> A new national economic expansion driven by information-intensive technologies and the extension of global business services should be under way by the early 1990s...Cities will be more polynucleated, with the development of more multiple-use megastructures, and medium-density planned housing unit developments...Increasing leisure and use of telecommunications will facilitate increasing low-density developments in which residential, work, leisure activities...and other local life-support activities are integrated, and increased emphasis on lifestyle and quality will ensure an increasing range and diversity of these developments within, at the periphery and beyond the urban area. The new affluence (for some) created by new technology will further add to this diversity and to the range of spatial development activities including global networks and virtual (global) cities...The cities of an advanced industrial society—the future metropoli —will be primarily engaged in indirect, and partially abstract, transactional activities, and may be hungry for collective rites to offset social fluidity, economic transience and electronic isolation.

This 'high technology regional form' notion, although normally tacit rather than explicitly articulated as formal theory, suggests that there is a typical pattern in the way urban/regional spatial patterns and the patterns of high technology industries coalesce. Walker (1985, p.226) observes this thematic development as follows:

> [T]echnology has come to be viewed by the public as the key to the magic kingdom of regional development and national competitiveness...It is not surprising, therefore, that various kinds of technological determinism have found their way into the regional debate, such as the notion that high tech industries have a unique locational pattern, that R&D centres are crucial to local growth because of their innovative function, or that the product cycle dooms older industrial regions to imminent stagnation.

Thus, high technology industries seemingly emerge in particular types of regions exhibiting a particular spatial structure, with the development of those industries subsequently exerting influence on the region and reinforcing the spatial features which first led to the flourishing of those industries. The result of these mutually reinforcing tendencies is that once a region becomes established as a high technology region, it develops an international competitive advantage. Conversely, cities or regions which lack the appropriate structure find themselves increasingly bypassed by advanced technology industries and employment.

There are practical policy implications of this perspective.

- Civic authorities and their advisers in high technology regions may adopt such a perspective in planning the 'urban' infrastructure most fitting to the evolving industrial base of their economy (e.g. transport facilities, housing developments, zoning requirements, project development regulations, educational institutions, communications facilities).
- Managers of advanced technology firms may be indirectly influenced by such theory when making decisions about the location of their activities.
- Regional and city policy makers wishing to improve the economic prospects of their region may look to such ideas to guide the adoption of policies aimed at altering their comparative economic advantage.

It is therefore important for this incipient theory to be closely examined in the light of empirical evidence.

Much of the research upon which this 'high technology regional form' theory rests has been based on a small sample of supposedly paradigmatic regions (e.g. Silicon Valley in California and Route 128 in Massachusetts) or on multi-region studies of high technology in general, or information technology in particular. *Is a simple 'high technology regional form' notion defensible when a wider diversity of regions and industries is considered, or is it largely a reflection of the limited research base from which it has emerged? More generally, in exploring whether there are single or multiple regional forms which advanced technology industries might adopt, might it be more productive to focus on internal-intangible aspects of regional milieux rather than external-tangible aspects of regional environments?*

This last question is of great importance, for example, in discerning prospects for cities in the Pacific Rim region. The Pacific Rim is home to a great diversity of cities and countries, each with different economies, demographics, cultures, resources and historical experiences. If multiple urban forms are possible with advanced technology industries, then the unique features of each Pacific Rim city become critical in the formulation of policy for the development of competitive metropolitan economies. If, on the other hand, there is only one 'high technology regional (or urban) form' then attempted imitation of the leading regions, such as Silicon Valley, would appear to be the most advisable policy option.

The literature on high technology and regional form

The recent debate over factors determining the location of high technology industry emerged against the backdrop of traditional location theory for manufacturing industry. This body of theory, 'Weberian location theory' (following Weber 1929), points to transportation costs as the key determinant of optimal industrial location decisions, with firms weighing the relative transportation costs of access to raw materials, labour and markets. Within this framework certain regions emerge as most economic for certain industries or firms because of their apparent capacity to minimise net transport costs. Once firms cluster in one of these optimal locations, agglomeration economies emerge, thereby reinforcing the existing economic advantages of the location for the particular industry in question. Variants of this type of theory have held sway until quite recently and have been reinforced by the observation that, both in Europe and in North America, the dominant trend in industrial location has appeared to be spatial concentration (Aydalot & Keeble 1988b, pp.1–2).

During the 1970s the capacity of traditional location theory to comprehensively explain industrial location patterns was progressively questioned in the face of the decline of traditional industrial regions and the rise of new regions linked with emerging industrial forms. This was symbolised through reference to the rise of 'sun-belt' cities based on 'sunrise' industries (Parry & Watkins 1977; Weinstein & Firestine 1978; Bluestone & Harrison 1982; Sawers & Tabb 1984). In contrast to the perceived general pattern of the previous half century, spatial dispersion emerged as the new emphasis in industrial geography. It appeared that throughout the industrialised world dispersion was superseding *concentration* as the key trend in industry location, and that this new trend also extended beyond the boundaries of the main industrialised countries into the newly industrialised countries of the Pacific Rim (Keeble 1976; Castells 1986; Breheny & McQuaid 1987).

Technological change emerged as a variable intimately linked with these economic and industrial-geographic changes. The development of new technological products and processes (particularly in the area of information handling and communications) was seen to provide the means for overcoming traditional physical or economic constraints to the spread of industrial activity, both between cities and within cities. Some commentators have sought to explain this by minor modifications to traditional regional growth and industrial location theories (see for example Rees 1986). Others have sought to introduce new concepts, such as that of the 'informational city', whereby 'space' is construed as the flow of information rather than as a geographical place (Castells 1984). In other words, the use of advanced technology is argued to enable decentralisation of many industrial activities from the core to the periphery, while still maintaining the possibility of control and coordination from the centre. Castells (1985, p.12) summarises the new perspective as follows:

> The most direct impact of high technology on the spatial structure concerns the emergence of a new space of production as a result of two fundamental processes: on one hand, high technology activities become the engine of new economic growth and play a major role in the rise and decline of regions and metropolitan areas, according to their suitability to the requirements of high tech production; on the other hand, the introduction of new technologies in all kinds of economic

activities allows the transformation of their locational behaviour, overcoming the need for spatial contiguity.

Thus, by the use of information technology, a firm is able to concentrate functions of the organisation while simultaneously dispersing the total organisation by locating various parts of its activities in geographical locations best suited to each respective function or the organisation's overall strategic goals. Some scholars have applied this insight to inter-metropolitan location decisions (Gordon & Kimball 1986), and some to intra-metropolitan location decisions (Scott 1983a; Blakely & Fagan 1988).

Despite the purported 'footlooseness' of high technology industries, such industries have in fact emerged in certain key geographical regions, the most famous of which is in Santa Clara county in California (Silicon Valley). Worldwide, the development of high technology regions has been rather uneven, with the result that much debate has emerged over just how feasible it is for more than a small number of such regions to thrive (Glasmeier *et al.* 1984). The phenomenon of high technology regions has once again raised the theme of industrial *concentration* into prominence (Swyngedouw 1989; Scott 1988c). Given the evidence of some urban areas emerging as clear leaders in high technology, and given that early entry into such activity may provide a competitive economic edge to those places, some commentators have argued against the view that the wide uptake of high technology will diminish the importance of geographic location for industries. The idea that new technology is likely to entrench the dominance of a handful of principal world cities is now quite established in the literature (e.g. Moss 1987). Thus, along with the theme of concentration has come the recognition that high technology regions tend to be located in urban areas.

Many attempts have now appeared to create profiles of high technology regions in the hope that they might form the basis of fruitful policy initiatives by city and regional governments (e.g. Herbig & Golden 1993). Saxenian (1989a, p.2) reports that the following features generally emerge from such studies as definitive parameters of high technology regions:

1 a high calibre research university to ensure a science base and a supply of scientists and engineers;
2 an ample supply of venture capital to fund new firms;
3 public investment devoted to research and procurement of new technologies;
4 a quality of life able to attract and retain footloose highly qualified professionals;
5 the absence of trade unions;
6 an industrial park to house start-up firms; and
7 adequate infrastructure to ensure efficient transportation and communication linkages.

Saxenian (1989a, p.2) wryly observes:

> The underlying message—though rarely stated—is that once these pre-requisites are assembled innovation and growth will follow. Like a soufflé which exceeds

the size of the initial ingredients, a region endowed with the proper mix of institutional and economic resources will be the lucky recipient of rapid high-tech growth.

Much of the literature on the nature of these high technology regions also appears to tacitly presume that there is such a thing as a *typical* high technology region; or, that with enough research, it might be possible to develop a single universal law of high technology development, capable of accounting for the evolution—or non-existence, as the case may be in some places—of high technology regions. One of the most ambitious large scale statistical–empirical studies of the location of high technology industry in the US, however, was unable to find evidence for such a general law, beyond the observation that the location of military spending appeared to be significant (Markusen *et al.* 1986).

Notwithstanding the difficulty of the challenge, some helpful contributions towards general theory in this field have been published. One line of research seeks to explain the geographic concentration and dispersion of high technology industry in a dynamic way by using product–profit cycle theory; thus, during early stages of the cycle, high technology firms need to cluster in high technology regions to take advantage of services which they are unable to provide internally, but this requirement declines in importance as the industry or firm matures, and dispersion takes place to enable firms to take advantage of regions which offer lower costs (Markusen 1985). Another line of research views high technology firms as involved in networks of transactions, with some firms highly disintegrated functionally and some highly integrated, reflecting the relative costs of internal and external transactions. Accordingly, high technology regional nodes ('technopoles') emerge as the spatial convergence of vertically disintegrated producers under conditions of uncertainty (Scott 1983a; 1988c; Scott & Angel 1987; Scott & Paul 1990). Another line of research has concentrated on the access of firms to financial resources; Florida & Kenney (1988) have demonstrated through their research in the US that the venture capital industry, which itself tends to agglomerate regionally for such reasons as the information-intensive nature of the investment process, appears to play an important role in facilitating agglomeration in high technology industries. In addition, through placing attention on the human and organisational processes by which technical knowledge is generated (in the case of the biotechnology industry in California), Willoughby & Blakely (1990; cf. Blakely & Willoughby 1990) have assembled evidence for the economic importance of concentrated local industry clusters.

Most of the recent research seeking general explanations of the spatial patterns of high technology industry has provided evidence for the importance of regional concentration rather than dispersion in advanced technology industries. It has also provided evidence, however, that individual industrial groups—whether based on high technology or otherwise—exhibit distinctive characteristics, with likely distinctive spatial tendencies. The theme of the variability of the spatial concentration of advanced technology industries, both between regions and between specific high technology industries, has thus emerged. For example, Felsenstein & Shachar (1988) have shown that metropolitan location is important for both small and large high technology firms in Israel, but for different reasons in each case. Davelaar & Nijkamp (1989) have discovered that spatial factors are associated significantly with the performance of high technology firms in the Netherlands, but with important distinctions in the importance of highly urbanised locations between whether the focus is on process innovations or product innovations. Amrhein & Harrington (1988) have assembled evidence that technological heterogeneity in an industry can lead to variations in the locational inertia of firms. Glasmeier (1988b), through a series of case studies in Texas, has observed that the development of high technology industry agglomerations, and the nature of their economic spin-offs, vary according to the product type and the organisational structure of firms.

To summarise, we may say that the dominant literature on technology and regional form which appeared during the second half of the 1980s contains a number of ambiguities and unresolved tensions. These may be thought of as dyadic themes, described in Table 13.1, lying along four theoretical dimensions: geographical contiguity, structural uniformity, locational determinism, and causal dynamism.

Within the first theoretical dimension, *geographical contiguity*, the literature exhibits tension between data and arguments in favour of the idea that new technology encourages the dispersion of industry within regions and beyond regions, and those in favour of the idea that technological change concentrates industrial activity into local geographical clusters. It is not always clear whether the subject of the debate is organisational control, manufacturing operations, intellectual capital, or human activity—and this ambiguity exacerbates and confuses the ongoing debate even further—but the tension between the two themes garners much attention.

Within the second theoretical dimension, *structural uniformity*, the tension in the literature is between the idea that there is one universal pattern to

Table 13.1 Theoretical tensions in the literature on technology and regional form

Theoretical dimension	Theme	Counter-theme
Geographical contiguity	Spatial dispersion	Spatial concentration
Structural uniformity	Homogeneous regional form	Heterogeneous regional form
Locational determinism	Single primary factor	Multiple factors
Causal dynamism	Direct linear causality	Indirect non-linear causality

the regional form of technological change and the opposing idea that the regional form may vary between places and industrial contexts. For example, amongst those who embrace spatial dispersion as the natural concomitant of technological advance, disagreement may remain over the question of whether or not this relationship may be observed uniformly across national or cultural boundaries; those who would argue in the affirmative would fit in the 'homogeneous regional form' category and those who would argue in the negative would fit in the 'heterogeneous regional form' category.

The third theoretical dimension, *locational determinism*, concerns the question of whether or not it is appropriate to describe one particular factor as the primary determinant of the spatial behaviour of high technology organisations or whether, instead, multiple locational factors ought to be seen as significant. For example, an argument to the effect that high technology firms relocate over time to low-cost areas fits within the 'single primary factor' category because it sees the drive for cost reduction as a singularly important determinant of the location of industries as they mature; conversely, those scholars who avoid the temptation to reduce industry locational dynamics to a simple universally observable homogeneous process would probably fit into the 'multiple factors' category.

The fourth theoretical dimension, *causal dynamism*, concerns the question of whether simple linear causality is even legitimate as the mechanism by which 'determining factors' influence the spatial behaviour of organisations and other actors in technology-based industries. It also raises questions about whether or not it is even appropriate to speak of an identifiable logic of industry location related to locational factors. This fourth theoretical tension is rarely, if ever, discussed explicitly; but most writings in the field tend to presuppose a bias towards one of the two theoretical positions. Those studies which seek to assemble a recipe of locational features conducive to high technology industry development tend to fit into the 'direct linear causality' category, and are differentiated by whether they look to a 'single primary factor' or to 'multiple factors' in explaining locational determinism.

In other words, they see factors in a location (external-tangible aspects of regional environments) as actually 'causing' the locational patterns observed. Research which abandons the direct linear causality presupposition is not common, but is emerging as part of the literature on 'innovative milieux' (to be discussed below). It is distinguished by the theme that even if particular features of regions could be associated with high technology industry locations, it would not follow that those features (or supposed 'locational factors') actually 'caused' the observed locational patterns. To the extent that some kind of locational logic might be observed to operate, those commentators fitting within the 'indirect non-linear causality' category would look more to internal-intangible self-reinforcing processes of the local milieu, than to external-tangible aspects of regional environments, as explanations of spatial behaviour.

It is not easy to attach particular scholars in a rigid manner to one theme only, as some appear over time to have subtly shifted their thematic stance and others may even exhibit both theme and counter-theme during the same period. It is perhaps also inappropriate even to speak of schools of thought based upon these themes. Rather, the themes run throughout the literature providing the fuel which has kept much of the debate going. An appropriate question to ask at this juncture is whether current research appears likely to resolve the tensions described above, or whether scholars will instead seek more productive debates during the 1990s, leaving the present theoretical ambiguities to be pondered upon by future historians of the field.

The future of today's high technology regions

It is not yet clear what the answer to the above question will be, but by the beginning of the 1990s several new themes have emerged in the literature, each representing a gradual shift in orientation from the top left corner of Table 13.1 towards the bottom right corner. That is, along with a recognition that heterogeneity of regional form may occur simultaneously with the spatial concentration of industry into local clusters, has come a recognition that the 'locational factors' approach to the industrial geography of high technology, with its focus on a narrow range of determining factors and its simplistic notion of system causality, may have been misdirected. Rather than seeking to identify 'what causes firms to locate' in a particular place, recent scholarship has come to focus more on the dynamics of the local industrial complexes. *The Holy Grail of the 'ultimate causal factor' has been superseded by the Round Table of the 'multi-dimensional evolutionary process'.*

The new themes in the literature emphasise organisational and institutional issues in industry, and utilise a dynamic perspective in theory about the connection between new technology and regional form—although this latter perspective, in itself, is not new (cf. Markusen 1985). These emphases appear to have been stimulated by two developments:

- widespread concern about signs of decline or loss of competitiveness of the hitherto pre-eminent high technology regions; and
- the emergence of the literature on flexible specialisation in manufacturing (Piore & Sabel 1984).

The first development has led to a shift from the search for general explanations as to why existing geographical-technological patterns have come about towards inquiries into the kind of organisational, institutional, political or legal actions which might be pursued to ensure a healthy future for 'our high-tech' industry (read Silicon Valley or Route 128 for the US, or Cambridge or M4 Corridor for the UK). In other words, intellectual activity has been redirected from explaining the past to understanding the present in order to influence the future. The second development has led to an interest in the role of inter-organisational and interpersonal relationships within and between industrial-technological-scientific complexes and, as a consequence, the forms of governance appropriate to the new flexible forms of production.

Most scholars agree with each other in recognising the importance of flexible specialisation, based on the application of advanced information technologies, as a critical aspect of the dynamics of high technology regions. There is much disagreement, however, over the long-term implications this has for the competitiveness of existing high technology complexes and over the business or public-sector strategies required to make them sustainable. At least four themes (which might even be considered as the focus for quasi schools of thought) have emerged as expressions of this disagreement.

Structured flexibility

The first theme, which we may label 'structured flexibility', using the term adopted by its leading protagonists Richard Florida & Martin Kenney (1990), is modelled on the Japanese approach to industrial restructuring, and emphasises the need for large corporations to provide certain system-governance functions throughout industrial complexes. Florida and Kenney argue that the highly flexible structure which has evolved in Silicon Valley and other American high technology regions contains a number of intrinsic externalities, such as separation of innovation and production and high labour turnover, which, they further argue, are not only undesirable in themselves but damaging to the international competitive position of the US. The structured flexibility they advocate as a remedy involves a kind of quasi-integration of small firms under the umbrella of large corporations who would manage the network for efficiency, stability and strategic focus.

Collective order

The second theme, 'collective order', is associated most closely with its leading protagonist, Alan Scott (Scott 1992; Scott & Paul 1990), and is based on a call for American industrial complexes to develop their own unique antidote to the problems and instabilities of flexible specialisation recognised by advocates of structured flexibility (à la Japanese keiretsu). Rather than look to quasi-integration of complexes of small firms within hierarchies controlled by large corporations, Scott advocates a kind of institutional collectivism (not state ownership) amongst members of new production complexes. In his own words, guidance as to the likely forms which such collective order might take may be found in 'novel social experiments involving interpenetrating structures of competition and cooperation and peculiar forms of collective action and governance' (Scott 1992, p.220). Scott's approach requires cooperation of both private sector and public sector actors in areas such as technology development, labour training, business services, manufacturing and land-use control.

Regional networks

The third theme, 'regional networks', is associated most prominently with the work of AnnaLee Saxenian (1989a; 1989b; 1990), although there are a number of well-established schools of thought united by their interest in the concept of networks (Billi 1992). This approach is less pessimistic than the previous two about the ability of small firms in high technology complexes to organise themselves to take advantage of the opportunities afforded by advanced technology in manufacturing, design and communication. Saxenian—whose ideas are based on detailed multi-year case studies of the semiconductor and related industries in and around Santa Clara, California, and Cambridge, Massachusetts—points to the gains in efficiency of product development and manufacturing which networks of small organisations and individuals may accomplish through building relationships of mutual learning and trust through joint activity and cooperation. Her approach stresses the effectiveness of self-organisation amongst small firms, while nevertheless recognising both the need for trans-firm institutions and the important role of large corporations in local industrial networks.

Innovative milieux

The fourth theme, which we will label 'innovative milieux', at present amounts to no more than a minor theme in the English language literature, represented recently by the conceptual syntheses by Hall (1990) and Maillat (1991), and the empirical work of Willoughby (1993). As summarised elegantly in Hall's essay, however, the theme has been much more strongly developed by several European scholars not working in the English idiom, especially Philippe Aydalot (1986) and Åke Andersson (1985) (cf. Aydalot & Keeble 1988b). The innovative milieux theme has much in common with the others (such as recognition of the local industrial complexes, based on flexible specialisation and vivified by network relationships) but, to a much greater degree than the others, it places emphasis on intangibles such as the attitudinal environment and the complex interplay of cultural factors (such as intellectual, aesthetic and practical creativity, propensity for political openness, or technically progressive values) as substrata for technological innovation. *More particularly, it sees innovation growing out of the local milieu as a self-renewing process which has no single cause other than the process by which it sustains itself.*

Proponents of the innovative milieu idea take little interest in the search for single factor determinants of high technology growth and seek, instead, to understand the complex multi-dimensional processes by which knowledge is generated and renewed within an industrial context. Finally, while the *local milieu* idea does relate closely to the notion of the *local environment* in which a firm operates, the two are conceptually distinct. In the words of Maillat (1991, pp.268–69):

> The milieu may be defined as a coherent area organised around its physical structures (territorial production system, regional labour market, regional scientific institutes) and around its non-material structures (culture and technical culture, and representation system—the collective way of perceiving events and responding to them). The milieu is thus an area integrated with elements, in particular resources; the environment, on the other hand, is a disparate complex from which elements have to be derived which are likely to enrich the milieu.

The balance of this chapter will be devoted to answering the questions raised thus far by developing the innovative milieux idea further through examining how it may be used to interpret the biotechnology industry in New York state.

Why biotechnology and why New York?

Biotechnology presents an interesting field through which to explore the themes raised in this chapter

- The biotechnology industry is a knowledge-based industry par excellence and, because much of the debate over technology and regional form relates to the knowledge intensity of new industry, a study of biotechnology should lead to significant insights on the difference that knowledge intensity makes
- Since much of the better literature on technology and regional form has been based on studies of the electronics and information technology industries, the charge might be raised that the conclusions reached therein might not be valid within other industries. Biotechnology is definitely a 'high technology' industry, but its scientific and economic context is quite different to that of the electronics industry.
- Interest in the field of biotechnology is widespread and, until recently, there has been a paucity of reliable empirically based research available on its economic and organisational dimensions.
- Preliminary work I conducted on biotechnology revealed the importance of the local milieu idea, and prompted me to explore it further.

As the lead in microelectronics held by the US appears to many observers to be slipping under the pressure of foreign competition, biotechnology has been promoted as the new economic 'golden goose' to replace the role previously played by defence-related industries. Changes in the US/Soviet strategic relationship, furthermore, have created uncertainty for American industries dependent upon defence subcontracting, and this has stimulated even greater interest in the commercial and industrial potential of biotechnology.

Since its emergence as a commercial activity during the 1970s (with the US being the lead nation) biotechnology has evolved from being an experimental outgrowth of modern biological science into a new industry. While at one level biotechnology is a collection of techniques (e.g. for recombinant DNA, cell culture, monoclonal antibody production, or microbial fermentation of enzymes) for application in existing industries (such as the pharmaceutical, chemical, agricultural and food-processing industries), the collection of firms and other organisations involved with these techniques constitutes an industry in its own right—which many communities have been seeking to cultivate as a regional economic asset.

In the wake of this trend, almost every federal and provincial government in the industrialised world has established some kind of agency or programme to facilitate the development of biotechnology. Competitiveness in biotechnology is seen as a key to future economic competitiveness. While it is not clear exactly how the economic benefits of biotechnology might be realised, and how they might be appropriated by investors, most industrialised countries (and a significant number of less industrialised countries) are competing with each other to develop a strong national biotechnology industry. While the US has been the clear leader in this industry, massive public sector investment and growing private sector involvement in biotechnology in other countries—particularly in Europe and Japan—threatens to undermine this lead.

In the above context, the policy challenge for state governments, local governments and local industry bodies is to identify strategies for building a competitive local biotechnology industry, and to gain regional economic rewards from the application of biotechnology in other industries. In order to effectively meet this challenge, it is necessary to understand the economic dimensions of biotechnology on at least two levels: the ways in which firms commercialise knowledge in biotechnology and make the transition to manufacturing activities; and the essential processes for nurturing the growth of clusters of successful firms in communities and regional economies.

The attention which entrepreneurs, investors, industrialists and government economic development specialists have given to biotechnology has been accompanied by a boom in academic research oriented towards the field. In addition to the growth of scientific work in modern biology and complementary fields, and a recent tendency for some institutions to place attention on relevant process engineering dimensions, there has also been an expansion of social-science, economic and policy studies associated with biotechnology. Most of these studies do not focus directly on the issues relevant to local economic *development* from biotechnology.

There are a great many biotechnology-related publications now available dealing with regulatory-cum-legal issues, ethics, public attitudes and responses, environmental impact issues, intellectual property factors, and public education and information. More recently, researchers from business schools have joined the fray with studies about the financial and organisational aspects of biotechnology firms; but on the whole, these studies take the firm itself as the unit of analysis, and do not yield useful results to aid in initiatives to facilitate the development of biotechnology industries in particular regions. Some business-oriented publications have appeared which adopt a much wider perspective than that of the individual firm, but instead tend to examine questions of international competitiveness in biotechnology (e.g. the growing threat to the US from Japan). While the insights gained from such work can be helpful to national policy makers, they do not offer much guidance for decision makers who are concerned with questions such as 'what can I do for my town?' or 'how can we stop jobs disappearing from our community and migrating to other states?'

I carried out the research from which the following results were derived during 1991 under the auspices of the Centre for Biotechnology, in Stony

Brook, New York, with the primary objective of constructing an accurate profile of New York's commercial biotechnology organisations, and documenting the size, scope, structure, character and competitive position of the state's biotechnology industry. The research, including its theoretical perspective and methodology, however, grew out of earlier work conducted by the author (together with Edward Blakely, Director of Berkeley's Biotechnology Industry Research Group) on the biotechnology industry in California.

California is home to the single largest concentration of biotechnology firms in the world, and it is also the focus for much of the popular interest in biotechnology because of the state also being home to many of the early high-profile companies in the industry, such as Genentech, Cetus or Chiron, and to path-finding scientific discoveries, such as that which led to the Cohen-Boyer patent for recombinant DNA. As a consequence, many policy analysts and economic development specialists have looked to California for insights into the economic potential of biotechnology. This has often involved a search for the elusive list of factors which would attract biotechnology firms to relocate from one place to another. The hope has been that if civic and commercial leaders from 'wherever' could identify the factors which could attract footloose California biotechnology firms to their town, then economic renewal would somehow follow. At the same time, policy makers and industry leaders in California have been seeking assurance that the lead held by the state in biotechnology would not be lost through inattention to environmental or regulatory factors inhospitable to the fledgling industry.

With these factors in mind, the research of the Biotechnology Industry Research Group aimed to both identify the factors which drive the locational dynamics of California's biotechnology firms, and the factors which determine the industry's competitive position in California vis-à-vis other US states and emerging international rivals. The Group's findings (see Willoughby & Blakely 1989; 1990; Blakely & Nishikawa 1989; Blakely & Willoughby 1990) have demonstrated that, in contrast to popular belief, biotechnology is not a footloose industry, but rather successful firms have tended to remain in certain, mainly urban, locations, in order to thrive. More specifically, the best firms have appeared to operate as part of strong local biotechnology industry clusters characterised by strong inter-organisational relations, and exhibiting distinctive technological and product competencies related to their localities. It has appeared, furthermore, that such clusters have evolved primarily from the emergence and growth of locally based firms rather than from industrial relocation. This has given support to the application of incubation-oriented policies rather than attraction-and-relocation policies as tools for gaining economic benefits from biotechnology.

The work of Willoughby & Blakely (1990) has also suggested that local biotechnology industry clusters containing firms making strongest use of local institutions and resources tend also to exhibit the strongest inter-national orientation. In other words, in the more successful biotechnology clusters, a local and global orientation tends to occur simultaneously. Interestingly, this phenomenon appears to occur even when locational factors, such as a stringent regulatory environment or high business costs (e.g. due to taxation, land prices or wages),make a location otherwise unattractive. Finally, the competitive strength of California in biotechnology was found to be particularly interesting in view of the fact that the state does not have a strong presence of complementary industries, such as pharmaceuticals or chemicals, often seen as a stimulus to biotechnology through both R&D funding and through the purchase of intellectual property and biotechnology end products.

Following the publication of the above findings it became apparent that, as a comparison with California, New York was an important region to study, vis-à-vis biotechnology competitiveness, because of its historical role as a leader in biomedical research. The New York region (especially when nearby New Jersey is included) appears to exhibit all the features which one would expect to lead to it being pre-eminent in industrial biotechnology. The state is replete with high-class medical and biological research centres, and universities with excellent teaching and research in the biomedical sciences. In addition, the concentration of pharmaceutical and chemical firms in the region should provide marvellous opportunities for rapid downstream development of New York's biotechnology R&D ventures into manufacturing and marketing; and the diversity and concentration of processing and manufacturing industries in the region provides a potentially rich set of opportunities for industrial linkages and spin-offs from biotechnology activities. Finally, the premier status of New York City as a financial centre should add an even greater fillip to the cash-hungry biotechnology industry. There is evidence that New York is not pre-eminent in biotechnology, however, despite the significant advantages the region appears to offer. This raises some important challenges for policy makers, managers, and those who provide professional and financial services to the industry. More particularly, however, it is theoretically interesting because it raises doubts about the 'access to resources' and 'locational factors' approaches to explaining high technology competitiveness.

Before proceeding further, some definitions are warranted. In my study of New York's commercial biotechnology industry I was careful to rigorously employ a very conservative definition of the type of organisation I was prepared to include in the study population. I called this a 'dedicated biotechnology business' and I defined it as a commercial organisation which:

- uses biotechnology in the manufacture of a product or the provision of a service, or
- produces biotechnology, or
- conducts R&D towards the production of biotechnology, or
- supplies specialised biotechnology inputs for biotechnology organisations; and which
- devotes at least half of its effort to such activities.

I defined biotechnology as:

technology in which biological systems, established and controlled through the application of molecular biology, cell biology or micro-biology, are employed as means towards the attainment of practical ends.

Specialised biotechnology inputs are:

- specialised technologies which are dedicated for use within biotechnology activities, or
- specialised materials which are dedicated for use within biotechnology activities, or
- other inputs which have been produced by the application of biotech-nology.

Specialised biotechnology inputs must be artefacts or tangible products, and may include specialised software dedicated to use in biotechnology activities, intellectual property of a scientific or technical kind, or biological materials such as enzymes, but may not include services such as technical consulting.

The term 'business' in 'dedicated biotechnology business', was used in preference to 'firm' or 'company' because it allowed for dedicated bio-technology groups (which might not be formally incorporated bodies) within larger companies to be included in their own right, without having to include the whole parent company.

Basic features of the New York biotechnology industry

Unless otherwise indicated, all of the data reported below is derived from the 1991 study described above (for an explanation of the methodology and for detailed results, see Willoughby 1993). By mid-1991 New York state contained 90 dedicated biotechnology businesses, with assets worth over $1.3 billion, and over 6000 employees worldwide. Despite trailing behind California, in both its absolute size and the public attention it receives, the New York biotechnology industry is substantial in scale.

Because an objective of the study was to provide insight into the potential for local economic benefits of biotechnology, it was necessary to develop a scheme for dividing up the state of New York into a number of subregions for analysis. Three criteria were used to set the boundaries of these 'biotechnology industry regions':

1 maximum consistency with the state of New York 'economic develop-ment region' boundaries;
2 maximum consistency with the regional boundaries used by other authorities for the collection and analysis of economic and demographic statistics; and
3 meaningful concurrence with the actual locations of clusters of dedicated biotechnology businesses (DBBs).

The solution which best matched the combined criteria is illustrated in Figure 13.1.

The first significant observation is that the industry is clustered across the state into nine local biotechnology industry regions (Willoughby and Blakely found 6, or perhaps 7—depending upon the criteria used— within California, a state roughly double the population of New York). Figure 13.1 indicates the number of DBBs in each region, together with the primary and secondary specialisations of each cluster in technology and market focus.

Table 13.2 lists some basic statistics about the scale of the industry clusters in each region relative to the local economic, demographic and geographic contexts. Figure 13.1 and Table 13.2 combined reveal that there is considerable diversity between the nine clusters in their economic and technological character and in their regional density.

Only 18% of the DBBs are located in New York City (the boroughs of Manhattan, Brooklyn, the Bronx, Queens and Staten Island), and the region with the single largest share is Long Island (with 28 businesses), accounting for one-third of the total. Metropolitan New York as a whole (New York City, Long Island and Lower Hudson—the latter including Westchester County—and the three regions together accounting for about 62% of the human population of the state) accounts for 62% of the whole population of biotechnology businesses, 57% of the total commercial biotechnology employment within the state, and 48% of the New York biotechnology industry's global employment.

The nine biotechnology industry clusters differ considerably in the average size of their businesses, ranging from 10 people per unit in the Southern Tier (the large region in the centre of New York surrounding Ithaca and Cornell University), to 108 people per unit in the north (the even larger region extending north-east from Syracuse). Long Island's businesses, with 30 people per unit employed within the local region, are smaller on average than the 'typical' business state wide (which has 45 people per unit).

The single largest cluster of employees occurs in Lower Hudson, and this is accounted for largely by the concentration of biotechnology R&D carried out there by groups within large pharmaceutical companies, such as Lederle Laboratories' Medical Research Division. Lower Hudson is also the home base for the largest number of biotechnology facilities globally (i.e. locations throughout the world where activities of New York biotechnology businesses are situated). The region with the lowest total commercial biotechnology employment locally is the Southern Tier (with 73 people, est. June 1991) followed by the Capital region (with 99, est. June 1991).

The urban part of metropolitan New York (i.e. New York City) contains only one-third as many biotechnology employees (just over 500 people) as the suburban part of metropolitan New York (i.e. Long Island and Lower Hudson), which are home to over 1800 employees. It appears that New York's biotechnology industry is something of a suburban phenomenon, rather than either an urban or a rural one. It should also be re-emphasised, however, that there is great diversity in the geographical character of the

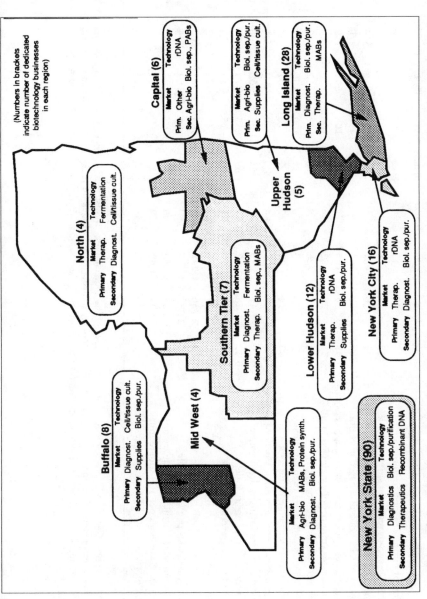

Figure 13.1 New York biotechnology industry, 1991; regional specialisation in market orientation and technology

Source: Dr. K.W. Willoughby, Centre for Biotechnology, June 1991.

Table 13.2 Employment and facilities of dedicated biotechnology businesses in New York regions, 1991

Biotechnology industry region	New York City	Long Island	Lower Hudson	Upper Hudson	Capital	North	Mid West	Buffalo	Southern Tier	Total New York State
Total population of businesses located in region	16	28	12	5	6	4	4	8	7	90
Number of businesses from region in survey sample	12	22	8	5	4	1	2	7	5	66
Estimated number of jobs (locally) in region	527	826	981	281	99	432	318	544	73	4 081
Estimated number of jobs (globally) based in region	781	957	1 264	361	107	432	978	1 270	113	6 263
Mean size of businesses (global number of employees)	49	34	105	72	18	108	245	159	16	70
Mean size of businesses (number of employees locally)	33	30	82	56	17	108	80	68	10	45
Estimated number of facilities (globally)	24	37	26	8	8	12	12	13	14	152
Percentage of businesses with more than one facility	42	25	50	60	25	100	100	14	80	41
Percentage of employees located outside region	33	14	22	22	7	0	67	57	36	35
Mean age of businesses (years at present location)	7.8	6.9	14.5	20.4	4.0	16.0	8.0	5.6	6.8	9.9
Biotechnology industry employment density index	0.29	1.47	3.73	1.69	0.54	1.25	1.02	2.00	0.49	1.00
Biotechnology industry firm density index	0.42	1.78	1.62	1.17	1.69	0.55	0.74	1.63	2.32	1.00
Composite annual rate of revenue growth (%)	143	21	75	31	277	15	16	20	31	70
Land area (square miles)	301	1 199	844	5 023	2 599	18 743	7 428	1 572	9 670	47 377
Population ('000 persons, 1990)	7 180	2 736	1 242	905	814	1 595	1 405	1 201	943	18 023
Population density (persons per square mile)	23 854	2 282	1 472	180	313	85	189	764	98	380
Number of local biotechnology jobs per square mile	1.75	0.69	1.16	0.06	0.04	0.02	0.04	0.35	0.01	0.09
Number of biotech. businesses per 100 square miles	5.32	2.34	1.42	0.10	0.23	0.02	0.05	0.51	0.07	0.19
Number of local biotechnology jobs per 1000 residents	0.07	0.30	0.79	0.31	0.12	0.27	0.23	0.45	0.08	0.23

Note: 1 sq mile = 2.59 km^2

Sources: K.W. Willoughby, Centre for Biotechnology, New York, Survey of the Biotechnology Industry, June 1991 (biotechnology data); New York State Statistical Yearbook (demographic data and geographic data). Biotechnology Industry Density Index for region 'n' = {(biotech units in region 'n')/(biotech units in New York)}/{(units of all industries in region 'n')/(units of all industries in New York)}; where units refers to either 'number of persons employed' or 'firms', as indicated.

industry. While it is spatially dispersed throughout the state, it does concentrate into local clusters, but the density of those clusters varies considerably.

A few other salient facts will provide a picture of the industry's present status. Almost 60% of the businesses employ no more than 25 people and a full 25% employ 5 or less people; 75% of the start-ups took place during the last decade, with 30% less than 4 years old. New York's commercial biotechnology industry consists primarily of organisations which are young and small. Nevertheless there is a significant minority of large, older businesses (12 were formed prior to 1970).

Despite the fact that biotechnology is widely thought of as a knowledge-based industry which concentrates on the generation of intellectual property rather than tangible products, and despite the fact that most of the businesses are relatively immature by normal conventions, 77% of the businesses in New York have begun to generate revenue from the sale of products or services based on biotechnology. An even higher majority (84%) claim to have commenced manufacturing operations already (and plan to continue manufacturing for the next 5 years) or plan to commence manufacturing within the next 5 years. Interestingly, only 14% say that they plan to rely exclusively on means other than manufacturing, such as the sale of intellectual property or services, for the generation of revenue; a further 5% are undecided about whether they intend to exclude manufacturing from their strategy for generating revenue.

The picture of the New York biotechnology industry which emerges from the above data conforms closely to the picture of the 'typical' post-Fordist high technology industry which has emerged from the literature reviewed earlier, including the pattern of 'flexible manufacturing' by local clusters of small firms.

The claim of the 'structured flexibility' advocates—that such complexes are beset by the separation of innovation and production—does not, however, appear to apply to this industry. The mixture of manufacturing and other activities in the one location by small firms (indicated above) is confirmed by the human resources function data in Table 13.3, which show that the businesses tend to maintain a mixture of research and technical operations activities in both primary facilities and branch facilities.

A rejoinder (by the proponents of the 'separation of mental and manual labour' view of flexible manufacturing complexes) to the presentation of this data might be that as the industry matured the pattern would change, with the older firms separating their functions more rigidly along spatial lines. Once again, however, the data suggests otherwise. Table 13.2 reveals that there is no apparent significant pattern across regions between the age or size of businesses and their propensity to establish branch facilities. Regions containing older businesses do not necessarily also contain businesses which are more spatially dispersed. In addition, Table 13.3 reveals that, to the extent that businesses *do* establish branch facilities, the activities of these are weighted more heavily towards R&D but no less heavily towards 'technical operations' than those of the main facilities.

Table 13.3   Distribution of personnel between organisational functions of New York's biotechnology businesses, 1991

	Mean proportion (%) of personnel in businesses devoted to each organisational function		
Organisational function	*Main facility only*	*All branch facilities combined*	*All facilities combined*
Management and marketing	24.1	16.8	21.5
Research and development	31.2	47.4	37.1
Technical operations	27.0	27.4	27.1
Other	17.7	8.4	14.3
Estimated total no. of personnel	3 972	2 291	6 263

Source: K.W. Willoughby, Centre for Biotechnology, New York, 1991.

Thus, there is little support here for either a simple 'life cycle theory' approach to explaining the industry's dynamics, nor for the pessimistic view associated with the 'structured flexibility' theme.

Locational dynamics

While much of the literature on the location of high technology industries is couched in the language of 'decision making by firms about where they intend to locate', my own research suggests that biotechnology locational dynamics have very little to do with such decision making. Over three-quarters (79%) of the total sample of businesses in the survey commenced operations at their present primary location in New York. Only 60% of the founders of the businesses have ever engaged in a significant planning exercise involving investigation of alternative locations for the business and the articulation of locational criteria. Of those who have engaged in such an exercise, only 38% have actually relocated the business (21% of the total sample), and of those, 69% relocated from *elsewhere in New York state*. Hence, less than 8% of all the businesses (or 31% of those which did relocate) relocated from out of state. Most of the relocations were relatively local, aimed at such objectives as improving facilities rather than gaining fundamental changes in the environment.

When asked where the geographical location of the primary founder (or founders) was at the time of the formation of the business, almost 80% responded that it was at or nearby the present primary location of the business, and another 6% indicated a location elsewhere in New York state. Of the remaining founders, 8% were located elsewhere in the US and 6% were located outside the US.

These results are interesting because much of the policy making of state governments intended to create local 'high technology' industries (such as biotechnology), and much of the promotional activity of private (or quasi-private) regional industry organisations with similar policy objectives, is based on the premise of seeking to entice 'footloose' high-tech/biotech firms to relocate to 'our' state or region. While such initiatives may indeed pay off from time to time (as may have been the case with the few firms mentioned above which did relocate from out of state), this study suggests that other approaches are more likely to be fruitful. Biotechnology businesses are basically not footloose. While a minority may make geographical shifts, it is hardly a strong enough phenomenon to form the foundation of an industrial development strategy.

As indicated by the above data and as suggested by other studies (Willoughby & Blakely 1990), *the prime determinant of the location of biotechnology businesses is the prior location of the founders of the business.* Biotechnology companies *emerge* out of places in which suitably motivated and suitably competent people (where 'competent' includes competence in getting access to necessary resources) *already exist.* This suggests that policies aimed at nurturing the pool of such people within a region would be more likely to lead to industrial growth in biotechnology than would industrial attraction efforts.

In the course of this study, a couple of ex-New York firms were identified which have relocated to other states (e.g. Marrow-Tech to California, American Diagnostic to Connecticut), but these are exceptions to the rule, and while there appeared to be good reasons in these two cases, relocations generally also involve considerable cost. Despite inadequacies which may be associated with the original location, most firms will relocate only when there are significant advantages intrinsic to the new location beyond those which may be 'packaged' by governments as inducements, and sufficient to outweigh the transitional costs (human, organisational and financial) associated with relocation (Blakely & Willoughby 1990).

The growth of the biotechnology industry in New York continues apace, despite widely discussed stringent circumstances. From 1990–1991 employment levels increased by an average of 51%, and the annual rate of employment growth averaged 53% over the previous five years. The annual rate of employment growth globally (i.e. internationally) was greater than the rate of employment growth within New York, from 1990–1991, for 12% of the biotechnology businesses, and was equal for about 85%. If measured over the previous 5 years instead, the annual rate of employment growth globally was greater than the rate within New York for 13% of the businesses. This suggests that as the industry continues to grow it is gradually becoming more international in scope, while remaining firmly rooted locally in the places where it has emerged.

Eighty-six per cent of the organisations interviewed in this study were dedicated biotechnology businesses at the time they were formed. The balance were mostly previously either hospital laboratories, medical diagnostic laboratories, or consulting/contract research organisations. When asked what kind of employment status the primary founder had at the time the business was formed, 19% responded that they were a member of another biotechnology business, 39% were university academics or members of a research institution, 22% were members of a business other than a biotechnology business, 8% were employees of a government agency (other than an educational or research institution), and 12% were self-employed.

Most of New York's dedicated biotechnology businesses (76%) were formed as a completely new business, but 18% were 'spin-offs' from an existing business, and 5% were formed by merging more than one existing business.

The picture of the industry which emerges here is of local clusters which have grown organically out of the existing fabric of institutions and social networks in each of their respective local regions, and which remain rooted in their locality, even as they become more national and international in their outlook and behaviour. This, in essence, is the innovative local milieu at work.

The study also involved detailed analysis of the factors which make locations attractive to businesses, and a detailed assessment of the advantages and disadvantages of each business's primary location compared with other potential locations outside New York. The results demonstrated, in short, that while New York is widely perceived as being disadvantageous from the point of view of costs and the regulatory environment, these disadvantages are outweighed on the whole by the advantages which New York offers to firms in access to good senior staff, access to good support staff, and the proximity of universities, research institutions and other biotechnology businesses. The managers of New York's biotechnology businesses are faced with a dilemma: cost-related factors, and to a lesser extent, regulatory factors, would lead them to prefer other locations, yet at the same time these factors are counterbalanced by several other factors (to do with knowledge, people, institutional resources and inter-organisational proximity) which would lead them to prefer their present location (for details see Willoughby 1993). The spatial inertia created by the local founder effect combined with the benefits of participation in local industrial networks reduces the importance of external locational factors as determinants of the geography of commercial biotechnology.

How does the local biotechnology milieu work?

If external locational factors are not key forces underlying the dynamics of the biotechnology industry, what then are? Drawing upon the evidence from New York, the following explanatory perspective is suggested.

If biotechnology businesses have emerged locally in New York, created from within existing local institutions by local people, with roots in local institutions and local networks for communicating knowledge and expertise, then those institutions and networks may be seen as the essential foundation for the industry. The emergent biotechnology businesses therefore draw upon those local institutions and networks for knowledge, people, specialised facilities, business and scientific contacts, and other kinds of resources; and the local milieu which acted as a seedbed for the new enterprises needs to continue to provide a fertile soil within which they may be nurtured. In other words, the capacity of a local biotechnology industry to thrive depends upon how well it maintains the vitality of its relationships with the community of people and organisations from which it emerged in the first place.

An implication of this perspective is that if a biotechnology business is strongly embedded in the local biotechnology milieu, and if it has cultivated human and institutional links within that milieu which are both rich and deep, then it is likely also to have cultivated capabilities which are strong enough to enable it to transcend the constraints imposed by the 'problems' of the region or locality.

For example, its communication with university research groups, its liaison with other biotechnology businesses, its ability to employ specialised instrumentation or software due to association with local technical suppliers, or its links with customers such as pharmaceutical companies, may give a business the capacity to innovate more rapidly than its competitors. This higher speed of innovation may be translated into a reduction of the total cost of developing a product (despite paying high wages or high rent), and an early positive revenue stream. In the same way, a business may be able to locate sources of investment capital on better terms and more rapidly than others. The total effect of these advantages may mean that strong 'embeddedness' in the local biotechnology milieu can lead to enhanced performance and enhanced access to resources sufficient to ensure that locational problems (e.g. costs and regulations) may be treated as non-critical, even though they may be very real. Some firms manage to become competitive in the face of severe locational problems, and some do not.

The cardinal management skill, according to this perspective, is therefore not necessarily that of applying the 'accountant's razor' to cut costs, but that of finding ways in which a business may cultivate competitive capabilities in the face of high resource costs and regulatory constraints. One way to mobilise such competitive capabilities is through facilitating better communication and collaboration within the local biotechnology milieu.

At the heart of this perspective—the 'technological milieu' perspective—is the idea that the relative performance of biotechnology businesses is dependent primarily upon their patterns of inter-organisational communication and collaboration, rather than upon external locational factors such as the cost of doing business or the attractiveness of the regulatory environment.

In order to explore the role of inter-organisational relationships as influences on the commercial performance of biotechnology businesses, each of New York's businesses interviewed in this study was asked a series of questions about both its informal communication and formal collaboration during the previous year with various types of organisations: other biotechnology businesses; universities and other research institutions;

278 Cities in Competition

hospitals and other health care institutions; pharmaceutical corporations; specialised suppliers of instrumentation, equipment, software and technical services; and other types of organisations. Data in response to these questions was collected for organisations located in three geographical categories: the same local region within New York state as the business; elsewhere in the US (including non-local parts of New York); and in foreign countries.

This data was then aggregated to construct 8 indices of inter-organisational communication and collaboration, and these in turn were plotted against a performance measure. The indices are:

- Index of informal communication
- Index of formal collaboration
- Index of local informal communication
- Index of local formal collaboration
- Index of national informal communication
- Index of national formal collaboration
- Index of international informal communication
- Index of international formal collaboration

For each of the eight indices the set of dedicated biotechnology businesses with an above-average score for the index was selected and the mean annual rate of revenue growth for that set was calculated. This figure was then compared with the mean annual rate of revenue growth for the whole sample of businesses in the survey. These calculations allow exploration of the possible relationship between commercial performance and propensity for communication or collaboration. The results of this exercise are summarised in Figure 13.2.

Figure 13.2 reveals that businesses with above-average levels of informal communication with other organisations (of a variety of types) achieve above-average rates of revenue growth. It also reveals that higher performance levels are associated with higher levels of informal communication with other organisations more than with higher levels of formal collaboration with other organisations.

The figure also shows that the international and local arenas are more significant than the national arena in this respect. Those biotechnology businesses with above-average levels of local informal communication with other organisations (i.e. within the same subregion within New York State) on average achieve the highest annual growth rates in their revenue. Likewise, those businesses with above-average levels of local formal collaboration achieve above-average revenue growth rates. A similar pattern exists for those businesses which exhibit above-average levels of both international informal communication and international formal collaboration with other organisations.

The results in Figure 13.2 therefore accord with the insights presented above as part of the 'technological milieu' perspective. The counter-

Ratio of Rate of Annual Revenue Growth for "above average interaction" set over Rate of Growth for Whole Sample

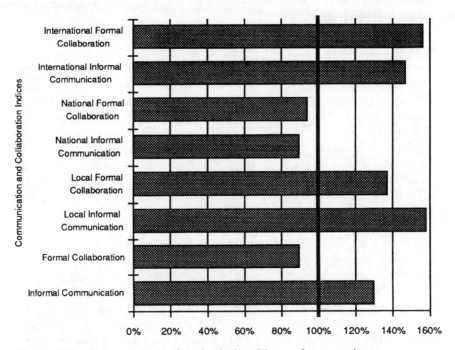

Figure 13.2   Inter-organisational relationships and economic performance

intuitive result, however, is the concurrence of high levels of inter-organisational relationships (and associated strong performance) at the international and local levels, while those businesses emphasising national linkages tend to exhibit lacklustre performance. These results accord fortuitously with the ideas of Matteaccioli & Peyrache (1989) published in French but paraphrased into English eloquently by Maillat (1991, p.274):

> Innovation is thus conditioned by the characteristics of the milieu, that is, the degrees of extroversion and of integration within the milieu. When the milieu succeeds in reconciling openness (acceptance of new ideas which trigger innovation) and closure (coherence of the socioeconomic fabric), it is able to stimulate and support the creativity of firms and, in return, to be enriched by the innovation they achieve. If it is too open it shatters or disintegrates; if it is too integrated it becomes to inward-looking and loses its competitiveness.

Strong biotechnology businesses tend to be simultaneously global and local in their outlook. Local embeddedness and global connectedness do not need to be traded off against each other as alternative strategic orientations of firms, but appear to be associated with each other. What then are some of the policy implications of this perspective?

The local economic development potential of biotechnology

Biotechnology may produce both direct and indirect economic development impacts. The direct impacts consist of the employment and wealth created by dedicated biotechnology businesses themselves. The indirect impacts consist of the 'multiplier' effects in other industries and in the wider economy which flow on from the activities of dedicated biotechnology businesses.

The indirect economic impacts are potentially much larger than the direct impacts. The dimensions of the indirect economic impacts of biotechnology are not uniform across the industry, however, because the multiplier effects depend upon the nature of the relationships between the dedicated biotechnology businesses and organisations in complementary industries.

Biotechnology, as well as being the basis for an industry in its own right, is also an *enabling technology* for other industries such as health care, pharmaceuticals, chemicals, agriculture, environmental management or food processing. Because the techniques produced by the biotechnology industry are so pervasive in the scope of their potential applications, the future competitiveness of these complementary industries in particular communities will be limited by the degree of their mastery of biotechnology. Access to the knowledge, products and expertise of biotechnology businesses will be a key to the mastery of biotechnology by organisations in complementary industries. The richness of links between biotechnology businesses and organisations in other industries will therefore be a key factor limiting the scale of multiplier effects from biotechnology.

Inter-organisational links are important for biotechnology-based economic development in two directions. The better the links between organisations in complementary industries (such as pharmaceuticals) and biotechnology businesses, the greater the capacity the former have for developing novel commercial applications for the new techniques. At the same time, the stimulus for technological innovation experienced by biotechnology businesses will be enhanced through richer exposure to the commercial opportunities and problems of other industries amenable to biotechnology solutions.

In principle, a company in a 'downstream' industry (such as the waste-treatment industry, food processing or horticulture) may gain access to biotechnology from anywhere in the world, and from any kind of organisation involved in biotechnology, whether commercial or in the public sector. The transfer of technology is not as simple as this suggests, however, because technology cannot be completely encapsulated in a 'blueprint', a patent, or in a set of operating codes—or even in a physical product. Technology involves a tacit dimension, which may be embodied in the routines, unspoken traditions or maintenance procedures of an organisation, or in the informal knowledge inside the heads of the people who developed the technology. It may also require specialised resources, materials, equipment or skills available to the company which developed the technology but not readily available to the licensee of the technology, and of which few people are consciously aware (e.g. the materials from which storage containers are made, a customised item of instrumentation or a supply of specialty biological inputs).

The successful transfer of the 'unarticulated' and informal aspects of biotechnology will probably require extensive human interaction and extensive organisational interaction. Such interaction is normally facilitated by close physical proximity. From the point of view of 'downstream' economic development from biotechnology, it is therefore important for complementary industries to have access to *local* biotechnology organisations. In addition, as suggested earlier, the capacity of a company to utilise information from international sources appears to be mediated by its capacity to participate in *local* networks for the exchange of information.

From the point of view of *local economic development* in biotechnology, what matters is that there is a strong *local* presence of both dedicated biotechnology businesses and businesses in complementary industries, and that these interact in a healthy way to transfer both technology and the stimulus for the development of new technology. Thus, the stronger the demand by local complementary industries for biotechnology innovations from local biotechnology businesses the stronger those local biotechnology businesses are likely to be. The stronger the local biotechnology businesses become, the higher the chance will be that businesses in complementary industries will be able to make significant advances in their activities based on the application of biotechnology.

Economic growth 'in general' is not of much use to particular communities—such as Long Island, Buffalo, or New York City—unless there is a *local manifestation* of that growth. *The local development of self-sustaining economic capacity (linked to biotechnology) requires a local network of dedicated biotechnology businesses and complementary organisations, including technical suppliers, customers and organisations providing specialised support services.* The stronger and more vigorous such networks become the greater the *direct* local economic impacts of biotechnology are likely to be and, consequently, the stronger the *indirect* local economic impacts of biotechnology are likely to be. A healthy local biotechnology industry network nurtures a virtuous circle of feedback between direct and indirect local economic benefits from local biotechnology activities.

The chief policy implication of this perspective is that efforts directed at the development of local biotechnology milieu through enhancing the quality and intensity of interactions amongst organisations within local biotechnology industry clusters (including complementary industries upstream, downstream and laterally) is more likely to succeed than efforts based primarily on attempting to attack unattractive locational factors such as the high cost of land, labour or taxation.

Because the constraints on improving the performance of biotechnology businesses are largely intangible, it means that future competitiveness of a local industry complex can only be increased indirectly rather than directly.

This means, first of all, that short cuts to local industry development by trying to 'import' ready-made firms from outside the region, with financial or other inducements, are unlikely to yield impressive results. The strategy of attracting firms to relocate from outside the region will generally only succeed if a strong local biotechnology industry has already been built through the local generation of new firms and new growth based on local people, local knowledge and local institutions. Put another way, there is no 'instant gratification' in biotechnology, only long-term pay-offs to sustained investment in the kind of environment in which biotechnology businesses thrive.

It also means, secondly, that it is generally not possible to build a competitive biotechnology industry through direct action aimed at particular firms or through particular methods or inducements alone (such as tax breaks, isolated regulatory reforms, technology parks, research grants or other kinds of subsidies). Rather, a whole range of measures need to be adopted to nurture the development of a strong local biotechnology milieu from which local biotechnology industry clusters emerge. Put another way, there is no magical 'key' to turn which will guarantee success. It is not possible to just create the desired end product (a sustainable collection of competitive biotechnology firms), it is only possible to employ a range of techniques to create the conditions which may indirectly lead to the desired end product.

The principle of indirect rather than direct support for biotechnology industry development may be discussed using a biological allegory. The controlled production of biological materials may be manipulated through the use of enzymes as catalysts in fermentation processes. Similarly, various policy instruments and industrial support measures for biotechnology need to be viewed as 'enzymes' in the biotechnology industry milieu 'fermenter', rather than as direct means of producing local biotechnology firms. These ideas may be illustrated by a simple model portrayed in Figure 13.3 for interpreting the process of economic development in biotechnology.[1]

The primary feature of the model is that the regional industrial process associated with biotechnology involves three main dimensions: local biotechnology industry clusters; a regional biotechnology milieu; and regional development factors.

This way of construing the industry accords with the evidence assembled in this study that biotechnology businesses tend to emerge in local clusters which exhibit distinctive characteristics (e.g. market focus, locational preferences, pattern of inter-organisational linkages, human resource requirements, technological specialisation). Locality is a fundamental aspect of the industry's dynamics, not just a convenient perspective from which geographers and planners may approach industry analysis. Economic development in biotechnology is a matter of *local economic development*, and not just of 'development' in general.

The model also embodies the notion that local biotechnology industry clusters do not emerge in isolation, but rather within a regional bio-technology milieu. There are two levels at which the concept of 'region'

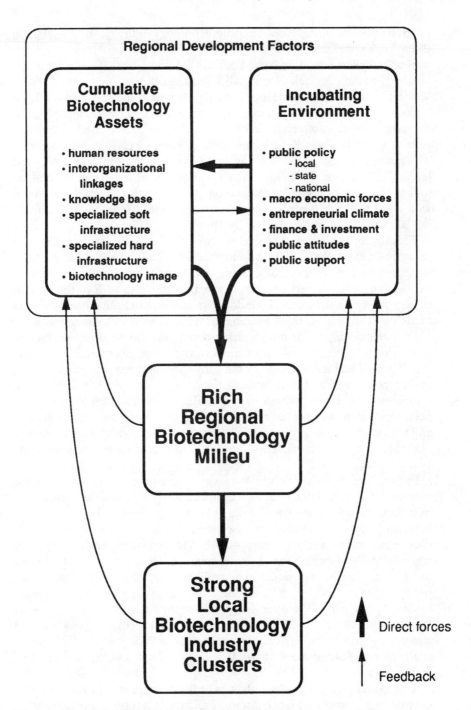

Figure 13.3   Model for local economic development in biotechnology

might be applied to the New York biotechnology industry: the *mega region* (New York state) and the *local region* (represented by the 9 'biotechnology regions' labelled herein as New York City, Long Island, Lower Hudson, Upper Hudson, Capital, North, Mid West, Buffalo and Southern Tier). While each of the biotechnology industry clusters is located within a local region, drawing upon the peculiar features and assets of that local region, the mega region in which each of the local regions is located provides a context in which each of the biotechnology industry clusters has emerged. The concept of the biotechnology milieu applies to both of the regional levels. Local clusters of firms emerge within a local region biotechnology milieu and a mega region biotechnology milieu. This observation accords with other research carried out on the biotechnology industry, particularly in California (Willoughby & Blakely 1990).

This use of the regional biotechnology milieu concept stresses that the growth of a strong local biotechnology industry cluster cannot generally be explained by the existence of any one particular locational factor. Rather, various locational factors contribute together to the growth of the milieu; there is no direct tight causal relationship between an individual locational factor and the emergence of a strong local biotechnology industry cluster.

Despite pointing to the milieu rather than to individual locational factors as a source of biotechnology firm clusters, the model does specify 'regional development factors' relevant to the formation of a regional milieu. Two main types of regional development factors may be specified: cumulative biotechnology assets (which are specialised resources in a *community* necessary for the emergence and flourishing of a biotechnology industry); and the incubating environment (which is necessary to both feed the assets and facilitate their mobilisation for the purpose of nurturing a regional biotechnology milieu).

In principle there are probably forces acting between each one of the elements in the model and all of the others, and this could be represented by a complex web of two-way arrows linking each one of them. Figure 13.3, however, seeks to discriminate between those forces which are significant (both empirically and for the purposes of policy initiatives) and those which, while formally identifiable, are not of great interest. The most significant forces included in the model are symbolised by the thick arrows, and represent the processes which most directly lead to the development of local biotechnology industry clusters. The thin arrows represent the important feedback processes by which a rich regional biotechnology milieu and strong local biotechnology clusters, once established, may in turn nurture the regional development factors which were preconditions for their emergence.

The model presents public policy as only one element of the incubating environment necessary for the assembly and mobilisation of a community's cumulative biotechnology assets, but it also presents the implications for the way in which public policy for economic development in biotechnology ought to be pursued if it is to be successful. Policies, taking on different forms at the various levels of government, should not aim directly at the establishment of biotechnology firms and clusters, but rather should have the following two objectives:

1 the building up of cumulative biotechnology assets in regions where the development of a biotechnology industry is desired; and
2 the management and mobilisation of those assets as a total system of resources for the nurturing of a regional biotechnology milieu.

The milieu itself, combined with serendipity and exogenous forces, will lead to the creation and strengthening of local biotechnology industry clusters. The clusters will produce feedback to nurture the regional development factors which undergird the regional biotechnology milieu, but this is not something that policy makers necessarily need to direct their attentions towards.

The model evokes the need for additional research to catalogue in detail the ways in which each of the different levels of public policy making ought to function within this schema. For example, the maintenance of a substantial and quality knowledge base in a community (fed through such means as universities and public research institutions) may best be dealt with at a state or national level, while the building up of specialised hard infrastructure (e.g. a supply of environmentally and scientifically appropriate physical plant and buildings) or combined specialised soft and hard infrastructure (such as a biotechnology research park, incorporating both professional services and physical facilities) might be more suitably handled at the local level.

Such analysis is beyond the scope of this chapter, but two important policy implications may be reiterated at this stage. First, public initiatives aimed at producing economic development from biotechnology should concentrate on the objectives of establishing and sustaining cumulative biotechnology assets in a region for the purpose of nurturing a regional biotechnology milieu. Second, such initiatives should recognise that industrialisation in biotechnology tends to take on a regionally specific character, reflecting characteristics of each incubation region, and therefore that policy initiatives ought to take into account the strengths, weaknesses and distinctive features of the local regions.

This chapter has demonstrated that biotechnology, as exemplified by the New York biotechnology industry, is not an amorphous, homogeneous, spatially diffuse economic form. Rather, it is highly differentiated in its spatial pattern, character, strength and relationship with the economic and geographic environment. The industry cannot be adequately understood unless it is analysed from a regional point of view.

NOTES

[1] This model was developed by the author of this study during a recent study of the biotechnology industry in California. Figure 13.3 and part of the accompanying explanation are taken from the report of that study (see Willoughby & Blakely 1990).

14 Developing the Regional Potential for Creative Response to Structural Change

Klaus R. Kunzmann

Introduction

LIVING in a traditional industrial region, such as the Ruhrgebiet in Germany, and monitoring the continuous public and private efforts to cope with the manifold repercussions of structural change, one is tempted to conclude that there is hardly any chance to guide or to master this change in any systematically planned or comprehensive way. The influencing factors at the various levels of decision making are too complex to be forecasted and the need to coordinate all strategic policy actions is met with hardly any of the required political commitment. Moreover, and despite all available modern communication, the value systems of regional and local opinion leaders and the attitudes of regional and local actors towards change are extremely difficult to influence.

Assuming that the waves of structural and technological changes will continue to shorten in the coming decades, and knowing that the chances of achieving regional consensus are continuing to diminish in an increasingly entropic civil society, the prospects for planners to intervene strategically in the process of change seem to diminish. Does this mean that the future planner is condemned to just react to the system and to developments which he can hardly influence? Will he just be a skilled monitor, using GIS for managing and visualising the abundance of space-related regional and local information? Or is he rather a moderator in the process of political negotiations on land-related matters, as recent literature on the subject suggests?

But the problem goes beyond the planner's own role in the process of structural change. The pivotal questions remain how the capability of a region for creative response to regional structural change can be maintained over longer periods and whether a sustainable creative milieu can be shaped and fostered in any region. As structural change is not a single event, but a continuous process caused by technological and social progress, such questions are crucial for regional economic development policies.

On a more general level, similar questions have repeatedly been asked over time: why are certain nations more successful than others? Why are certain regions successful at one time and in decline at another? Answers vary over time and depend on ideologies and academic fashions. Usually they are built around a central hypothesis, e.g. the importance of defence industries or the crucial role of cities in regional economic development (Kennedy 1987; Bairoch 1988). Others have explored the reasons why innovations happen at certain locations, or why not (Andersson 1985; Aydalot 1986; Maillat 1990; Saxenian 1989a). It is not the purpose of this chapter to analyse the respective academic literature. Rather, based on experience drawn from the Ruhrgebiet in Germany (an old industrial region in the heart of the former West Germany with a population of about 5 million), I will explore some of the factors which seem to be crucial for enabling a region to sustain its capability for responding creatively to waves of structural changes occurring over shorter or longer time periods.

The questions for which I am searching to find answers are:

1 Why are certain regions more creative than others, why are they more successful in adapting to structural change than others?
2 What are the factors driving regional innovative changes?
3 Can innovative milieux be created, are they sustainable?
4 Can synergies be planned to create innovative milieux?

I have not yet found a consistent framework to answer these questions. Hence I can present only a few initial deliberations on the subject matter, which has many cultural and psychological, social and economic, and political as well as administrative dimensions.

Some lessons from restructuring efforts in the Ruhrgebiet

Since 1945 the Ruhrgebiet in Germany has repeatedly been confronted with the various economic and social implications of developments weakening the established regional economic base. The public and private efforts to master the respective crises have been more or less successful. Usually it has been an all but easy balancing act of the state and local governments, between the support of the declining coal—and steel-related industrial base and cushioning the social impacts of plant closures, and the promotion of innovative policies triggering off new, future-related industries which are competing in an international market. In accordance with the influential headquarters of the political parties in power, and in close cooperation with the unions and the chambers of commerce, varying sets of regional policies, strategies, programmes and plans were developed by the State government of Northrhine-Westphalia following the respective mainstream regional economic development paradigms. The more serious the respective crisis of the regional coal and steel production complex had been, the faster consensus was found among the various regional actors on what to accept, what to do and what to initiate. In this process, as a rule, elections to local, state and federal governments played a crucial role. At least they speeded up the decision-making process of the politico-administrative complex prior to election dates.

When, in the early 1960, the post-war recovery period had come to an end and the first economic crisis had hit the region, it became obvious that the development of a better, more highly qualified regional labour force was essential for any longer term development of the Ruhrgebiet, where 5 million people lived and worked. Subsequently, the State government undertook an impressive effort to establish five large public universities in the region which, prior to that policy and for various historical reasons, was void of any major institutes of higher education. It was an explicit shift to developing the human potential of a region. However, it took almost a generation until the universities had any major impact on the regional economy. The old corporate structures and the established politico-administrative networks still dominated the regional decision-making machinery. And in subsequent years it was learnt that the transfer of ideas and competences from the universities to regional industry requires sophisticated and highly subsidised transfer mechanisms to trigger off new economic activities in the region.

Since the mid-1980s the universities in the Ruhrgebiet have become a pivotal factor for regional economic innovation. This coincided with the third or fourth crisis within the traditional regional coal- and steel-production complex. Unemployment rates in the Ruhr, particularly in Dortmund, had reached almost 20%. And there was little hope that the conservative Federal Government—in contrast to the socialist government of the State now dominated by the conservative party—would be as generous as earlier governments in subsidising the ailing coal and steel industries by taking over the costs of economic restructuring and of the social cushioning. This fact had mobilised the more innovative regional forces. In the city of Dortmund, for example, the economic crisis had brought about a hitherto unknown consensus of the major local political & economic forces, which subsequently led to a series of innovative projects aimed at modernising and diversifying the obsolete local economic base. Soon the partial restructuring of Dortmund, driven mainly by the activities of the Dortmund technology park, was labelled (and widely marketed) as a success story of innovative action (Hennings & Kunzmann 1990; 1993).

However, this short period of innovative action now seems to be stagnating. Since the reunification of the two German states, the focus of industrial policies has obviously shifted from West to East Germany, and a new steel crisis is hitting the whole region. Now it turns out that the innovative milieu in the region is still fragile and not yet sufficiently well established to master the new crisis. One might argue that the new crisis is not severe enough to cause the more creative forces of the region to act.

Experience shows that political action in a region is not caused by well-founded scientific arguments and insights resulting from continuous regional monitoring, but is provoked by spectacular events which find wide regional or national media coverage, such as militant squatting of derelict housing, the breakdown of the public transport system following a series of accidents, the unexpected lay off of thousands of workers in a certain industry, an extreme smog situation caused by week long inversions, or water pollution caused by human failure.

Although there is some truth in the belief that crisis situations are instrumental in triggering off political focus and reaction, it would be too cynical, politically not acceptable and, above all, too costly to wait for a regional crisis before any fundamental action is taken, particularly because there is some evidence that a region, once it has been labelled a declining region (e.g. Merseyside in England, the Saar in Germany, the Wallonie in Belgium), has extreme difficulties in getting rid of the negative image and in recovering economically. History shows that, for various historical and political reasons, some cities and regions never again reach the economic importance they once had. Augsburg, a world city in the 16th century, is such an example; Lorraine in France, Liverpool, Venice and Istanbul are others.

Hence the crisis argument does not lead any further. One can neither wait for the next regional economic crisis, nor even provoke one to be able to take action or to lay the foundations for a longer term strategic concept. Moreover, given the speed of the globalisation of the economy and the new uncertainty of political development, it will become even more essential in the future to strengthen those forces in a region which are capable of responding to any regional changes occurring. This will also be necessary for a number of other reasons which, in the future, will render planned regional development in Germany more and more difficult:

1 The time has passed when a few ideologies characterised the value systems and aspirations of a regional populace. With the growing fragmentation of value systems and interest groups it is becoming more and more difficult to agree and to find political consensus on consistent goal systems for long-term regional development.

2 Traditional political parties which, due to their established consensus-producing mechanisms, have been the guarantors of longer term political stability, are more and more eroding and losing their former power. Forced by the media into short-term populist action, they are less and less capable of pursuing the longer term regional development visions and strategies they once stood for.

3 The gap between knowledge and action in regional development is widening daily. Convincing the growing number of regional institutions and actors to accept substantial paradigm shifts (e.g. accepting additional, environmentally justified land-use restrictions, or even a shift from growth to non-growth policies) and to move in a certain direction and

coordinate and synchronise their activities has become almost impossible. Hence it is more and more difficult to initiate any change to ongoing mainstream developments based on longer term visions.

4 The growing polarisation of society, and its many cultural, social and economic implications, has weakened the social consensus, eroded solidarity mechanisms and increased incidences of local social conflicts. This polarisation is caused by both the devastating deregulation paradigms of the 1980s and the inability of the established state apparatus to stimulate continuous innovation and regeneration. It makes it more and more difficult to compromise. It also makes any intra-regional cooperation and functional division of labour more strenuous.

5 The exploding abundance of available information and the complexity of interrelationships seems to force the planner to focus just on single problems and short-term projects, while neglecting their comprehensive context and longer term frameworks, which obviously are much more difficult to imagine and to change.

6 The growing difficulty in defining consistent boundaries of urban regions for coordinated action, as varying functional requirements (e.g. for public transit, waste disposal or water provision, or for economic promotion) require different regional boundaries.

To my knowledge, these and other trends which make it daily more and more difficult to intervene into regional development, are not occurring only in the Ruhrgebiet or in other old industrial regions. They can be found in most regional conurbations of the Western Hemisphere and they will most likely continue unless unforeseeable political events change the global political and economic agenda.

Given such trends, it may be essential to consider a strategic effort to foster and to maintain an interrelated set of longer term strategies to foster and maintain the innovative milieu of a region so that the various institutions and actors in a region can better respond to any short and longer term developments. It is the endogenous regional creativity potential which has to be developed and sustained.

The regional creativity potential

After observing public and private efforts towards regional restructuring in the Ruhrgebiet over a period of almost 20 years, I have come to the conclusion that the potential of a region for creative response to all the problems it faces over time depends mainly on seven elements (Kunzmann 1990). These elements of the regional creativity potential are:

1 A diversified and differentiated set of educational institutions offering a wide choice of educational programmes targeted at the regional human capital, but also at international target groups. Obviously the universities and their innovation potential and competence have a pivotal role in building up this key component of the regional knowledge complex.

2 A diversified system of public and private research and development institutions, incorporating centres of excellence of international importance as well as a broad innovative R&D base for the regional industry, reflecting the whole product chain from basic research to the consumer.

3 A cultural environment which serves both the cultural needs of an elitist clientele, demanding cultural events of high international standards, as well as those of the fragmented multicultural society in a region, who wish to have local socio-cultural community and neighbourhood centres.

4 A positive international image for the outside world, documented in continuous international media interest in the region's activities, achievements and events, and a regional identity which is deeply rooted in the regional population. Both the external image and the regional identity are mutually reinforcing.

5 The quality of regional and local natural and built-up environments which has to be protected, conserved or yet created.

6 A broad regional information base comprising print and audiovisual media, media both reporting regional news to the outside world as well as selecting international information as to its relevance for inspiring regional development efforts.

7 A socio-political environment which is open to innovation, supports grassroots experiments, and considers criticism to be a starting point for creative response.

Activities in the respective 7 areas must combine:

- public and private responsibilities for the region and
- encompass both the local and global dimension of regional actions and changes.

Developing the regional creativity potential

One fact seems to be crucial. The development and, above all, the continuous maintenance of the various elements of the regional creativity potential, as described above, cannot be left to market forces alone. Their promotion requires consistent and visionary creative public management to establish long-term policy guidelines for regional actors and actions. Whereas in the past the development of physical infrastructure has been the predominant development task of the public sector, it is the sketched set of elements of the creativity potential which will be similarly essential for successful and sustainable regional restructuring.

From experience in the Ruhrgebiet, the following suggested actions are crucial for fostering regional innovative milieux:

1 A vision of the region's future, serving as a guideline (and a filter!) for the incrementalist day-to-day decision-making process is an indispensable tool for regional spatial and economic development. Such a 'Leitbild' both supports the strengthening of regional identity and the

regional image. Already the process of designing such a regional vision and finding consensus among the regional actors, evokes essential mutual information flows and new communication opportunities.

2 Skilled moderators may be needed to identify and bridge intra-regional communication barriers between the various institutions and actors. This bridging is essential for creating the necessary synergetic actions of regional restructuring and development. Such moderators could be based in newly established (lean!) agencies within local or regional governments or in jointly funded public–private centres of regional change.

3 Allowing or even promoting intra-regional competition for state support and among the various local governments grants can mobilise the creativity of the respective local bureaucracies. Experience from assessing grant programmes based on intra-regional competition, such as the ZIM-programme and the IBA-International building exhibition in the Ruhrgebiet, or the City Challenge Initiative in Great Britain, may be useful for designing appropriate competitions.

4 A multiple choice of communication opportunities has to be created and sustained in the region. Such opportunities range from conferences to workshops, from standing commissions to ad hoc task forces. Such venues where people from various public and private institutions and interest groups can meet and exchange ideas, where they have a chance to overcome their respective prejudices and to bridge the gaps between what are usually rather isolated sections of society, are pivotal to create and to innovate communication networks in a region.

5 With the growing complexity of the regional environment, synergies in regional action will become a key to the likely success of regional policies. Such policies, however, are not easy to achieve. They may best be experimented within carefully selected small flagship projects, acting as experimental innovative catalysts. Only by successfully implementing such projects will the various actors and institutions involved learn about the interests and hidden aspirations of their project partners. Only then might it be possible for biased opinions and value systems to be altered.

6 In successful regional development the linking of single projects and sectoral policies is desirable, even indispensable, to strengthening and mutually reinforcing the mental and physical networks within a region. Hence, in designing single projects and programmes, the potential interfaces to other projects should be considered and made explicit, in order to encourage other actors and institutions to link up.

7 Regional innovative action requires careful monitoring of local and regional grassroots projects and initiatives by a specialised centre of innovation. Such a centre, as a focal point for local as well as international information exchange on such projects, would have to document all the relevant efforts in an ideas bank, and disseminate their achievements (and their failures) to a wider public.

8 The regional media also play a crucial role in this process. They should be incorporated in regional restructuring and reminded of their positive and influential role for regional development, and of their great ethical responsibility, which should guide their almost monopolistic monitoring, information diffusion and mobilisation potential in regional development and restructuring.

9 In a world where investments in real estate are planned for shorter and shorter time spans, where industrial production complexes of the 1960s have been dismantled after 20 years at the most as technologies have changed, where shopping centres or business parks are becoming obsolete after only 15–20 years, where housing, with few exceptions, will no longer survive the third generation of tenants, traditional land-use planning may have to become more flexible. We have to accept or even to plan for 'shifting urbanisation' within a region. Hence the innovative centres of a region may well move around over the whole territory of the region, following the respective life cycles of real estate investments. This in turn suggests that land uses may change over longer time periods, just as three different crops are grown on a field to regenerate and enrich the soil.

10 One final point has to be made. In the past, regional and local bureaucracies have not been very active and effective with regard to continuously retraining their employees and reviewing their actions. This fact has been one of the major reasons for discrediting the public sector and for justifying the quest for deregulation. It has also led to the mushrooming of new and more innovative intermediate regional institutions which, as lean regional and local management units, have been more flexible in their negotiation and cooperation efforts. Either the public sector must be willing to learn from this experience and to attach much more weight to internal qualification strategies and the flexibilisation of its development-related activities (and on the related matter of individual career promotion), or it will have to limit its tasks to formal control, leaving the development tasks to the more flexible intermediate agencies. My plea is for more and better qualification and the reorganisation of its immobile hierarchical structures, along the lines of the deconcentration models some larger private corporations have already successfully implemented.

Conclusions

What are the conclusions of these deliberations on building up a sustainable regional creativity potential? One lesson to be learnt from observing restructuring efforts in traditional industrial regions, like the Ruhr, over a longer period is that the mental infrastructure, that is the capability for creative response to the complex challenges of subsequent waves of structural change, is at least as important as the physical infrastructure or consistent public sector led industrial policies. There is also some evidence that this is even true for advanced industrial regions (e.g. Southern California or the Stuttgart region in Germany) where the continuous innovation of the regional creativity potential to respond to economic crises has been neglected during economic boom times.

The building up of a sustainable innovative mental infrastructure in a region, however, requires continuous investment in human capital and in a wide variety of facilities and opportunities to establish or to improve regional information exchange and communication networks.

NOTES

This chapter has been written while the author benefited from the hospitality of the Graduate School of Architecture and Urban Planning at the University of California in Los Angeles. Graham Cass at the University of Dortmund was so kind to look through the text and to polish up the English where it was necessary.

15 Techno-Sustainability in Taiwan

Edward J. Blakely

THE multifunction polis or MFP discussed in Steven Hamnett's chapter is the international metaphor for technological sophistication. Hamnett's chapter concerns the creation of technology innovation in Australia as the vehicle for economic renewal as much as the development of an industrial form. In other words, governments are anxious to demonstrate that their nations are technologically competitive. This used to be called 'smokestack chasing'. Today, the smokestack has become the technology city. Almost every newly industrialised nation in Asia is designing its own version of a *techno-city* based on the Japanese technopolis model and loosely fashioned after the various US technology nodes.

Taiwan is among the last of the Asian Tigers to establish a technology city presence, attempting now to alter its national economic development agenda by creating a Silicon Valley. This small nation has emerged as a leading producer of the world's computer and information systems, producing 10% of all microcomputers, 17.5% of all terminals, and 67.4% of all motherboards, mice and similar peripherals. Historically, labour-intensive export-oriented manufacturing of computer equipment has accounted for Taiwan's phenomenal growth in earnings and its huge international trade surplus, creating the world's largest foreign reserves with over $US 82 billion on hand. The country's incredible productivity has, however, also led to social and environmental dilemmas such as increasing auto dependence and air and water pollution.

The combination of money and technological sophistication is requiring the nation to rethink its economic and political destiny. The most difficult challenge, according to Taiwan officials, is that the country must develop its technological leadership by retaining and attracting its own highly sought-after engineers and professionals now residing in the United States and Europe. This strategy is required for Taiwan to become the technology nerve-centre for the emerging China. As a consequence, the Taiwanese government is embarking on a national economic strategy that embraces new high-tech-oriented growth. The backbone of the new high-tech Taiwan is a rebuilding of its urban form to accommodate future industries and to recapture as well as retain the intellectual resources that leave the country. Thus, the success at transforming Taiwan's economy hinges, in part, on balancing sustained economic development against the sometimes conflicting goal of environmental protection. In order to meet this challenge, Taiwan is establishing one of the world's most ambitious new cities and regional development building programmes. The Taiwanese government plans to spend over $300 billion over the next 6 years to modernise the nation. A major component of this modernisation is the design and development of six new cities dedicated to science-based economic development. These new cities range from communities of 30 000 attached to existing metropolitan areas to a new science region built around the city of Hsinchu with a current population of 600 000. The Hsinchu area is being redesigned as a technology region of over 1.3 million. The scope of these developments and their economic consequences are enormous. The Taiwanese government anticipates that they will become one of the most productive high-tech research and development nations in the world, using economic superiority as a major international bargaining chip. Taiwan has elected the Silicon Valley regional model over a single tech-park approach, partly because land assembly is so difficult. However, a high-tech lookalike is not enough.

Concentrating Western-style amenities in only a few locations is also strategic, creating a large enough critical mass to make Taiwan's technology area easily recognised. Moreover, Taiwan needs to develop these new communities to handle its modernisation drive in an environmentally sound and economically productive way. The country is densely settled, with severe environmental problems. These new cities are intended to provide a better physical and social environment for the high-tech economic growth of the next century's industries and cosmopolitan population.

In the past year, a project has been undertaken that combines the California Silicon Valley and the North Carolina Research Triangle approaches to design a techno-region for the Taiwan government. These two models were merged because each offered appealing ingredients. The Silicon Valley research park offers the link between university and private sector activity throughout a wide area, while the North Carolina model provides an excellent example of using a large land mass and central coordination of marketing and infrastructure development.

The high-tech issues and images

Just as it remains for Australia's MFP, it has been the subject of considerable controversy what urban form a technology community must possess. Some writers suggest that any metropolitan area is or can become technologically sophisticated (Rees & Stafford 1986). Others argue that technology communities have particular characteristics, and clear choices can be made to alter a community's technology development potential.

Blakely & Willoughby (1990) argue that there are social, cultural and physical dimensions that can be assembled to develop a node or community conducive to technology development. These normative choices, outlined in Figure 15.1, are interrelated with other attributes such as the quality of the social and physical amenity environment.

In essence, there are two considerations. First, the community must be located where the surroundings offer high physical amenity. Physical

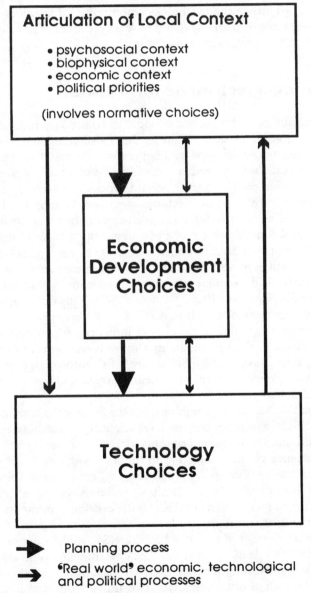

Figure 15.1   Economic development planning and technology choice

characteristics are as important in the technology era as the presence of iron ore, rivers or good natural resources used to be when the economy was based on manufacturing and agriculture. Thus, it is unlikely that places too cold, too hot or in unattractive venues will become 'tech-nodes'.

Second, it is important for a community with technology potential to assemble the right set of internal attributes that match the technology industries' knowledge-based requirements and other 'human resource attractors'. The combination of physical amenity and social/knowledge resources must attract the technology community. In fact, it is fairly clear that the area must be marked off and have a special set of features that give the community a clearly separate identity.

Designing a tech-node for Taiwan

Taiwan has adopted the Japanese model of technology inducement to a specific geographic location. That is, the Taiwanese have attempted to attract or create Japanese-style 'technopolises' or zones. These tech-nodes, as I prefer to call them, are either a single city, a group of cities, or a region specifically redesigned to attract certain types of technology firms. The national government is a prime actor in identifying, funding and providing the special infrastructure for these tech-nodes. In most instances, this means either the establishment of a new university with dedicated research resources to support a predetermined technology or the repositioning of one or more universities to act as the support base for newly emerging industries in a designated field. Another approach is the 'centrally planned model' of technology development (Blakely 1987, p.735). This differs from the US market-led approach inasmuch as it relies on government intervention to create a stimulating climate for certain industrial forms rather than the random chance of market forces. In a more direct 'government-guided' approach, best illustrated by Singapore's IT (information technology) strategy, the government selects research areas and invests directly in certain technology arenas.

In Taiwan's case, central planning takes the form of the National Science Council (NSC), which not only initiates scientific research, but acts as the founder and sponsor of one and possibly more new science cities or regions. The first science city is being built on the framework of the existing city/county structure of Hsinchu, approximately 60 km north of Taipei. In this respect, it is very much like the Japanese technopolis (city of technology) which are based on reshaping or designating existing communities as high-tech communities or regions.

Hsinchu has been well resourced by the NSC and the national government to play the role of a tech-node. The government has invested heavily in Hsinchu, establishing a major national basic research laboratory that employs more than 6000 scientists, two universities and a network of applied research facilities ranging from a new cyclotron to special tele-communication facilities.

In addition to this public infrastructure, the NSC funds and operates the nation's largest technology park, with the most advanced state-of-the-art physical facilities for technology firms in Taiwan and perhaps in all of Asia. By the end of 1992, Hsinchu Science & Industrial Park had over 135 firms which employed more than 23 000 people. These firms can be basically subdivided into computer and peripherals, integrated circuits, telecommunications, opto-electronics, automation, biotechnology, environmental technology and energy groups. They are primarily first generation manufacturing establishments, meaning they have moved beyond research and development and become established in the international market with a set of products or services. These firms use research rather than produce it.

The best example of HSIP's firm profile is the fact that Acer Computer is the park's largest tenant. In many respects, the science-based industrial park is aptly named. It is not presently focused on technology or research which presents one of the NSC's challenges. While HSIP is successful in housing-developed technologies, the NSC wants the park and the community to become the host environment for the creation of new technologies.

HSIP has attracted firms, but it has not provided the intersection between science development and new product design that Saxenian and others describe as the basis for a strong technology milieu (Saxenian 1990). In fact, in a paper describing regional linkages, Wu & Leung (1992) assess HSIP as not having much of a relationship with the surrounding scientific infrastructure or many of the region's firms.

NSC's goal is very clear. Taiwan must become a technology exporter and not merely a low end high-tech merchandise producer. Thus, HSIP is not enough. A total environment must be created, stimulating creative activity that results in both new science and new firms.

Looking for a better mouse trap

From the NSC perspective, HSIP can only be a success if it attracts the kind of people and firms that create new technologies. An oft-stated goal by the NSC leadership is that Taiwan must reattract its lost scientific brains trust from the US and Europe. The subscript of this is that a verisimilar living and working environment will achieve this end as well as induce European and American entrepreneurs to be attracted to Taiwan. In sum, if 'we build it, they will come'.

It is not always clear from the literature or from industrial folklore what the right environment is and what components it must possess. There are several different approaches to 'tech-node' building. At this juncture, the NSC has taken on a science-city/region approach by default rather than careful assessment. Table 15.1 is an attempt to articulate the features of the various choices NSC examined before making its selection. Each approach can be examined against a common set of variables.

Table 15.1 Major research centre paradigms

	Basic Components	Organisational Structure	Goal	Examples
Science City/Region	• International urban design • Multi-university & research centres • Gov't research centre • Magnet science facility	• Local/regional gov't • National gov't involvement in regulation process & designation as tech city • Development authority manage parks & facilities	• International presence • Agglomeration of technology (node) • Concentration of human resources	Technopolis, Japan Multifunction Polis, Australia
University Research Park	• Single university base • Concentrate on single sector (biotech, electronics) • Specialised university institutes	• University managed sometimes with private developer	• Promote university business interface & technology transfer	Stanford Research Park MIT
Non-Profit Research Centre	• Multiple university base • Control resource centre • Shared science facility with local business/ industry	• Joint university & regional or state gov't managed • Special legislation	• National presence • Technology firm attraction	N. Carolina Triangle Wisconsin Florida
Private Tech Park	• Special infrastructure • Incubator areas • Business offices & related facilities	• Private developers • Joint local gov't & private development	• Real Estate Development • Marketing of facilities to the private sector	Rt. 128, Boston Harbor Bay, Alameda Silicon Valley, San Jose

The regional/science city approach emphasises a total living environment conducive to technology firms and attractive to the technology workforce. Science regions or cities can be created either by government or private-sector actions, or merely evolve into their form à la Silicon Valley. The Japanese have attempted, with varying success, to create technology areas. Sophia-Antipolis and the Australian Multifunction Polis are currently the world's best illustrations of this approach.

Another more modest effort is the University Park, based around a university. The idea here is to create a direct tie between the research generators and the research consumer/producers. This is a US style that has become enormously fashionable throughout the industrialised world. The success rates of US research parks have been recently examined by Luger & Goldstein (1992). They describe the mixed record of these endeavours.

Non-profit research parks have emerged as an attempt to create a base that bridges the university and other research resources in the public and private sector. The North Carolina Research Triangle is the world model for this type of approach.

Finally, private developers have emerged to fill the market for research-oriented space as well as design links with universities, national research laboratories and private research facilities.

These approaches have aspects in common and, in one way or another, form a conceptual base for forging a model. The issue for Taiwan is how to use them to design technologically sophisticated places for new town or community developments in Hsinchu as well as several other communities in the planning stage.

The techno-sustainability challenges

Taiwan is very densely settled along its west coast, where the land is relatively flat, in contrast to the centre of the island and most of the east coast, which is covered by mountainous terrain. As a result, agricultural and urban living are in close proximity in most parts of the country. Moreover, as the island has industrialised, it has placed extreme pressures on the small land inventory. Air pollution levels, in Taipei as well as the other major cities on the west coast, are among the world's worst. Vehicle exhaust contributes most to this pollution, with the island expanding its auto registration by more than 6000 vehicles a day. Motor scooters lacking anti-smog devices are the dominant means of daily short-haul transport; many are in poor repair and emit clouds of blue smoke. The particulate matter in the air is so severe that an observable number of Taiwanese wear face masks outdoors at all times.

Water pollution is also well beyond World Health Organisation standards. Much industrial waste is poured directly into streams and rivers, and there is no tertiary treatment of sewage before it reaches the ocean from any city or village on the island. Finally, other pollutants and toxins are not yet controlled to the degree they are in Europe, Japan and the US. As a result, soil, air and water contaminants emitted from establishments are only slightly above Third World levels.

None of this is to suggest that Taiwan is oblivious to any of these problems. To the contrary, a vigorous new Environmental Protection Agency has been formed that is rapidly moving the country to world standards in all pollution and environmental protection areas. In the interim, Taiwan views this problem as one of the prices of rapid industrialisation. Furthermore, new technologies are providing cheaper means to eradicate these problems. As a result, Taiwan is pursuing a strategy of both cleaning up existing conditions and rebuilding the island nation infrastructure as pollution free through large-scale transportation, water and sewage, and new-town developments. In summary, Taiwan wants to build its way out of its current environmental problems.

The building programme to meet these goals is ambitious. A new high speed rail is proposed to traverse the island. A multi-billion dollar underground and above ground light rail system is being built in Taipei, and numerous new highway and bridge building projects are underway. But the basis of this new infrastructure is 6 to 10 new settlements modelled on Hsinchu as high-tech communities. These communities will take advantage of modern technology and will also become the nodes for new research-based technology economic development.

Taiwan has both economic and environmental sustainability problems/opportunities in several respects. First, rebuilding the existing city structure is a challenge. Second, a culture clash is occurring as the nation merges into a globally competitive system. The global dimensions of growth are reflected in the building styles and city organisation as well as changing tastes and preferences. For example, the large-lot single family house is becoming a desired commodity in a small island with very little available space.

There are new tensions on the natural resource base as well. As new settlements are proposed, agricultural lands, foothills and sensitive watersheds are encroached upon. The direct and indirect dimensions of these changes cannot be effectively measured in an environment already so disturbed.

Finally, there is the technology sustainability issue. Can design or organisation create a high-tech milieu? In effect, can Taiwan create the physical, social and research environment that will attract overseas scientists and retain local ones so they will create new products?

Each of these issues is discussed in the balance of this chapter. Obviously, there are no easy solutions. We will use Hsinchu as the reference point for this discussion, evaluating it as a surrogate for a set of newly planned communities.

Settlement sustainability: land use and jobs, housing balance

Hsinchu is planned to grow, based on the HSIP park, which will increase fourfold to become a city within the city. The organisation of the new science city is still under discussion. However, the basic structure calls for a blended working and living environment. The rapid growth in Hsinchu even without the park can be seen in Figure 15.2.

This figure also shows the region's anticipated growth as a science city area. Rapid growth has its own special problems no matter where it occurs. In the case of Hsinchu, growth is being induced into a pattern aimed at serving two goals. First, it will be concentrated on expansion of the park and its immediate periphery. Second, the style of growth will be western; i.e. more low-rise than the usual Asian pattern. As a result, science city planners are attempting to design housing into the park setting as well as develop housing in a concentrated new town area adjacent to the park. To accommodate higher level executive housing, private land is being released for housing developments in the surrounding hills.

Although the planners are proposing higher density development that connects with a circular municipal light rail, the park is designed primarily for the auto. This design form is a direct replica of Route 128 and Silicon Valley, intended to simulate other high-tech environments. But inherent conflicts in this form and the desired environmental outcomes have not been well articulated.

Another issue is that the surrounding hills are not suited for development of any type; accessibility by public service vehicles will be a severe problem. The hillside vegetation is also very flammable and represents an extreme hazard during some periods. However, to restrain growth there would limit

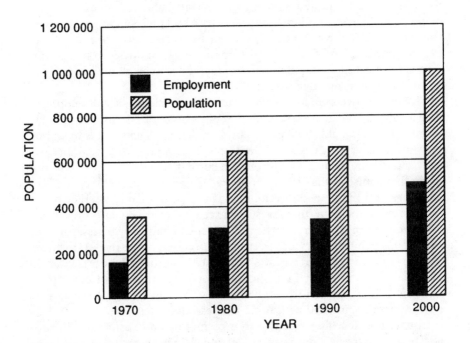

Figure 15.2 Population and employment, Hsinchu

housing style choices and thereby make Hsinchu less competitive on the basis of housing amenity levels.

The issue of housing style is more than architecture. It represents an entire organisation of life. The lower density style is not only auto-oriented, it is also more amenable to smaller family living arrangements. This has implications for the extended family lifestyle of the Taiwanese as well as the organisation and delivery of retail services. Shopping malls and hypermarkets are already being planned at highway intersections to cater to the single family home owners, which may affect current community shopping centres.

Cultural sustainability

Hsinchu and Taiwan in general are culturally colonised. Western building styles dominate the skyline. Western food and films and other culturally homogenising influences are widely available. On some levels, this provides comfort and familiarity for western families or families that have lived in the west. However, on other levels it dilutes local cultural expression by reducing cultural variety and style. The Hsinchu HSIP development epitomises this transformation. The park looks and feels like many similar technology parks in the world, although it has some unique features. Housing, for example, is directly incorporated into the park grounds, which are closed in behind security gates (an increasingly common feature in the US).

Building styles have merged into the 'international box' and lack any form of cultural uniqueness or vernacular articulation. As a result, one could be looking at a building of similar style and presentation anywhere in the world. A few buildings have been provided with 'pseudo-Asian' expression on the roof gables and trimmings. However, even this passing nod to the local surroundings is so pale that it has to be pointed out by park employees or it is readily missed.

The park is becoming the style setter for the science city development. As a result, the organisation and design of its buildings is pushing out older forms. In fact, the old city hall is to be abandoned. A new one is to be built near the core of the new HSIP developments along with a new shopping mall and cultural centre, away from the current central city area. This may mean the demise of the historic core area. Already, encroachments of new buildings and unplanned development have permanently altered the old city system to the point that few historic remnants are intact.

Building form is only a superficial depiction of cultural expression. In the case of Hsinchu, its previous identity is precisely what the 21st century science city is intended to remove or displace. This new identity is intended to assist the community in attracting overseas Chinese and other internationals. The idea is to present them with a primarily western visual environment augmented with some Chinese character.

In addition to selling Hsinchu to a cosmopolitan constituency, it must also be sold to locals. Currently, professionals in most sectors live in Taipei

and drive the 60 km daily to Hsinchu. Figure 15.3 presents an occupa-
tional/ sectoral distribution of the Hsinchu population.

Informal discussions suggest that there are two reasons for this. One is
that Hsinchu schools are not perceived as 'good'. Since schooling is so
important to social and economic progress, the perception that Hsinchu
schools will not lead to the best universities is damning.

Another factor limiting Hsinchu is the fact that it is the 'backwater'.
Important people, cultural activities and other issues of significance occur
in the national capital. Hsinchu has to alter its image, to remake itself into
a techno-pole similar to the familiar icons of Silicon Valley or Cambridge,
Massachusetts. Of course, the down side of this approach is that it may turn
out to leave Hsinchu with no image at all or one alien to its host environment.

Natural systems sustainability

As mentioned previously, Hsinchu is in an environmentally delicate area
with several picturesque settings. There are several rivers and streams that
serve the province flowing down from the surrounding hills and mountain
sides, some of which have been marked off as sensitive. However, one
particularly sensitive area is already under threat because of private sector
anticipation of the science city.

Expansion of HSIP will directly threaten the area's wetlands. However,
since most of this land has been farmed for many years, this in itself is not a

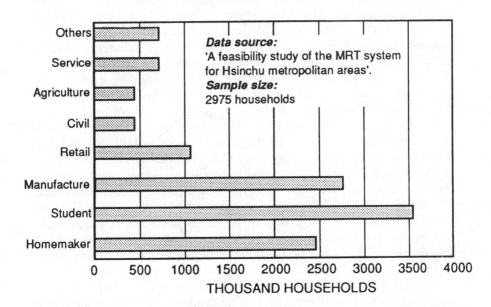

**Figure 15.3   Occupation distribution, survey of Hsinchu metropolitan
areas**

serious issue. The most important thing for Hsinchu is to preserve views and open/recreational space to serve the needs of the potential new immigrants.

One environmental asset not being developed for the science city is the nearby seashore area. The central city is about 10 km from the ocean. The Hsinchu dock area has nice views but no beach. It is smelly but has a general old-world charm. There are adequate areas for housing and private boat docks. However, local prejudice against the sea makes development along the seashore unlikely in the short term.

There is insufficient data for real environmental analysis on the cumulative impact of the development of the scale proposed. Clearly, the added population, autos and resource requirements could alter the physical environment substantially. It is possible to model these impacts when better environmental data are available. This process is under way.

Techno-sustainability

The science city is an infrastructure but it does not produce science. Thus the real argument for sustainability is whether the national investment will result in new science and new products. The evidence to date is mixed. HSIP is certainly a successful park which has attracted technology manufacturing firms. Some of these firms are new start-ups, but the majority are existing Taiwan expansions or overseas branch plants. Since land is in short supply and quality industrial layouts almost non-existent, this early success might be anticipated. The real question is whether the growth rate of the Hsinchu HSIP can be sustained from internal spin-offs, or new joint ventures, coming from its own firms or from the local university research centre complexes. Table 15.2 shows the composition of HSIP to date. As the table indicates, most of the firms form two types: computer and peripherals, and integrated circuits. The other companies occupying the park are primarily in related markets like biotech.

Hsinchu is a manufacturing centre, as Figure 15.3 shows clearly. According to Wu & Leung's (1992, p.10) research, only 5.25% of HSIP firm expenditure was on research in 1992. This indicates that these firms still view research as modifications of known or found technologies and not path-breaking new science. Creativity is difficult to induce. However, despite Wu's data, it does appear that relationships are being forged between industrialists and scientists. While Taiwan is anxious to promote new science by concentrating human resources in a locus like Hsinchu, it will take decades to know whether this community or any other will reach the creative take-off point.

The other issue is to what extent the HSIP firms, as internationals, use local resources as suppliers. The existing firms are not yet well integrated into the regional economy. A great deal would have to occur before suppliers would give up current venues for Hsinchu. On the other hand,

Table 15.2 Support systems required for advanced technology centres

Support System	Science City	University Research Park	Non-profit Research Park	Private Tech Park
Information Communications Strategic relationships	Teleport & international major communication facilities	Conference facilities	Innovation centres	Publications Media support Commercial marketing
Human resources orientation	International research scientists	University faculty as consultants	Skilled technicians	Top management skilled and flexible
Education and research facilities	University or research institute	Research labs	Research labs	University/technical college
Environmental quality	High quality houses and recreation areas	High quality recreation and cultural areas	High quality living and working environm.	High quality work facilities and equipment
Business and community institutions	National policy and process	Venture start-up firms	Entrepreneurs and local social network	High-tech specialists as catalyst Close ties with large firms
Local Government Organisation	Local and national gov't collaboration	Joint-venture university & private companies	Information services centre	Local gov't provide support infrastructure
Finance	Heavy gov't finance	Research grants	Investor consortia	Institutional finance
Physical infrastructure	Specialised infrastructure in support of technology Transportation Infrastructure	Research facilities accessible to local investors	Fibre optics or related transmission services	International airports Toxic and scientific waste systems
Enterprise facilities	Research institutes, universities	Incubator buildings	Innovation centres	Technology and/or business and science parks
Community facilities and support	Good community facilities	Regional recreational, cultural and social facilities (e.g., golf, symphony)		Specialised commercial and professional services
Technology image	International status	Environmental quality reputation	Designation/or reputation for specific technology	Top quality high-tech firms
Planning and development regulations	National directed regional technology strategy	Provision of research and testing areas	Designated mixed use areas	Controls related to industrial requirements

new suppliers might emerge from the process. As Wu & Leung (1992) note, there is a high degree of polarisation in the high-tech development process. The Hsinchu-Taipei region is the growth pole for the high-tech industrial base and pulls all such activity to it. In fact, the accelerated growth of Hsinchu may retard development of other technology nodes elsewhere in Taiwan. So, it is very important for all of the nation's regions to develop complete technology bases rather than remaining concentrated around Taipei. This seems unlikely on the basis of Wu & Leung's (1992, p.12) data.

A decision-support system for tech-nodes

The decision to build HSIP and other technology based communities has already been made. As these communities come on line, policy makers are asking what the mix of ingredients needs to be to insure success. Clearly, the physical design is not enough. An understanding of separation of functions within the city, such as manufacturing, housing and retail, is required to guide the technology community building process. Historically, industry and community were only tied through natural resources and hard infrastructure like railways and highways. The technology is different, inasmuch as a set of 'soft infrastructure' requirements are now central to the process. Taiwanese policy makers, like many others in the world, have not been able to integrate the variety of soft and hard infrastructure into a single paradigm that bridges the various tech-node forms. Taiwan decision makers need a decision-support matrix to clarify the range of options. The matrix (Table 15.2) provides a decision guide. It is not the answer but it is useful in thinking about the mix of physical, social and community infrastructure that can form the basis for designing a technology community or area of any scale.

In addition to this support matrix, an environmental decision-support system has been devised to forecast the environmental consequences of various development alternatives for the Hsinchu area. The model links existing economic, visual and geographic computer modelling technologies that have not yet been used for city planning, forecasting and monitoring. Finally, planners have been able to see how merging data into visual form can influence decision making. The initial project is coming to an end just at the time when the pioneering computer simulation and modelling system could yield important information for policy makers on the integration of economic and environmental factors. The team is in the process of devising the protocols and operating systems for an integrated visual, economic and geographic modelling process. This new modelling effort would incorporate three simulation technologies into a common system that could measure the cost, environmental and visual impacts of any proposed new city based on real data and in real time.

Geographic information systems or even visual and economic models are not new. What is new is the attempt to allow decision makers to see the physical dimensions and economic requirements of a project as large as a city in real time. Hsinchu science city provides a unique opportunity to utilise some of the world's most sophisticated information technologies in the design, assessment and monitoring of a new international high-tech node. A complete city with several thousand new technology firms based around two national universities and a national research lab will be the template for the plan. While focusing on Hsinchu, the concepts will be applicable to a far wider set of places. Concepts like jobs–housing balance, high density, and travel intensive design systems, along with Asian versus western housing styles and design approaches, can be examined with geographic, economic and visual modelling. The project's goal will be to design a computer modelling and visualising system based on Hsinchu with applications anywhere. The computer modelling and visualising system can be transferred to other locations and opportunities for analysis once it is developed.

Conclusion and discussion

Technology chasing has definitely replaced smokestack chasing in the economic development lexicon. Building new physical forms for new technologies does not differ that much from the industrial development approaches. The primary difference in these strategies is well illustrated in Taiwan, where the nation is building communities rather than factory buildings 'to suit'. The goal is to follow the lessons of Silicon Valley, amplified in the literature by Saxenian (1990) and Blakely & Willoughby (1990), which show conclusively that the quality and presentation of the place or venue matters. Thus, Taiwan is basically changing its settlement form, visual appearance and social instruments to accommodate or induce creative productivity. Balancing the physical, cultural and social environmental challenges in this process is an enormous challenge.

The 'jury is still out' on Taiwan's attempt to transform its economic destiny by recreating itself as a technology developer versus a high-tech manufacturer with low-end products. However, the evidence available suggests few real alternatives, if the country wants to remain internationally competitive. Will a set of new towns with technology ingredients pave the way to a new future of the island nation? Who knows?

What we do know is that attempting this approach in Taiwan leads to physical and social tensions with the existing environment. Because it is a clear physical form and operating entity, a great deal can be learned from the Hsinchu science city experiment. New technologies are being pioneered to forecast as well as measure the economic, physical and visual impacts of various alternatives. This decision-making support system represents a powerful resource for not just Taiwan but all countries and regions interested in repositioning themselves for technology-based economic development.

16 The Adelaide Multifunction Polis in 1993:
From Technopolis to Enterprise Zone

Stephen Hamnett

A quite remarkable amount has been written about Adelaide's multifunction polis, or MFP, and especially about the genesis of the proposal, real or hypothesised, and about its significance to the development of Australia–Japan relations. For detailed accounts of the origins and early stages of the MFP, and of the broader evolution of Australia–Japan relations, readers are directed to the bibliography and to the works of McCormack (1990; 1991a; 1991b), Mouer and Sugimoto (1990), James (1990), Inkster (1991), Rimmer (1989) and Hamilton (1991) in particular. This chapter summarises briefly events between the first announcement of the MFP in 1987 and the decision to locate it physically in Adelaide, South Australia, in mid-1990. Its principal focus, however, is on the subsequent progress of the Adelaide project itself and on the status of the MFP at the time of writing in mid-1993.

The MFP before 1990

Two specific antecedents of the MFP are commonly identified. The first of these is the Japanese 'technopolis'. The second is the short-lived proposal, hatched by the Japanese Ministry of International Trade and Industry (MITI) in 1986 under the name of 'Silver Columbia', for encouraging retired Japanese to settle abroad, and the related ELSA ('Extended Leisure Stays Abroad') scheme which sought to provide younger Japanese with opportunities to enjoy longer stays in foreign countries. The 1987 multifunction polis proposal was seen to combine the leisure and resort—'high touch'—elements of these schemes with the 'high-tech' elements of the technopolis.

The MFP proposal was first revealed in January 1987 and set out in detail in a 65-page 'basic concept' document in September of that year. It involved the construction somewhere in Australia of a new city of about 100 000 people. This city would be a prototype for cities of the 21st century, combining leading-edge industries such as computer and information technology, biotechnology and health sciences, with activities based on leisure, resorts, conventions and tourism. The proposal was enthusiastically accepted by the Australian government as a means of assisting with the restructuring of Australian industry and of building a special relationship with the most dynamic economy in the world. As for Japan's interest in the project, McCormack (1991b, p.41) has identified as key factors Japan's desire for an enhanced international role; the expansionist thrust of Japanese capital (and the desire, in particular, for the Japanese construction industry to extend its already significant activities in Australia); and the Japanese tendency to utopianism and the search for alternatives to the alienating characteristics of Japanese urban life, already expressed in Japan in a succession of projects with names like 'Green City', 'Super City', 'Sunshine City' and 'Teletopia'.

Following the Japanese presentation of the MFP concept in 1987, agreement was reached in January 1988 to conduct a joint feasibility study. MFP national committees were set up in both countries, incorporating major national companies, each paying an annual fee of A$10 000 which, together with matching funds from the Australian and Japanese governments, funded a Joint Secretariat and Joint Steering Committee to guide the project.

At the end of 1987 the Australian government endorsed 9 principles to guide the feasibility studies (Brown 1989):

1 The development of an MFP based around internationally traded information, education and training, leisure and tourism, and research and development activities, should be in Australia's interest, with particular emphasis on the pursuit of scientific and technological excellence. The MFP should be developed as a way of assisting structural change in the Australian economy geared towards the development of an internationally competitive and export-oriented industry structure.

2 Fundamental to the competitive advantage of the concept will be the development of leading-edge infrastructure in areas such as telecommunications, information and education.

3 The MFP should be truly international in terms of its links with the world economy, its investment sources and the people participating.

4 The MFP should be developed as an entity which is not an enclave but is linked with the remainder of the Australian economy, providing a leading-edge test bed and technology transfer.

5 Further work should be undertaken on the assumption that the proposal will only proceed to fruition if it can mobilise significant private investor support, particularly in Japan and other countries, which results in a net addition to available capital resources in Australia.

6 The MFP should not be financed through the provision of special location-specific Commonwealth and State subsidies.

7 The feasibility study should investigate a range of urban development options, including those involving multiple sites, all of which assume that the MFP would not be a cultural enclave, but would be integrated within the remainder of Australian society.

8 The Commonwealth government should have carriage of all negotiations with the Japanese government for the implementation of the MFP principles. The States may discuss commercial proposals and provide information to Japanese government representatives.

9 The Commonwealth and State governments are committed to examining the regulatory environment with a view to facilitating investment in the MFP. In particular, the Commonwealth government will examine the climate for the movement of people, money and goods in a positive way so as to enhance the MFP proposal.

Feasibility was assessed in a series of parallel studies, of which the most significant were those carried out by the consultants Arthur Andersen between November 1988 and December 1989, dealing with commercial aspects of the project, concept development and international marketing, in association with Kinhill Engineers who carried out background studies designed to assist in defining the 'spatial attributes' of the project and criteria for eventual site selection. The Joint Steering Committee and the Australian Department of Industry, Trade and Commerce (DITAC) also commissioned work on economic analysis, urban development, governance and social issues, the last leading to a report by Professor David Yencken of the University of Melbourne which drew attention to the lack of emphasis on social considerations in the MFP project at that time and the dangers attendant on not having a well-thought out public information and consultation strategy (Yencken 1989; 1990). It was in this period that the MFP derived a reputation for secrecy and a highly selective approach to the presentation of information (see, for example, Huxley 1990, Nittim 1990 and Peace 1992) which it has struggled ever since to shake off, despite subsequent attempts to repair its image, most notably through a national community consultation exercise which took place in 1990 and 1991 after the selection of the Adelaide site[1].

Yencken (1989, p.25) was also one of a number of commentators who expressed serious concerns about the principle 'that the MFP should not be financed through the provision of special location-specific Commonwealth and State subsidies', and about the lack of any clear or convincing statement about the 'essential role of the public sector in any large polis project'. Peter Self (1989, p.29), writing at the same time, was another, drawing on a review of international new town development to warn that

> a genuine new town requires 'patient money' because up-front investment is high, holding costs may be expensive and profits from land and rents take a long time to materialise fully. Experience with private new towns suggests that for financial reasons the new town will usually need to be fairly small and priority will have to be given to maximising the development of upper and middle income housing.

A key part of the process of seeking to establish the feasibility of the MFP project at this stage was the attempt to identify Australian commercial and industrial activities that might form part of an MFP. This process involved 18 industry 'think tanks', comprising leaders from a range of business sectors as well as academics and government representatives. Out of this process 7 industries were identified as potential core activities of an MFP:

• information and telecommunications;
• construction and design;
• advanced transport;
• environment and agriculture;
• education;
• health; and
• leisure and entertainment.

The final Andersen–Kinhill report in 1989 stated that 'the MFP concept was compelling as it met the requirements of its Australian sponsors, the aspirations of Japanese proponents and the demands of potential investors'. It also concluded that the MFP concept was financially, economically and socially viable. Some observers were not convinced (McCormack 1991b, p.49):

> Public confusion and concern over the project, attributed by its promoters to deficiencies in the national character such as 'insular pessimism' or a negative, defeatist approach to national problems on the one hand, or racist anti-Japanese sentiments on the other, actually reflected nothing so much as confusion at the heart of the project itself. The core documents rang hollow with utopian hype and the shifting image they presented suggested phantasmagoria, behind which there might be no substance at all.

Overlapping with the feasibility study, State governments were invited to submit detailed site proposals. Four States, New South Wales, Queensland, South Australia and Victoria, eventually made final submissions and a fifth proposal was lodged by a private syndicate in north Queensland. The Commonwealth government was given responsibility for the selection of the preferred site, based on the advice of the Australian members of the Joint Steering Committee. As the site selection process went on, the notion of an MFP based on multiple sites lost support and a clear preference for a single MFP site emerged. Final proposals were submitted on 14 May 1990 and on 14 June a decision was reached in favour of the Queensland Gold Coast submission, even though much of the proposed 4 700 ha site was in private ownership. The Queensland government was required to purchase the site at short notice or, failing that, to guarantee that the 'betterment' resulting from the development did not go to the existing landowners. After a brief and rather farcical interlude, the Queensland government expressed itself unwilling or unable to meet these requirements, whereupon the MFP project was awarded to Adelaide on 19 June.

The choice of Adelaide was formally endorsed in the final report of the Joint Steering Committee to the governments of Australia and Japan the following month (Joint Steering Committee 1990) .

The whole of Adelaide was nominated as the MFP, but a specific site within the existing urban area, a crescent of degraded and contaminated land centred on Gillman and Dry Creek, 12 km north of the city centre, was designated as the core site for new urban development (Figure 16.1).

Unlike the Queensland site, just inland from Australia's Gold Coast tourist mecca, the chosen Adelaide site had little obvious potential for resort or tourist development. But it was in public ownership and adjacent to a city with good universities and with some significant high-technology research and development enterprises already established. It was plausible to suggest that MFP developments in areas such as information technology, health and medical science could build upon and be integrated with these

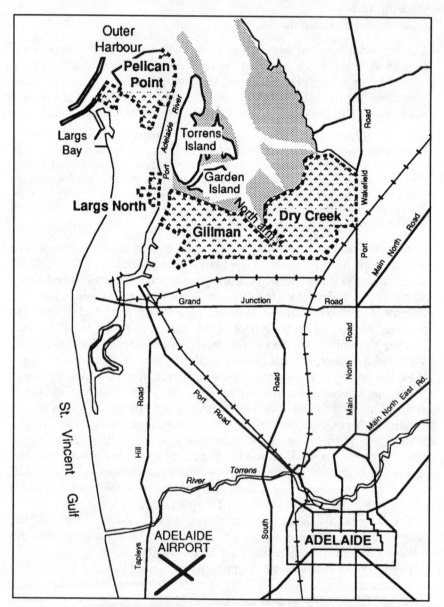

Figure 16.1   The MFP core site in relation to the Adelaide city centre

existing resources. A key notion in the 1990 submission was that of a 'world university', a new type of networked educational institution linked to other universities in South Australia, interstate and overseas. This university was to comprise elements of existing universities linked with 'five new centres encompassing advanced learning technology, environmental management, Asian/Pacific languages, information technology and health' (South Australian Government 1990, p.4–6). The university's administrative centre and several of its new advanced institutes would be at the Adelaide core site. Another important proposal in the submission was the creation of an 'information utility'—a public utility providing a computerised information system for Adelaide which would sell electronic mail, data interchange and brokering facilities and which would be accessible to government, private sector and community users, including those working from home. And there was an intriguing proposal, separately generated but incorporated in the MFP submission, to integrate air, sea and land freight handling in a new 'transport hub' which had the potential to turn Adelaide into the Rotterdam of Australia (South Australian Government 1990, p.4–8).

The Adelaide submission went into considerable detail about the urban development aspects of the MFP. The Andersen-Kinhill report had assumed an MFP with a population of between 100 000 and 200 000 people, of whom 60–80% would be 'highly skilled international workers who take up residence in Australia'. The Adelaide proposal spoke of a population 'up to 100 000' but argued that the particular location of its proposed core site had the potential to avoid any risk of an enclave of overseas workers developing. The core site within the Adelaide metropolitan area adjoins suburbs with many characteristics of social and economic disadvantage, as well as environmental health problems (Bunker 1991, p.151).

> An MFP settlement in the crescent, and connected with these surrounding communities, is obviously one of the means for the amelioration of these social, economic and environmental circumstances...Siting the polis in this location could not be criticised either as establishing an enclave or as appropriating some of Adelaide's desirable real estate; rather it would be an important contributor to a better future for those communities.

Integration with the existing metropolitan area and the advantages which this could offer through connections with existing urban infrastructure were central elements of the submission. The MFP was to be (South Australian Government 1990, p.3–1):

> a large, new and surprising urban development that will become an integral part of the existing metropolis. Its key features will be visible from Adelaide and from the hills. It will develop an identity of its own. The settlement will also be part of Adelaide, both because of its proximity and because of the deliberate social, cultural and economic connections that are an integral part of the design concept.

The core site was close to Adelaide's Technology Park and the Levels Campus of the University of South Australia, on its eastern side, and to the recently established submarine construction plant at Osborn on the western side of the Port River. While seriously contaminated in parts, the Adelaide site also included substantial and significant stands of mangroves and important fish-breeding grounds. The problems of reclaiming and rehabilitating the degraded parts of the site and preserving its remaining mangroves and the adjacent marine environment, threatened by urban pollutants, in part borne by stormwater run-off, were fundamental to the rationale for proposing such a difficult 'brownfield' site. It was seen as an opportunity to demonstrate that such environmental challenges could be overcome and, in the process, to develop new environmental management techniques and expertise which could find commercial application elsewhere (McCormack 1991b, p.53).

> ...the disadvantages of the site, if overcome, could become its long-term strengths and the plan to build upon and refine existing scientific and technological strengths rather than wait for some grand cornucopia of technological and investment blessing to be poured in from outside, was in accord with the lessons to be learned from the few examples Japan had to offer of successful Technopolis development.

In 1990, as part of the first round of feasibility studies, the National Capital Planning Authority in Canberra had been commissioned to develop some urban design principles for an MFP. Its report set the MFP's urban development in the broader context of contemporary debates about the future of Australia's low density cities (NCPA 1990, p.5):

> Today Australia is facing several crises. These are primarily about economic performance, but equally vital is the debate about cities, the cost of living in them, and a future in a world concerned as much with sustainability and environmental quality as with economic growth...Australian cities are exceptionally large relative to its small population. Indeed, more Australians live in cities (over a 1.0 million population) than in most other nations. Yet increasingly, Australian metropolitan areas are chaotic and wasteful. To provide for a growing population, scarce capital is freely expended on housing and infrastructure. Our suburban lifestyles epitomise the 'lucky country'. Australians consume energy and use mobility as voraciously as we consume land. Not only are the resulting costs high, but increasingly Australia is entering an era where energy shortages are of concern and significant increases in national wealth are necessary to sustain present forms of urban development.

Against this background the NCPA report suggested that an MFP urban development should embody higher densities than normal in contemporary Australian urban developments, reduced dependence on the private car and more emphasis on public transport, cycling and walking. It also advocated that long-standing objective of 20th century planners, a balanced distribution of living, working and other activities, in order to reduce the need for people to travel. These considerations led to a suggested 'polis' population of at least 100 000, with 45 000 jobs to be available within the MFP, 50% of these within half a kilometre of home. The NCPA argued that these aims for the MFP were most likely to be achieved through the adoption of a linear plan with a central activity spine containing the most important urban activities and with strong provision for public transport. Residential areas of about 4 000–5 000 people were suggested, differing from 'traditional suburbia' in higher building densities—160 people or 60 dwelling units per hectare—and mixed uses—'home-based activities, corner shops, studios, galleries and small restaurants/cafes' (NCPA 1990, p.49).

The NCPA report also noted that the idea of a privately funded MFP was unrealistic and that

> The postwar experience of private sector development of new communities in the United States of America indicates that large-scale developments involving more than 1000 hectares or requiring more than six years to build have been exceptionally risky (NCPA 1990, p.65).

The Adelaide urban design proposals had much in common with the principles expressed by the NCPA report, but they included some additional ideas and were also shaped by the specific characteristics of the core site at Gillman. The South Australian submission described an indicative urban form made up of 'villages' of varying sizes up to 10 000 people, disposed about a grid at varying densities of development, making creative use of the water features of the site and including the establishment of extensive 'urban forests'. A 'polis' based on separate 'villages' was seen as providing an overall framework within which there would be considerable flexibility for development to be staged and for the character of settlements to evolve over a long period of time (South Australian Government 1990, p.3–1):

> in a creative combination of the past and the future; high technology and human scale; the age-old theme of villages, separate but linked in a new expression of a city.

While the way in which the villages might evolve was inevitably left open-ended, it was envisaged that each might develop a distinctive function (South Australian Government 1990, p.3–2):

> One will be the site of the World University Centre, others the sites for the new advanced research and training centres, the teleport, the environmental management complex, one of the target industries, or a totally new industry yet to be defined.

Some—especially disaffected proponents of submissions from the other States—were unimpressed, dismissing the Adelaide urban design proposals as 'Arcadian', 'naive and vague' and as 'a cross between Noddyland and the background to Thomas the Tank Engine' (Inkster 1991, pp.61–63). Certainly the Adelaide submission was not free of the hyperbole which had characterised much MFP material to that point and, in particular, its designers' notions of 'thematic axes' representing 'land–humanity', 'water–environment' and 'monument–technology' attracted widespread derision.

But most of these urban design concepts were only indicative and representative—how could they be otherwise at that stage? Some of the difficulties which the MFP project has experienced since then are perhaps attributable to the way in which these preliminary indicative notions of how the site might be developed came to be given more weight than they really merited. On the other hand, the Adelaide submission did take the MFP project a significant step forward by beginning to provide some tangible notions of what the elements of an MFP might be and by suggesting a coherent framework within which its physical form might develop.

MFP Adelaide

With the nomination of Adelaide as the site, another period of feasibility assessment began. The Joint (Australia–Japan) Secretariat and Steering Committee were dissolved. A new MFP Adelaide Management Board was set up in August 1990 and an international advisory board was created. There had been negligible Japanese input into the Adelaide submission and, from being the original proponent of the MFP in 1987, Japan's role had now become only advisory. Investigations to determine the viability of the site, to detail the urban design proposals, the infrastructure costs and broader economics of the project, the degree of community support and the prospects for attracting Australian and international investment in the Adelaide project now began in earnest.

The Board completed an interim 'status report' in December 1990 and presented its final report to the Commonwealth and State governments in May 1991. The report was positive and in July 1991 the Federal government formally approved the Adelaide project and provided an initial allocation of $A12 million, primarily to fund the MFP's staff and administration costs. At this time it was hoped that on-site construction would commence in 1992, with the full development estimated to take 20–30 years. The target population of the Gillman core site and associated sites was now 50 000 people, with a further 50 000 possibly attracted to other parts of the metropolitan area.

The May 1991 report reaffirmed the broad vision of the MFP.
MFP Adelaide would become:

- an international centre of innovation and excellence in urban development and in advanced technology to serve the community
- a leading centre of innovation in science and technology, education and the arts
- a national focus for economic and technological developments of international significance
- a model of conservation and management of resources and the natural environment
- a focus for investment in international business development based on new and emerging technologies that will form the basis of the economics of the 21st century
- a social model for the 21st century based on equitable social and economic development.

The World University had by now become the 'MFP University or MFP Academy'. Space industry proposals, identified at an early stage of the MFP process, were now revived, drawing together the activities of British Aerospace in Adelaide, the University of South Australia's research in the area of signals processing, a proposal to build a new radio telescope to the north of the city and plans to revive the Woomera rocket launching facility in the far north of South Australia. Other projects specified at this stage included a 'telemedicine' centre and a tourism and leisure complex at Pelican Point on the tip of the Lefevre Peninsula on the western side of the Port River.

The 1990 submission had identified, rather loosely, an area of over 3 500 ha for the MFP. In the 1991 report the core site was redefined more precisely as 4 discrete parcels of land totalling 2 343 ha—the main Gillman/ Dry Creek site of 1 840 ha and 3 smaller sites, 2 of which were on the Lefevre Peninsula, west of the Port River and the third at Garden Island, north of the main site. The notion of 'a mosaic of villages separated by forest, lakes and open fields' from the 1990 submission was broadly unchanged, although a significant increase in the area of land required for village settlements at the core site was proposed—800 ha as opposed to the original 370 ha—in response to feasibility studies carried out by consultants Kinhill–Delfin. The 1990 NCPA report had suggested a target density for the MFP of 60 dwellings to the hectare and this had formed the basis of the 1990 Adelaide submission's figure of a population 'up to 100 000' at the core site. The Kinhill–Delfin studies were much more conservative, based on densities of 30 dwellings to the hectare—still higher than recent medium-density developments in Adelaide, but not dramatically so. On the basis of these assumptions, it was concluded that the Gillman/Dry Creek site could eventually house about 42 000 people in 17 000 dwellings in a larger number of villages than originally suggested, each nominally of 36 ha in size. The balance of the site was made up of 416 ha of lakes and canals and 624 ha of forests and open space. Kinhill–Delfin's analysis indicated that the development of the Gillman/Dry Creek site could provide an internal rate of return on investment in excess of 24%. This conclusion was arrived at by estimating the return from the development and sale of serviced allotments on the Gillman/Dry Creek site, after allowing for some sharing of incomes and costs between private developers and the government. The analysis assumed that 64% of buyers would be from Adelaide and that the attractiveness of the site would increase as land suitable for development elsewhere in the metropolitan area became scarce. The direct costs to the public sector of the development at Gillman/Dry Creek were estimated at $A251 million, partly offset by contributions made by developers to public works and by the deferral of the costs of providing services at other locations where development would now take place at a later stage because of the development diverted to the MFP core site. The costs to the public sector were thus reduced to $A202 million, or $A105 million in net present value terms (MFP Adelaide Management Board 1991). These assumptions were regarded as rather heroic by local observers, who questioned whether the Adelaide housing market would take up the sort of housing proposed at Gillman at the rate suggested and at the price implied.

MFP in 1992

At the end of 1991 'MFP Adelaide' was renamed 'MFP Australia' as a symbolic reminder that the MFP was still intended to be seen as a national project, although interest in the project in other Australian States was by now lukewarm at best. In the course of 1992, the most obvious focus of

MFP development work continued to be the core site at Gillman/Dry Creek. A draft Environmental Impact Statement (EIS) for this site was released in February (PPK Consultants 1992). During the preparation of this EIS, further modifications were made to the development proposal for the site, principally to take account of more detailed information about site conditions as this became available. Further work also continued on the waste, water, energy and sewage management requirements of the proposed urban development and into the provision of a fibre-optic network to serve the first 5 or so villages, now intended to be built over a five-year period to house about 10 000 people.

In March 1992 the Commonwealth government gave some indication of its continuing commitment to the project by announcing a further $A40 million injection into MFP through its 'Building Better Cities Programme', although this received a mixed response since it was believed by some that these funds had been diverted from worthwhile projects with a strong social justice focus elsewhere in the metropolitan area which would otherwise have been more generously funded.

Then, on 28 April 1992, legislation was passed by the Parliament of South Australia providing for the establishment of an MFP Development Corporation to manage the further progress of the project. The MFP Development Act empowered the Development Corporation to invest and borrow money, acquire land and enter into partnership and joint venture arrangements. The Act also vested Crown land at the core site in the corporation and provided for the integration of South Australia's Technology Development Corporation, the body responsible for the development of Adelaide's successful Technology Park and its fledgling Science Park, with the MFP Development Corporation. The MFP Development Corporation was seen by the South Australian government as an essential vehicle for restoring momentum to the project, and the appointment of its Board members and chief executive officer were seen as matters of high priority.

As the year went on it became clear that there was widespread and growing concern with what was perceived as an over-emphasis on the detailed design of the first settlements at Gillman at the expense of the broader vision of the MFP and its economic and technological purposes. A new Metropolitan Planning Strategy for Adelaide was released in July 1992 and raised questions about the priority which the Gillman site should have in relation to the staging of other residential development projects in northern Adelaide. And the following month a detailed report on *New Directions for South Australia's Economy*, prepared by consultants Arthur D. Little International, made the following comments about MFP (Little 1992, pp.62–64):

> The MFP suffers from a weak public image at present because, while its fundamental purpose and objectives may be understood by a small group of public policy makers, there is insufficient understanding of its role among the wider business community and the public. After discussions with MFP's own management and a wide cross-section of the South Australian community, our own perception of the potential of MFP is as follows:

- MFP's strategic objective is to stimulate economic development in Australia through the imaginative application of new technologies
- To do this, MFP is engaged in three complementary sets of activities:
  - supporting the development of business clusters
  - providing a physical environment for technology demonstration
  - developing a new community.

We believe that our perception of the MFP accords with that of the MFP's management team, although there are differences of emphasis. The most obvious of these concerns the role of the Gillman site: we see it as a means to an end whereas the small MFP management team, faced with the prospect of developing an experimental community from scratch has, perhaps inevitably, been pushed into giving it an undue emphasis. Developing the business clusters is a more subtle task and we believe it deserves greater emphasis and more public recognition...The business projects are central to the MFP's contribution to South Australia's and Australia's economic development and offer the prospect of more rapid payback than the more risky and costly technology demonstration community development activities.

In the same month the local Murdoch-owned newspaper, the *Adelaide Advertiser,* gave voice to concerns about the MFP project in rather less measured language (Hopkins 1992):

> You could be forgiven for losing faith if you ever had it in the MFP. It's still looking for a permanent Chief Executive, it has searched Australia for a Chairman and a Board. It doesn't have the support of the Prime Minister or the Federal Cabinet, with the possible exception of Industry Minister Senator Button; it doesn't have the support of any other Australian State, although it is supposed to be a national project; it doesn't have the support of Japanese investors, or Korean investors, or any other type of major investor, with the notable exception of BHP in a fairly unspecific sort of way; and, most important of all, it doesn't have the support of the people of South Australia who remain uncertain about what the MFP is supposed to be...The Gillman site , no matter how much it reflects the ideals of the MFP by taking a degraded site and restoring it, is not the MFP. It was only ever meant to be part of the MFP, one of several nodes. But that larger vision of the MFP has been lost.

The same newspaper article quoted Professor John Lovering, the Vice-Chancellor of the Flinders University of South Australia, describing the apparent demise of the World University concept:

> We expected to be invited to participate after we won the bid, on the basis that higher education was a major focus of the MFP...but that didn't happen and if the universities aren't involved with the MFP, then it's unlikely it is going to happen.

In October 1992, the appointment of a Board of Directors for the MFP Development Corporation was announced under the Chairmanship of Alex Morokoff, a previous Chairman of Australia's Telecom and, amongst other things, the deputy chairman of the Australian and Overseas Telecommunications Corporation.[2]

And then, at the end of November 1992, the final version of the Environmental Impact Assessment for the core site at Gillman/Dry Creek was released. Its findings were equivocal. It gave a cautious go-ahead to urban development on the site, while warning that there could still be undetected contamination and that earthworks could stir up toxic waste. The *Adelaide Advertiser* drew the conclusion that (Pearce 1992):

> the report adds to rather than defuses many misgivings about the core site. The development of business clusters based on existing centres of technological development is preferable in the current economic climate to the high risk strategy of persevering with Gillman. The core site concept simply is not viable without a major investor commitment. Even then, as the latest report clearly shows, it will remain a leap into the unknown.

By the end of 1992 it was already clear that it would now be 3 or more years before any initial development could be completed. A project update from the MFP office in late 1992, sensitive to the continuing concern about the Gillman site, sought to play down its significance by pointing out that[3]:

> The greatest benefit from MFP to the people of this State will not be the activity stemming from the building of a city to house up to 50 000 people at Gillman, but the economic outcomes from new businesses and activities established as a result of the project.

One of the first tasks of the recently appointed MFP Board was to assess the timing and extent of commitment to the first stage urban development. Perhaps to the surprise of some, the Board reaffirmed that the Gillman site was its first priority. Further significant changes to the 1991 urban development concepts were endorsed, however, after yet another feasibility study commissioned to advise on the cost, timing and appropriate location of the first housing to be built on the site.

On the positive side it must be said that the same 'project update' document of late 1992 contained an impressive list of MFP activities under way, in addition to the urban development at Gillman. The core areas on which MFP was then focusing were confirmed as education traded services, communications technology and environmental management, with ancillary 'industries' including health, media, transportation and the space industry. Specific projects mentioned included a national and international tele-medicine network linking remote locations in Australia and some locations in Asia to a major centre based in Adelaide. The main applications of this are intended eventually to be in the provision of diagnostic and treatment services, the education of health workers and the administration of remote area health services. Another proposal was for a multilingual software development centre as part of a broader programme to create a

> multi-cultural and multi-lingual learning environment to enable students from the Asia–Pacific region to develop their computer-based skills, together with broad international communications and cultural skills.

Also in late 1992 the establishment of an 'MFP services company' was announced: a consortium of 16 entities which includes a major French utility company (Lyonnaise des Eau) and a large Japanese engineering firm (Chiyoda), formed to investigate the commercialisation of new technologies in solid and liquid waste management, energy systems and water services for the MFP site and surrounding urban and industrial centres. The objective is to develop specific technologies for the MFP's urban development which can then be marketed as products and services to other cities in Australia and overseas.

Some of the more convincing conceptual work to have come out of the MFP office in the past year or so has related to the innovative environmental management practices which will need to be introduced as part of any development at Gillman. At present the site is a major storage basin for stormwater from Adelaide's northern suburbs. The MFP's environmental scientists believe that this water can be captured, treated and used to recharge the natural aquifers which are becoming depleted through heavy usage by industrial and horticultural activities. Artificial wetlands will be used to trap the stormwater, heavy metals will be allowed to settle and eventually the cleansed water will be returned to the aquifers. There is also a proposal to convert sludge from the nearby Bolivar sewage treatment works for use as fertiliser or in combination with waste products from a nearby soda ash works to make bricks and paving materials.

Some time ago the Commonwealth government announced that it would locate a major operational arm of its environmental protection agency at the MFP core site and an Environmental Management Centre involving other Commonwealth research organisations is still being studied.

In October 1992 South Australia was successful in securing the Australian Open Learning Technology Corporation for its Science Park site. The University of South Australia is now the home of a new Signal Processing Research Institute, and an Advanced Engineering Centre which draws on the resources of the three South Australian universities is also being funded. The MFP can claim to have had some involvement in all of these projects. And there are some other more speculative projects said to be under way—a project to create artificial heavy oils from eucalypts, using sewage effluent as a source of nutrients, for example, and an International Business Management Institute intended to link South Australia's universities with a number of large Japanese companies.

The Federal election of March 1993

At the beginning of February 1993, the Australian Federal Parliament was dissolved and a general election was set for 13 March. During the previous Federal election campaign, in March 1990, the MFP became a major electoral issue when the then Federal opposition leader, Andrew Peacock, came out against the MFP, first of all on grounds of immigration policy in an attempt to make electoral capital out of fears of a foreign enclave and then, several days later, on the rather different grounds of economic viability. This time the opposition Liberal–National coalition, the favourite in the opinion polls right up to the election, adopted the following policy, announced by its Shadow Minister for Industry and Commerce on 12 February 1993:

A coalition government would refocus South Australia's MFP Gillman to become the Adelaide technopolis based on existing centres of excellence in the State...The $52m already earmarked for MFP Gillman would be redirected to areas such as the Waite Institute, Technology Park and the Universities. Since Gillman was chosen as the site for a MFP several years ago, the Federal coalition has been careful not to condemn the concept so that it had the best chance to attract private investment, both locally and internationally. It is now clear that MFP Gillman was a grandiose concept of the '80s and there is now a need for a more realistic and sensible plan, building on what is already succeeding. There is no doubt in my mind that MFP Gillman has hit a brick wall and it has no future in the short-term, so that Adelaide is sitting off the pace while Gillman goes nowhere.

To the surprise of many, the Liberal–National coalition did not win the election. Instead, the Federal Labor government was returned with a narrow majority for a fifth term. The Gillman development was reprieved.

Prospects

In mid 1993 preparations for urban development at Gillman are proceeding and detailed designs are being drawn up. The environmental arguments which were used to justify the choice of the contaminated Gillman site in the first place are still persuasive and are likely to lead, at least, to the construction of some sort of model urban development at Gillman which exemplifies some of the advances in environmental management and technology which the MFP can fairly claim to have made. The continuing commitment to Gillman seems to reflect the MFP Board's view that it needs urgently to demonstrate progress and that Gillman is the only place where actual development can be undertaken in the short term. The success of this development is likely to be judged on whether it really does appear innovative or whether it turns out to be, as cynics have long predicted, another Australian suburb with a slightly higher density.

From a broader perspective, of course, whether or when development occurs at Gillman is not the principal issue. In November 1988 the then Federal Minister for Science, Barry Jones, addressed a seminar on the MFP concept in Canberra in the following terms (Jones 1989, p.7):

> I strongly urge you, before you rush to dust off drawings of new city plans, to give a lot of thought to the conceptual possibilities of the proposal. The emphasis in the MFP is very much on development of the right 'soft' infrastructure rather than the hard infrastructure of bricks and mortar.

Jones was in the forefront of those who, throughout the MFP project, have cautioned against overemphasising the MITI notion of a new utopia on a single site—the 'City of the Fifth Sphere':

> ... new cities are a late nineteenth century idea; new organisational and electronic networks are a late twentieth century idea that is changing the nature and role of the city (Hepworth 1987, quoted in Mandeville 1991, p.241).

The 1990 Adelaide submission was certainly concerned to make Adelaide a 'systems city', with its 'information utility' as a key element. There is precious little to show from this initiative yet, but perhaps this ought not to be too surprising. David Yencken, writing of the MFP in 1989, noted that (p.22):

> the project will take many years to plan and build and many more years to come to fruition. It will have to work well at many levels, not least financially and commercially and that at best will take many years to achieve. Any very large project will inevitably progress through periods of intense opposition, debate and uncertainty.

Members of the MFP's small and committed staff argue, with some considerable justification, that the project has scarcely started and that a great deal has been accomplished when one remembers that it is only about two years since the Adelaide Management Board completed its feasibility report and the Adelaide project was endorsed by the Commonwealth government. They claim also that interest in the project remains strong in Japan, especially on the part of members of the 85 companies which are members of the MFP Co-operation Association, and that there is also investor interest in Korea and Taiwan.

But turning interest into investment remains the challenge that has to be overcome if the MFP is to realise its potential. And it still seems that more evidence of Australian commitment to the project will be required before any serious prospect of overseas investment is likely. Stuart Fysh, corporate manager of BHP, the one large Australian company to have made some commitment to MFP, was quoted late in 1992 as saying that there needs to be more government support to entice other companies—especially the Japanese—into the project (quoted in Moodie 1992):

> The Federal government's $40 million Better Cities allocation to the MFP was a step in the right direction, but the South Australian government needs to lobby harder to make MFP the focus of national information technology develop-ments...During discussions with 15 Japanese companies in March, I found a great deal of enthusiasm for the MFP, but also a lot of confusion over what the project is all about. They felt there was no structure, no positive effort to make MFP Adelaide an attractive proposition. And until that happens, and they can see Australian companies getting behind the project, the Japanese will stay away.

The prospects of greatly increased government investment in MFP infrastructure, hard or soft, are not promising however. At Federal level, economic rationalism has prevailed on both sides of the political fence for most of the past decade and there is little acknowledgment of the wide-spread evidence that new technology-based industries have done best in managed economies—like Japan!—where governments have played a leading strategic role in encouraging and supporting the development of such industries (Dosi et al. 1988; Brander 1987).

The South Australian economy has serious structural problems, exacer-bated by calamitous losses suffered by its State Bank in the past few years, which led to the resignation of State Premier John Bannon in 1992, and the capacity of the South Australian government to provide further support to the MFP is clearly limited. It did, however, seek recently to encourage investment in the MFP by declaring Gillman and adjacent areas as an 'enterprise zone', within which investment will be exempt from various State taxes and charges.[4] Given the intellectual origin of the 'enterprise zone' notion, there is some irony, not to say conceptual confusion, in the simultaneous designation of MFP as both a technopolis and an enterprise zone. But the need to encourage investment is clearly critical. Denis Gastin, a former senior trade commissioner in Tokyo, former Chief Executive of the MFP Joint Secretariat in the project's early days, and someone present at the meeting in January 1987 when the MFP was first mooted by the Japanese, has summed up the MFP's position as follows (quoted in Hopkins 1992):

> The failure to indicate substance to the project is forming a national attitude that it's gone...Positive things are happening but only major commitments will turn this sentiment around and there are no major commitments in the pipeline. A lot of good work has been done on the MFP but it's coming together too slowly. The Japanese delegation which came through late last year summed it up when they said 'We'll believe in it when you do'.

Postscript

In December 1993, South Australia elected a new Liberal state government. This government moved quickly to play down the importance of the Gillman site and is proceeding, instead, with a refocused MFP, to be built adjacent to South Australian Technology Park and the University of South Australia's northern campus. Both federal and state governments remain committed to the MFP and early development of the new site is anticipated.

NOTES

The author would like to acknowledge the helpful comments made on an earlier draft of this chapter by Associate Professor Raymond Bunker of the University of South Australia, and also the generous assistance provided by Neil Travers and Bruce Harper of the MFP Australia office. The observations and interpretations contained in the chapter are the sole responsibility of the author.

[1] See 'Report of MFP Adelaide Community Consultation Panel', August 1991, Australian Government Publishing Service, Canberra.

[2] The MFP appointed a chief executive officer in April 1993—Ross Kennan, an Australian who came from a senior executive position with the Honeywell Corporation in the USA.

[3] This quotation comes from the transcript of an unpublished address by Neil Travers, Manager, Public Affairs, MFP-Australia, to an economic development seminar of the South Australian Multicultural and Ethnic Affairs Commission, 1 December, 1992.

[4] South Australian Economic Statement, 22 April 1993.

PART 4 Sustainable Patterns of Living and Working

OVER THE LAST few decades or more, the structure of major cities in the developed world has been transformed. It has changed from essentially single-centred to a form in which the majority of commercial and industrial activity is now located in the suburbs. In the major UK, US and Australian cities considered here, about two-thirds of employment is now located outside the central city in suburban centres, special-use zones and even more dispersed. *Commuting patterns* have changed accordingly.

This change may be viewed as part of a broader *transition from an industrial to an information economy*. In *pre-industrial cities* most employment was dispersed in cottage industry or in the surrounding fields, and access was by foot. The *industrial revolution* produced a reversal of this pattern with the new industry concentrated near the centre of the city, enabling the development of mass transit on radial rail networks. With the *information revolution* this pattern has again reversed, with dispersal of employment to the suburbs and beyond and with a shift again towards personal transport to access these dispersed destinations from scattered origins and routes. The scale of metropolitan land use patterns is now such that only a small percentage of trips is short enough for walking, and their dispersed nature is such that only a similarly small proportion can be served by public transport.

Working hours and conditions underwent similar changes. The industrial revolution represented a peak of formality in working conditions and rigidity of hours. The information economy is providing a return to less formal conditions and more flexible hours.

There is a view (Marchetti 1992) that there are also *anthropological invariants* in human daily travel behaviour which influence commuting and urban scale, that people have territorial instincts and budget a reasonably constant proportion of their day to movement about that territory; and that this time for travel is essentially independent of any travel technology used. Normally much of that time would now be devoted to the journey to and from work. Early, pre-industrial cities were apparently limited in size by the return distance that could be covered on foot within that time. Horse-drawn vehicles extended that distance. Mass rail transit allowed an expansion of that scale by almost one order. The motor car has allowed a further increase assisted by employment dispersal, and the potential *population capacity* of cities has expanded accordingly.

The proportion of central city jobs is still diminishing. About one-third of employment currently remains in the central city and includes company headquarters and various producer and other specialist services (legal, finance, insurance, advertising and travel). Its labour market extends throughout the metropolitan area and beyond. Many of these activities are locating in large suburban centres also. Jobs greatly outnumber residences in the central city, with homes generally outnumbering jobs in the suburbs.

Commuting patterns show some variation between the countries and cities considered, and over time. In the chapters following, Cervero, and Gordon and Richardson discuss the major US cities, Spence and Frost analyse London, Manchester and Birmingham in the UK and Brotchie *et al.* focus on Melbourne, Australia. These next four chapters analyse the extent of *dispersal of employment and housing* and the impact on travel, particularly *journey-to-work*. They indicate the relative size of *labour markets* for central city and suburban jobs by analysing the average commuting distance to each. Commuting distances from various residential locations are also evaluated. Changes in average commuting trip time over recent years, particularly for major US cities, are analysed by Cervero (chapter 17) and by Gordon and Richardson (chapter 18). Different aspects and interpretations of these trends are presented.

The net changes in *travel time* are relatively small and the trends unclear. An apparent impact of *job dispersal* is to divert trips to less congested suburban destinations and routes, thereby limiting inner urban congestion, and to private transport—generally increasing average travel speed. In this way it appears to stabilise travel times—consistent with the concept of *travel time budgets*—and to restrain the level of urban operating costs, contributing to *urban competitiveness* and *sustainability*. Dispersal also provides a path for urban structural change which could lead toward lower *energy use*. It produces an urban structure which, under a scenario of energy constraints, could change its internal travel pattern via destination or origin changes to transform the city from integrated and multi-centred towards a set of *self-contained centres*, with *telecommunications* providing much of the interaction among them—enabling shorter trips by more energy-efficient modes. Dispersal can lead to both sustainability and *robustness* in this sense and provide an alternative to *urban containment*. However, it can also be inequitable—for low income groups—in that it reduces access via public transport to these dispersed jobs. It can also lead to greater vehicle travel distances and emissions, and drive-alone vehicle travel, in the US case at least. Cervero examines these aspects. He links high vehicle travel to dispersal of employment and to the lower residential densities in the suburbs—an argument developed by Newman & Kenworthy (1989) and an issue widely promulgated and discussed among planners—leading to broad support for policies of urban containment.

An evaluation of this containment policy is provided by Breheny,in chapter 21, with surprising results. Breheny addresses the UK (and international) debate on urban containment and policies for its achievement. He considers all regions in England and Wales and all urban travel (journey-to-work represents only about a third of this, although it is a major part of peak hour travel and an essential core in any low energy future) and examines current policies of urban containment in the light of present *deconcentration* trends. Using national data on travel, including travel modes and their energy use, in relation to city size and type he evaluates the primary energy use in total urban travel under two scenarios. The first is for the present land-use pattern. The second is for the pattern resulting if the outward migration of urban population and jobs over the last 30 years had been entirely prevented. The results throw serious doubt on the feasibility of present policies of urban containment, in the UK at least, and on their effectiveness if they could be implemented.

In the final chapter in this section (22) an alternative to commuting is explored. Moss and Carey examine *telecommuting* as an option for some metropolitan workers. In an experiment in New York, a sample of workers from several organisations telecommuted about once a week—with largely positive results. The organisations reported increased productivity, the individuals claimed more free time and lower costs and the city gained environmentally with less vehicle miles travelled and less emissions. The longer term implication of its wider introduction could be a further flexibility of locational choice for households and firms. The electronic cottage and virtual city—extending beyond city boundaries—may be coming closer for some workers as telecommunications capabilities increase, emission controls are tightened and as teleworking receives wider social and organisational acceptance.

17 Changing Live–Work Spatial Relationships: Implications for Metropolitan Structure and Mobility

Robert Cervero

Economic restructuring and metropolitan form

THE spatial structures of most large US metropolitan areas have changed over the past few decades from a monocentric form to a more dispersed, and in some cases polycentric, form. Decentralisation of jobs has been the dominant force behind this transformation. Between 1976 and 1986, employment in the suburbs in the 60 largest US metropolitan areas rose from 16 million to 24 million jobs. By 1990, around two-thirds of all jobs in US metropolitan areas were outside central cities, up from 45% just a decade earlier (Hughes 1992). While suburban employment growth has been most pronounced along America's sunbelt crescent (Cervero 1986a), the trend has been truly nationwide in scope, occurring even in older industrial areas. In greater Philadelphia and St Louis, for instance, suburban employment grew by 8% and 17% respectively between 1982 and 1986, contrasted with a loss in central city jobs over the same period (Urban Land Institute 1987).

The movement of jobs from the metropolitan core to the metropolitan periphery and beyond has been spurred by post-industrialisation—the restructuring of America's economy from a predominantly manufacturing base to a service and information processing orientation. Among the pull factors that have lured corporate America to the suburbs have been the availability of cheaper land, easier access to labour (particularly married women seeking clerical positions), lower taxes, improved telecommunications links and closer proximity to regional airports. Push factors, like rising property taxes and deteriorating inner-city conditions, have also led to suburban job gains. While many decentralising jobs have involved back-office support functions, increasingly corporate headquarters and entire companies in fields like finance, retailing and wholesaling are relocating in the suburbs (Stanback 1991).

In addition to jobs, the majority of America's population growth has taken place in the suburbs as well. In 1990 over half of the nation's population lived in the 39 metropolitan areas containing over one million residents (Hughes 1992). The suburban population in these areas increased 55% between 1970 and 1990, while the traditional, central city population increased only 2%. Retailing, of course, has followed both residents and jobs, with many US metropolitan areas today containing over five times as much retail space in the suburbs as in core cities. The New York metropolitan area mirrors the spatial changes that took place in many large US regions during the 1980s. From 1980 to 1990, Manhattan added 54 million square feet of office space. The suburban ring, including Long Island, north-east New Jersey and Westchester County, added 173 million square feet (equal to the entire Chicago metropolitan office market). Thus, suburban counties captured two-thirds of the region's office growth during the 1980s. Overall, Manhattan still accounted for 56% of all office space in the region, but its market share fell from 85%. Population in the core cities increased from 8.21 to 8.40 million (2.3%) during the 1980s, however in the suburban ring it rose from 6.87 to 9.62 million (40%). New York's suburban ring now has 48 fully enclosed regional malls, encompassing 49 million square feet of retail space (Hughes 1992).

Few of America's suburbs today resemble the Levittown bedroom communities of the 1950s. Except for their relative newness and locations, the very largest suburban centres are virtually indistinguishable from many traditional cities. Such places have proven difficult to label, giving rise to a variety of names like 'suburban downtowns', 'edge cities' and 'technopolises' (Hartshorne & Muller 1986; Scott & Angel 1987; Garreau 1991).

The transformation of suburbia from well-manicured residential enclaves to America's principal place of employment has had a particularly dramatic impact on commuting. Already by 1980 suburb-to-suburb trips accounted for 42% of all intra-metropolitan journeys to work, up from 30% 20 years earlier. Between 1960 and 1980, moreover, suburbs received over three-quarters of the increase in metropolitan worktrips (Pisarski 1987). These percentages have no doubt increased, perhaps significantly, over the intervening decade given the rapid rate of suburban job growth during the 1980s.

This chapter examines how metropolitan growth trends during the 1980s have influenced the spatial relationship of residences and workplaces and, accordingly, regional mobility. For the most part, changes in 1980 to 1990 journey-to-work census data are used in this pursuit. The analysis is necessarily preliminary in that, at the present time, national journey-to-work statistics are incomplete. A second caveat is also in order. There has been a tremendous slow-down in new office construction in nearly all American regions since 1990, due to the glut in available office space as well as tight credit (brought on by changes in federal tax laws and the savings and loans crisis). Moreover, a number of companies are in the midst of downsizing. The combination of tight credit, overbuilt commercial-office markets, and white-collar restructuring virtually guarantees that the development trends of the 1990s are going to be far different from the 1980s. This does not imply that the decentralisation trends of the past decade will be reversed; rather it is likely that much of the pattern that held in 1990 will hold true for the remainder of this decade as well.

Industrial organisation and decentralised growth

The effects of decentralisation on live–work spatial relationships and mobility depends, in part, on the degree to which this growth is orderly or spatially well defined. The two extremes are dispersed patterns of growth, or 'scatteration', versus subcentering or a polycentric form.

Research by urban geographers has shed important light on this question. Changes in industrial organisation in response to increased global competition have dramatically changed the spatial morphology of many regions, especially those that have successfully developed a modern, high-technology industrial base, whether Silicon Valley or Massachusetts Route 128 in the US or Singapore or any other Asian NIE. Many such successful regions feature a close-knit network of firms that practise 'flexibly specialised' forms of production—a term used to characterise businesses that make a wide and changing array of customised products using flexible, general-purpose machinery and skilled, adaptable workers (Hirst & Zeitlin 1990). In contrast to Fordist mass production, such firms employ small but technologically sophisticated units to craft rapidly changing, innovative product lines under short production cycles.

The characteristics of many flexibly specialised firms—i.e. strong inter-firm linkages, frequent subcontracting, reliance on specialised skills, the relative cleanliness of the industry—favour spatial agglomeration, however not in central cores but rather in smaller, more numerous centres spread across the metropolis. Some research suggests that whether a polycentric or dispersed form emerges depends, in part, on industrial structure and divisions of labour. Scott (1985; 1989) found a tendency toward clustering among small, flexibly specialised firms within industries which were horizontally integrated. Such industries were as varied as women's dress making and printed circuit production. More mature, vertically integrated industries with oligopolistic structures and practising routinised mass production, on the other hand, were more spatially dispersed. In the study of printed circuit producers in Southern California, Scott (1983b) found that spatial distribution changed from clusters to dispersion as plant size increased. Saxenian (1990) has similarly attributed Silicon Valley's economic success to the emergence of a dynamic network of innovative, nodally clustered firms that specialise in critical central components like silicon wafers and software. In contrast, Route 128 in Massachusetts and Los Angeles' aerospace complex, both of which have experienced recent downturns, are comprised of larger, vertically integrated firms that are spatially dispersed and that focus on end-use products (e.g. minicomputers and computer-system controlled fighter-bombers).

Changing production processes, like flexible specialisation, suggest many large metropolitan areas will continue to decentralise, however with concentrated employment nodes—i.e. they will become more polycentric in form. Factors that could inhibit this trend include: the durability of the existing built environment (i.e. current capital stock of offices and housing); frictions to movement like restrictive zoning and local no-growth moratoria; and telecommunications advances that reduce transportation costs. Castells (1991) sees continued and increasing concentration of the control of production in a few global cities because this arrangement best protects the power of corporate entities. Telecommunications advances, he and others argue, will eliminate any benefits from agglomeration since any locale will be able to plug into a ubiquitous telecommunications network, allowing, in the language of information sciences, 'virtual' agglomeration economies anywhere. While it is questionable whether even information-handling firms will become completely footloose, it may be the case that any clustering wrought by flexible specialisation and other new-age forms of production will be diluted over time by telecommunications advances.

Empirical evidence on subcentering and polycentric growth

A number of empirical studies have documented the emergence of sub-centres in the US. Using minimum thresholds for office and retail floor-space and jobs, analysts have identified 13 subcentres in greater Washington, DC (Garreau 1991), 17 in greater Atlanta (Atlanta Regional Commission 1985), and 22 in the Houston area (Rice Center 1987). In a national study, Cervero (1989a) found 57 suburban employment centres located at least 5 radial miles (8 km radius) from a CBD and containing over 2 000 full-time workers and over one million square feet (93 000 m^2) of office space. Three separate studies of the Los Angeles area have identified anywhere between 6 and 54 subcentres (Gordon *et al.* 1986; Heikkila *et al.* 1989; Giuliano & Small 1991). While large suburban downtowns and edge cities have gained recent media attention, in many areas, a far more dispersed, less structured form of suburban office development has taken form. In a study of 6 large US metropolitan regions, Pivo (1990) concluded that most office jobs were located in small and moderate-sized, low intensity clusters along freeway corridors. Gordon *et al.* (1986) and Giuliano & Small (1991) have likewise found that, except for several large concentrations, small-scale clustering best characterised Los Angeles' form of subcentering. These findings suggest that the decentralisation process in contemporary urban America is complex and spans a continuum ranging from dispersal at one extreme to more orderly polycentric forms on the other.

Decentralisation: a blessing or curse?

The debate over whether decentralisation is a positive or a counter-productive trend is centuries old. Early commentators viewed decentral-isation as a mark of social progress, helping to alleviate the overcrowding and miserable living conditions in many cities. The aggressive attacks against decentralisation as environmentally damaging and class motivated have been relatively recent, postdating World War II.

Any normative assessment of decentralisation's impacts depends on some expression of an objective function. On environmental and social equity grounds, many might argue that decentralisation in the United States has imposed high costs. In terms of industrialisation and economic productivity, others might counter that resulting benefits have far exceeded costs. Since these outcomes are often only indirectly related to and difficult to unambiguously correlate with decentralisation, most assessments to date have focused on spatial interaction and transportation impacts. Even here, however, any evaluation begs the question: what is the objective function? If assessed in terms of changes in travel distances, experiences in the US show that contemporary decentralisation trends have had modest impacts. In terms of travel times and average trip speeds, though, the impacts have been clearly positive. However, as will be argued in this chapter, the greatest transportation impact of recent decentralisation trends in the US has been with reference to mode of travel, specifically the increasing reliance on private automobile travel, which in turn has in many instances had serious environmental repercussions. Only when a multi-dimensional objective function is used in evaluating the transportation impacts of decentralisation, one which weighs impacts on modal splits in addition to trip distances and speeds, can a balanced perspective be gained on the public policy implications of recent urbanisation trends in the US. This chapter aims to provide such a balanced portrait.

Decentralisation and commuting

A body of research has emerged on the impacts of polycentric-like structures and employment decentralisation on commuting patterns, though research findings have reached conflicting conclusions and the phenomenon is still only partly understood. Gordon et al. (1986; 1991) have argued that polycentric structures reduce urban commuting. They estimated that average commute travel times fell in 18 of the 20 largest US metropolitan areas between 1980 and 1985, concluding that 'polycentric metropolitan structures are especially favourable to short commutes' (Gordon *et al.* 1991, p.419). Several studies have explored this question by comparing actual commuting to the amount of commuting theoretically necessary under the monocentric model, defining any differences as 'wasteful' or 'excessive'. Using this approach, Giuliano & Small (1992) recently concluded that Los Angeles' suburban centres averaged shorter commutes than the CBD. A subsequent study of commuting in Los Angeles which corrected for many of the deficiencies of earlier work on excess commuting reached the opposite conclusion that 'the polycentric required commute is considerably larger than that required by the monocentric model' (Song 1992, p.18).

Several detailed case studies of individual employment centres suggest that the built environment characteristics of suburban centres might have some bearing on commuter distances. Of the greater Houston area's 11 suburban centres, the Rice Center (1987) found that workers in West Houston's Energy Corridor, the farthest centre from downtown and the one with the lowest densities and strongest automobile orientation, averaged longer trips than workers from any other employment centre— 9.8 miles (15.8 km, even further than the 9.2 mile/ 14.8 km average for CBD workers). For two of the San Francisco Bay Area's largest and fastest-growing suburban employment markets, Pleasanton-Livermore and the Golden Triangle of Santa Clara County, average one-way trips exceeded 15 miles (24 km) in the mid-1980s, over one-quarter longer than the regional average (Cervero 1986a). The distribution of trip lengths was found to be more concentrated for suburban centres than other metropolitan workplaces, with fewer short distance and fewer very long distance journeys (Figure 17.1).

Among America's 57 largest suburban employment centres, those characterised as office parks averaged 11.9 miles (19 km) travel compared to an average of 9.5 miles (15.3 km) for workers at denser suburban downtowns (Cervero 1989a). A Rice Center (1989) study of 62 US employment centres showed that average commuting distances to CBDs (11.6 miles/ 18.7 km) were markedly longer than average distances to suburban downtowns (10.1 miles/ 16.3 km) and all other types of suburban centres except office parks (12.2 miles/ 19.6 km). Virtually all studies agree

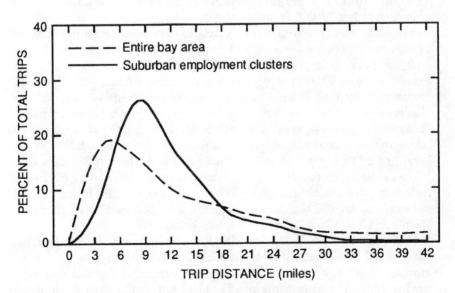

Figure 17.1   Worktrip distributions for suburban employment centres and the entire San Francisco Bay Area, 1980
Source: Cervero 1986b.

that travel times are shorter for workers heading to suburban centres, especially those with lower densities, than to CBDs, confirming that those travelling in outlying areas average higher speeds.

Jobs–housing imbalances

Among the benefits that might be expected from the decentralisation of employment is the transformation of America's suburbs into more 'balanced' communities, thus reducing commute distances and congestion levels. Cervero (1986b; 1989b) documents overall regional balance of jobs and housing in the San Francisco Bay Area in 1980, however within 7 of the 22 largest communities, jobs exceeded housing units or vice versa by more than two to one. More surprising was the fact that for four of the communities with approximate balance (any jobs–housing ratio between 0.75 and 1.25 was considered to be reasonable local balance), fewer than 20% of their workers resided locally. This suggested that the jobs–housing balance has both quantitative and qualitative dimensions—balance requires a concordance of worker earnings and local housing prices, not just numerical parity. A gravity model analysis of journey-to-work census data suggested that imbalances could be at least partly explained by high housing prices relative to workers' earnings and shortages of residentially zoned land, suggesting that restrictive policies like fiscal and exclusionary zoning could be impeding locational choices.

Other research suggests that imbalances could reflect choice (in the Tiebout tradition) more than barriers to movement. In a provocative article, Hamilton (1982) raised doubts over the reliability of the traditional monocentric model of urban economics that explains residential location as a trade-off between housing and commuting expenditures. For 14 US and Japanese cities studied, he concluded that actual trip distances exceeded those predicted by the monocentric model with dispersed employment by 87%. While a number of subsequent studies concluded that Hamilton overstated the extent of excess commuting when other factors are accounted for,[2] these studies nonetheless consistently show that not all commuting can be explained by the sheer geographical imbalances in the locations of housing and jobs. Other factors besides job access that influence the locational choice of residence (and thus commuting distances) include quality of schools, neighbourhood amenities, mixes of community services, and familial ties and attachments to neighbourhoods (Quigley & Weinberg 1977; Clark & Burt 1980; Ley 1985).

Giuliano (1991) and Downs (1992) question whether jobs-housing balance will ever be an achievable policy target due to such factors as the growing number of two-earner households, regulatory barriers that keep worker earnings and housing prices out of kilter, and high rates of job turnover and residential mobility. Richardson & Gordon (1989) note that since non-work travel is the fastest growing category of travel, jobs-housing balance will over time be less and less effective at reducing traffic congestion.[3] Moreover, Lowry (1988) and Downs (1992) argue that regional balance is a natural evolutionary process brought on by market conditions. In an unrestricted land market, Giuliano (1991) maintains that subregional imbalances caused by rapid growth will erode over time, noting that Orange County has steadily moved toward balance over the post-war period, from a ratio of jobs to population of 0.21 in 1950 to 0.46 in 1985. Nowlan & Steward (1991) demonstrate how the market worked toward balance in downtown Toronto. There, traffic gridlock from central Toronto's office building boom in the 1970s and 1980s was averted through accelerated downtown housing construction that was later occupied by people working downtown. Wachs *et al.* (1992) recently traced changes in journeys-to-work over time for over 8 000 hospital workers throughout southern California, finding that the average distance travelled actually decreased slightly from 10.0 miles (16.1 km) in 1984 to 9.7 miles (15.6 km) in 1990.

The spirit of jobs–housing balance is less one of government fiat and more one of breaking downtown regulatory and exclusionary barriers to movement so that the marketplace can indeed work toward greater balance (Cervero 1991a). When combined with pricing schemes that more closely reflect true social costs and any externalities associated with peak-hour travel, the erosion of locational barriers like fiscal and large-lot zoning would clearly yield important mobility dividends.

Changing commuting patterns and distances

Regardless of whether America's metropolises are truly polycentric in form or regionally balanced, it is clear that decentralisation of jobs and residences has produced a complex web of cross-town and lateral trip-making. With many origins and destinations, today's commute paths resemble Brownian motion—seemingly random movements in all directions. Suburb-to-suburb commuting captured 58% of the total increase in US commuting between 1960 and 1980; by 1980, intra-suburban trips accounted for around 40% of all metropolitan commuting, approximately twice as much as suburb-to-downtown commuting (Pisarski 1987). Schnore (1959, p.205) foresaw these trends three and a half decades ago, predicting that decentralisation would lead to 'a confusing and asymmetrical compound of variously oriented threads of traffic, overlaying the older centre-oriented pattern'.

The trend toward suburb-to-suburb commuting does not square well with the nation's metropolitan road and rail networks. Most major thoroughfares and expressways were laid out in a hub-and-spoke pattern designed to funnel suburbanites to downtown offices and central city factories. Those making lateral and cross-town journeys are often forced into circuitous trips on out-of-the-way beltways and secondary highways. Whether this mismatch between the geography of commuting and the geometry of traditional highway networks has significantly worsened traffic conditions would be difficult to prove. Ambient levels of traffic congestion, however, are clearly on the rise in most metropolitan areas. Using data on freeway and highway density levels, Hanks & Lomax (1991) document a trend toward increased congestion from 1982 to 1988 in 17 of 20 US metropolitan areas. The greatest increases in congestion were in San Diego, Nashville, the San Francisco Bay Area, Sacramento and Los Angeles, all of which experienced suburban population and employment growth rates of between 25% and 60% during the 1980s.[4]

Commuting distances

Recent national statistics on commuting distances have surprised many. From 1977 to 1990, the average commuter trip length in the US increased by 15%, from 9.2 miles to 10.6 miles (14.8 km to 17 km, Hu % Young 1992),[5] which according to Bookout (1992, p.10) reveals 'an even poorer relationship between jobs and housing than experts expected to find'. Total vehicle miles travel (VMT) increased by 55%, compared to an increase in population of only 12% for the same period.[6] Longer distance trips accounted for 38 per cent of the growth in VMT.

Overall, these data suggest that more and more American households are living farther from their workplace less by choice and more because of an absence of affordable or suitable nearby housing (Federal Highway Administration 1991). On balance, the national census shows that decentralisation of employment has not placed the typical American worker closer to his or her job.

Inter-county and external commuting

During the 1980s there was a significant gain in inter-county commuting within metropolitan areas, further suggesting a trend toward longer trips (Pisarski 1992). The share of MSA (Metropolitan Statistical Area) employees working in their county of residence fell from 79% to 76%. Those working downtown declined in share, while those working in the remainder of the MSA and outside the MSA area increased, in some cases significantly.

In the San Francisco Bay Area, the fastest growth in intercounty commuting has been in the fringe counties, most notably Solano County where 38.6% of resident workers commuted to another Bay Area county in 1990, up from only 11.8% in 1960 (Purvis 1992). The most significant growth in inter-suburban commuting has been along the East Bay-to-Santa Clara County corridor, which nearly doubled from 30 000 commuters in 1980 to 59 000 in 1990. The Bay Area's three largest suburban employment concentrations—Bishop Ranch and Hacienda Business Park in the East Bay and Silicon Valley in Santa Clara County—lie along this corridor.

The Bay Area's transformation from a monocentric to a polycentric structure is reflected by San Francisco's decline in the share of regional commutes from 20.5% in 1980 to 18.0% in 1990. Only 2 of the 9 Bay Area counties—Napa and Solano—increased their number and share of total commuter trips to San Francisco during the 1980s. Over this period trips across the Golden Gate Bridge between Marin County and San Francisco actually declined 10.5%. At the same time, reverse commutes from San Francisco homes to suburban jobs increased 53%.

External commuting also rose sharply during the 1980s. Americans commuting to workplaces outside their MSA of residence rose by 3.5 million trips during the 1980s, from 5.4% to 7.6% of all commuter trips. Extended commuting (from non-MSAs to MSAs) likewise increased significantly, continuing a trend documented two decades earlier by Berry & Gillard (1977) and Fisher & Mitchelson (1981). In northern California, the largest increase in external commuting was between the burgeoning bedroom communities of the Central Valley, historically an agricultural belt, and the Bay Area. Commutes along this axis over the Altamount Pass increased from only 5 600 trips in 1980 to 32 000 in 1990.

Commuting times and speeds

Nationwide, average commute times increased by only 40 seconds during the 1980s, from 21.7 to 22.4 minutes. Given that the number of persons driving alone increased from about 62 million to over 84 million (35%) over this period, Pisarski (1992, p.3) calls this 'an extraordinary comment on the flexibility and capacity of the Nation's highway system'. This would appear to lend credence to travel time budget theories that hold the time an individual devotes to transportation is

> close to an anthropological constant, ranging from 1 to 1.5 hours per day, both in rural agricultural and urban-industrial societies (Grubler 1990, p.2).

Thus Americans have adjusted to longer distance commutes by increasing average speeds, such as using limited access highways for inter-suburban trips, travelling more during the off-peak, and switching from slower-moving buses to faster single-occupant automobiles.

Table 17.1 reveals an association between decentralisation and commute times in 20 of the largest metropolitan areas in the US.[7] Metropolitan areas are ranked in order of their 1990 populations. In all 20 regions, population and employment growth in the suburbs[8] outpaced overall metropolitan growth. Atlanta had the fastest suburban population growth and the New York region gained the most suburban jobs (on the strength of central New Jersey's mid-1980s office building boom).

The table shows that average travel times increased in 18 of the 20 metropolitan areas. For suburban residents, average trip times actually fell in 7 of the 20 MSAs. Larger metropolitan areas generally posted the largest increases in commuting times. Overall, there was little difference in commuting times among residents in these 20 large MSAs—suburban residents devoted about the same amount of time commuting in 1990 as their central-city counterparts.

From Table 17.1, the correlation between 1980–90 changes in suburban employment and changes in suburban commuter times was calculated at

Table 17.1 Trends in worktrip commuting times for suburban residents of 20 large US metropolitan areas, 1980–90

CSA or MSA[b]	1980–90 % change pop. Metro.	1980–90 % change pop. Suburb	1980–90 % change emp. Metro.	1980–90 % change emp. Suburb	Metropolitan area 1980	Metropolitan area 1990	Metropolitan area % diff.	Suburban residents[a] 1980	Suburban residents[a] 1990	Suburban residents[a] % diff.
New York	19.0	39.9	41.6	144.9	26.7	30.6	14.6	26.1	31.8	21.8
Los Angeles	26.4	29.6	32.0	37.5	23.3	26.4	13.3	22.9	25.1	9.6
Chicago	2.5	10.3	14.4	21.7	23.0	28.1	22.2	24.4	30.8	26.2
San Francisco	16.5	26.2	27.2	29.3	22.9	25.6	11.8	23.4	23.3	-0.4
Detroit	1.0	6.7	6.4	21.7	21.1	23.4	10.9	21.2	24.3	14.6
Washington	28.2	36.9	35.3	45.2	28.5	29.5	3.5	28.0	29.9	6.8
Dallas	30.6	44.2	42.1	69.3	22.8	24.1	5.7	22.5	24.0	6.7
Boston	22.7	28.6	19.1	12.8	21.9	24.2	10.5	23.8	24.0	0.8
Houston	19.7	41.2	23.3	39.4	23.8	26.1	9.7	26.4	29.7	12.5
Miami	20.8	26.7	40.4	60.1	23.2	24.1	3.9	24.0	24.3	1.3
Atlanta	39.6	52.0	57.3	80.1	25.9	26.0	0.4	23.3	25.8	10.7
Cleveland	-2.6	2.1	2.6	14.7	20.8	22.0	5.8	22.6	22.3	-1.3
Seattle	22.3	29.6	36.6	59.5	22.3	24.3	9.0	24.9	25.6	2.8
San Diego	34.2	40.7	55.6	74.7	19.6	22.2	13.3	20.7	23.0	11.1
Minneapolis	16.6	23.9	23.4	33.2	21.0	21.1	0.5	23.0	21.7	-5.7
Baltimore	9.6	18.7	14.5	30.9	26.5	26.0	-1.9	25.1	23.7	-5.6
Pittsburgh	-0.9	1.8	3.5	6.9	23.1	22.6	-2.2	22.6	22.2	-1.8
Denver	14.0	18.9	23.3	30.3	21.1	22.4	6.2	24.9	24.3	-2.4
Cincinnati	5.1	17.2	9.0	30.0	21.0	22.1	5.2	23.6	22.7	-3.8
Milwaukee	2.4	3.6	5.5	18.2	18.2	20.0	9.9	21.0	21.7	3.3
Average[c]	16.4	24.9	25.7	43.0	22.8	24.5	7.6	23.7	25.0	5.4

[a] Mean commute times for those residing in the suburbs (i.e. outside of the central city).

[b] Data are for Consolidated Statistical Areas (CSAs) except for metropolitan areas that only have a Metropolitan Statistical Area (MSA) designation.

[c] Unweighted (i.e. not weighted by metropolitan population).

Source: US Bureau of Census, Summary Tape Files 1A for assorted metropolitan areas.

+0.40. At one extreme was the New York region, where suburban employment more than doubled in the 1980s and the average trip of suburban residents increased by 5.5 minutes. At the other extreme is Pittsburgh, where suburban jobs increased only 7% and suburban residents got to work 30 seconds faster. Overall, suburban residents of America's largest regions appear to devote about the same amount of time to commuting as their central-city counterparts. Additionally, average commute times seem to be increasing fastest in those regions that are most rapidly decentralising, particularly in terms of jobs.

Some of the longest increases in commute times during the 1980s was in traditional bedroom counties on the fringes of large MSAs. In the Los Angeles area, average commutes for Riverside County and San Bernardino residents rose from 22 to 28 minutes (Pisarski 1992); in Solano County midway between San Francisco and Sacramento, the change in averages was identical—22 to 28 minutes (Purvis 1992). Lee (1992) documents increases in very long commuter trips among residents of California's Central Valley. Fewer than 2% of Modesto workers commuted more than an hour each way in 1980; in 1990 nearly 10% did. And in Patterson, a traditional farming community of 8 600 residents, one in six workers spend over 3 hours per day commuting. Such statistics suggest the Bay Area's labour-shed extends well into the vast Central Valley as more and more young families and first-time home buyers trade off long travel times for cheaper housing. Fulton (1990) cites similar instances of 2–3 hour trips among residents of the Antelope Valley set in the high desert of northern Los Angeles County and in Moreno Valley in Riverside County (the fastest growing city in the fastest growing county in the country).

Changing modes of commuting

Far more dramatic than changes in trip distances and travel times have been recent changes in the modal composition of travel. The increase in Americans driving alone to work, about 22 million, exceeded the number of new workers, meaning all job growth during the 1980s was absorbed by solo driving (Pisarski 1992). During this period, drive-alone commuting rose from 64.4% to 73.2% of all commutes. Significantly, from 1980 to 1990, all alternatives to the drive-alone automobile fell both in shares and absolute numbers of commuter trips: carpooling, 19.7% to 13.4%; and transit, 6.4% to 5.3%. Despite the passage of dozens of trip reduction ordinances and transportation control plans in US metropolitan areas throughout the 1980s, average vehicle occupancy fell from 1.15 to 1.09.

For the 20 large MSAs, Table 17.2 shows that drive-alone shares of commuter trips made by suburban residents rose in all areas. Greater New York had the largest gain in drive-alone commuting by suburbanites, made up of many former transit commuters who switched to solo commuting when their job sites moved from Manhattan to Long Island, northern New Jersey and Connecticut. Transit's share of trips by suburban residents, on the other hand, fell in 17 of the 20 MSAs. Of the three MSAs that had increases in the share of suburban residents commuting by transit, two (San Francisco and San Diego) have rail transit systems; the other MSA, Houston, began phasing in the nation's most extensive HOV/busway facility during the 1980s. Two other rail cities, Washington DC and Miami, more or less held their transit market shares among suburban commuters. Since all urban rail systems in the US focus on downtown areas, it follows that transit's singular growth market has been radial, suburb-to-downtown trips, at least in large rail cities.

The switch-over by suburban residents from transit to drive-alone commuting was most prevalent in metropolitan areas whose jobs were decentralising the fastest. For the 20 MSAs, the correlation between 1980–90 employment changes in the suburbs (from Table 17.1) and percentage point change in transit's share of commutes by suburbanites was –0.60.

Table 17.2   Trends in work trip modal shares for suburban residents of 20 large US metropolitan areas, 1980–90

		Per cent of trips by suburban residents[a]				
		Drive alone			Transit	
CSA or MSA[b]	1980	1990	% pt change	1980	1990	% pt change
New York	55.5	77.5	22.0	27.5	10.1	–17.4
Los Angeles	75.9	79.4	3.5	3.3	2.6	–0.7
Chicago	72.0	81.1	9.1	9.1	7.1	–2.0
San Francisco	71.4	73.9	2.5	7.2	10.9	3.7
Detroit	80.7	89.6	8.9	1.5	0.4	–1.1
Washington	62.0	71.2	9.2	10.8	10.6	–0.2
Dallas	76.1	84.8	8.7	1.3	1.0	–0.3
Boston	67.5	80.2	12.7	11.7	8.1	–3.6
Houston	73.3	82.8	9.5	0.7	1.7	1.0
Miami	74.2	80.0	5.8	3.8	3.6	–0.2
Atlanta	73.4	83.0	9.6	4.7	2.9	–1.8
Cleveland	76.6	86.3	9.7	6.1	3.3	–2.8
Seattle	72.9	82.5	9.6	4.5	3.9	–0.6
San Diego	72.8	78.4	5.6	2.5	2.8	0.3
Minneapolis	72.0	84.8	12.8	4.5	2.7	–1.8
Baltimore	70.5	82.3	11.8	3.8	2.9	–0.9
Pittsburgh	69.5	80.3	10.8	8.4	5.5	–2.9
Denver	72.2	82.8	10.6	4.2	3.0	–1.2
Cincinnati	76.1	86.0	9.9	3.3	2.1	–1.2
Milwaukee	76.0	87.7	11.7	2.7	1.6	–1.1
Average[c]	72.0	81.7	9.7	6.1	4.3	–1.7

[a] MSA or CSA residents residing outside of the central city.
[b] Data are for Consolidated Statistical Areas (CSAs) except for metropolitan areas that only have a Metropolitan Statistical Area (MSA) designation.
[c] Unweighted (i.e. not weighted by metropolitan population).
Source: US Bureau of Census, Summary Tape Files 1A, 1992.

Employment decentralisation has clearly had a far greater impact on mode choice, and transit usage in particular, than on commute distances or travel time in America's largest metropolitan areas.

Telecommuting

Working at home, or telecommuting, is perhaps the ultimate form of jobs–housing balance. Nationwide, there was a 56% increase in the number of Americans working at home during the 1980s, bringing the share of telecommuters to 3% (Pisarski 1992). Among large MSAs, Atlanta had the largest increase in telecommuters, from 29 500 in 1980 to 54 800 in 1990 (an 86% increase). Among suburban Atlantans, telecommuting rose 113% during the 1980s.

Mass telecommuting would, over time, reduce the need for firms to spatially concentrate. However research shows that telecommuting's potential is likely to be limited. Saloman (1984) found that 45% of computer professionals would not work at home if given the opportunity. According to Gurstein (1990), many homeworkers feel cut off from office social life and promotion opportunities. Research by Mokhtarian (1990) nevertheless suggests that telecommuting could yield major mobility benefits: it is most appealing to long-distance commuters and, contrary to the popular view, telecommuters do not make more non-worktrips on their days off, but rather make more non-worktrips when they physically commute to the office (i.e. they chain trips). A middle ground between working at home and at the office is perhaps neighbourhood telework centres equipped with computer and facsimile technology. A network of telework centres would most likely lead to a more dispersed and hierarchical pattern of metropolitan subcentering. The mobility advantage, however, could be superior to many transportation strategies available, including road pricing.

Finer-scale analyses of job decentralisation and commuting

Several studies provide a far more detailed perspective on the commuting impacts of job decentralisation by focusing on how the commuting characteristics of specific workers change when their work address changes from downtown to a suburban locale. In his study of office relocations in greater London, Daniels (1972; 1981) found that most employees experienced a longer travel time after their jobs relocated. He also documented a dramatic switch in commuting from public transit to the drive-alone auto. Similar research by Wabe (1967), O'Connor (1980) and Ley (1985) found average commuting distances changed little after firms moved to the suburbs, but like Daniels, their work showed a tremendous drop-off in transit usage.

A more recent study of workers whose jobs moved from downtown San Francisco to several large suburban workplaces in the Bay Area found that while average trip distances remained virtually unchanged, transit modal splits plummeted from 58% to 3% (Table 17.3).

Their average trip times fell from 50.2 minutes to 36.5 minutes. However, total VMT produced by the relocated workers rose an estimated 280%—accompanied by a comparable increase in fuel consumption and tailpipe emissions. To the extent that relocated workers switched from electrically powered BART rail trains and Muni trams, the energy and air quality impacts of these mode changes were likely to be even greater.

Table 17.3 also shows that impacts varied significantly depending on whether workers resided in the suburbs or in San Francisco. Suburban residents (whose commuting patterns changed from radial to inter-suburban) averaged significantly shorter and faster trips, and virtually abandoned mass transit as a travel option. San Franciscans, on the other hand, averaged distances five times as long and travelled 24 minutes more once their jobs were located to the suburbs. These new reverse commuters switched over dramatically from transit to drive-alone commuting. When working in the city, over 75% of San Francisco residents rode trains, trams and buses to work; following the change in workplace, transit's modal share fell sharply. Ethnically, around 54% of the San Francisco residents whose jobs moved to the suburbs were non-white (compared to only 28% of all relocated workers) and many worked in service positions that paid only moderate salaries. Thus, those most disadvantaged by job relocation tended to be ethnic minorities who, for whatever reason, chose to maintain a San Francisco residence. This finding is consistent with the 'spatial mismatch' literature that links the flight of businesses from central cities, coupled with relocational barriers faced by many minorities, to rising inner-city unemployment (Jencks & Mayer 1990; Ihlandfeldt & Sjoquist 1990; Holzer 1991).

Table 17.3 Changes in commuting due to suburban office relocation, San Francisco Bay Area, 1987–89

	All workers		Suburban residents		San Francisco residents	
	Before	After	Before	After	Before	After
Average distance (miles)	12.7	12.5	14.5	12.9	3.5	20.0
Average Time (min)	50.2	36.6	55.6	37.7	32.1	56.3
% trips by:						
Drive-alone	22.8	74.9	20.5	74.3	7.7	50.0
Carpool	16.9	21.5	18.2	22.2	7.7	37.5
Transit	58.1	2.8	59.8	2.1	76.9	12.5

Note: 1 mile = 1.61 km
Source: Cervero and Landis (1992).

While the commuting impacts of job decentralisation remain somewhat fuzzy due to many conflicting research results, one finding is consistent and unequivocal: that transit ridership and, to a lesser extent, carpooling fall off sharply as a result. The built environment of central cities and traditional suburban work settings clearly accounts for much of the difference in transit and ride-sharing modal shares. High densities are universally necessary to support frequent and convenient mass transit services (Pushkarev & Zupan 1977). Figure 17.2 underscores the importance of density in influencing mode choice in the Bay Area.

The graph plots 1990 drive-alone commuting shares versus residential density for 33 super districts in the 9 county Bay Area.[9] The negative exponential function[10] fit to the graph confirms what others have found—every doubling of density is associated with a 25–30% decline in auto-commuting (see Newman & Kentworthy 1989b; Holtzclaw 1990). Statistically, this relationship seems to hold whether the scale of analysis consists of sites, cities or even countries.

Stark differences in the amounts of mixed land uses and subsidised parking between downtowns and most suburban work centres also account for dramatic differences in modal shares. Many suburban jobs are in campus-style office parks that are surrounded by free available parking (Cervero 1986a; 1989a). Office space typically accounts for 90% or more of all floorspace. While suburban downtowns are denser and potentially

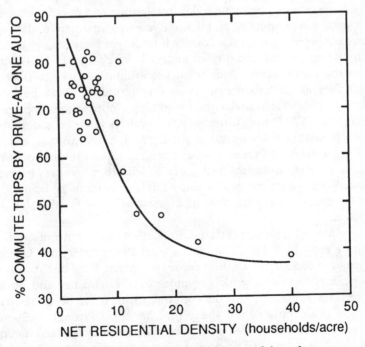

Figure 17.2   Influence of residential densities on drive-alone commuting, San Francisco Bay Area (33 superdistricts), 1990

more supportive of frequent transit services, such places as Tysons Corner in northern Virginia and Uptown in west Houston are laid out on super-blocks with vast expanses of paved parking separating buildings and few continuous sidewalks and pedestrian amenities. In traditional downtowns, by contrast, workers can carpool or commute by transit and still easily access restaurants, banks, shops and other services. A recent study of commuting in greater Washington, DC, underscores just how important the built environments of employment centres are in shaping commuting choices. Douglas (1992) found transit modal shares for worktrips to be four times higher in downtown Washington than in a suburban downtown (Bethesda, Maryland), and four times higher in suburban Bethesda than in a suburban office park (Rock Springs Park, Maryland).

Conclusion

Powerful economic forces have given rise to a far more decentralised and sometimes multi-centred pattern of urban form in most US metropolitan areas. A vast body of research has emerged which has focused on the question of whether decentralisation has reduced commuting distances and times. To date, dozens of varying and sometimes conflicting findings have been published. The absence of any clear consensus on how these trends have affected commuting has unavoidably impeded public policy making. While the impacts of employment decentralisation and polycentric growth on trip lengths are important and deserve research attention, to date many studies have been silent on the equally if not more important question of how modal splits and equality of access have changed. This chapter suggests that, on balance, Americans are around as close to their jobs whether they are in urban centres or on the metropolitan periphery. Impacts on mode choice are far greater and have more serious implications for changes in VMT and, ultimately, natural resource consumption. It would appear that worsening air quality and suburban congestion in many US metropolitan areas has been related less to suburban workers commuting longer distances and more to suburban workers becoming increasingly reliant on the private automobile and, perhaps, being forced to use the same limited number of cross-town road facilities as other commuters.

In conclusion, the more serious implications of decentralisation and polycentric growth in the US are less spatial and more modal. Changing transportation conditions, I would argue, are affected less by the distances that separate home and work and more by the land use and physical characteristics of both ends of the commuter trip. It has been the charac-teristics of place rather than space that has most strongly influenced changes in modes of travel. Whether suburban densities and land use mixes that are conducive to transit riding and other commuter alternatives are best planned for or brought about through market strategies like road pricing is largely a political decision. What is important is that decision makers become acutely aware that America's emerging pattern of suburban growth is creating greater levels of automobile dependency than ever, and that the net impacts associated with such dependency could be substantial.

NOTES

[1] The New York metropolitan area's core cities are New York, Newark, Jersey City, Yonkers, New Rochelle, Elizabeth and Mt Vernon.

[2] These factors include the effects of road configurations (White 1988), two-earner households and housing characteristics (Cropper & Gordon 1991), varying density gradients (Suh 1990), frictions to mobility (e.g. high transactions costs) (Dubin 1991), and greater data disaggregation (Small & Song 1993).

[3] Recent national statistics show that work trips are accounting for a growing share of total VMT, however, increasing from 30.1% of VMT in 1983 to 32.8% in 1990 (Hu & Young 1992). Work trips remain the single trip category that contributes the most to regional traffic congestion.

[4] Percentage growth in population and employment outside of the central cities of these metropolitan areas were, respectively: Sacramento (51% and 59%); San Diego (41% and 56%); Los Angeles (30% and 32%); San Francisco (26% and 27%); and Nashville (26% and 30%). Source: US Bureau of Census, STF-1A, 1992.

[5] For work trips, the average vehicle trip length rose from 8.5 miles in 1983 to 11.0 miles in 1990, nearly a 30% rise.

[6] Average annual VMT per household rose from 3 815 to 4 853 over the 1977–90 period, a 27.2% rise.

[7] Statistics on suburban commuting were not available at the time of writing for the Philadelphia, St Louis and Phoenix metropolitan areas, which were among the 20 largest MSAs in the country in 1990.

[8] Defined as outside of central city but within the MSA or CSA.

[9] The downtown San Francisco district was omitted: at 130 dwelling units per acre, it had the lowest share of worktrips by solo commuters in 1990—only 20%.

[10] $Y_i = 146 \exp(-0.12X_i)u_i$, $r^2 = 0.80$, where Y_i = per cent of commuter trips by drive-alone auto and X_i = net residential density in households/acre; $\exp$ = natural exponent; and u_i = disturbance term.

18   Sustainable Congestion

Peter Gordon and
Harry W. Richardson

Introduction

PRESUMABLY, the concept 'sustainable cities' derives from the related idea of 'sustainable development.' The latter is open to a wide range of definitions but it usually implies, inter alia, economic development compatible with preservation of environmental quality. In addition, it may imply 'stability' and 'tolerability.' The link of 'productive *and* sustainable cities' suggests the notion of combining increasing productivity, economic stability and environmental balance. We employ the phrase 'sustainable congestion' in these contexts to suggest congestion levels that can be tolerated by the travelling public, that are compatible with metropolitan efficiency, and that do not generate substantial environmental damage.

Our analysis is based upon several premises.

1 Some degree of congestion is not only inevitable, it is desirable. If all travel took place on uncongested roads, we would have been guilty of over-investment in highways. Indeed, Downs (1962) suggested equilibrating behaviour by individuals that equalised travel times on all superior routes between every O/D pair; this implies some congestion on potentially high-speed routes such as freeways.

2 The commonly used measures of congestion (e.g. distribution of traffic among roads classified from A to F according to the degree of traffic flow, road congestion indices based on ratios of VMT [vehicle miles travelled] to lane-miles of highway, volume-to-capacity ratios, estimates of hours of delay) are severely deficient. For example, the FHWA (Federal Highway Administration) estimate that congestion had increased because 6.4% of freeway travel was on level of service F roads in 1988 compared with 5.2% in 1985 means nothing in terms of overall congestion levels in cities. Similarly, the TTI (Texas Transportation Institute) road congestion index's critical value of 1.0 implies a speed in excess of 40 mph (65 km/h), but urban commuting speeds of 30 mph (48 km/h) would be regarded as acceptable. A better measure would be the average

ratio of VHT (vehicle hours of travel) to VMT, which is the simple inverse of our measure: average speed.

3 If average travel speeds remain constant, there cannot be system-wide worsening congestion. Of course, travel speeds could decline, and probably will decline, along specific arteries. But the critical test is the average degree of congestion throughout the metropolitan area.

4 If average speeds remain constant, any deterioration in air quality can be traced to increases in VMT. But increases in VHT are much more closely associated with worsening air pollution than increases in VMT (Bae 1993). The contribution of increasing VMT to declining air quality can easily be offset by technological advances affecting emissions per vehicle (e.g. preheated catalytic converters to avoid cold starts, reconstituted gasoline, improved emissions testing techniques).

There are opposite and competing views on what is going on in American cities. One view emphasises 'encroaching gridlock'[1], which it associates with 'urban sprawl' and 'uncontrolled growth'. Another view emphasises the fact that cities have always expanded outward and that suburbanisation is much more the solution than the problem. Economic activities have followed the labour force into the suburbs (Gordon & Richardson 1989) with the result that most commuting is now suburb-to-suburb, allowing many people a relatively quick commute and relieving pressure on the traditional downtown. Only a small fraction of commuters still work downtown. This process explains the remarkable stability of trip speeds and durations in US cities. Consider that:

1 Comparing results from the 1977 and 1983 Nationwide Personal Transportation Studies, we found that trip durations did not increase with city size, nor did they systematically deteriorate in the six-year time span (Gordon *et al.* 1989a).

2 Comparing 1980 Census with 1985 American Housing survey (also conducted by the Census Bureau) results on commuting times, we found that the averages in all of the top 20 US metropolitan areas either improved or remained about the same (Gordon *et al.* 1991).

3 We have found data for Los Angeles county that go back further. Comparing local survey results for 1967 and 1976 with the more recent 1980 and 1985 data, we found no perceptible changes in county-wide commuting time averages (Lobb 1979).

4 Evidence presented by Ira Lowry (1988) for Pittsburgh shows that the distribution of commuting times for that city has not changed since 1934.

Recent findings

The availability of the 1990 Nationwide Personal Transportation Study (NPTS) makes it possible to retest our hypothesis.[2] In the discussions that follow, the only work trips considered are those that are non-stop. This controls for the increase in trip-chaining; as reported by Liao (1993), 14.7% of the work trips were interrupted by other errands in 1983 but this proportion had increased to 19.2% by 1990. Tables 18.1a and 18.1b show average trip durations and distances for the two most recent NPTS surveys (1983, 1990).

As before, we aggregated trips by purpose (work and other),[3] place of residence of the respondent (central city or outside central city; only metropolitan area residents are included in our compilations), time of day (our peak periods are 6–9 am and 4–7 pm; weekend trips are included with off-peak trips), and metropolitan area size class. We only report data for travel by privately owned vehicles. Table 18.2 (p.352) summarises the comparisons between the two NPTS surveys.

The table shows a mixed picture of net time and distance improvements and deteriorations. For Inside Central City residents, trip durations improved during the morning peaks for 3 of the 5 Metropolitan Statistical

Table 18.1a   Mean trip times and distances, 1983 and 1990 by trip purpose, time of day, metropolitan size, place of residence (inside central cities—private vehicles)[a]

MSA population size		Year	a.m-peak[b]		p.m.-peak[b]		Off-peak	
			Work	Other	Work	Other	Work	Other
Residing inside central cities								
Below 250 000	T[c]	1983	15.2	15.7	17.2	12.2	13.6	16.0
		1990	15.0	11.5	16.8	13.6	13.0	14.3
	D[c]	1983	6.7	7.6	7.6	5.0	6.2	7.8
		1990	7.8	6.4	9.6	7.3	7.0	8.3
250 000–499 999	T	1983	15.1	13.9	15.2	11.7	15.7	13.3
		1990	14.8	10.1	15.7	11.8	14.2	14.8
	D	1983	6.1	6.0	7.7	5.0	7.5	5.6
		1990	7.6	3.7	7.7	5.6	7.8	8.8
500 000–999 999	T	1983	17.3	17.9	20.8	12.3	14.9	17.2
		1990	17.9	12.5	17.9	14.1	16.1	14.8
	D	1983	8.5	8.5	9.3	5.0	6.9	9.0
		1990	10.2	5.7	9.2	7.5	9.2	8.0
1–3 million	T	1983	18.3	16.4	20.8	14.0	17.9	15.1
		1990	19.5	12.4	21.3	14.1	17.8	14.8
	D	1983	8.7	–	8.3	5.5	8.7	6.8
		1990	10.4	6.1	11.1	7.1	9.7	8.0
Over 3 million	T	1983	28.8	15.1	29.4	17.8	23.0	17.3
		1990	22.9	15.7	24.6	14.7	21.7	16.1
	D	1983	12.7	5.3	12.3	6.7	10.4	7.4
		1990	11.7	7.4	11.8	6.9	12.1	8.5

[a] All work trips refer to non-stop one way trips.

[b] In this and subsequent tables, the a.m.-peak is defined as 6–9 a.m.; the p.m.-peak is 4–7 p.m.

[c] T refers to time in minutes, and D to distance in miles.

Table 18.1b   Mean trip times and distances, 1983, 1990 by trip purpose, time of day, metropolitan size, place of residence (outside central cities— private vehicles)

MSA population size		Year	a.m.-peak		p.m. peak		Off-peak	
			Work	Other	Work	Other	Work	Other
Residing outside central cities								
Below 250 000	T	1983	18.4	9.8	20.2	12.9	16.6	13.3
		1990	19.1	19.5	20.0	17.5	20.4	17.5
	D	1983	9.9	4.3	9.9	6.6	8.8	6.9
		1990	11.7	12.3	12.2	11.1	13.2	11.4
250 000–499 999	T	1983	19.2	18.2	19.7	14.9	16.9	14.1
		1990	19.3	13.1	21.9	12.5	19.4	17.2
	D	1983	10.6	12.4	9.9	7.9	8.8	7.6
		1990	12.0	6.9	13.6	6.5	12.5	10.2
500 000–999 999	T	1983	22.5	17.9	25.5	13.5	21.7	15.8
		1990	21.1	16.9	23.0	13.3	20.8	15.9
	D	1983	12.1	10.4	13.2	6.5	11.1	8.6
		1990	13.1	10.3	13.9	7.5	13.2	9.8
1–3 million	T	1983	22.1	14.9	23.2	17.7	19.5	16.1
		1990	21.5	13.2	22.8	15.1	21.0	16.6
	D	1983	11.2	7.4	11.2	9.3	10.7	8.8
		1990	12.5	7.2	12.1	8.1	12.9	10.5
Over 3 million	T	1983	22.3	15.9	25.5	14.1	18.3	17.3
		1990	24.3	13.2	26.4	14.3	21.7	15.9
	D	1983	11.2	7.6	11.5	6.9	9.3	9.0
		1990	13.5	6.8	14.0	7.1	12.9	9.0

Area size groups (including the 3 million-plus group). There were trip time improvements for 3 of the 5 groups for the afternoon peak. For Outside Central City residents, trip durations also improved for 2 of the 5 groups. For the pm-peak, work trip times improved for 3 of the 5 cases.

Though these results suggest further corroboration of our hypothesis, they require analysis of their statistical significance. As in our previous reports on NPTS data, we converted the survey results on durations and distances to data on trip speeds. The latter are more likely to be normally distributed and, therefore, more appropriate for standard statistical testing. Data on trip speeds are shown in Table 18.3 (p.353). *1990 trip speeds are significantly higher for 57 of the 60 comparisons shown.*

We also conducted ANOVA tests on the null hypothesis that average trip speeds were independent of city size. Table 18.4 (p.353) shows F values and significance levels for the 12 categories (two possible places of residence, three time slots, two trip purposes) for which the tests were conducted.

Whereas the null hypothesis that city size makes no difference when comparing average trip speeds was not rejected (at the 95% confidence level) for 1983 work trips (while sustained for 1983 non-work trips), the findings for 1990 are somewhat different. Table 18.4 shows that city size

Table 18.2 Comparison of mean trip times and distances, 1983 and 1990 by trip purpose, time of day, metropolitan size, place of residence (private vehicles)

Population Size		a.m. peak		a.m. peak		Off-peak	
		Work	Other	Work	Other	Work	Other
Residing inside central cities							
Below 250 000	T	Down	Down	Down	Up	Down	Down
	D	Up	Down	Up	Up	Up	Up
250 000–499 999	T	Down	Down	Up	Up	Down	Up
	D	Up	Down	n/c	Up	Up	Up
500 000–999 999	T	Up	Down	Down	Up	Up	Down
	D	Up	Down	Down	Up	Up	Down
1–3 million	T	Up	Down	Up	Up	Down	Down
	D	Up	–	Up	Up	Up	Up
Over 3 million	T	Down	Up	Down	Down	Down	Down
	D	Down	Up	Down	Up	Up	Up

Population Size		am-peak		pm-peak		Off-peak	
		Work	Other	Work	Other	Work	Other
Residing outside central cities							
Below 250 000	T	Up	Up	Down	Up	Up	Up
	D	Up	Up	Up	Up	Up	Up
250 000–499 999	T	Up	Down	Up	Down	Up	Up
	D	Up	Down	Up	Down	Up	Up
500 000–999 999	T	Down	Down	Down	Down	Down	Up
	D	Up	Down	Up	Up	Up	Up
1–3 million	T	Down	Down	Down	Down	Up	Up
	D	Up	Down	Up	Down	Up	Up
Over 3 million	T	Up	Down	Up	Up	Up	Down
	D	Up	Down	Up	Up	Up	n/c

makes no difference for Inside Central City a.m. peak and off-peak work trips but is significant for the ten other categories.

Inspecting data for the 20 Consolidated Metropolitan Statistical Areas (CMSAs, typically the largest metropolitan areas; Table 18.5), however, it can be seen that there is no simple relationship between trip speeds and metropolitan area size.

The middle-sized cities often show the highest speeds (as in 1983), with generally slower speeds seen in the larger and the smaller metropolitan areas. Whereas the 5 smallest CMSAs show the shortest commuting times, the relationship between metropolitan area size and trip duration is, at best, weak. More importantly, relative or absolute CMSA population growth does not correlate with commuting times. The large metropolitan areas appear to absorb large numbers of newcomers without significantly lagging in average commuting times. The p.m. peak work trip for central city-based residents is the basis for comparison because some observers think that this

Table 18.3   Comparison of mean trip speeds, 1983 and 1990, place of residence (private vehicles)

Population Size	Year	a.m. peak		p.m. peak		Off-peak	
		Work	*Other*	*Work*	*Other*	*Work*	*Other*
Residing inside central cities							
Below 250 000	1983	25.0	20.6	22.9	20.3	24.6	20.9
	1990	29.6*	26.4*	31.4*	26.8*	30.2*	27.4*
250 000–499 999	1983	23.5	19.3	25.0	23.1	24.9	21.2
	1990	29.9*	22.1**	28.7	26.0*	31.5*	28.8*
500 000–999 999	1983	27.4	19.4	25.4	21.0	25.4	23.0
	1990	31.6*	24.2*	29.3**	26.6*	32.4*	27.1*
1–3 million	1983	27.6	31.7	24.2	20.5	27.5	22.4
	1990	30.7*	24.3	30.4*	26.8*	31.7*	27.7*
Over 3 million	1983	25.9	15.9	25.5	18.0	25.9	19.2
	1990	30.0*	23.3*	28.0	25.0*	30.6*	26.5*

Population Size	Year	a.m. peak		p.m. peak		Off-peak	
		Work	*Other*	*Work*	*Other*	*Work*	*Other*
Residing outside central cities							
Below 250 000	1983	30.3	20.9	27.3	24.5	28.2	25.8
	1990	35.2*	30.3*	34.1*	29.6*	36.6*	32.6*
250 000–499 999	1983	30.7	24.8	27.7	24.6	28.4	27.1
	1990	34.8*	28.3**	35.1*	29.4*	34.9*	30.7*
500 000–999 999	1983	30.6	24.5	29.6	26.3	28.6	25.9
	1990	35.2*	30.6*	34.9**	30.7*	35.9*	31.5*
1–3 million	1983	28.4	23.1	27.0	25.3	29.9	28.3
	1990	33.5*	26.6*	30.7*	28.5*	34.7*	31.0*
Over 3 million	1983	28.1	20.9	25.8	23.8	27.3	23.7
	1990	31.8*	25.5*	30.7*	27.3*	33.1*	31.8*

* Significantly greater than 1983 at the 99% confidence level.
** Significantly greater than 1983 at the 95% confidence level.

Table 18.4   ANOVA f-values for null hypothesis that city size does not affect average speeds (private vehicles)

Residence	a.m. peak		p.m. peak		Off-peak	
	Work	*Other*	*Work*	*Other*	*Work*	*Other*
Inside central cities						
1990	1.05	3.70	2.65	2.71	1.48	10.74
	(0.3797)*	(0.0052)	(0.0321)	(0.0287)	(0.2067)	(0.0001)
Outside central cities						
1990	8.11	11.61	9.00	10.16	5.77	72.35
	(0.0001)	(0.0001)	(0.0001)	(0.0001)	(0.0001)	(0.0001)

* Significance levels are shown in parentheses.
Source: Compiled from 1990 Nationwide Personal Transportation Study data.

Table 18.5 Work trip durations and speeds (private vehicles) compared with CMSA growth (1980–1990)

CMSA	1990 pop. ('000)	Pop. change ('000) 1980–1990	% Pop. change 1980–1990	Central city pm-peak worktrip duration 1990 (min.)	Worktrip duration, 1990 (mins.) Residing Inside central city	Outside central city	Worktrip speed, 1990 (mph) Residing Inside central city	Outside central city
Los Angeles	14 532	3034	26.4	25.5	23.7	26.0	31.7	33.6
Dallas	3 885	954	32.6	25.5	21.0	18.8	33.0	36.1
San Francisco	6 253	885	16.5	16.6	19.7	21.9	29.6	33.9
Houston	3 711	611	19.7	24.7	20.2	24.5	29.2	33.9
Miami	3 193	549	20.8	19.8	19.7	23.5	32.8	28.6
New York	18 087	547	3.1	26.1	23.0	23.4	26.7	31.5
Seattle	2 559	466	22.3	19.8	20.1	30.1	32.3	29.5
Denver	1 848	230	14.2	24.7	21.2	20.5	31.6	32.2
Philadelphia	5 899	218	3.8	22.3	22.6	22.1	34.8	30.8
Boston	4 172	200	5.0	24.3	21.3	20.8	26.9	33.2
Portland	1 478	180	13.9	18.3	16.8	21.6	26.7	35.0
Chicago	8 066	129	1.6	30.4	27.8	23.3	32.5	28.1
Cincinnati	1 744	84	5.1	19.2	17.4	22.0	31.5	34.8
Hartford	1 086	72	7.1	19.2	17.1	22.2	29.6	32.6
Providence	1 142	59	5.5	16.7	12.5	19.1	39.0	35.1
Milwaukee	1 607	37	2.4	21.5	19.8	19.1	29.9	35.1
Buffalo	1 189	−54	−4.4	16.7	17.7	24.1	35.5	34.3
Cleveland	2 760	−74	−2.6	19.1	19.8	20.3	27.1	30.4
Detroit	4 665	−88	−1.9	24.2	20.9	22.8	29.4	36.8
Pittsburgh	2 243	−180	−7.4	28.8	22.2	17.7	25.6	29.5

Sources: 1990 and 1980 Census Population and NPTS data.

is the type of trip that is more vulnerable to congestion in conditions of growth; however, analysis of other trips yields similar results. Yet CMSAs experiencing substantial population growth (including Los Angeles which added more than 3 million citizens over the decade) fared about as well as the other metropolitan areas. It is likely that spatial adjustments helped.

In fact, not only are there more people living and working in large cities, but daily trips per capita continue to increase; Pisarski (1992) reports that the 1983–1990 growth in trips per capita was 7%. Much of this is accounted for by the continuing growth in non-work trips which now make up 75% of all metropolitan area person-trips.

The key point is that modern cities avoid congestion by spreading out; they remain competitive, by avoiding high land costs (and high export prices) by taking advantage of agglomeration economies that are apparently available at comparatively low densities and throughout each metropolitan area. Spatial structure adjustments stave off most of the traffic calamities that many predict for the largest US cities.[4]

It is well known that congestion will occur as long as pricing is avoided. The important question is 'how much'? Because suburbanisation occurs for almost all activities, what levels of road and highway congestion are sufficient to push residences and job sites away from congested areas? Apparently, not great. The key to explaining limited congestion in US cities is the apparent ease with which activities have suburbanised. The decentralisation story in US cities is well documented. In previous research (Gordon & Richardson 1991a), we have shown that similar patterns of decentralisation are occurring in all major US cities regardless of the location (region) or the age of the city: there does not appear to be a 'sunbelt–frost belt' difference, for example. Table 18.6 (p.356) summarises the results for the 12 major CMSAs that were studied.

Many more people subscribe to the 'doomsday' view of urban growth than to the more benign 'waves of development' perspective. The former view is usually supported by anecdotal evidence. Yet the large data files now available consistently tell an entirely different story.[5] Because the doomsday rhetoric provides the justification for expensive rail transit plans (in spite of the sorry record of these systems; see, for example Pickrell (1992)), there is much more at stake than discussions about the most appropriate urban development model. Our vision of how to manage cities hinges on our understanding of how they decentralise in order to compete and succeed. Since many have argued that we are at the beginning of a new age of telecommunications, the dynamic processes of urban spatial adjustment are likely to be even greater and more auspicious in the future.

Technology and travel behaviour

Given the focus of this conference on technological change and urban form, it is pertinent to ask whether, and to what extent, our arguments are affected by ongoing and prospective technical progress. As suggested above,

Table 18.6 Sectoral employment trends in 12 major CMSAs*, 1982–87

Sector	Central City			Ring I			Ring II			CMSA
	1982 share	1987 share	annual growth rate 1982–87	1982 share	1987 share	annual growth rate 1982–87	1982 share	1987 share	annual growth rate 1982–87	annual growth rate 1982–87
Manufacturing	0.302	0.258	-0.0388	0.349	0.377	0.0075	0.349	0.365	0.0012	-0.0078
Retail	0.281	0.258	0.0215	0.368	0.379	0.0454	0.351	0.364	0.0468	0.0394
Wholesale	0.377	0.319	-0.0029	0.336	0.360	0.0445	0.287	0.321	0.0535	0.0304
Services	0.416	0.361	0.0397	0.300	0.327	0.0885	0.284	0.313	0.0904	0.0698

* Includes New York, Los Angeles, Chicago, Philadelphia, San Francisco, Detroit, Houston, Miami, Cleveland, Milwaukee, Cincinnati, Seattle.
Source: Richardson and Gordon (1993), 'L.A. lost and found', Tables 4 and 5.

we believe that auto-related air pollution can be largely relieved by improved technology with respect to emission controls and fuel advances. The validation of this belief is important to our analysis because it casts doubt on the criticisms of suburbanisation based on its modest contribution to increases in VMT and on social engineering efforts to improve 'jobs-housing balance' or to induce changes in travel behaviour (e.g. modal shifts, fewer trips). If technology can deal effectively, either now or in the foreseeable future, with the automobile emissions problem, some of the environmental issues in the 'sustainable cities' debate evaporate.

A more speculative set of issues relates to how the telecommunications revolution will impact upon traffic congestion. Particularly in the information-processing industries (a large and expanding component of the services sector), both the relocation of back-office functions to smaller towns (often outside metropolitan areas) and increased telecommuting within the big cities offer an additional form of adjustment by firms and households to increases in metropolitan congestion. How important these trends might become has been a subject of great controversy. Back-office relocations have been driven more by wage cost concerns than by diseconomies of congestion, although they could have a congestion-relief impact if they become extensive enough to affect city-size growth rates on any scale (however, we should remember that city-size variations in speeds remain very modest). The barriers to more telecommuting remain more social and managerial than technical; many workers are reluctant to abandon the pleasures of socialising at work while some managers remain leery of the lack of direct supervision of working-at-home employees. Interestingly, although some analysts have projected increases of telecommuting in the millions over the next decade or two, the American Housing Survey shows that working at home declined between 1985 and 1989 from 3 to 2.5% of the labour force (Pisarski 1992, p.38). Although there have been shifts within the working-at-home categories, in favour of suburban technical workers and away from urban home-based jobs (e.g. family day care providers) and rural small holders, the growth in telecommuting remains a potentially important safety valve against worsening commuting congestion.

Conclusion

Our argument that congestion remains sustainable has strong policy implications. It suggests that many currently favoured transportation policies have to be viewed with suspicion on the grounds of cost-effectiveness. The most obvious examples are heavy capital investments in fixed rail transit and mandatory measures to change travel modes and behaviour. Instead, the focus of transportation interventions should shift towards those less costly measures that help to reinforce the voluntary adjustments that travellers are willing to make to avoid congestion. Examples include the introduction of road congestion pricing and deregulation of route carrier restrictions to foster para-transit alternatives. Sustainable congestion implies sustainable transportation policy: measures that work and that do not require more tax revenues.

Notes

[1] A typical example is the Los Angeles County Transportation Commissions' recently released 30-year plan (1992). The document suggests that under its 'no-build' alternative, average rush-hour freeway speeds and average rush-hour commuter speeds will fall from the current 29 mph (46 km/h) to 17 mph (27 km/h).

[2] The 1990 data are from a telephone survey of 22 300 households (including 48 400 persons) throughout the US. The Research Triangle Institute conducted the interviews for the Federal Highway Administration (US Department of Transportation).

[3] Because work trips and non-work trips are occasionally combined in trip-chains, this explains some of the peak period growth in non-work trips (being examined in related research).

[4] An alternative coping mechanism that has been suggested by some observers is peak-spreading. In previous work, we found only limited peak-spreading between 1977 and 1983 (Gordon et al. 1990). Preliminary results for the 1990 survey corroborate this finding. Whereas 39.1% of all trips in the largest metropolitan areas took place in the 6–9 am and 4–7 pm peaks in 1977, the corresponding number for 1990 was 38.3%.

[5] Pisarski's report of the 1990 NPTS takes a view fully consistent with our explanation. Referring to all commuting trips he states: 'The average commute trip length increased by 7% from 1983 to 1990, from 9.9 miles to 10.6 miles. Yet the commute time declined by 3 % during the same period. This observation might be partially due to the fact that a greater number of suburban and exurban residential areas and employment centres were developed. The resulting commutes are longer but are travelled at faster speeds. The decline in travel time is also influenced by changes in commuting modes, with a decrease in transit and car pooling and an increase in driving alone' (Pisarski 1992, p.24).

19 Work Travel Responses to Changing Workplaces and Changing Residences

Nigel Spence and Martin Frost

Introduction

USING comparable information for three large cities in Britain—London, Manchester and Birmingham—this study attempts to answer some basic questions about changing patterns of workplaces, residences and work travel. Such patterns are highly distinctive in these cities, especially when reflected in density terms. The numbers of people and jobs being lost from within the areas of these cities in recent times is substantial on any measure yet it seems to have done relatively little to alter the fundamentals of the distributions. Work travel is still predominantly radial and London is the prime example of such. Where local centres of employment exist away from the cores of cities then patterns of work travel adjust accordingly but these tend to be minor diversions from the main pattern. Over time the relative distribution of homes and jobs is slowly changing. Workplaces are becoming increasingly separated from residences and work travel trips are becoming somewhat longer. But given the scale of the total net shifting of population and employment the characteristics of work travel seem enduring. It is salutary to reflect on the complexities of the processes which give rise to change in the work journey for individuals. Changes in the surface of employment opportunities within a city have a role, as do the differential changes that in and out migration produces. To these must be added changes in the facilitating infrastructure both in terms of the urban built form and the pricing structures evolved for its use.

Changing patterns of work travel reflect the outcome of some complicated processes involved in finding places to live and to work. The aims of this chapter are to present some results of research which attempts to describe such changing patterns. It will be shown that cities have been undergoing significant changes in recent times with substantial numbers of people and jobs decentralised (Frost & Spence 1991a; 1991b; 1993). Yet the basic density profiles retain their traditional character in the main. The distribution of workplaces relative to residences is changing, but only slowly as travel distances between the two modestly increase. The patterns of work travel too appear to be enduring, although some significant features of change specific to certain parts of the city are to be seen. From the following descriptions it will be demonstrated just how complex a set of processes changing patterns of worktravel represent.

The empirical context for this research is England in the 1970s (Dasgupta *et al.* 1989; 1990a; 1990b). In all, some 27 cities made up the full research sample and for each one a centre was selected and a radial extent demarcated to coincide with consistent population density thresholds (10 persons per hectare). All large cities were present and these were combined with a selection of 13 smaller cities and towns. This chapter will focus emphasis on the results for only the three largest cities—London, Manchester and Birmingham—those for which some detailed change over time is available (Frost & Spence 1981).

To facilitate analysis, a methodology was developed to aggregate information collected for small areas (enumeration districts and wards) into a series of regular concentric annuli around city centroids. Three points are worthy of note here.

- All information relating to these small areas (usually wards) is assumed implicitly to be locationally represented by the centroid of the area.
- It is also assumed that these small areas lie wholly in the annuli wherein their centroid is located.
- Distance between wards is measured as the straight-line distance connecting centroids.

This measure has great computational advantage but clearly makes no allowance for the network characteristics of routes connecting the wards. One further important complicating factor about the spatial referencing of these data concerns the information for 1971. For this date statistics are available only on a restricted and more spatially aggregate basis and as a result best-fit solutions to the adopted ring and sector spatial systems were developed.

All of the data used in this chapter are derived from population censuses. The principal source is the special workplace statistics data set which is divided into three tables. The first two present the characteristics of residents and workplaces respectively. The last presents the work travel flows where the number of people moving are classified by mode of travel, occupation, industry and age of the employed person. This data file presents flows within, into and out of each ward. All work travel data are based on 10% sample information. A second, similar but considerably smaller, data set used in the study was the special workplace statistics for 1971. These statistics were not collected on such a regular basis. Flows in Greater London are presented as flows between traffic zones; other cities are presented with residence classified by enumeration district and workplace by local authority specified traffic zone.

Patterns of residences and workplaces

The spatial distribution of employment and its distribution relative to residences are crucial factors in determining work travel patterns and trip characteristics. If the distribution of homes and jobs are both centralised under conditions of high urban density, then workplaces are easily accessed by foot and work journeys are shorter. Where the distribution of employment is strongly centralised and population is decentralised within a city, naturally travel patterns are radially oriented to the city centre and the longer work journeys are well supported by public transport. Conversely, a decentralised distribution of employment attracts out-commuting and orbital trips; these journeys are of varying lengths and are more easily made by car. This section examines the distribution of residences and workplaces as well as the relative distribution of the two.

Population change and the distribution of residential densities

It has been well documented that during the 1970s the larger British cities experienced a loss of total residential population, whilst smaller towns grew. Of the 27 towns and cities in the study, all those with a population over 100 000 in 1981 showed losses over the decade, whilst almost all settlements with less than 100 000 showed population gain. On average in the group of growing settlements with populations in 1971 of less than 75 000 there was an increase of over 20%, although individual towns within that size class had varying rates of growth. The sample of towns in the 75–100 000 size class also grew (by 6% on average). Among the larger settlements there is a weak relationship between increasing size and greater population losses. However, there are considerable differences between cities that do not relate simply to size differences.

The pattern of population losses and gains within the three largest cities can be seen in Figure 19.1 (p.362).

These density profiles for 1971 and 1981, using 2 km annular bands, show that there has been population loss at all radial extents, although the magnitude of this loss is much greater in the inner areas of the cities, producing significant reductions in the population densities in these areas. By contrast, in the medium-sized cities there has been some population gain in the outer areas but not enough to compensate for the inner area losses, hence these cities have experienced overall decline. Many smaller cities are also losing population in the inner area, although to a lesser extent than in the larger cities, but their outer area gains more than compensate for these losses.

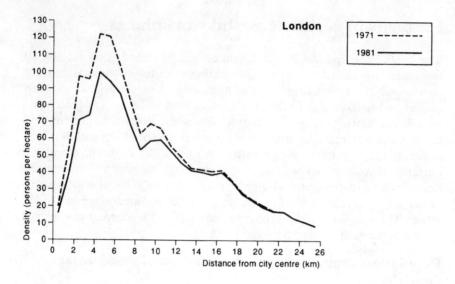

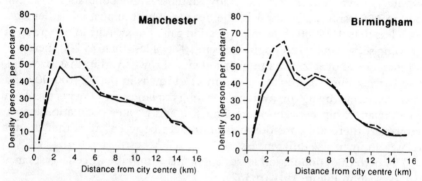

Figure 19.1 Population density profiles within cities, 1971 and 1981

Employment densities within cities and employment change, 1971–1981

Employment is of course unevenly spread over the city area. The city centre, by virtue of its accessibility, has been the favoured location for specialised retail outlets and for financial, legal, administrative and other office-based activities. The demand for centrally located office and retail space has maintained high rent and land values and available land is therefore intensively used. In most cities, employment densities fall sharply beyond this area and then much more gradually as distance from the city centre increases.

The largest cities of the study conform to this expected employment density profile, as shown in Figure 19.2 where employment density is

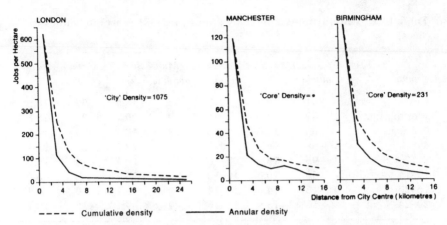

Figure 19.2 Distribution of employment densities within cities, 1981
* indicates problems in urban core statistical definition

calculated for each 2 km annular band from the centre and recorded both separately and cumulatively.

'Core' densities (within 1 km of the city centre) are recorded wherever the ward data permitted reasonably comparable definitions—Manchester was problematic in this respect. When averaged across annular bands, employment densities fall rapidly over the first 4 km from a central peak and then more gradually to reach low employment densities in the urban periphery.

Employment densities in London are of an entirely different magnitude to those elsewhere. In London, the highest density is recorded in the 'city' (1075 jobs per hectare) and the annular density falls from around 600 in the cental area to 100 over the first 4 km. In Manchester and Birmingham, central densities reach about 120 jobs per hectare and 'core' densities are about 200. In all three major cities, the annular densities fall sharply to between 20 and 30 in the 2–4 km band.

Across the full sample, with few exceptions central densities broadly decline with the position of a city in the urban hierarchy. Across the sample of towns and cities, central densities range from 10 jobs per hectare in the smallest towns (population size 30–40 000) to around 40 in the smaller cities (population size 100–250 000) to 50–80 in medium-sized cities (below half a million population) and to above 100 jobs per hectare in the million plus cities.

Like population, employment is also being lost throughout the cities considered here. Table 19.1 (p.364) shows job losses by annuli bands within the cities.

For the capital, around 120 000 jobs were lost over the decade from the dual central cores of the 'city' and Westminster, and these double figure percentage declines are typical of most of the rest of the annuli. For both Manchester and Birmingham the declines in the central annuli in relative terms are greater than those for London. Curiously the 8–10 annuli in

Table 19.1 Employment change by annuli, 1971–81 (absolute values in '000s)

Annuli	London absolute	%	Annuli	Manchester absolute	%	Birmingham absolute	%
City	−41	−12.2	0–2	−33	−18.5	−32	−18.5
Westminster	−82	−15.0	2–4	−19	−20.7	−27	−18.8
0–4 rem	−63	−13.6	4–6	−18	−17.1	−13	−12.6
4–8	−98	−16.7	6–8	−26	−22.3	−28	−21.1
8–12	−93	−15.7	8–10	23	19.7	−18	−15.6
12–16	−56	−10.3	10–12	−22	−13.7	−8	−6.9
16–20	−18	−4.0	12–14	−9	−8.1	0	−0.1
20–24	−55	−13.9	14–16	−9	−9.1	−16	−18.7

Manchester does manage to record employment gains but elsewhere the pattern is one of almost consistent heavy decline.

The relative distribution of jobs

The relative strength of the city centre (i.e. the proportion of city-wide jobs which are located in the central area) is an important component in accounting for work travel patterns in different cities. One way of measuring the relative strength of the city centre is to compare the cumulative proportion of jobs at different radial extents from the city centre (Table 19.2).

London, although of a much larger radial extent than the cities in the next level of the urban hierarchy, has a third of its jobs concentrated within a 4 km radius of the centre. Comparing Manchester with Birmingham, although the proportion of jobs is identical within the central area, clearly there is a higher concentration of jobs in the 2–8 km rings in Birmingham.

Table 19.2 Cumulative percentage of jobs, 1981

Annuli	London	Manchester	Birmingham
0–2	21.6	17.1	17.0
0–4	33.0	25.7	30.9
0–6	40.7	35.9	41.7
0–8	46.8	46.4	54.1
0–10	53.6	62.6	66.0
0–12	61.0	78.7	79.7
0–14	67.4	90.1	91.4
0–16	75.1	100.0	100.0
0–18	82.3		
0–20	87.3		
0–22	91.6		
0–24	96.7		
0–26	100.0		

Table 19.3 **Relative distribution of workplaces (W) and residences (R) within cities, 1981—containment levels by area in square kilometres with percentage areal extent in parentheses**

		Containment levels				
		20%	40%	60%	80%	100%
London	W	12.4(1)	107.3(5)	435.4(21)	945.2(45)	2079.6(100)
	R	167.2(8)	414.0(20)	740.7(36)	1175.8(57)	2079.6(100)
Manchester	W	20.9(2)	152.2(18)	292.9(34)	477.5(56)	851.1(100)
	R	107.2(13)	240.5(28)	390.0(46)	574.1(67)	851.1(100)
Birmingham	W	13.9(2)	92.2(12)	241.7(30)	480.9(60)	801.1(100)
	R	98.1(12)	180.0(22)	284.3(35)	485.9(61)	801.1(100)

Table 19.3 gives a measure of the relative strength of the city centre and the level of employment dispersion.

The table shows the areal extent which contains given percentages of the total employment (and residences which are discussed later). Areal extents are expressed in square kilometres and as a percent of the total area of each city. The figures are calculated by cumulating the areas of wards, each having been ranked by distance from the city centre. The principal feature as far as workplaces are concerned is the remarkable concentration of jobs in central cities. The 20% employment containment level is attained in all cities in less than 2% of the city area. In London, some 40% of the jobs are located within 5% of the city's area. For comparison, in Manchester and Birmingham, the 40% containment level occurs within 18% and 12% of the total area respectively. In both cities, the 60% containment level for workplaces requires about a third of the city wide area compared to the much lower areal proportions (21% and below) in London and most other large cities.

Distribution of jobs relative to homes

It is important to consider not only the distribution of jobs but also the relative distribution of jobs and homes. The information in Figure 19.3 (p. 366) and Table 19.3 provides a description of the relative distribution of jobs and homes of people in employment in the largest cities of the study.

That all cities have a higher concentration of workplaces than residences is naturally true, but some cities, for example London, are relatively more centralised than others. In Manchester and Birmingham the distribution of homes and jobs are roughly similar although there are interesting local variations. In Manchester there is evidence of secondary concentrations of workplaces between 8 and 12 km from the city centre, whereas in Birmingham the secondary concentration is adjacent to the core area from about 2–6 km from the centre. It is here that the gap between homes and jobs is widest, contrasting with Birmingham's outer zones where there is hardly a gap at all. It should be remembered that this information records only the aggregate numbers of homes and jobs within the urban area and

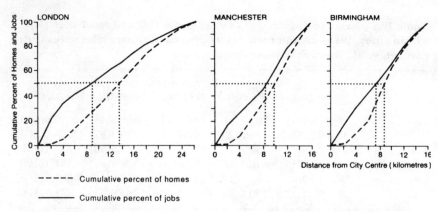

Figure 19.3   The relative distribution of homes and jobs of employed persons, 1981

does not provide a true indication of work travel flows either within, into or out of the city system.

Table 19.3 gives an alternative comparison of the relative levels of centralisation. The principal feature of this table as far as employed residents are concerned is the expected contrast with the distribution of workplaces. The distribution of residents is, of course, much more dispersed. For all the cities in the sample, some 20% of employed residents live within 8–16% of the total areal extent. At the 40% containment level, Manchester is one of the most decentralised. Over the remainder of the containment levels for each of the cities in the sample, the deviations between the workplace and residence concentrations are still apparent although less marked than in the centre. In general the higher the containment level the lower the difference in areal extent between workplaces and residences.

Another measure of the attractive capacity of the city centre, as far as worktravel is concerned, is provided by the 'work-ratio' (i.e. the number of jobs per employed resident in any area). These are given in Figure 19.4.

As expected, the central areas have many more jobs than employed residents and these areas therefore experience a net inflow of work trips. As in the case of employment densities, there is a clear relationship between the work ratio in the central areas and the overall size of the city, ranging from 29 jobs per employed resident in central London to about 4 jobs per employed resident in the smaller cities of the full sample. Interestingly there is much more variation in the core areas of cities. In London, the city commands a work ratio of 91. In Manchester, the work ratio (at 68) is much higher than in Birmingham (at 20).

Distances between workplaces and residences

In terms of urban structure, work travel distances are determined by a number of factors. For instance, it might be expected that large cities with their greater radial extents would be characterised by longer distance trips

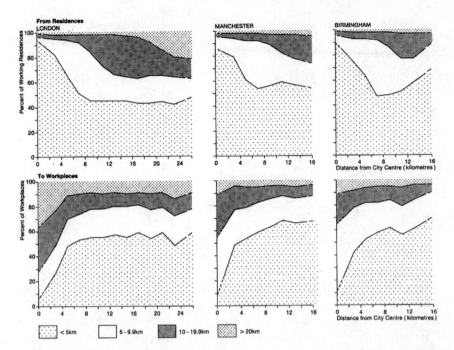

Figure 19.4   Work ratios within cities, 1981 (number of jobs per employed resident)

than smaller cities. But this simplification could be clouded by the extent of the segregation of homes and jobs within each city. In cities where workplaces are clustered in or near the city centre and central employment densities are high, the city centre would naturally attract long distance commuting. If some of this labour force was being drawn from satellite towns, the latter would be generators of long distance trips. The internal structure of cities (in terms of the spatial segmentation of labour markets) is also an important determinant. So for instance, if there was a good local match between various types of jobs and different groups of working residents, then journey distances would be shorter than otherwise. Urban residential densities are also a contributing factor, longer journeys being associated with lower densities.

Travel distances from residences and to workplaces

Accurate measures of travel distances spatially disaggregated to a small area level are not generally available. For the first time, the 1981 census provided tables giving distance to work at both the residence and workplace ends of the work journey. The data provide five categories of distance bands—less than 5 km, 5–9 km, 10–19 km, 20–29 km and 30 km or more. Distances are calculated from the grid reference of the centroid of the enumeration district of residence and that associated with the post code of workplace. The measure is a straight-line distance. A summary of travel distance data (for totals in employment) is shown in Table 19.4 (p.368) for London, Manchester and Birmingham.

Table 19.4 Distance travelled to work as percent of all distances, 1981

(a) to workplaces	Less than 5 km	5–9 km	10–19 km	More than 20 km
London	42.1	22.3	19.2	16.5
Manchester	55.0	25.4	14.2	5.5
Birmingham	50.3	29.5	14.1	6.2
(b) from residences	Less than 5 km	5–9 km	10–19 km	More than 20 km
London	48.4	25.3	20.7	5.6
Manchester	59.0	25.9	12.2	2.9
Birmingham	55.6	30.7	11.6	2.2

The basic feature of this information is that it shows journeys from place of residence by distance band and in turn journeys to the place of work again by distance class. Of course the largest component of flows from both ends of the work trip are those less than 5 km. In all cases the proportion of such trips is higher from residences than to workplaces. Generally, small towns and cities have largest percentages of such short-distance journeys. The second category of distance, 5–9 km, is particularly well represented in the medium-sized and larger towns in the system. The proportion of trips in this category tend to be similar in the case of flows from residences compared with flows to workplaces in most large and medium-sized cities, as is shown in London, Manchester and Birmingham. Small towns, however, tend to attract a much larger proportion of medium length trips than they generate. The 10–19 km category follows next in general proportional importance for most large and medium-sized cities.

Contrasts between the flows calculated 'from residences' and 'to workplaces' are especially apparent for these longer distance trips. The relationship with city size is not straightforward. In large cities the proportions are roughly similar, whereas medium-sized places attract a higher proportion of long-distance trips to the workplace end than from residences. Large and medium-sized cities attract a much higher proportion of trips which are more than 20 km in length than they generate, as the information in Table 19.4 demonstrates. The converse is the case in the smaller towns and cities. Longer distance commuting from small towns is an important feature especially for those towns which are well linked to London by rail.

The phenomenon of long-distance commuting is best illustrated in the London data: some 16.5% of people who work in London travel more than 20 km to do so. Persons, then, living in small towns around metropolitan areas, especially London, are prepared to commute longer distances in return for the advantages of living in such locations. But many of these small towns are also considerable attractors of long-distance trips, suggesting a fairly open labour market beyond the London green belt.

Distance distributions within cities

Figure 19.5 shows worktravel flows from residences and to workplaces located in different parts of the three largest cities of the study.

These illustrations confirm that the central parts of large British cities attract long-distance commuting but generate a large proportion (over 80% in most cities) of short-distance trips. At the residence end, as distance from the city centre increases, the proportion of short-distance journeys declines over the first 6 km to about 40–50% of journeys and then levels off to the outskirts. Deviations from this general pattern occur in Birmingham where the proportion of short-distance trips rises again beyond the 12 km radius. This reflects the fact that outer Birmingham has a number of employment nodes which provide job opportunities for the resident population.

Short- to medium-length trips (5–9 km) usually peak at around 6–8 km from the centre. Medium to long journeys (10–20 km) peak further out at about 12–16 km in the larger cities and generally increase in importance with distance from the city centre in medium-sized cities. The only large city which generates a significant proportion of long-distance trips is London, where the proportion of long-distance trips (of over 20 km) reaches about 20% beyond the 22 km radial extent. This pattern of successive trip length bands peaking at increasing distances from the city centre is of course a reflection of the fact that a substantial proportion of trips generated in any part of the city is directed to the city centre.

At the workplace end, as distance from the city centre increases, short-distance trips increase at first and then level off at around 55–65% of all trips. This feature is most pronounced in large cities where short-distance trips are but a small component of trips to the centre. Short- to medium-distance trips are generally evenly distributed through the city, except in Birmingham, where there is a higher than average representation of these

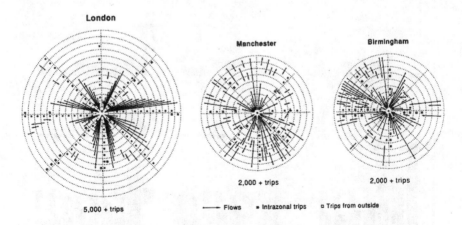

Figure 19.5 Trip length distributions from residences and to workplaces, 1981

trip lengths in the city centre. The 10–20 km trips play a more important role in central workplaces than elsewhere in the city, although in some cities the proportion of journeys in this distance band rises again in the outer annuli. Long-distance commuting (greater than 20 km) usually terminates in the city centre, especially so in London where it accounts for more than 30% of trips.

Patterns of work travel

This section presents a summary of the patterns of work travel. The features highlighted are the geographical orientation of work journeys; the degree to which the residents of different parts of a city are dependent on the central versus 'local' (defined here in a limited way as those living and working in the same annulus) job opportunities; the catchment area for jobs in the central area; and the changing scale of work travel distances.

Geographical orientation of commuting

There are significant variations in the orientation of work journeys in London, Manchester and Birmingham. An overall view of commuting patterns (analysed for a ring and sectoral representation of city structure) is given in Figure 19.6. For ease of visual interpretation, only heavy flows are given.

London is characterised by high volumes of long-distance in-commuting flows to the city centre as well as to the inner and some outer areas. Remarkably, there is little by way of long-distance orbital journeys except in the western sector near Heathrow (not shown, less than 5000 trips).

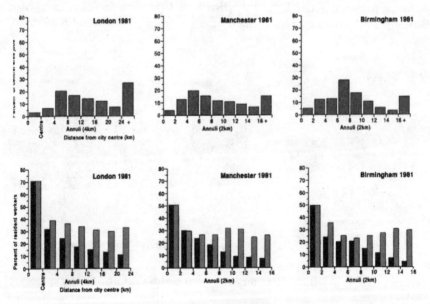

Figure 19.6 Work journey flows for 2 km annuli within cities, 1981

Overwhelmingly, the trips attracted by outer area employment centres are short-distance trips. The overall circularity of London is well reflected in its commuting patterns. Circularity is also a characteristic of Manchester, though, unlike London, the principal attraction centres seem to lie either in the central area or at radial extent of 8–12 km from the centre. This is not surprising given the history of urban development in Manchester, where a ring of industrial centres have remained substantial attractors in their own right despite the fact that newer residential spread now extends well beyond these older centres. Commuting patterns in Birmingham are quite different. There appears to be short- and long-distance commuting flows in the western and north-western sectors reflecting the city's asymmetric geographical structure and its proximity to neighbouring towns.

Changing patterns of worktravel flows within cities, 1971–1981

The losses of employment workplaces within cities has already been discussed. Table 19.5 (p.372) provides information on the equivalent losses of residents in employment (provided by the totals row for each city). With few exceptions most annuli show declines, with those near the centre being most marked.

The 29% loss from annuli 4–8 km in London amounted to a quarter of a million employed residents. Indeed it seems that this zone of cities (say between 2 and 8 km) has provided the largest worktravel flow reductions and this is entirely consistent with the patterns of population loss. For Manchester the 2–8 km band lost around 66 000 employed residents, and for Birmingham the losses amounted to 78 000.

The diagonal part of the origin–destination matrix records the intra-annuli flow changes. Most of these are negative, indicating both reductions in the population resident in these areas as well as employment decentralisation. Declines of over 30% in these flows are common in the inner annuli of the three cities.

Below the diagonal are recorded the changing flows that reflect exchanges from inner to outer annuli. For all three cities, the absolute flows that these percentage changes represent are small. Much more important are the changing exchanges between the outer annuli and the inner annuli and these are depicted in the area of the matrix above the diagonal. The cities discussed in this report differ somewhat in this respect but the similarities are more noticeable. London shows declines for all of the annuli sending employed residents to the dual cores of the city and Westminster. The heaviest declines are from the areas near to the centre. Some minor gains, or at least not so dramatic declines, are shown to occur in flows to the 4–8 km annuli from other areas of the city outside this zone. Manchester shows the same clear profile of reduced flows to the central area, the greater the reduction from nearer the centre. Again some increases in work travel are shown in the matrix and these primarily involve flows into the 6–10 km

Table 19.5 Percentage changes in worktravel flows between annuli, 1971–1981

	City	Westminster	0–4 rem	4–8	8–12	12–16	16–20	20–24
			London origins destinations					
City	*	−27	−32	−30	−19	−13	−12	−16
Westminster	*	−34	−31	−27	−11	−8	−11	−16
0–4 rem	*	−31	−29	−29	−6	*	−3	−14
4–8	*	−37	−15	−30	−9	2	5	−6
8–12	*	*	−15	−31	−22	−11	−3	−13
12–16	*	*	−27	−32	−24	−16	−5	*
16–20	*	*	−39	−26	−8	−2	−9	−11
20–24	*	*	−77	−76	−33	−22	−15	−21
Total	*	−34	−27	−29	−15	-8	−5	−11

	0–2	2–4	4–6	6–8	8–10	10–12	12–14	14–16
			Manchester origins destinations					
0–2	−49	−39	−32	−15	−17	−8	−6	14
2–4	−59	−33	−25	*	*	*	−18	−28
4–6	−64	−57	−42	−34	−30	−28	−29	*
6–8	*	−25	*	*	−7	31	*	*
8–10	*	57	25	17	−2	9	36	98
10–12	*	−29	−32	−10	−29	−15	−16	4
12–14	*	*	*	*	−13	−18	−17	6
14–16	*	*	−34	*	−24	−22	−26	*
Total	−51	−34	−26	−6	−11	−6	-9	5

	0–2	2–4	4–6	6–8	8–10	10–12	12–14	14–16
			Birmingham origins destinations					
0–2	−58	−43	−27	−24	−17	13	50	90
2–4	−45	−38	−24	−18	−10	7	53	*
4–6	−30	−40	−21	−17	−5	26	99	62
6–8	−69	−55	−34	−26	−21	11	19	82
8–10	*	−19	−21	−13	−24	−8	*	25
10–12	*	*	*	15	−12	−5	−15	*
12–14	*	*	82	76	28	*	−11	−8
14–16	*	*	*	*	*	−14	−36	−18
Total	−40	−39	−22	−17	−14	2	−2	−2

* Changes in worktravel flows between annuli of less than one per cent (plus or minus).

annuli from areas of the city outside this zone. Birmingham is a little different in its pattern of changing flows. Reductions in employed residents flowing to the central area are apparent as elsewhere but from annulus 10–12 km outwards increases, modest in absolute terms, are to be seen. Furthermore such increased numbers journeying to work from these zones are also shown to be the same for flows into all of the inner annuli.

Local versus central dependence for jobs

The above description of commuting patterns does not adequately portray the balance between central and local commuting. Perhaps the most important component of the spatial structure of work travel is the dominance of the city centre for the provision of jobs. The extent to which the work force resident in different parts of a city commute into the city centre or work locally is shown in Figure 19.7 (p.374).

Some general patterns emerge. A large proportion of central area residents tend to live and work in the central area (defined as the high density employment core) although the absolute number of residents tends to be small.

It is clear that the proportion of working residents who travel to the central area falls as distance from the centre increases. In the case of London this means that, whereas about a quarter of working residents in the 4–8 km band work in the centre, this figure drops to 10.7% in the 20–24 km band. In Manchester and Birmingham, the fall is slightly steeper. Here only 4–7% of the outer area work force commute to the centre.

The declining importance of the central areas (with increasing distance from the city centre) is only partially offset by an increase in the importance of local working, and in some cases this does not occur at all. In the case of London, the proportion of residents living and working in the same annulus also shows a decline away from the centre in all but the outermost annulus. In Manchester and Birmingham the profiles are more irregular but in neither case is there a systematic increase in local working to offset the declining importance of centrally orientated flows. Rather, it can be seen that the sum of the two columns shows declines as distance from the city centre increases. This is a result of an increase in the proportion of residents who cross an annular boundary but who do not travel into the central area of the city. Towards the outer areas of all these cities the non-central, non-local movements constitute a majority of all work travel journeys.

In spite of substantial changes in the quantity of employment demanded and in the industrial structure of employment (with a shift away from traditional industries to service sector jobs), the basic structure of work travel within the cities has remained surprisingly constant. There are some minor changes in patterns of work travel. Overall, in all three cities, there is a marginally lower dependence on either central or 'local' jobs in 1981 when compared with 1971 (in general, the differences are no more than two percentage points). The consistency of the trend does, however, suggest that the patterns of origins and destinations may be becoming more diffused.

Labour catchment for central area jobs

The previous section considered to what extent workers resident in different areas of a city depended on job opportunities in the central area. This section looks at the extent to which employers in the central area draw their

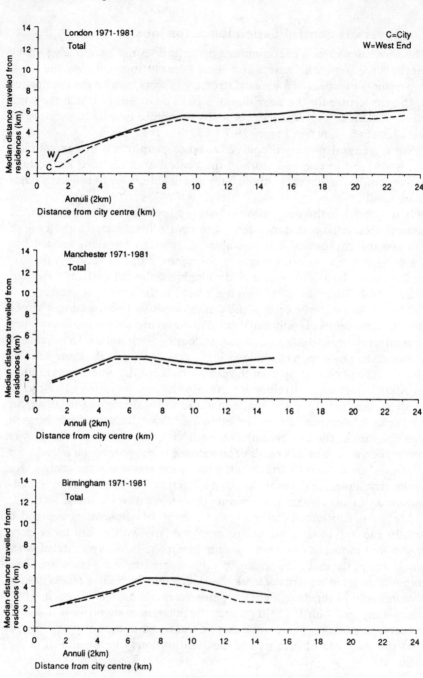

Figure 19.7 Local versus central demand for jobs and labour catchments for central area jobs within cities, 1981

labour force from different parts of a city and from beyond the city boundary. The labour catchment areas for central area jobs are also given in Figure 19.7.

Labour catchment areas seem to 'peak' twice: once quite close to the city centre and then again beyond the city boundary. This probably reflects the nature of the jobs market in the central areas. Typically central areas are characterised by a higher than average proportion of 'high quality' professional and managerial workers, many of whom prefer to live in the more pleasant environment of outer suburban areas and rural districts. On the other hand, the bulk of the central area work force comprises intermediate and junior non-manual workers drawn from the densely built-up residential areas surrounding the urban centre. In London the importance of long-distance commuting into the centre is clear with over a quarter of the central city work force being drawn from beyond 24 km. Birmingham and Manchester are much more 'self-contained'; in both cities, only 14% of the central area work force is supplied from beyond the contiguously built-up urban area. The labour catchment patterns within the cities reflect, to an extent, the distribution of population: the high-density areas supply a large proportion of the central area work force. The highest proportion is drawn from the 4–8 km bands.

The impact of changes in urban structure on the spatial spread of the central area work force is evident when labour catchment areas in 1971 are compared with those in 1981. Two general features are prominent. First, the inner city areas which, in 1971, were the peak suppliers of labour to the central zones, have been reduced in importance by 1981 as their populations have declined over the decade. Second, the decline in importance of these areas can be set against the relative growth in commuting from beyond the city limits. The most prominent example of this is London where, in 1981, more than one in four workers in the central area originated from beyond 24 km from the city centre in comparison with about one in five in 1971. This relative dispersal of workers with respect to central area employment is also seen in Manchester and Birmingham. It must be stressed, however, that these are relative changes and do not reflect absolute volumes of commuting. Indeed, the steep declines in overall levels of employment in all three cities has resulted in absolute reductions in the volumes of in-commuting to the central area during the 1971–1981 period.

It is clear that the changing structure of the city affects patterns of work travel movement as population and employment redistribute within the urban form. In many cities these effects are concentrated into certain critical areas of change as the population of inner areas declines and the relative flow of longer distance commuters increases. However, within this pattern of change, the broad pattern of work travel for the residents of the less affected areas of the city maintains a significant level of stability over the 10 year period.

Changing median distances travelled to work

Variations in median travel distances in these major cities (shown in Figures 19.8 and 19.9 for journeys from residences and to workplaces respectively) reflect the patterns of work travel observed in the previous sections. It was seen earlier that workers living near the city centre were heavily dependent on either central or local jobs. This is reflected in the short distances travelled by central area residents (Figure 19.8).

The median distances are barely 2 km in length in central Manchester and Birmingham and are even lower in London where both population and employment densities are higher than in the other two cities. Trip lengths initially increase as distance from the city centre increases. In London, trip lengths rise to about 6 km in the 8–10 km annulus and remain at that level throughout the rest of the city. In Manchester, median trip lengths peak at just over 4 km in the 4–8 km annuli and then decline in the 8–12 km annuli. The latter clearly reflects the commuting patterns observed in Figure 19.2 where it was seen that these annuli are characterised by the presence of older satellite towns which provide local job opportunities. The pattern in Birmingham is somewhat different in that trip lengths continue to rise (to about 5 km) further into the outer area.

Trip lengths attracted to different areas of a city (Figure 19.9, p.378) contrast with the above.

The labour force required for central area jobs is drawn from a wide catchment. Indeed, in the major metropolitan areas, 20–25% of the central area work force commutes into the city centre from outside the city. As a result, trip lengths to central areas tend to be markedly higher than elsewhere. This is particularly noticeable in London where trips attracted to the 'city' averaged over 16 km in length in 1981. Trip lengths drop sharply to about 4.5 km in the 4–6 km band and level off thereafter. In Manchester and Birmingham the decline is less marked, from 8 km in the centre to about 4 km in the 4–6 km band.

Between 1971 and 1981 there has been a general increase in travel distances as expected, given the decentralisation of population and employment between these dates and the consequent lowering of urban densities. For trips originating from residences, all three cities show a similar pattern of change. The values for 1981 are consistently higher than those for 1971 but this difference is very small within the inner area of the cities, increasing a little but rapidly levelling off towards the fringes of the cities.

The comparable workplace profiles highlight the 'central city' effect. Particularly in London, where there has been a greater decentralisation of population relative to central area jobs, travel distances to the city centre have risen sharply. As distance from central London increases there is a diminishing difference between the median values for the two years. A similar pattern of change, but on a much subdued level, can be seen in the case of Manchester. In Birmingham, however, there is a more even distribution of change over the whole area of the city, but this is consistent with the smaller levels of change in the catchment patterns for central area jobs in Birmingham when compared with the other two cities.

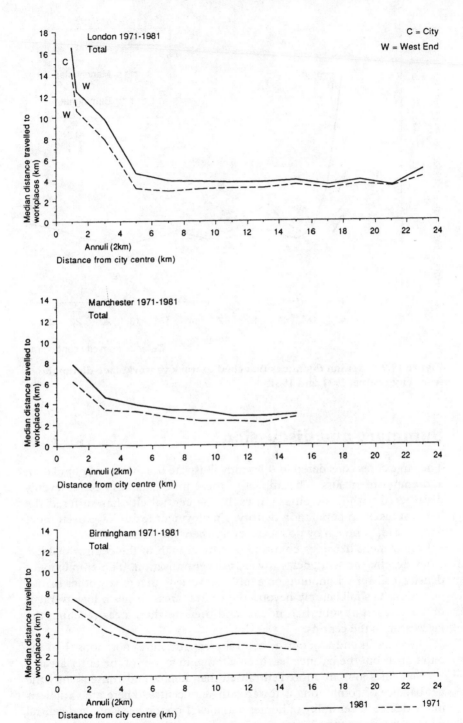

Figure 19.8   Median distances travelled to work from residences, 1971 and 1981

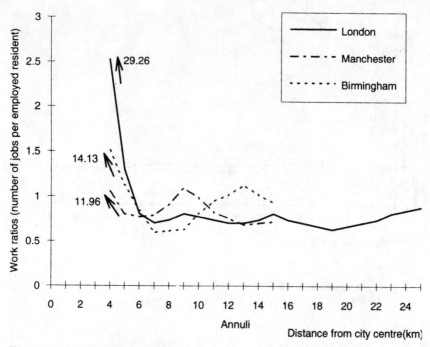

Figure 19.9 Median distances travelled to work vs workplace distance from city centre, 1971 and 1981

Summary and discussion

The large cities considered in this study illustrate the themes of population and employment loss. The impact of these processes has been unevenly distributed within the cities. In each, the central city has suffered the heaviest losses in population density. Employment losses have been more widespread, although by no means everywhere the same.

Employment densities continue to remain high in the centres of large cities despite the tendencies towards decentralisation. The employment density of Central London is on a different scale to that of any other British city. Densities fall sharply beyond the central areas to reach low averages only exceeded at suburban nodes, and then nowhere near reaching the levels seen in the centres.

The areas in and around the central areas still have more jobs than the outer areas but the balance has been altering in favour of the outer areas.

In all cities jobs are more centralised than homes and consequently the net balance of commuting is towards the city centre. There are variations in the degree of centrality of homes and jobs. London has a marked central concentration of jobs but a relatively dispersed distribution of homes. Manchester and Birmingham, by contrast, have relatively high levels of dispersion of both jobs and homes.

The city centres attract long distance commuting but generate short work journeys. At the residence end of the work journey there is a clear progression; short trips peaking at the centre, medium-length trips being well represented at about 6–8 km from the centre and the longer journeys coming from the outer areas. This is a consequence of a major proportion of trips generated in all parts of the city ending in the centre.

Broad similarities exist in the spatial structure of work travel, although in each city urban structure plays an important role in determining its precise nature. Between the cities, there are variations in

- the balance of importance between central and local movement;
- longer distance commuting;
- peripheral movement around the city, and
- distances travelled.

Long-distance commuting is dominated by radially oriented flows to the city centre. In London and Manchester these flows are more radially symmetrical than in most other large cities where long-distance flows tend to be concentrated into specific corridors. Even in the larger cities, there is a surprising absence of long-distance out-commuting and orbital travel. There are a few exceptional cases where employment centres in the outer area (such as Heathrow) attract long-distance commuting. The bulk of work trips attracted to outer area destinations are short journeys.

The dominance of the city centre in attracting long-distance trips is highlighted in London where, in 1971, 20% of the central area work force travelled further than 24 km to work. Between 1971 and 1981, as population has decentralised relative to employment, the proportion of workers commuting to the city has risen to 25% of the central area work force.

Changes in the patterns of work travel flows within cities are of three main types.

- Reduced flows to the centre are characteristic of most parts of the city, but more especially from the inner areas.
- Almost all annuli in the cities surveyed had reduced internal flows, but again these are more pronounced in the inner areas.
- Some parts of the outer city have seen increases in numbers travelling to work to local employment nodes and, in the case of Birmingham, additionally over longer distances to more central nodes.

In Manchester and Birmingham the labour catchment areas for central jobs are more evenly spread, and, although there has been a relative increase in long distance in-commuting over the 10 year period, the proportion of in-commuters remains low (less than 15%). For the resident work force in the three largest cities, the changes between 1971 and 1981 show a marginally lower dependence on either central or 'local' jobs. The consistency of the trend in all parts of the city suggests that the patterns of origins and destinations may be more diffused in 1981 than in 1971.

These differences in work travel structure are reflected in the spatial distribution of trip lengths: the longest journeys are to central area destinations (most markedly in London), from outer area origins. The decentralisation of people (relative to jobs) is also reflected in longer travel distances over the 10 year period in all three major cities for both the residence and workplace ends of the journey.

Between 1971 and 1981 there was little change in median distances travelled by residents. In many instances the values for the two dates are almost indistinguishable. Trip lengths to workplaces, however, show a different picture. There has been an increase in trip lengths by train and by car to the central area (most noticeably in London). This is consistent with the observation that long distance in-commuting has increased over the period.

As descriptions of the changing patterns of work travel response become more sophisticated, answers to more probing questions as to the *causes* of such changes become not only more pressing, but also more surprising by their absence. The reasons for this may simply be due to the complexities of the processes that produce the patterns of work travel and, even more so, their changes over time.

Both the supply of and the demand for labour within the city change over time, as this chapter demonstrates.So also does the provision of transport infrastructure and the cost of travel. Vickerman (1984) has argued that work travel change comprises a variety of components related to these changing forces.

1 There is the differential effect of in- and out-migration on the characteristics of people making work journeys. Some migrants moving to an area may not have the same work travel characteristics and propensity to change workplaces as the receiving population. Furthermore, some out-migrants may again be untypical of the population left behind in these respects.

2 Businesses prosper and wither, and sometimes move, giving rise to changing patterns of demand for labour in the city. Existing residents and in-migrants necessarily must respond, especially in tight labour market conditions, to such changing opportunities.

3 Sometimes alterations to the accessibility surface of the city, perhaps due to a new transport link or a new transport pricing structure, may facilitate or constrain the matching of the supply and demand for labour.

In effect residents of an urban area make decisions related to the work journey in a number of specific markets—those for housing, labour and transport being the most important. The fact that decisions are often made in several markets at the same time illustrates how potentially complex these processes can be. Some five types of change have been identified again by Vickerman (1984):

1 a change in workplace without a change in residence. Certainly decisions in the labour market will be needed and some change in transport might result.

2 a change in residence without changing jobs. Here housing market decisions are required and the journey to work may well change.
3 a change of either residence or workplace, in the expectation that in the near future a change for the better will be made in the other location.
4 these changes are actually brought about, rather than just expected from the initial decision to move workplace or residence.
5 the change which results from the joint moves of homes and jobs.

The structure and changes reported upon in this chapter reflect the outcomes of these processes of urban change. Future research should attempt to link these outcomes with the processes that bring them about.

20 Changing Metropolitan Commuting Patterns

*J.F. Brotchie, M. Anderson
and C. McNamara*

Introduction

L AND use, transport and communications are key components of the urban system and key factors in determining the extent to which it meets objectives such as increased urban productivity, social justice and ecological sustainability. Changes in one of these key factors with techno-logical change have impacts also on the other two, and on urban system performance against the objectives above.

The increasing integration of telecommunications and computing, and their interactions with production, service activities and transport, are changing land-use patterns and the resulting transport task. Service industries are increasingly information-based, and now provide by far the major share of urban employment in Australia. Many of these activities are choosing suburban locations with lower factor costs and, in some cases, higher amenities as a result of increased freedom of location enabled by information and communication technologies. Jobs are following people into the suburbs and because of the benign nature of many of these activities they can locate close to, and in some cases within, residential areas. This is substantially changing urban transport patterns, particularly commuting, shopping and freight movements. Cities are changing from essentially single-centred to be increasingly multi-centred. Transport patterns are changing accordingly, with multiple centres as destinations and multi-directional flows.

Global influences

The developed world is in transition from an industrial to an information economy in which resources are shifting from material- and energy-based industries to information- and knowledge-based activities increasing their outputs while production in the industrial sector is maintained. This transition is gaining strength with the confluence of various information technologies, particularly telecommunications and computing. Their increasing integration with other technologies such as production and transport is changing industrial locations, processes and products, and the nature of work itself. It is also changing the urban transport task and the means for meeting it.

Expanding global electronic networks are creating wider markets for many goods and services, and for industry location. Service industries are becoming increasingly information-based and are increasing their share of the labour force. Manufacturing is beginning to utilise information technology to increase productivity through automation, flexible manufacturing systems, and inventory management including just-in-time processing and production- and distribution-on-demand. In this way, new technology is also linking supply and demand and extending this link over broader spatial areas.

There is increasing interest in coordination of production with distribution, communication and computing, and of service industries with travel, telecommunications and computing. These linkages are creating the new 'engines' of the economy (Figure 20.1, p.384).

Each is being coordinated more closely with consumption and consumer preferences. Service industries now provide 76% of Australian GDP. They employ about 83% of the urban workforce in Australia, and manufacturing only 17%.

The global competition created by expanding markets is emphasising the need for the restructuring of national economic activities, so that all components can operate at full efficiency. In Europe and Japan the regional coordination of transport and telecommunications with service systems has been assisted by the introduction of fast rail networks for intercity travel, while coordination of production systems has been assisted by fast highway and rail networks for the carriage of goods and for travel by knowledge workers and their expertise (Nakamura & Ueda, 1989; Taniguchi *et al.*, chapter 9 this volume).

Metropolitan employment and commuting patterns

At a metropolitan level, activities are increasingly linked via telecommunications, assisting the spatial expansion of organisations and the establishment of suburban centres and industrial parks. In Melbourne, many large manufacturing plants have located on the metropolitan periphery with access to intercity transport links. Service activities are following people and locating in various suburban centres. The central region now has only one-third of employment and the CBD only 15%. The remainder is in large and small suburban centres, industrial areas and special-use zones. Nearly 4% work from home. This home-based activity can be expected to increase as technological support capabilities improve,

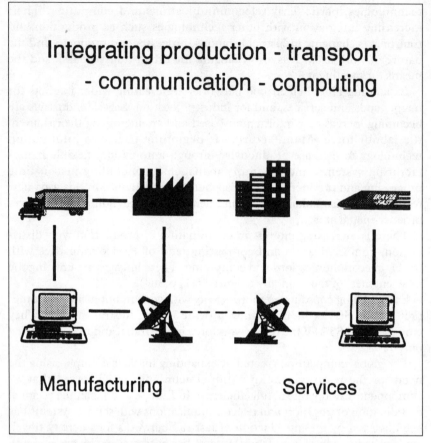

Figure 20.1 The integration of production, distribution, communications and computing in manufacturing, and of commerce, travel, communications and computing in the service industries and the further linking of these activities to provide the 'new engines' of the economy

economic benefits are realised, and social and institutional acceptance broadens. Suburban employment is highest in the middle ring of suburbs (Figure 20.2). Only the central region has more jobs than dwellings, and it has one order more. The relative distributions are shown in Figure 20.2.

There is a consequent shift in transport demand including commuting patterns, shopping trips, recreational movement patterns and their times and frequencies, with consequences for urban transport and its modal shares. These changes in urban form and transport patterns are reasonably representative of other major Australian metropolitan areas, and of major cities in the USA (Gordon *et al.* 1989a; Cervero, Gordon & Richardson, chapters 17 & 18 this volume) where dispersal of employment appears to have developed earlier.

Commuting trips are increasingly to suburban employment, in the same or another suburb, so that a closer suburban destination is substituted for a

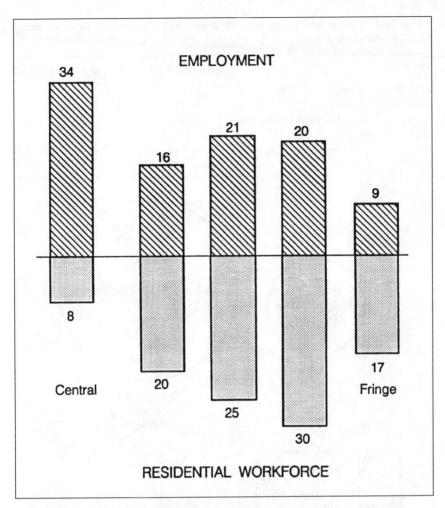

Figure 20.2 Distribution of employment and employees' residences between rings of zones indicating imbalances in each ring

more distant central one, with less congestion en route (Melbourne, see Table 20.1 and Figure 20.3).

The proportion of commuters with Melbourne LGA jobs decreases with increasing distance from the CBD (Figure 20.4 p.387).

As a consequence, trip distances are substantially (nearly 40%) less than for an otherwise equivalent single-centred city (Figure 20.5) and trip times are even further reduced (over 50%, Table 20.2 p.388).

Average trip lengths from each zone show only about one-third the variation that they would for a single-centred city. Average trip distances and times are generally increasing in absolute terms but not in relation to dispersal of homes and jobs (Table 20.2).

Interestingly, travel times from each origin zone averaged over all destinations are almost constant, but their variation between destinations

Table 20.1   Commuting patterns, Melbourne, 1986, divided into central plus 15 suburban zones

Trip type	Per cent of trips	Ave. trip distance (km)
Suburban to central	28	21
Intra-suburban zone	30	8
Intersuburban and central to suburban	36	18
Within central	6	6

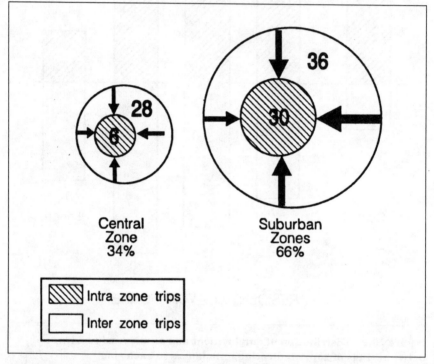

Figure 20.3   Distribution of commuting trips, 1986: (a) within and to central core (34%), (b) within, between and to suburban zones (66%)—indicating eccentricity of labour markets in each case

increases with distance from the CBD (Figure 20.6 p.389). Changes over the period 1976–86 are shown in Tables 20.2 and 20.3 (p.388). Preliminary data for 1991 are also included in Table 20.2.

The changing pattern

The central core and inner middle rings were stable but declining in population and worktrip origins. The central core was also declining in jobs and hence worktrip destinations, with a small increase in the inner and middle rings which together now contain the largest share of the jobs. There was a net migration of people and jobs to the outer and fringe rings. The

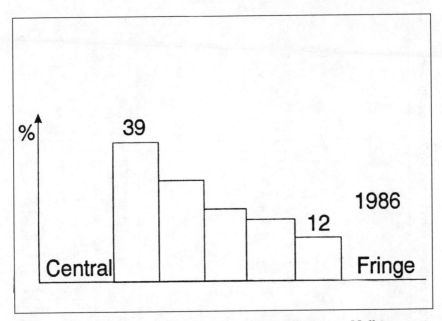

Figure 20.4   Percentage of trips from each ring of zones to Melbourne LGA

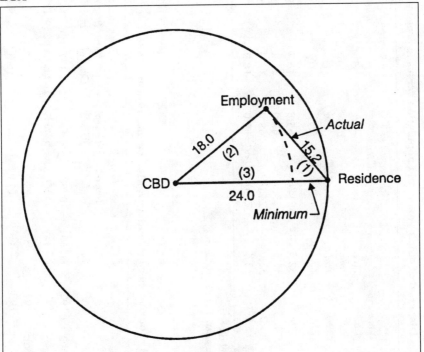

Figure 20.5   Average distances of: (1) worktrip length (15.2 km), (2) employment from CBD (18.0 km), (3) employees' home from CBD (24.0 km)

Table 20.2 Dispersal of employment and population versus average worktrip length and travel time, Melbourne, 1976–91

Year	Work trip		Melbourne LGA trip from home		Job dispersal from Melbourne LGA		Ratios			
	Ave. dist. (km)	Ave. time* (min.)	Ave. dist. (km)	Ave. time* (min.)	Ave. dist. (km)	Ave. time* (min.)	(2)/(4)	(3)/(5)	(6)/(4)	(7)/(5)
(1)	(2)	(3)	(4)	(5)	(6)	(7)	= (8)	= (9)	= (10)	= (11)
1976	13.8	32.5	21.9	65.8	16.4	54.2	0.63	0.49	0.75	0.82
1981	14.5	32.6	23.0	67.0	17.2	55.5	0.63	0.49	0.75	0.83
1986	15.2	33.0	24.0	68.0	18.0	56.5	0.63	0.49	0.75	0.83
1991**	15.5	32.4	25.0	69.0	18.9	58.2	0.62	0.47	0.76	0.84

* Peak travel time
** Preliminary data from 1991 Census

Table 20.3 Journey to work patterns, Melbourne, 1976 and 1986

From (living) To (working)		Central ring			Inner and middle			Outer and fringe			Total Melbourne
		Central	Inner/middle	Outer/fringe	Central	Inner/middle	Outer/fringe	Central	Inner/middle	Outer/fringe	
Trips (number)	1976	80 601	16 777	4 449	193 877	266 742	44 569	88 936	103 096	190 845	989 892
	1986	64 429	15 993	5 253	180 682	239 409	56 633	111 004	132 905	251 855	1058163
Trip dist. (km)	1976	5.9	13.7	24.3	16.7	8.8	18.3	28.1	18.1	10.9	13.8
	1986	6.0	14.0	26.5	16.8	9.4	19.4	29.0	19.6	12.6	15.2

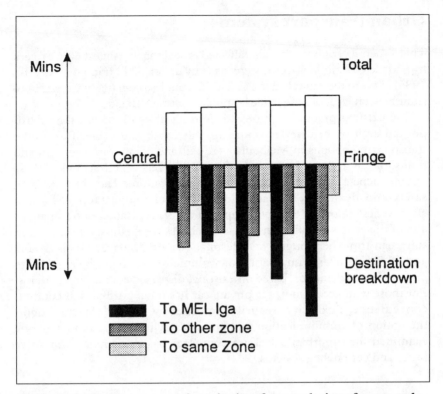

Figure 20.6   Average commuting trip time from each ring of zones and their variations with trip destination

outer and fringe rings have increased their share of population and of trips to all three destinations, with the smallest share to the central core and the major share to jobs in the outer and fringe zones. The outer and fringe contained nearly half the population and almost a third of jobs (Table 20.4 and Figure 20.2 on p.385). This share increased further in 1991.

Table 20.4   Distribution of jobs and workers' homes, Melbourne 1986

To jobs *(destinations)*	*From homes (origins)*			*Total jobs* *(destinations)*
	Central	*Inner/middle*	*Outer/fringe*	
Central	64 429	180 682	111 004	356 115
Inner/middle	15 993	239 409	132 905	388 307
Outer/fringe	5 253	56 633	251 855	313 741
Total workers (origins)	85 675	476 724	495 764	1 058 163

Note: Central and inner/middle are stable regions. Outer/fringe is developing and contains essentially all population and job growth.

Urban transport systems

This dispersal of industry has also affected modal shares. Almost 60% of work trips to the CBD in Melbourne were made by urban public transport (see ABS 1985), 34% to the central region as a whole, but less than 6% of work trips to suburban job locations are made by public transport (Figure 20.7).

The existing practical options for journey-to-work within the metropolitan area by private transport are car, walk and cycle. For public transport the options in Melbourne are rail, including some light rail and trams, or bus and taxi in the case of road. The relative attractions of these options depend on trip distance, and its relative time and cost, comfort, safety and flexibility, by each mode. The relative attractiveness and effectiveness of public transport depends on concentration of destinations and of trip routes. Public transport is much less competitive for the average suburban trip where origins and destinations are dispersed (Figures 20.3 and 20.5) and a large number of low volume routes would be required to meet potential demands, based on existing, dispersed commuting patterns. For those with access to it, the private car has other attributes of comfort, convenience, flexibility, security, informality, in-car entertainment, the option of communication and goods carrying. Other options being examined for suburban travel are computer-based, demand-responsive buses and car pooling systems.

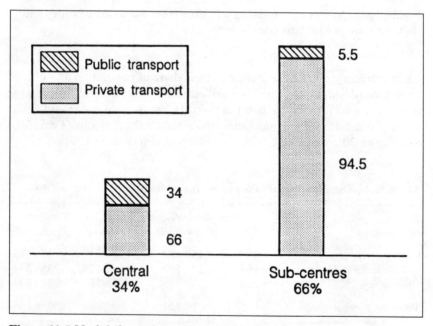

Figure 20.7 Modal shares between public and private transport for commuting trips to central zone, and to suburban employment in other zones

A framework for examining these changing patterns of land use and resulting transport/communication interactions is now presented as an urban land use/transport interaction model and options for increasing urban transport efficiency will be examined.

An urban land use/transport model

The simplest urban land use transport interaction model has only two parameters (Brotchie 1984; 1992). The transport parameter is the ratio of average worktrip length to the average trip distance from housing to the CBD, which is the ratio of sides (1)/(3) of the triangle in Figure 20.5 (p. 387). The land-use parameter is the ratio of average employment dispersal (the average distance of jobs, from the CBD) to average housing dispersal (the average distance of housing from the CBD), i.e. the ratio of sides (2)/(3) in Figure 20.5. This simple two-dimensional land use/ transport interaction model is illustrated in Figure 20.8 (p.392), with transport the vertical ordinate and land use the horizontal one.

The limits to average trip length, and to average employment dispersal (i.e. to land use/transport interaction) are defined by the triangle ABC. A single-centred city is the point A. Complete employment dispersal, or a job next to every house, is the line BC. The line AC represents shortest trip length if every worker chooses the house closest to his or her job (or the job closest to home). The line AB represents the longest average trip length, i.e. where the choice is made without regard to distance. Melbourne in 1986 was at point D in Figure 20.8 with a trip length ratio of $15.2/24 = 0.63$ and an employment dispersal ratio of $18/24 = 0.75$ (and 0.62, 0.76 in 1991, Table 20.2). The triangle ABC has been termed the urban triangle. The current urban transport task is seen to be only about 60% of that for a single-centred city—indicating a path to ecologically sustainable development through further trip shortening as later discussed (ESD 1991a; see also Roy 1992).

Trajectory over time

The trajectory of the point D over time to the present is indicated in Figure 20.9 (p.393). The pre-industrial city with cottage industries and pedestrian travel was near the point C. Industrial development caused a concentration of production activities and hence a movement towards the point A, enabling mass transit systems on radial routes. The development of services, the motor car and telecommunications has seen a dispersal back towards BC, with telecommunications separating from travel to provide additional agglomeration economies and new network linkages over the city as a whole. These trajectories could continue—with relative reductions in trip length, travel time and energy use. Movement towards the point C could result in the worktrip length stabilising and perhaps shortening in absolute terms as the vertical ordinate D reduces. The data for major US

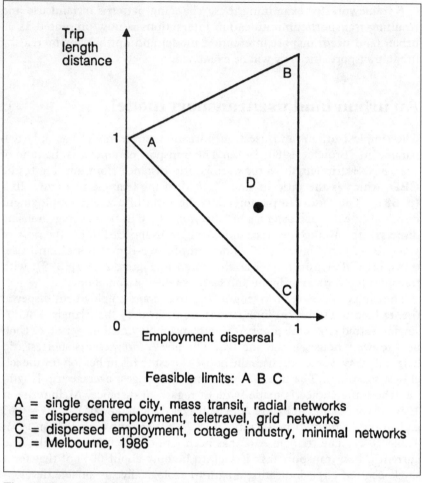

Feasible limits: A B C

A = single centred city, mass transit, radial networks
B = dispersed employment, teletravel, grid networks
C = dispersed employment, cottage industry, minimal networks
D = Melbourne, 1986

Figure 20.8 Employment dispersal versus worktrip length, and feasible limits ABC—normalised by average distance of housing from CBD. ABC is the 'urban triangle'. Melbourne in 1986 is point D.

cities suggest this stability for travel time at least (Gordon *et al.*, 1990; chapter 18 this volume; Cervero, chapter 17 this volume; Pisarski, 1992) and it may be close for the major Australian cities also. Policies of containment of residential development with continued industrial dispersal could assist this stabilisation. However, the evidence suggests that market forces are assisting this stability, particularly for travel times and energy use, as indicated by Breheny in chapter 21.

Further insight is provided by the concept of human territorial instinct and of time spent in daily movement about this territory as an anthropological constant (Marchetti 1992). Marchetti claims a mean of about an hour per day is devoted to this movement independent of any travel technology adopted; much of this time is now used for commuting. This is

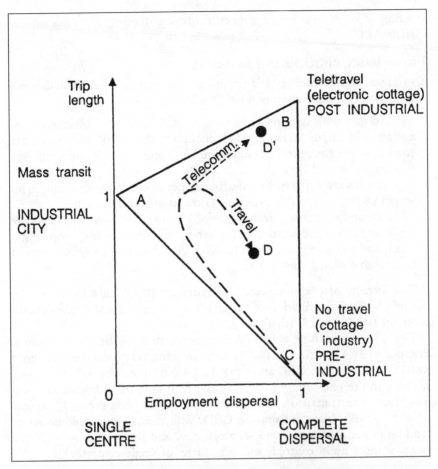

Figure 20.9 Urban development trajectory showing transition from pre-industrial cottage industry with no commuting (C) towards a single-centred industrial city with radial mass transit network (A), turning back as employment diperses with commuting (D) moving towards C and diverging from telecommuting (D¹) which is notionally moving towards the post-industrial electronic cottage (B).

consistent with the notion of travel time budgets, and has implications for land use patterns and urban scale. In early pre-industrial cities, movement was on foot and the distance covered limited employment opportunities and urban scale. Horse drawn vehicles extended this distance. In the industrial city, mass rail transit increased it by almost one order. In the post-industrial city the motor car has extended it further and telecommunications and employment dispersal are extending it still further, thereby increasing population capacity, and the range of services and employment opportunities. Further technologies such as intercity freeways and high speed rail allow further urban extensions along these corridors and the merging of cities into larger concentrations, extending the scale of the triangle ABC.

Three basic commuting patterns

An analysis of commuting (and shopping) patterns shows that Melbourne acts in several complementary ways (Table 20.1):

1 28% of commuting trips are to the central region of Melbourne of average trip length 21 km (or 34% including the 6% of trips which are entirely within the central zone, Figure 20.3) and to this extent it still acts as a single-centred city;
2 36% of trips are between 15 suburban zones or regions with average trip length 18 km, and to this extent it is now multi-centred;
3 30% of trips are entirely within one of 15 suburban zones with average trip length 8 km and 6% within the central zone with average trip length 6 km, and to this extent it now behaves as a cluster of 16 single-centred cities with reduced energy use.

This breakdown is illustrated in Figure 20.10 which also shows the triangle $A^1 B^1 C$ for division of the city into 16 equal-sized single-centred cities (of 100% type (3) trips).

The 28% commuting to the Melbourne central region live an average distance of 21.4 km from it. The 36% commuting between suburban zones live 25.9 km from the CBD and work 22.2 km from it. The 30% living and working in the one suburban zone reside 27.6 km out and work 27.3 km out. Thus the transition from trips type (1) to trips type (3) largely progresses with distance from the CBD, with central-inner-middle zones tending to act more as a single-centred city, and the middle-outer-fringe zones as more multi-centred, and as a cluster of single-centred cities.

Potential for future trip shortening

The potential for future trip shortening within the existing built environment is evident from Figures 20.8, 20.9 and 20.10. Some potential for movement in this direction is provided by the increasing home services sector, part-time employment, the informal sector and working from home. The growth of small businesses may be a further factor. Further potential would result from future fuel price rises. However, it could also result from increased locational mobility of households. There are presently a number of institutional and financial impediments to residential relocation. Home mortgages are not readily transferable. Stamp duties, agents' fees, legal fees and conveyancing can amount to $15 000–$20 000 for the householder on an average house sale and repurchase within a major Australian city. A reduction in these charges would enable increased locational mobility and lead to further trip shortening in an environment of energy conservation and higher transport costs.

This future behaviour can be considered in terms of the multi-centred city. Presently the city is behaving as multi-centred with CBD trips from each zone and interzonal trips between suburban zones. Trip shortening

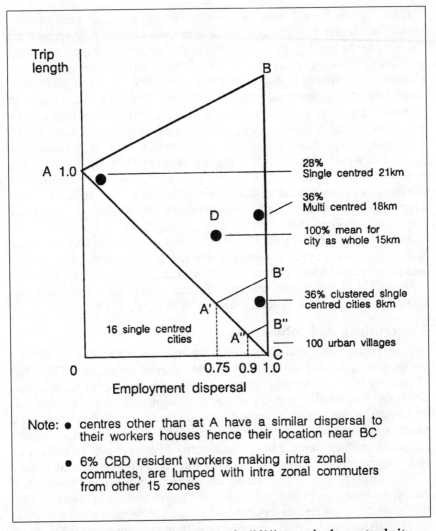

Figure 20.10 **Melbourne operates partly (28%) as a single-centred city, partly (36%) as a multi-centred city and partly (36%) as 16 clustered single-centred cities. The reduction in trip length with a cluster of 100 urban villages is also indicated by the triangle A[11] B[11] C.**

would occur in transition to behaviour as a contiguous cluster of single-centred cities. Further trip shortening would occur with the formation of new centres (further dispersal) and transition towards behaviour as a larger number of smaller single-centred cities or self-containment at the local level. Behaviour as a set of self-contained urban villages would reduce the transport task even further (Figure 20.10). Telecommunications (and freight movements) would need to provide the agglomeration economies to maintain productivity while reducing urban operating costs.

If all workers were to select a house in the same zone as their employment, the average trip length with 16 zones would reduce to about 8 km (to A^1 or to the line AC, Figure 20.10) i.e. to almost half the present average trip length. The potential for further trip length reduction by increased employment dispersal is also evident in Figures 20.8, 20.9 and 20.10. If all employment were dispersed into, say, 100 urban and suburban villages, all of similar size, and workers selected a house in the same village as their employment, the average trip length would reduce to about 2.5 km (to A^{11} in Figure 20.10). With average trip lengths of this order, a substantial proportion would be made by walking or cycling, and low energy consuming vehicles would be appropriate for the rest. New cities and extensions to existing major cities could be developed in this form. The creation of new centres in existing cities would also allow trip shortening, subject to the constraints outlined above.

The concept of travel time budgets does not rule out trip shortening but it does suggest the need for attractive low energy consuming travel options and for enhanced telecommunications which can partly substitute for larger daily trips. Otherwise only under high future transport costs and travel constraints would these potential savings be realised.

Uncertainty and robustness

An increasingly important feature in future cities is uncertainty of their social, economic, technological and environmental conditions. The range of possible futures is increasing with increasing technological change, with economic and social changes, and with increasing pollution of the land, air and water, making environmental conditions less predictable. Future cities must be robust to be capable of operating effectively and efficiently in a range of possible futures. They must be capable of internally reconfiguring their trip patterns to meet each of these futures. Multi-centred cities have this capability. The point D (Figure 20.10) represents present commuting behaviour based on present land uses and travel costs. The distance DA^1 in Figure 20.10 represents a number of points D^n defining potential urban travel patterns with lower energy usage than the present pattern D. It represents a measure of robustness of the present land use pattern as previously noted: and the capacity to change to meet various future scenarios of lower energy supply and higher unit travel costs. The distance A^1C represents a range of future land use and travel changes to enable even lower energy use futures. Urban and suburban villages, and lower energy-consuming vehicles are among these futures.

Thus the urban system must be capable of adapting to a range of possible futures. Multi-centred cities where jobs, homes and recreational facilities are in close proximity enable shorter commuting trips and provide flexible and robust urban systems capable of operating effectively under a range of possible energy futures. Urban villages take this trip shortening one stage further. The transport system required is one which is capable of operating efficiently at low energy use levels, ensuring an ecologically sustainable future.

Telecommunications and a public transport network linking centres would offer further options—of mode choice.

Urban consolidation and transport interaction

The conventional wisdom is that higher residential densities support public transport, particularly rail transport systems on dedicated tracks, and in this way lead to urban energy savings. However, this appears to be a significant factor only for CBD trips from the data above. There is also some US evidence (Gordon *et al.* 1989b; 1990; chapter 18 this volume) that lower densities may assist road travel by reducing the congestion en route, particularly for suburban destinations, enabling faster trips and possibly shorter trip times. This could be a second route to ecological sustainability. (In this same way, of course, higher densities would increase road congestion, favouring rail travel on dedicated track.) Gordon *et al.* show for US cities that commuting time and congestion are dependent on a number of factors. The principal factors are interpreted here as:

1 employment density—concentration of trip destinations;
2 residential density—concentration of activities along the route ;
3 city area—overall city size;
4 multi-centredness—dispersal of employment from the CBD.

Increases in the first three factors increase travel time, while increasing the fourth reduces travel time. The reduction in commuting time reported over the 20 major US cities over the period 1980–85 apparently resulted from this dispersal of industry, as the other factors would have generally increased or remained relatively constant over the 5 year period. Dispersal of industry in which jobs are following people into the suburbs can reduce travel time by substitution of a suburban destination and a less congested suburban route, for a CBD destination on congested radial routes, enabling a shorter and faster trip, and hence a reduction in commuting time as reported above. The evidence over the period 1980–90 is less clear in direction, as discussed herein by Cervero and by Gordon & Richardson, and stabilisation of trip time may be a more appropriate description of the changes—both positive and negative—that have occurred over the longer term.

Further transport technology options

Coordination of public transport modes, and provision of real-time information on them, provides further avenues for transport improvement.

Fast intercity rail systems running into metropolitan CBDs offer a further option for commuters on their stopping trains, enlarging the scale of daily urban movements, but providing a faster alternative for CBD commuting (Hall 1991a; Brotchie 1991).

Real-time information systems providing in-vehicle advice to motorists, or optimally controlling traffic signals, provide further options for reducing congestion and improving traffic flows. There are opportunities here for optimal route selection and for demand-responsive centrally controlled multi-hire taxis and buses and for computer-based carpooling systems. Such systems represent a further example of the integration of commerce, travel, communications and computing and are the subject of increasing attention from transport and information technology researchers. They offer the possibility of system optimisation at the metropolitan-wide level.

The US Federal Government has earmarked substantial funds for research and development of 'smart highways' which can price travel on congested routes, provide information on optimal routes and speeds, or even guidance for vehicles perhaps on dedicated lanes, making them a potential option for the 21st century city, although the potential for accidents in an automated system is being recognised.

The increasing capabilities of broad-band and mobile telecommunications and of computing, and their integration, offer an alternative to travel for some purposes—particularly for movement of documents, and possibly for commuting by information workers (teleworking). Telecommunications will no doubt also stimulate some travel and freight movement and, as indicated above, will provide support for all transport services.

Paths to ecological sustainability

There may be more than one pathway to future sustainability of urban development. The presently accepted path is towards increasing urban densities, particularly around urban centres and linking these by existing and, where necessary, new public transport routes. Fixed rail is the accepted mode. The concept is that public transport needs higher densities for viability; and higher densities mean a smaller urban area—and through this a reduced transport task with less energy use and less greenhouse emissions. It also means less encroachment of urban development into the surrounding hinterland and other land uses such as agriculture and recreational space. Rail transport causes minimal air pollution, less energy use, and greenhouse emissions per passenger km of only two-thirds that for private car travel, with present occupancy rates for each (ESD 1991b). This option means concentration of new jobs in a restricted number of urban centres on rail transport nodes, and concentrating new housing in and around these centres, and around rail stations. The rate at which this can change urban behaviour depends in part on the rate of development of new housing—about 2.5% pa—and the success of directing and concentrating this around urban centres and transport nodes. The following chapter by Breheny throws doubt on the feasibility and effectiveness of this path.

The alternative path to sustainability aligns more closely with recent trends. It is to allow jobs to continue to disperse towards housing, and housing towards jobs. This can occur in urban and suburban villages, mixed-use zones and the integration of commercial, residential and recreational activities to minimise daily movement patterns. Market forces and a reduction in planning constraints can assist this process.

The current level of dispersal of employment in Melbourne is such that the transport task is reduced to just over 60% of what it would be for a single-centred city. This means that its greenhouse emission levels are already comparable to those of a single-centred city in which public transport is the dominant mode. Continued dispersal of industry towards residential areas could reduce this transport task and emission levels further, providing an alternative path towards sustainability. For those trips which reduce substantially in distance and time, other, more energy- and emission-efficient modes of transport become feasible, such as lower energy consuming vehicles, car pooling, demand-responsive public transport (e.g. small buses), electric and hybrid vehicles and, for even shorter trips, e.g. in suburban villages, cycling and walking.

Working from home, and telecommuting, are other increasingly viable options as telecommunications capabilities increase and costs decrease. Among these capabilities are multi-media transmissions and electronic data interchange—computer-aided design (EDI CAD and CAD conferencing).

At the urban level, these telecommunications networks would be required to provide interactions enabling the economies of scale and agglomeration necessary for urban productivity and competitiveness. These properties are increasingly necessary to support viable and competitive industries and to attract new ones, particularly sustainable information- and knowledge-intensive industries. It is these latter industries that enable urban areas to be clean and green.

Under this scenario, the metropolitan area would behave increasingly as a cluster of single-centred cities from a transport viewpoint, and these centres would increase in number with new ones of smaller size.

A more likely scenario is a combination of these two. The city would behave in part as a single-centred city or with a small number of centres containing higher level activities connected by transit links. However, it would behave also as a multi-centred city with additional centres, and with transition to further centres and towards self-sufficiency in activities and passenger travel within each; with telecommunications linking these centres providing city-wide agglomeration economies. This would enable further informality of work activities and their merging with residential and recreational activities and spaces.

Comparison between cities

A comparison of Melbourne with London, Manchester, Birmingham and a sample of US cities in regard to employment dispersal and journey-to-work may be attempted from results presented in the previous chapters and elsewhere, using the urban triangle of Figure 20.8.

For a sample of 14 US cities, (1) the mean commuting distance, (2) the mean distance of jobs from the centroid of the city, and (3) the mean distance of houses from this centroid, are each estimated by Hamilton (1982). These are the parameters of Figure 20.5 (p.387)and their ratios [(1)/(3), (2)/(3)] define the coordinates of D for the city in the urban triangle of Figure 20.8 (p.392). They are plotted for the mean of the sample of 14 US cities, and for various cities in this sample in Figure 20.11. Los Angeles is also included (Small & Song 1992). The three UK cities, London, Manchester and Birmingham (Spence & Frost, chapter 19 this volume) and Melbourne are also included in Figure 20.11.

Figure 20.11 indicates a consistency of levels of employment dispersal among the various US, UK and Australian cities, with the UK cities and Melbourne comparatively less dispersed at about 0.75, than the US with mean 0.87.

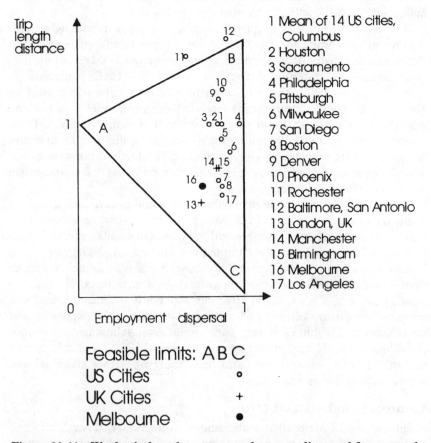

Figure 20.11   Work trip length versus employment dispersal for a sample of 14 US cities (Hamilton 1982), Los Angeles (Small and Song 1992), London, Manchester and Birmingham, UK (1981), and Melbourne (1986).

The trip length ratios, however, vary substantially from about 0.6 to 1.5 (Hamilton data), indicating very different levels of transport cost and network development, spatial layout of activities, and degree of diversity and specialisation of employment. The cheaper the travel, the better the transport network, and the more diverse the employment tasks, the closer is the city in theory to the line AB. The dearer the travel, the poorer the transport infrastructure, and the more homogeneous the employment, the closer it is to the line AC.

The UK cities (0.55–0.85) and Melbourne (0.63) are lower than the US cities (mean 1.0), indicating dearer transport and possibly more congested networks and slower average speeds in the UK and Australia. The trip length ratio D also represents the potential for trip shortening which could result from changing travel behaviour, and increased job dispersal under futures which substantially constrain transport energy use.

Conclusion

The continued rapid development of communication and computing technologies and their integration, together with their linking to the production of goods and services and to distribution and travel, is changing the nature and location of employment and of the transport task. Telecommunications and information technologies are interacting with transport—not so much directly as through their interactions with land use. Information and communication technologies are increasing location options, allowing industry and commerce to seek lower factor costs and, in the case of knowledge-intensive industries, higher amenity, as jobs follow people into suburban or fringe locations. In this way these new information technologies are interacting indirectly with transport, substantially affecting the nature and size of the transport task. This task has changed from one in which the destination was concentrated centrally, and the routes were radial, allowing the development of fixed rail systems, to one in which the large majority of destinations are dispersed, as are origins, and the routes between them are randomly aligned. Thus, the key interaction between telecommunications and transport is not direct but indirect via land use. Human behavioural instincts, including travel time budgets, provide a further influence, allowing additional opportunities to be realised through a combination of telecommunication and travel. New models linking land use, transport and communication interactions are therefore required to simulate and analyse this behaviour and to determine optimal planning policies.

21 Counter-urbanisation and Sustainable Urban Forms

Michael Breheny

Introduction: a 'Canute' policy?

The major contribution of planners and human geographers to the environmental sustainability debate has, to date, been focused on the question of what types of urban form might be the most energy efficient. Conclusions have been drawn—some based on research, others on assertion—that point to the need for greater containment of cities and towns. In some cases an extreme variant on this view is taken, with proposals for high density, compact cities contained largely within existing urban boundaries (e.g. Commission of the European Communities 1990; Sherlock 1990; Elkin *et al.* 1991; McLaren 1992). Other, less strident views acknowledge the need to plan for a variety of forms of containment (e.g. Banister 1992, Newman & Kenworthy 1989b; Newman 1992; Orrskog & Snickars 1992; Yanarella & Levine 1992). The growing support for the containment solution is now being voiced by politicians as well as academics. For example, the European Commission has proposed the 'compact city' solution on energy efficiency and quality of life grounds. The Dutch government has put forward a range of urban form policies, as part of its environmental programme, in the 'Extra' supplement to the Fourth National Physical Plan (National Physical Planning Agency 1991). The UK government has been less explicit, but does now require planners to adopt policies that reduce the need to travel (Department of the Environment 1992). Even in the United States, urban 'sprawl' is now derided, and growth management approaches are proposed which focus development within existing urban areas.

There is, then, a broad consensus of academic and political views which favour some form of compact city as one answer to the demand for more sustainable lifestyles. There have been a few dissenting voices. Some, such as Green & Holliday (1991) and Robertson (1990), represent a contrary view which favours decentralised living—geographically and institutionally—and a focus on 'rural values'. Others, including Gordon & Richardson (1989b; 1991b), argue that the energy-efficiency case for the compact city is not proven, and that market mechanisms will best resolve urban problems. Breheny (1992) suggests that the compact city proposal has contradictions that have not been addressed seriously by its advocates. At the moment these dissenting views are very much in the minority. The compact city solution has quickly gained momentum, and is now accepted as a legitimate policy objective in many countries.

Although the merits of this containment solution have received considerable attention, there is one rather obvious question that has not been posed seriously: is it feasible? In particular, is it feasible in the light of the powerful forces of urban decentralisation evident in virtually all western countries since 1945? These forces—of 'counter-urbanisation' in their most extreme form—may be so strong that the assumption that they can somehow be reversed—almost overnight—may be very naive. The picture of King Canute comes to mind: furiously whipping the irresistible waves. In this case, the waves are of people and jobs flowing from cities and towns.

This neglected question is addressed here. Clearly, the compact city proposal implies the need to reverse existing counter-urbanisation trends. But the possibilities of actually doing this have not been considered. If the case for the compact city and the evidence on counter-urbanisation are put together, what is implied about the prospects for the former being implemented?

This chapter is structured around three sections. Section two reviews the evidence on the relationship between urban form and energy consumption, including some new analysis. It poses the question:

> what evidence is there to suggest that compact city forms can significantly reduce urban energy consumption?

Section three reviews the available evidence on the patterns and causes of counter urbanisation. It addresses the question:

> is it possible that counter-urbanisation can be reversed?

When this chapter was started, it was intended that any novel contribution would come from answering this second—the hitherto neglected—question. However, it was found that consideration of this question also cast some light back on the first. Thus, section four relates the two sets of evidence and considers the additional question:

> is counter-urbanisation exacerbating the problem of urban energy consumption by increasing significantly overall travel (whether or not it can be reversed)?

Sustainable development and urban form

Of the areas in which planners can make a contribution to environmental sustainability, the most significant seems to be the whole issue of the need for travel in urban areas, and the degree to which changes to land uses and to transport systems can affect travel and hence energy consumption and pollution. Thus, the issue that has received most attention from researchers to date is that of urban form and transport. The general assumption behind this research is that certain urban forms, if they can be identified, will both reduce overall travel by private transport and induce greater use of mass transport. The question is: what urban forms will produce these effects?

Transport makes a significant contribution to environmental problems. It contributes to oil consumption, which is set to rise steeply at current consumption rates. Road vehicles in 1988 accounted for 18% of world output of carbon dioxide, the major cause of global warming; 85% of carbon monoxide; 45% of nitrogen oxides; and 50% of lead pollution (Banister 1992). Road transport accounts for 80% of all transport emissions (Rickaby *et al.* 1992). Transport is likely to be responsible for an increasing share of total CO_2 emissions over the next decade, and hence must be a focus of action if national governments are to meet their target of reducing such emissions to 1990 levels by 2000. It seems then, that if planners can reduce, or even stabilise, travel levels then they can make a significant contribution to the alleviation of environmental problems. If the prospects for using the planning system to reduce domestic and industrial energy consumption are added, then the role of planning becomes even more obvious.

In looking at the evidence for the role of changes in urban form and transport systems in bringing about greater urban sustainability, it may be useful to categorise contributions into:

1 empirical studies;
2 hypothetical modelling studies;
3 policy formulation studies; and
4 synthesis studies.

As usual, the problem is that such categorisation hides overlapping areas of work; but in this case the benefits of clarity probably outweigh the disbenefits of excessive compartmentalisation. Within these four categories only a fraction of available work will be reviewed; in order to give an indication of the nature and direction of evidence, rather than a comprehensive assessment.

Empirical studies

The basic aim of empirical studies is to ascertain whether there is evidence in practice that certain types of urban form are correlated with low levels of travel and hence with energy efficiency. Two related approaches to this question concern urban densities and urban size. In the former case, the question is whether high urban densities reduce the need for travel and facilitate the provision and use of mass transport systems. In the latter case, the question is whether larger urban areas—because of the benefits of higher densities—are more energy efficient than smaller areas.

Recently, the empirical debate on *densities* has focused heavily on the major investigations of Newman & Kenworthy (1989b). They have produced an analysis of the relationship between gasoline consumption per capita and a series of other characteristics of cities. The analysis has been carried out for 32 cities around the world, including 10 in the United States. Having correlated gasoline consumption with other characteristics, they have identified urban density as the only variable that varies systematically with gasoline consumption.

Newman & Kenworthy's (1989b, p.25) analysis of the 10 US cities led to the following conclusion:

> The relative intensity of land use in the ten US cities is clearly correlated with gasoline use overall and in the inner and outer areas. The strongest relationship is with the population density in the inner area...These patterns suggest that urban structure within a city is fundamental to its gasoline consumption.

The most densely populated US cities in the sample, New York and Chicago, with approximately 40 people and 20 jobs per acre, had gasoline consumption rates of 335 and 367 gallons per capita per year respectively. In contrast, the least densely populated cities of Houston and Phoenix, with 8 people and 10 jobs per acre, had consumption rates of 567 and 532 gallons respectively.

The analysis was also carried out for intra-urban differences in consumption as well as for whole cities. Table 21.1 shows Newman & Kenworthy's (1989b) consumption and urban density figures for different parts of New York.

It demonstrates very clearly the decline in consumption with density. They contrast the 90 gallons per capita in Central New York with the most extravagant consumers, in ex-urban Denver, who accounted for 1043 gallons each in 1980.

The policy conclusion from the analysis of US cities is that if the low density cities such as Houston and Phoenix were to develop urban densities similar to those in New York or Chicago, there would be an overall fuel saving of 20–30%.

Within the worldwide sample of 32 cities, Newman & Kenworthy (1989b) identify much greater variations in gasoline consumption rates than between the US cities. Asian cities show the lowest rates, followed by European cities. Australian and Canadian examples tend to fall between the European and US rates. Generally, high inner city densities and high levels

Table 21.1 Gasoline use and urban density in New York by region

Area	Gasoline use (gallons/capita)	Urban density (persons/acre)
Outer area	454	5.3
Whole urban area	335	8.1
Inner area	153	43.3
Central city	90	101.6

Source: Newman & Kenworthy (1989b).

of public transport provision correlate with low per capita gasoline consumption. There are interesting exceptions to the general findings. For example, although in the non-US cities there is a clear positive relationship between low consumption and low average work trip lengths, as we might expect in principle, one of the puzzles of the American sample was that there appeared to be a negative relationship between consumption and average work trip length. Another exception was the case of cities with strong, and reasonably dense sub-centres. European cities and Toronto exhibit this feature, and consequently have low consumption rates. This may add weight to the arguments of Owens (1991) and Rickaby (1987) that 'deconcentrated concentration' may be an urban form that is just as efficient as central concentration.

Newman & Kenworthy (1989b, p.33) conclude that planners can make a significant contribution to energy efficiency by:

- Increasing urban density;
- Strengthening the city centre;
- Extending the proportion of a city that has inner-area land use;
- Providing a good transit option; and
- Restraining the provision of automobile infrastructure.

The most substantial response to the Newman & Kenworthy (1989b) work has come from Gordon & Richardson (1989b, p.342). They argue that their '...analysis is faulty, that the problems are wrongly diagnosed, and that their policy and planning prescriptions are inappropriate and infeasible'.

Newman & Kenworthy (1989b) are accused of advocating '*Maoist planning methods*' and of wanting the '*Beijingisation of US cities*' (Gordon & Richardson 1989b, p.344). Gordon & Richardson (1989b) argue for economic—that is, pricing—polices to determine the consumption patterns of scarce resources, including petrol. If petrol has to be conserved—the need for which they doubt—the answer is a fuel tax, not a new set of restrictive planning policies. They argue that the variations in gasoline consumption noted by Newman and Kenworthy may be a function of different lifestyles in different cities rather than of densities. They also contend that the demands for greater investment in public transit systems is misguided, given the very poor record of such investments across the world.

Gordon & Richardson (1989b) also point out that a focus on work trips to urban cores is misleading in two respects. Firstly, many work trips are no longer focused on the core; many are from suburb to suburb. In an earlier paper, Gordon *et al.* (1988) have argued that the joint decentralisation of people and jobs has reduced average work trip lengths. They point out that average work trip lengths are shorter in the '*gas guzzling*' cities than in the apparently efficient New York and Chicago. Secondly, Gordon & Richardson (1989b) point out that work trips are becoming an increasingly small proportion of all trips. Non-work trips are short and have been facilitated by decentralised lifestyles. Policy measures which focus on core area work trips are thus misguided.

An alternative perspective on the urban energy consumption issue is to focus on *urban size*, rather than urban density. Banister (1980; 1992) and ECOTEC (1993) have attempted to gauge the energy-consuming characteristics of different urban sizes. The problem with Banister's work is that his 1980 exercise was related only to a small number of villages in Oxfordshire, and the 1992 exercise, based on the UK National Travel Survey (Department of Transport 1988), had crude size categories that placed all urban areas less than the size of London and greater than 25 000 population in the same category. Banister's conclusions are nevertheless clear. Large urban areas have higher levels of energy consumption than smaller areas, with remoter rural areas having particularly high levels of consumption. He did find that the largest city that he studied—London—was not as efficient as other, smaller cities. However, his urban size bands did not allow conclusions to be drawn on the relative merits of different sized towns and cities.

The special tabulations from the National Travel Survey used by ECOTEC provide the very detailed classification of urban size that was not available to Banister (1992). Thus, they were able to determine distances travelled by residents in finer urban size categories. The information is available by work and non-worktrips, and also for mode (but not mode by type of journey).

Table 21.2 shows distances travelled per person per week by mode, for each of nine urban sizes. In this table, metropolitan areas are treated as one group; in the ECOTEC report details are given for each of the major metropolitan areas in the UK.

It is also clear that average distances travelled are generally longer with decreasing size of urban area. However, the largest urban areas—Inner and Outer London—do not exhibit the shortest distances travelled. O⁺her metropolitan areas, and other cities with more than 250 000 population, appear to induce the least travel. This result for London is consistent with

Table 21.2 Distance travelled per person per week, by mode, by urban size

	Car	Bus	Rail	Walk	Other	Total
Inner London	76.1	12.0	34.1	2.5	16.6	141.3
Outer London	113.3	8.9	23.3	2.6	18.5	166.6
Metropolitan areas	70.6	16.9	4.7	3.4	17.1	112.7
Other urban over 250 000	93.6	11.2	8.3	4.2	23.9	141.2
100 000–250 000	114.8	8.6	11.3	3.2	22.6	160.5
50 000–100 000	110.4	7.2	13.0	3.7	20.2	154.5
25 000–50 000	110.8	5.7	12.5	3.7	18.2	151.0
3 000–25 000	133.4	7.2	8.0	3.0	24.1	175.7
Rural	163.8	5.7	10.9	1.7	28.9	211.0
Total	113.8	9.3	11.3	3.2	22.0	159.6

Source: adapted from ECOTEC (1993) (Table 9).

Banister's (1992) findings for the capital relative to other urban types. Table 21.2 also shows that people in large urban areas, with the exception of Outer London, use cars less than those in other areas. Concomitantly, they use bus and rail more. Rural residents travel considerably further by car than those in other areas. ECOTEC's (1993) conclusion from this analysis is that smaller urban areas, and particularly rural areas, are more fuel inefficient—and hence relatively unsustainable. Large urban areas—by virtue of higher urban densities, shorter travel distances, and mass transit facilities—are much more sustainable.

Table 21.2 gives a clear pointer to the relative roles of different sized urban areas in the consumption of energy in transport. However, it does not give actual estimates of energy consumption. If energy consumption could be measured for each of the sizes of urban area, would the differential between the most and least efficient types be greater still than Table 21.2, based solely on distances travelled, suggests?

ECOTEC (1993) provide rates of energy consumption by mode. Strangely, they do not multiply these through by the journey length information that is available by urban size. Advantage can be taken here of their oversight. Table 21.3 presents the primary energy consumption rates by mode for different occupancy levels.

These were originally derived from a European Commission study (Commission of the European Communities 1992c). These energy consumption rates can now be applied to the information on distances travelled in Table 21.2 to give the overall energy consumption levels shown in Table 21.4.

Unfortunately the two sets of definitions of modes do not match. Thus a simplified set of consumption rates has to be used. For present purposes, the following rates, assuming 50% occupancy, are used: car, 1.31; bus, 0.58; rail, 0.59; walk, 0.0; and other, 0.82 MJ. The last category is a problem. The 'other' category in Table 21.3 includes 'two-wheeled vehicles, taxis, domestic aeroplanes, other public transport and other types of bus'. The question is what overall consumption rate to use for this 'other' category? The solution here is to use the average of the car/bus/rail rates; that is 0.82 MJ per kilometre per person per week.

Table 21.4 shows that metropolitan areas have the lowest levels of energy consumed per person, with the levels rising as urban areas get smaller. Thus, rural areas have consumption levels 109% higher than the most energy-efficient metropolitan areas. This compares with a difference of 87% when distances travelled are measured. Thus, the effect of measuring energy consumption, as in Table 21.4, rather than distance travelled, as in Table 21.3, is to widen the differences between the most and the least energy-efficient urban types. This widening results, of course, from the use of modes of travel to arrive at the energy-consumption measures. The greater dependence of rural and small-town dwellers on the car is the major source of their greater energy consumption.

Table 21.3   Consumption rates for transport modes at different occupancy levels (in MJ primary energy/passenger km)

Mode	25%	50%	75%	100%
		Occupancy rate		
Gasoline car				
<1.4	2.61	1.31	0.87	0.62
1.4–2.0	2.98	1.49	0.99	0.75
>2.0	4.65	2.33	1.55	1.16
Diesel car				
<1.4	2.26	1.13	0.75	0.57
1.4–2.0	2.76	1.38	0.92	0.69
>2.0	3.65	1.83	1.22	0.91
Railways				
Intercity	1.14	0.57	0.38	0.29
Super Sprinter	1.31	0.66	0.44	0.33
Suburban electrical	1.05	0.59	0.35	0.26
High speed (300 km/h)	2.86	1.43	0.96	0.72
Bus/car				
Double-decker	0.70	0.35	0.23	0.17
Bus	1.17	0.58	0.39	0.29
Minibus	1.42	0.71	0.47	0.35
Express car	0.95	0.50	0.33	0.25
Air				
Boeing 727	5.78	2.89	1.94	1.45
'Soft' transport				
Cycling	–	–	–	0.06
Walking	–	–	–	0.16

Source: ECOTEC (1993) (Table 5).

Table 21.4   Primary energy consumed (MJ) per person per week, by mode, by urban size

	Car	Bus	Rail	Walk	Other	Total
Inner London	99.7	7.0	20.1	–	13.6	140.4
Outer London	148.4	5.2	13.7	–	15.2	182.5
Metropolitan areas	92.5	9.8	2.8	–	14.0	119.1
Other urban over 250 000	122.6	9.8	4.9	–	19.6	156.9
100 000–250 000	150.4	5.0	6.7	–	18.5	180.6
50 000–100 000	144.6	4.2	7.7	–	16.6	173.1
25 000–50 000	145.2	3.3	7.4	–	14.9	170.8
3 000–25 000	174.7	4.2	4.7	–	19.8	203.4
Rural	214.6	3.3	6.4	–	23.7	248.0
Total	149.1	5.4	6.7	–	18.0	179.2

Source: adapted from ECOTEC (1993).

This analysis, which is consistent with other empirical evidence, simply relates urban structure—that is, size—to energy consumption. It is, of course, possible that other factors are determinants of the variation in energy consumption: income, social structure, lifestyles, car ownership, etc. Ideally, an attempt should be made to isolate the contribution of each of these factors to energy consumption. Unfortunately, lack of data makes this very difficult. It might be reasonable to assume, however, that if these factors could be accounted for, they would tend to reduce the difference between urban and rural consumption rates. Incomes and car ownership, for example, which are associated with higher energy consumption, are generally higher in small town and semi-rural areas.

One further aspect of the debate on which some empirical evidence is available concerns the question as to whether planned settlements are any more energy efficient than other towns. The European Commission's Green Paper on the Urban Environment (Commission of the European Communities 1990), which advocates high-density, compact cities, also comes out very strongly against the idea of new towns or new settlements. It argues that these are unacceptable on two grounds: they are energy inefficient, and they fail to create the quality of life of the urban core. However, there is little evidence presented to support this view.

The Commission's work coincides in Britain with a debate about the merits of proposals for small, privately funded new settlements. One element of this debate has concerned the sustainability merits of this form of development (Breheny *et al.* 1992). Are these new settlements likely to be more energy efficient than increments to existing settlements? Is there a minimum size for such energy efficiency?

Breheny (1990) has attempted to assess the energy efficiency of existing new towns in the UK by analysing their degree of journey to work self containment, on the assumption that high degrees of self-containment entail lower levels of fuel consumption. In this analysis, he has built on work carried out by Cresswell & Thomas (1972). They calculated an 'independence index' (internal work trips divided by the sum of in and out work trips) for new towns and a comparable set of other towns in southern England. They found that average levels of self containment peaked—well above that for 'natural towns'—in the period 1961–66. Breheny (1990) has demonstrated that this advantage has now been lost, and that new towns are unlikely to be more self sufficient, and hence energy efficient, than other towns. He draws tentative conclusions about the potential energy efficiency of new settlements from his updating and extension of Cresswell & Thomas's (1972) work:

> ...in order for new settlements to have low average journey to work lengths, and hence energy efficient, they should be:
>
> • Large and relatively isolated, or
> • Small and relatively isolated, or
> • Small and close to existing urban areas.

Hypothetical modelling studies

An alternative approach to the question of appropriate urban forms is that of modelling. Here, the aim is to build mathematical models of cities that represent the interrelationships between the major land uses and transport systems. Once a satisfactory model has been developed, it can be used to forecast the effects of certain changes; either from changing behaviours in the system or from policies. There is, of course, a long tradition of modelling of this kind. However, one group of researchers has focussed their work on modelling the energy-consuming characteristics of cities. In a series of studies (e.g. Rickaby & de la Barra 1989; Rickaby 1987; 1991; Steadman & Barrett 1991; Rickaby *et al.* 1992) they have attempted to identify those hypothetical urban forms which minimise travel and hence energy consumption.

These researchers have produced a model of an 'archetypal' English town. This is based on a synthesis of the characteristics of 20 actual towns. Having set up the model, they have then experimented with various changes to land uses and the transport system in order to observe changes in energy consumption. Rickaby (1987) worked with a version of the model that placed the town in a city region setting. Of the policy options modelled, those that showed the lowest levels of energy consumption either focused activity at the city centre or involved village dispersal around the edge of the city region. A more recent exercise with the model (Rickaby et al. 1992), without the wider region being modelled, showed no particular advantage for any form of development.

Despite this recent set of results, Rickaby (1991) suggests that overall the modelling work points to some form of 'decentralised concentration' as being the most likely, under varying circumstances, to minimise energy consumption. This solution has the merits of maintaining overall high densities, but without the reliance on a single, often congested, centre. A city with a core and a number of high density sub-centres, based around public transport nodes, may be relatively energy efficient.

Policy formulation studies

One proposed solution to the urban energy problem that has received considerable support is that of the 'compact city'. The most important proponent of this solution is the Commission of European Communities (1990). In its 'Green Paper on the Urban Environment' it advocates that all future development should be within the boundaries of existing urban areas. The resulting high density compact city, the Commission argues, will be superior on two grounds: environmental sustainability and quality of life.

Two major problems are identified by the Commission: urban 'sprawl' and the spatial separation of functions. Together, these reduce urban creativity by undermining the intensive milieu provided by the compact city. Thus, for the Commission, the one solution—the compact city, with mixed land uses—could solve both the sustainability problem and quality of life problem in Europe's cities.

The compact city solution also receives support from other commentators. A more carefully articulated advocacy of the idea than that put forward by the Commission comes from Friends of the Earth (Elkin et *al.* 1991) and the Council for the Protection of Rural England (Owens 1991). Sherlock (1990) has taken a similar line. The Friends of the Earth study (Elkin *et al.* 1991, p.8) comes to the same conclusions about high density and mixed land uses as the European Commission:

> We have outlined alternative transport, but the built form of the city dictates the overall demand for transport, which can be reduced. The maintenance of high 'urban' densities of population alongside integrated land use can achieve this while providing the social interaction that makes cities desirable.

Breheny (1992, p.3) suggests that this line of reasoning, while it may prove to be entirely valid, has been accepted a little too readily:

> we have a solution being proposed by the Commission before the problem has even been articulated properly, and certainly before optional solutions have been assessed.

He suggests that the compact city proposal contains a number of possible contradictions that need to be assessed. For example, he suggests that proposals for urban greening may be inconsistent with higher urban densities. Also, in the UK at least, high urban densities are not assumed to correlate with high urban quality of life. Indeed, there are now concerns over the consequences of 'town cramming'.

The Commission of European Communities' (1990, p.19) Green Paper points out that in addition to sprawl, the greatest enemy of 'urbanity' is the separation of land uses. What is required, it is argued, is *'the efficient, time and energy-saving combination of social and economic functions'*. Elkin *et al.* (1991), Owens (1991), and Cervero (1991b) concur with this advocacy of mixed uses. Such mixed uses are regarded as desirable because they both create a richer urban milieu and reduce journey lengths. The former reason is difficult to contradict. However, the superior performance of mixed uses on energy grounds is more questionable. Again, this superiority is asserted rather than demonstrated. For certain basic, non-specialist goods and services, local provision may reduce journey lengths and facilitate walking and cycling. But for work trips and many other trips originating from home, the need or desire for specialist destinations—particular jobs, specialist goods, particular recreational facilities, for example—makes local provision impossible. Thus, it is doubtful that segregated or mixed land uses will necessarily produce different trip lengths or patterns.

Synthesis studies

A number of commentators contribute to the debate on urban sustainability, not by providing empirical evidence, by modelling, or by the assertion of policy, but by considered synthesis of all of these. Owens, for example, concludes that there is a need for 'energy conscious planning'

(Owens 1991, p.30), but does not follow the Commission of European Communities (1990) in its promotion of compact, centralised cities:

> Appropriate planning policies would therefore include discouragement of dispersed, low density residential areas or any significant development heavily dependent on car use; some degree of concentration, though not necessarily *centralisation*, of activities; integration of development with public transport facilities and the maintenance of moderately high densities along transport routes.

Rickaby (1987) also concludes that some form of 'decentralised concentration' of cities may be relatively efficient. This solution has the merits of high densities, but avoids the congestion that undermines the efficiency of the single-centred large city. However, Owens (1991) warns that the effects of particular urban form will be heavily dependent upon other prevailing policy conditions. For example, the multi-centred city envisaged by Owens (1991) and Rickaby (1987) would only be efficient if the propensity to travel was low. Thus, if fuel taxes were high, people would be inclined to use their nearest centre. If the propensity was high, as with low fuel taxes, people would continue to travel to multiple, often distant, destinations. Under the latter conditions, the single-centred city with good public transport may be the more efficient solution. Koushki (1991) has investigated empirically the elasticities of demand for travel, and has demonstrated that with high rises in fuel prices the number of trips does decrease.

Although a lot of the research on the relationship between urban form, transport and energy consumption is persuasive, a note of caution is in order. Assertions that there can be a direct relationship between changes in urban form and environmental improvement must be treated warily. The forces that determine energy consumption, urban change and environmental degradation are complex. They are political, social and economic, as well as physical. Just as market responses to changes in energy prices are difficult to predict, so too are responses to planned physical change. In the circumstances, the best strategy, as Owens (1991) and Rickaby (1987) suggest, might be to find urban forms that are robust; i.e. that are energy efficient in a variety of circumstances.

These conclusions are consistent with the findings of the ECOTEC (1993) study of the role of planning in reducing emissions, that was initiated in the Environment White Paper of 1990 (Department of the Environment 1990). Theirs is the most thorough of British studies that deal both with empirical evidence and policy prescriptions. They conclude that urban containment and higher urban densities are consistent with lower travel demand; that larger urban settlements are likely to be more energy efficient; that urban decentralisation has increased travel demand; that towns where there is a balance of employment and houses will be more fuel efficient; and that both centralised and polynucleic (multi-centred) urban forms may be efficient.

ECOTEC's (1993) recommendations for action follow these findings closely. Central government is urged to adopt, among other things,

- targets for the reduction of CO_2 from transport;
- arrangements that enable land use and transport planning to be coordinated; and
- a development control system that acknowledges the travel demand characteristics of different uses.

Local government is urged to:

- give priority to reduction of travel demand when making decisions;
- explore the travel demand implications of alternative locations for development;
- use travel demand implications as a criterion in making development control decisions; and
- prepare 20 year plans which have travel demand reduction targets.

A number of these recommendations have been taken up in government guidelines to planners (Department of the Environment 1992; 1993).

An emerging consensus?

Despite some dissenting voices, and calls for caution in drawing conclusions, it is clear that there is now considerable momentum behind the containment or compact city solution to urban sustainability. Much of the research points to lower levels of energy being consumed by travel in large, dense urban areas, where overall private travel seems to be relatively low and where the prospects for the provision and use of public transit systems are high. These research findings appear to have been accepted by national and local governments anxious to make—and be seen to make—their contributions to greater environmental sustainability. There now seems to be a demand for action in many countries that places urban containment— the compact city—at the top of the planning policy agenda.

The implication of this is, of course, that the urban decentralisation— indeed counter-urbanisation—trends of the last 50 years will have to be reversed. The problem is that the profundity of this implication has not been appreciated. No one seems to have asked the second of the questions posed in the introduction to this chapter: 'is it possible that counter-urbanisation can be reversed?'. To answer this neglected question it is necessary to determine the strength of the counter-urbanisation forces currently shaping the economic and demographic geography of the UK, and, indeed, other western nations.

The counter-urbanisation evidence

There is little doubt that the major change to the economic geography of most of the western industrialised countries in the post-war period has been the generally pervasive process of urban decentralisation (Cheshire & Hay

1986). This appears to have affected all major urban areas of the UK, although its effects have been most pronounced in and around the south east region. There has been a 'wave' of economic growth, and hence population growth and development pressures, spreading apparently from the south east and into adjacent regions. Thus, in the last decade, population and employment growth have been fastest in counties such as Cambridgeshire, Northamptonshire and Dorset. The south east region, plus these adjacent, growing counties, has been referred to as the 'greater south east' (Hall 1989).

Before proceeding to provide evidence on this decentralisation process, a clarification of terms may be useful. 'Decentralisation' is used here in a general sense, to refer to change which appears to be focused away from established large centres. 'Suburbanisation' refers to continued growth at the edge of existing urban areas. This may take the familiar form of direct extensions to urban areas, but it might also take the form of discontinuous suburbanisation. In this case, the process is one of suburbanisation, but where continuous growth is hindered, for example by a green belt. 'Counter-urbanisation' is a specific term. It suggests that growth is not simply focused on urban extensions—that is, extended suburbanisation— but is occurring as a distinct antidote to existing urbanisation; with the focus of growth being in the least urbanised areas. Much of the debate has centred on the question of whether the phenomenon observed in the UK, and particularly in the 'greater south east', is extended suburbanisation (Gordon 1988) or counter urbanisation. The term 'counter-urbanisation' is used here because, as will be seen, the evidence does point to radical changes in settlement structure. However, it does not matter greatly to the concerns here whether the process is one of continuing suburbanisation or counter-urbanisation: the question is, will the process continue?

Measuring decentralisation

The UK Office of Population Censuses and Surveys (OPCS) uses a standard classification of areas in the UK, based on cluster analysis, that can be used to trace the decentralisation of population, because the classification moves generally from large urban areas to remote rural areas. Table 21.5 shows population changes in these standard categories for each decade since 1961.

The table shows a clear decentralisation of population over 30 years. Through all three periods, the largest cities, including Inner London, show the highest population losses. The areas of highest population gain do, however, move to successively smaller urban areas over the years. In all periods, the new towns have seen substantial population growth, as have the 'mixed and accessible' areas, both inside and outside the south east. Over the period, such areas outside the south east have gained an increasing share of growth, as the wave of decentralisation has moved beyond the south east regional boundary. 'Resort and retirement' and 'remote, mainly rural' areas have gained large increases over each of the three decades.(This latter group is particularly interesting. It is the gain in this category that has

Table 21.5   Population change, 1961–91, for OPCS urban types in England and Wales

	1961–71		1971–81		1981–91	
	'000s	%	'000s	%	'000s	%
Greater London boroughs						
Inner	−461	−13.2	−535	−17.7	−147	−5.9
Outer	−81	−1.8	−221	−5.0	−171	−4.2
Metropolitan districts						
Principle cities	−355	−8.4	−386	−10.0	-258	-7.4
Others	412	5.5	−160	−2.0	−327	−4.2
Non-Metropolitan districts						
Large cities	−41	−1.4	−149	−5.1	-98	-3.6
Smaller cities	38	2.2	-55	-3.2	5	0.3
Industrial districts						
Wales and northern regions	118	1.3	42	1.3	-72	-2.1
Rest of England	342	5.0	158	5.0	59	1.8
New towns	337	21.8	283	15.1	133	6.1
Resort and retirement	346	12.2	156	4.9	174	5.2
Mixed and accessible rural						
Outside SE	627	21.9	307	8.8	156	4.1
Inside SE	960	22.1	354	6.8	162	2.9
Remote, mainly rural	399	9.7	468	10.3	328	6.4
England and Wales	2 629	5.7	262	0.5	−57	−0.1

Source: Office of Population Censuses and Surveys (1992a), 1991 census.

led to the view that the changes are evidence of genuine 'counter-urban-isation' rather than simply suburbanisation. Those opposing this view have tended to dismiss change in this 'remote, mainly rural' category as insignificant on the grounds that high percentage change is on a small absolute base. However, as Table 21.5 shows, the change is large in both percentage and absolute terms.

The preliminary report of the 1991 census, which has been used in compiling the figures in Table 21.5, gives a graphical version of population change for the 1981–1991 period in each of the urban to rural categories of the OPCS urban types. These are shown in Figure 21.1.

The decentralisation pattern shows up very clearly in this form. It is clear that, despite claims in the mid-1980s that there was a 'return to the city' (Champion 1987; Champion & Congdon 1988), the decentralisation process is continuing.

An alternative, and slightly conflicting, view of population changes in the 1981–91 period is given by OPCS mid-year population estimates. Although

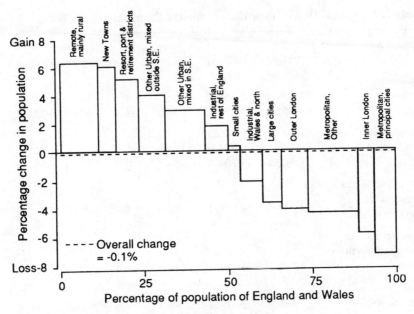

Figure 21.1 Population change 1981–91 by OPCS urban types
Source: UK Office of Population Censuses and Surveys (1992a).

estimates for mid-1991 are based on the preliminary results of the 1991 census, the results differ considerably from the preliminary census figures given here in Table 21.5. This is because the mid-year estimates have been adjusted to account for under-enumeration in the census.

Table 21.6 gives 1981–91 population change figures based on mid-year estimates.

The major difference between Tables 21.5 and 21.6 for 1981–91 change is at the national level, where the figure diverges considerably. The census results in Table 21.5 suggest a small loss of population, while the mid-year estimates of Table 21.6 show a considerable gain of 1.3 million. Inevitably, such large differences will be reflected in the figures for each urban type. It is estimated that under-enumeration occurred largely in the major cities. Thus, the preliminary results in Table 21.5 probably overestimate population losses in the larger cities. Table 21.6 shows smaller population losses in the larger urban areas, and larger population gains in the more rural areas; suggesting a weaker counter-urbanisation trend overall, but much larger rural gains, than Table 21.5.

One benefit of using the mid-year estimates in Table 21.6 is that they give natural change and migration components of the change. Table 21.6 makes it clear that the large population growth in the 'mixed and accessible rural', 'resort and retirement' and 'remote, mainly rural' categories has arisen from large migration moves, rather than natural increase. In the latter two categories there is a negative natural increase change, perhaps suggesting

Table 21.6   Population change 1981–91 for OPCS urban types ('000s)

	1981 pop.	1991 pop.	Change 1981–91	% change	Natural change	Migration
London						
Inner	2 550.1	2 566.4	16.2	0.6	112.3	−96.1
Outer	4 255.4	4 236.7	−18.7	−0.4	129.7	−148.5
Metropolitan districts						
Principle cities	3 550.2	3 401.6	−148.6	−4.2	59.0	−207.6
Others	7 803.5	7 704.5	−98.9	−1.3	155.1	−254.0
Non-Metropolitan districts						
Large cities						
Smaller cities	4 577.9	4 620.3	42.4	0.9	99.7	−57.3
Industrial districts						
Wales and northern regions	6 709.9	6 851.3	141.4	2.1	166.9	−25.5
Rest of England						
New towns	2 186.9	2 379.1	192.2	8.8	116.2	76.0
Resort and retirement	3 368.8	3 633.9	265.1	7.9	−158.5	423.6
Mixed and accessible rural						
Outside SE						
Inside SE	9 500.4	0 000.9	500.4	5.3	235.7	264.7
Remote, mainly rural	5 131.2	5 560.2	429.0	8.4	−30.9	459.9
England and Wales	49 634.4	50 954.8	1 320.5	2.7	885.2	435.3

Source: Office of Population Censuses and Surveys (1992b).

an older age structure and past migrants who are relatively elderly. By contrast, the 'mixed and accessible rural' areas show strong natural increase gains. In this case, this probably implies the opposite: a younger population and younger migrants. It is this 'mixed and accessible rural' area that has been the focus of the strongest economic growth in recent years. This economic growth may be associated with younger migrants in search of work.

The question of the source of population change under counter-urbanisation is important from a sustainability point of view. Where population gains result largely from indigenous change—that is, natural increase—then planners can have little prospect of influencing that change. Where in-migration is the cause of population growth, then it is possible to envisage constraints on such movements—by restrictions on housing and job increases, for example. The 'remote, mainly rural' category seems to be the category most susceptible to constraints of this kind.

If the figures in Table 21.6 are to be believed, then the counter-urbanisation process in the UK remains very strong. Over a 10 year period, four urban types—from new towns to remote rural areas—gained nearly 1.4 million people, 1.2 million by migration. The most rural type, 'remote, mainly rural' areas, gained the largest absolute number of migrants and the largest percentage overall gain.

Measured in population and employment growth terms, the decentralisation process in the UK can be seen most clearly in and around the south east. Table 21.7 shows population change in the five rings—Inner London, Outer London, the outer metropolitan area, the outer south east, and the south east fringe—used by Hall (1989) in his discussion of change in the greater south east.

The pattern shows that in each successive decade since 1945, the ring with the highest growth rate is successively further out, with the south east fringe—the area beyond the standard region—witnessing the highest rates of growth in the 1981–91 period.

Other measures of interest—and ones that help in moving towards explanations—are migration patterns and business relocations. A simple recent picture of migration flows—using records of patient movements between family practitioner committees (FPCs) in England and Wales, and area health boards (AHBs) in Scotland—is provided by Bulusu (1990). The major net effects are positive flows from London to the rest of the south east and from the south east to neighbouring regions such as East Anglia, the south-west and the East Midlands; generally, down the urban hierarchy.

A comprehensive study of migration to and from the south east region has been completed for the UK Department of the Environment. This also relies on FPC data. The report, by Fielding (1991), looks at data over a 20 year period. The general picture of migration flows around the south east is, however, consistent with Bulusu's (1990) 1989 picture. Although the geography of migration flows can vary considerably on a year-to-year basis, Fielding (1991) demonstrates a general tendency—consistent with other information on decentralisation—for the destinations of south east out-

Table 21.7 **Population change in the south east, 1961–91 ('000s)**

	1961–71	*1971–81*	*1981–91*
Inner London	−13.2	−17.6	−5.9
Outer London	−1.7	−4.6	−4.1
OMA	18.6	4.9	0.7
OSE	18.3	8.2	5.6
SE fringe	14.8	10.4	8.3
Greater London	−6.7	−9.9	−4.7
ROSE	18.4	6.4	3.0
Greater SE	6.9	0.1	1.1

Source: Office of Population Censuses and Surveys (1992a), 1991 census.

migration to be progressively further from the south east; a process of 'rolling back' the frontiers of the region.

The Department of the Environment has also commissioned work on business relocations to and from the south east. Prism Research (1991) identified company moves into or out of the south east, and within the south east, during the period 1985–90. The major net beneficiaries of company moves within the region are the 'Western Crescent' (Hall *et al.* 1987) counties of Berkshire, Buckinghamshire and Hertfordshire. London is the major loser. Prism Research's (1991) general conclusion is that at the national level there is a distinct 'distance decay' effect in company moves from the south east.

Explaining counter–urbanisation

The most comprehensive recent review of these issues is in Cross's (1990) book *Counter-urbanisation in England and Wales*. He surveys the literature and identifies five distinct sets of possible explanations for decentralisation:

1 longer distance commuting;
2 residential preference for non-metropolitan areas;
3 economic change in favour of peripheral areas;
4 employment decentralisation; and
5 integrated theories.

Those theories that focus on *commuting patterns* argue that the decentralisation phenomenon is simply a case of extended suburbanisation; and hence is not a new phenomenon at all. Certainly, Cross (1990) does provide evidence that commuting distances into the south east, and to London in particular, have lengthened over time, but doubts that changes in commuting patterns can explain overall decentralisation trends.

The argument that decentralisation results from *changing residential preferences* in favour of non-metropolitan areas, is based on both pull and push factors. The pull factors suggest that there has been a popular shift in favour of rural and small-town life, and that this has been met readily by the market. The push factors are the increasingly poor quality of life afforded by many major urban areas; with problems of congestion, crime, social deprivation, and physical decay increasingly being associated with cities. Barlow (1989) argues that house prices have a substantial influence on decentralisation patterns. He argues that through the 1980s, as house prices rose rapidly in the south east, they fuelled household moves to cheaper areas. With the subsequent collapse of the housing market in the south east, it will be interesting to see if decentralisation moves are slowed.

Cross's (1990) third cluster of explanations highlights *economic changes which have favoured peripheral locations*. He argues, however, that the obvious neoclassical explanation of migration moves from areas of high unemployment and slack labour markets to areas of growth and tight labour markets, is not borne out by statistics at the national level. He does not assess whether this is true, however, specifically for the south east. The possibility that decentralisation has been fostered by government regional policy is also investigated under this heading.

The most promising group of explanations, according to Cross (1990), relate to the role of *employment decentralisation*. Here it is suggested that fundamental shifts have occurred in the way that capitalism uses geographical space. Gone, it is argued, are the days of regional sectoral specialisation, in which regions specialised in a particular industrial sector, and held all of the operations of companies involved in that sector. This, it is argued, was replaced by a spatial separation of those operations, into new spatial divisions of labour.

This newer geography of economic activity tends to separate core functions of a company—administration, R&D, marketing—from routine production. The former, core functions have tended to locate in major cities and in decentralised but accessible locations. Production has tended to focus away from traditional urban locations to cheaper peripheral locations, aided also by demands for new industrial property at lower densities (Fothergill & Gudgin 1982). Similar trends are also discernible in the service sector, with the splitting of small, central functions from large-scale 'back offices' in decentralised locations.

Although this 'new' geography appears to provide a valid basis to understanding decentralisation, ironically it may be being superseded. There is now a view that the hierarchical divisions of labour developed in the post-war period are breaking down. The more recent theories, under the label of 'flexible accumulation', suggest that pressures to compete on the basis of production flexibility (including 'just-in-time' production) mean that companies will tend to locate previously separated functions together. Thus, the theory suggests new 'industrial districts' are being established; not in the former industrial areas, but in new 'clean' locations such as the south east fringe locations—which tend to equate to the 'mixed and accessible rural' urban type discussed earlier—that have witnessed substantial population increase in recent years. The preference of companies for these types of locations has been confirmed in survey work by Prism Research (1991).

The obvious geographical consequence of these changes is an increasingly sharp divide between the declining traditional industrial areas and the newly developing, largely service-based, rural and small-town areas.

The final set of explanations reviewed by Cross (1990) are those that attempt some *integration* of previous sets of separate explanations. One group of integrative theories suggests that employment decentralisation and housing market factors—such as the search for lower house prices away from major cities—together provide a plausible explanation.

The prospects for continuing decentralisation

Although there is a considerable body of literature dealing with explanations of the south east decentralisation phenomenon, there is very little that attempts to forecast future change. Cameron *et al.* (1990) have produced county-by-county forecasts of population and employment change for the period 1989–2000. These suggest that some counties that were to the fore in the last decade—for example, Buckinghamshire, Cambridgeshire, West Sussex, Berkshire—will remain so through the 1990s. Some counties with high ranks in the 1980s begin to lose them in the 1990s. Dorset, Surrey, Northamptonshire and Wiltshire, for example, drop down the rankings. Of particular interest are those counties that show a significant positive change of rank. Warwickshire's elevation (in employment but not population terms) is consistent with the idea of the 'boundary' of the south east extending further north. However, counties that might be expected to be candidates for inclusion within the wave, such as Leicestershire and Lincolnshire, do not figure.

The Cameron *et al.* (1990) forecasts are largely based on past rates of growth and industrial structure. An interesting alternative way of looking at possible future changes is via housing and planning policies. Some recently published research helps to identify those counties which have growth policies which intend substantial deviation from past trends; both positive and negative.

Table 21.8 is based on a report, *Rates of Urbanisation in England 1981–2001*, by Bibby & Shepherd (1991).

The report provides estimates of the rates of urban growth in each county—expressed as percentages of the county area to be converted to urban uses—under two assumptions:

• growth will be determined by projected household formation.
• growth will be determined by policies (from structure plans and regional guidance) for housing growth.

Table 21.8 Forecasts of urban growth 1981–2001: comparison of household projection-based forecasts and policy-based forecasts

County	Policy-based forecast (A)	Household-based forecast (B)	A–B
Greater London	0.27	1.57	−1.30
Hampshire	0.54	1.52	−0.98
Dorset	0.50	1.16	−0.66
Northants	1.13	1.42	−0.29
Buckinghamshire	1.95	2.21	−0.26
Oxfordshire	0.84	1.02	−0.18
South Yorkshire	0.97	0.39	+0.58
Greater Manchester	1.15	0.56	+0.59
Humberside	0.93	0.34	+0.59
Tyne and Wear	1.85	0.26	+1.59
Merseyside	1.88	−0.39	+2.27

Source: based on Bibby & Shepherd (1991).

The difference between these two estimates can be used as a reflection of the degree to which policy diverges from trends. Table 21.8 shows the two different estimates of growth for each county, and the difference between the two. Counties are ranked according to the degree of divergence.

It is interesting to compare Cameron et al.'s (1990) trend forecasts with the growth rates implied by Bibby and Shepherd's assessment of planning policies. Hampshire provides an interesting case for comparison. The Cameron *et al.* (1990) forecasts suggest a rise in its rank, both on employment and population growth measures. In the Bibby and Shepherd analysis (Table 21.8), Hampshire comes out as an extreme example of policy restraint. It has the second highest divergence between what trends suggest is required (1.52%) and policy will allow (0.54%). If this restraint policy holds, it suggests that Hampshire will have a growth rate somewhat below the Cameron *et al.* forecast. For other counties, the Cameron *et al.* and Bibby and Shepherd analyses are more consistent. Dorset, Northamptonshire and Oxfordshire, for example, fall down the Cameron *et al.* (1990) rankings and also exhibit strong restraint policies in the Bibby and Shepherd analysis.

The prevailing political attitude in southern lowland England (approximately the 'mixed and accessible rural' areas discussed earlier) is reflected in the Bibby and Shepherd analysis. After years of growth pressures, policies now aim to resist further growth. The success of this approach—pejoratively labelled 'NIMBY' ('not in my back yard') by some—will depend on the degree to which planning policies are strong enough. Demographic trends will make resistance difficult because these areas now have strong built-in household growth—as shown by government household forecasts (Department of the Environment 1991)—as a result of earlier rounds of in-migration. If resistance is to some degree successful, it is likely to push development pressures to more rural areas still—to the 'remote, mainly rural' category—rather than back into the large urban areas.

The Prism Research (1991, p.131) study of company decentralisation speculates about the likelihood of the recent pattern of decentralisation being continued:

> Of course, the future is uncertain, but there is nothing in recent data to suggest that this underlying cycle will not be repeated some years hence, once regional output returns to previous levels and once slack in the London premises and labour markets has been taken up...Equally, too, it would be forecast, that, other things being equal, the same effects are likely to emerge, with those patterns discerned in the 60s and 70s as well as in the recent study, continuing to characterise future movements...Indeed, there may be increased rationale for that repetition. Until those areas which have received net inflow from SE relocation themselves begin to generate the same pressures which caused those earlier waves of movement, they offer attractive new markets to other mobile firms. Thus a new cycle of business growth, with sufficient strength to catch up on ground lost in the recent recession, is likely to continue to push economic activity out of London and over the South East Regional boundary, in Western and Northern directions. There is no intrinsic business logic which will cause such movement...to seek more distant solutions.

In the circumstances, this trend scenario seems to be a reasonable prediction. But in accepting it, certain unknowns have to be borne in mind. For example, much of the decentralisation activity in the last decade has been driven by the service sector, which was itself growing enormously. This growth has now stopped and may not continue when the recession ends. If the 1980s service sector growth was a one-off phenomenon, then the major source of decentralisation may have disappeared. Clearly, it is possible that a stable services sector will still generate relocation, but it is unlikely to be on the scale witnessed in the recent past.

Notwithstanding this proviso, the likelihood then seems to be a continuation of the growth pressures witnessed recently in the outer south east and the south east fringe, once the recession ends. Given Prism Research's (1991) condition to this prediction—that it will hold unless past net recipients of growth begin to exhibit negative pressures—it is likely that the rolling back of the south east will continue.

The prospects for stopping counter-urbanisation

The answer then to the simple question posed at the beginning of this section—'*is it possible that counter-urbanisation can be reversed?*'—seems to be 'no'. The evidence suggests that the process is as strong now in the UK as at any point in the post-war period, with accessible lowland and remoter, rural areas receiving the greatest percentage and absolute levels of growth. These changes have taken place during a period when planning policy has favoured containment. It can be assumed reasonably that decentralisation would have occurred at a faster rate still without this planning regime. The evidence reviewed here casts serious doubt on the ability of central or local government to resist, still less reverse, counter-urbanisation trends as a device to reduce urban energy consumption.

Perhaps a more sensible approach would be to accept the inevitability of further urban decentralisation, but to channel it into a settlement pattern that is relatively energy efficient, if not as efficient as the preferred compact city alternative. Population growth in the 'mixed and accessible rural' areas will be the most difficult to halt because much of it will be indigenous. These areas are also the main recipients of economic decentralisation. In contrast, the 'remote, mainly rural' areas may be more susceptible to attempts to prevent or channel growth. Much of the growth will come from in-migration, and hence may be dissuaded. That growth which does occur could be focused on larger villages and small towns, rather than being allowed to filter down to the smallest, and least energy efficient, settlements.

Counter-urbanisation and sustainable urban forms

Having reviewed evidence on urban form and energy consumption and on the force of counter-urbanisation trends, the two can now be related in order to address the third question posed in the introduction: *is counter-*

urbanisation exacerbating the problem of urban energy consumption by increasing significantly overall travel (whether or not it can be reversed)?

In principle, this can be done by assessing the degree to which urban areas which are growing are also areas of high energy consumption. Unfortunately, the evidence on counter-urbanisation, in Table 21.5, is available by urban types, while that for energy consumption, in Table 21.4, is available by urban size bands. To carry out the assessment then, it is necessary to attempt a matching of the OPCS urban types and the ECOTEC (1993) size bands. Table 21.9 proposes such a matching.

The matching is necessarily rather crude, but is a reasonable approximation of the average size of settlements in each of the OPCS categories. This matching allows a comparison to be made of the degree to which the recipients' areas of population growth arising from counter-urbanisation are also high energy consuming areas. Figure 21.2 plots energy consumption rates on top of the counter-urbanisation details given earlier in Figure 21.1. This shows a close correlation.

The areas of population decline ('principal cities', 'Inner London', etc.) show the lowest rates of energy consumption per capita; the areas of largest population increase ('mixed and accessible rural', 'resort and retirement', 'remote, mainly rural') show the highest rates of energy consumption. Thus, it is clear that counter-urbanisation changes do exacerbate the

Table 21.9 A matching of OPCS urban types and ECOTEC urban size categories

OPCS urban types	ECOTEC size categories
London	
Inner	Inner London
Outer	Outer London
Metropolitan districts	
Principle cities	Metropolitan areas
Others	Other urban over 250 000
Non-metropolitan districts	
Large cities	100 000–250 000
Smaller cities	
Industrial districts	
Wales and northern regions	50 000–100 000
Rest of England	
New towns	25 000–50 000
Resort and retirement	
Mixed and accessible rural	
Outside SE	3 000–25 000
Inside SE	
Remote, mainly rural	<3 000

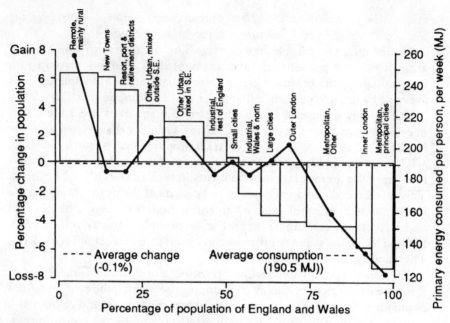

Figure 21.2   **Population change 1981–91, and transport energy consumption, for OPCS urban types**

problem of urban energy consumption. People are moving to areas that will induce more travel and hence higher fuel consumption. The biggest problem appears to be with 'remote, mainly rural' areas. These have much higher per capita energy consumption rates than other areas, yet are showing the highest rates of population growth—in percentage and absolute terms.

It seems clear, then, that counter-urbanisation trends have been increasing fuel consumption. But the questions remain: by how much, and is the amount significant? Given the estimates of energy consumption by OPCS urban type categories, it is possible to calculate a crude estimate of the additional energy consumption that arises from counter-urbanisation. This is calculated in Table 21.10.

The aim is to calculate total energy consumption in each area type at 1991, and then to calculate the equivalent energy consumption levels on the assumption that no counter-urbanisation occurred over the 1961–91 period. The difference between the two should then represent the urban energy consumption that occurs because of counter-urbanisation. The 1991 'non-counter-urbanisation' population is calculated by allowing the 1961 population in each urban type to grow by the national growth rate (+6.1%) from 1961–91. This, in effect, gives a 1991 population in each urban type as if no population movement had taken place over the previous 30 years.

Columns 1 and 2 of Table 21.10 give the 1991 census population and the per capita energy consumption rates, derived from Table 21.4, for each urban type. Column 3 shows the result of multiplying these together: the

Table 21.10 Estimates of fuel consumption arising from counter-urbanisation changes, 1961–91, by urban types (GJ per week)

Urban type	Pop. 1991 ('000s) (1)	Energy use rate/ cap. (MJ) (2)	Total energy use 1991 (GJ/week) (3)	Pop. non-counter-urban. 1991 ('000s) (4)	Total energy use non-counter urban. 1991 (GJ/week) (5)	(GJ) Difference (5)-(3)	%
Greater London Boroughs							
Inner	2 350	140.4	329 940	3 707	520 463	190 523	57.7
Outer	4 028	182.5	735 110	4 777	871 802	136 692	18.6
Metropolitan districts							
Principal cities	3 227	119.1	384 336	4 485	534 164	149 828	39.0
Others	7 424	119.1	884 198	7 959	947 917	63 719	7.2
Non-metropolitan districts							
Large cities	2 666	180.6	481 480	3 135	566 181	84 701	17.6
Smaller cities	1 691	180.6	305 395	1 808	326 525	21 130	6.9
Industrial districts							
Wales and northern regions	3 278	173.1	567 422	3 386	586 117	18 695	3.3
Rest of England	3 381	173.1	585 251	2 995	518 435	−66 816	−11.4
New towns	2 305	170.8	393 694	1 647	281 308	−112 386	−28.5
Resort and retirement	3 509	170.8	599 337	3 007	513 596	−85 741	−14.3
Mixed and accessible rural							
Outside SE	3 951	203.4	803 633	3 036	617 522	−186 111	−23.2
Inside SE	5 278	203.4	1 073 545	4 513	917 944	−155 601	−14.5
Remote, mainly rural	5 422	248.0	1 344 656	4 487	1 112 776	−231 880	−17.2
England and Wales	48 960	−	8 487 997	48 960	8 314 451	−173 546	−2.0

total energy consumption arising per week from travel in each area type. Column 4 then gives the 1991 population of each area type if no differential population changes had taken place over the 1961–91 period: in effect, if there had been no counter-urbanisation. Column 4 gives the 1961 population multiplied by the national population change (+6.1%) for 1961–91. Column 5 then gives this population multiplied by the energy consumption rates of column 2. Column 6 gives the difference in energy consumption between the estimated actual levels of consumption (column 3) and the 'non-counter-urbanisation' consumption estimates of column 5. This suggests that 173 546 000 MJ of energy are consumed per week in England

and Wales because of counter-urbanisation changes over the last 30 years. Column 7 gives the equivalent percentage changes. This suggests that the counter-urbanisation forces of the last 30 years now mean that weekly transport energy consumption is 2.0% higher than it would be had those forces not taken place.

This result is surprising. The weight of evidence in favour of urban containment as a way of reducing urban energy consumption implies that it would bring about significant savings: savings so significant as to warrant radical policies to reverse immensely powerful counter-urbanisation trends. Indeed, at the outset of this work, the assumption was that this significance would be demonstrated. Given the potential margin of error in these calculations, and the possibility that if other factors could be taken into account (income, car ownership, etc.) they would reduce the urban/rural energy consumption differences, the figure of 2.0% could be regarded as a minor difference. The calculations have been made on the basis of counter-urbanisation data taken from Table 21.5 rather than Table 21.6. The latter suggests more extreme counter-urbanisation changes over the 1981–91 period. But they are not likely to make a great deal of difference to the above result.

This result challenges the validity of the fundamental assumption upon which much of the emerging conventional wisdom on urban sustainability in the UK is based. This assumption is that urban decentralisation is responsible for a large proportion of rising transport energy consumption. The consequent policy stance—until now an obvious response to the evidence—favours stronger urban containment policies. The evidence presented here suggests that the consequent savings are likely to be trivial. There may be other environmental benefits from urban containment, of course, but the suggestion here is that energy savings should not be a prime motivation in promoting such a policy. Urban containment may be consistent with policies to reduce travel overall, because, for example, it provides the potential for greater use of public transport. However, significant savings must not be expected simply from a change in settlement patterns.

Conclusion

The original intention behind this chapter was to counter-pose the emerging academic and political consensus in favour of containment—or the 'compact city'—as a contribution to urban sustainability, with the apparent neglect of one important issue: is it feasible? In particular, the issue to be addressed was whether the proposed solution is feasible given the evidence that urban decentralisation forces are so strong. This issue has been addressed here by two questions:

1 what evidence is there to suggest that compact city forms can significantly reduce urban energy consumption?
2 is it possible that counter-urbanisation can be reversed?

Brief reviews of available evidence on both counts have been presented. On the first question, despite some doubters, the general consensus—academic and political—seems to be that large, high-density cities involve less private travel and induce greatest use of mass transit: thus reducing energy consumption and pollution. The result is a remarkably powerful movement in favour of urban containment policies in many countries.

On the second question, the weight of evidence suggests that counter-urbanisation forces remain very strong—even when the prevailing planning regimes favour containment—and show no signs of abating. This suggests, then, a conflict between the desire for stronger containment policies and prospects of actually implementing such policies. Even if containment policies are regarded as environmentally desirable, there is little chance that they can be effective. This implies that a rethink of the new-found consensus on the compact city is needed. It does not necessarily imply a complete abandonment of the idea. Perhaps an acceptance of counter-urbanisation trends is required, but a more specific channelling of such movements into environmentally acceptable, but decentralised settlements.

Thus, the original intention of this chapter has been met. However, a bonus has also emerged. The review of counter-urbanisation has provided further evidence on the merits of the containment solution itself. Thus, a third question has been posed:

3 is counter-urbanisation exacerbating the problem of urban energy consumption by increasing significantly overall travel (whether or not it can be reversed)?

It has been suggested tentatively that, contrary to expectations, counter-urbanisation movements in the last 30 years in the UK, although powerful, have made little difference to current energy consumption levels: that is, if the decentralisation movements had not occurred, energy consumption by transport would only be marginally lower than it is now. Thus, the answer to the third question is: apparently, no.

A chapter, then, that assumed at the outset that the environmental case for containment was basically sound—and that the only issue was whether counter-urbanisation forces would undermine it—finishes by casting doubt on the validity of that very case.

22 Information Technologies, Telecommuting and Cities

Mitchell L. Moss and John Carey

TELECOMMUNICATIONS has long been viewed as a threat to cities: with the decreasing cost of communications technology, many observers believe that cities will eventually become obsolete. Clearly, the spatial dispersal of economic functions from the central cities of North America to the suburban areas and so-called 'edge cities' has reinforced the belief that central cities have lost their traditional function as a headquarters and business centre. In recent years, the diffusion of the personal computer and facsimile machine has fostered the widespread view that telecommuting will eventually replace the office as the primary work site. This chapter examines the way in which telecommuting influences the pattern of work in central cities, by analysing data from a telecommuting project in the New York metropolitan region. The chapter points out that telecommuting is not about to eliminate the need for office activities, whether in the central city or other locations. However, the opportunity to work from home has important implications for the future pattern of development in cities and metropolitan regions.

Telecommuting has attracted a great deal of attention from government policy makers, employers and individual workers as well as in the popular press (Lueck 1992; Young 1993). It is viewed by many as an important component in the information age: a high-tech workforce that uses telecommunications to 'commute' to the office and conduct work (Page & Brain 1992). For more than a century individuals, especially crafts people and professionals, have conducted all or part of their work at home. Operating a business from home, full or part time, is increasingly feasible with advanced telecommunications systems. Many people work at home on an informal basis—reading reports, preparing for meetings, grading exams. This is not generally considered to be telecommuting. A telecommuter is someone who relies on communications technology to do much of his or her work at home, from the car, airplane, or even a hotel room. The distinguishing characteristic is that the work process involves the use of telephone lines and related communications equipment—from the home or a non-traditional work site.

Telecommuting offers the potential to solve a number of immediate, practical problems. For example, some research has indicated that telecommuting can help companies to meet the requirements of clean air legislation by reducing the number of employees who commute to work in a car (Boghani *et al.* 1991). It is described as a way to help reduce traffic congestion at peak periods (Hodson 1992). Many argue that it provides a viable use for an 'information highway' to the home, thereby supporting the construction of an advanced telecommunications infrastructure in metropolitan areas (Malech & Grodsky 1991). Earlier research studies have also indicated that telecommuting can increase productivity in an organisation (Bowman & Davis 1990), improve employee morale (Arizona Department of Commerce 1991), reduce the stress associated with commuting (Boghani *et al.* 1991), help companies recruit better employees (Nolan & Shirazi 1991) and help to develop a more competitive workforce (Malech & Grodsky 1991).

Several obstacles to telecommuting have also surfaced. Some concerns have been raised about the social isolation of people who work at home alone (Argyle 1990). There are many unsettled insurance and liability issues associated with working at home (Bowman & Davis 1990). Resentment by fellow workers has occurred in some cases (Bowman & Davis 1990) and some unions have objected to telecommuting (Page & Brain 1992).

In order to study these issues, the New York Telephone Company in conjunction with the New York City Department of Telecommunications and Energy, New York City Department of Transportation and New York Metropolitan Transportation Council conducted a 10 month study of telecommuting from March 1992 through January 1993. In the project, selected employees from three private organisations (Bankers Trust Company, Merrill Lynch and New York Telephone) and one city agency (NYC Department of Transportation) worked one or more days per month from their homes. They used telecommunications rather than transportation to get to work and communicate with fellow workers.

The purpose of the study was to understand the feasibility, potential benefits and obstacles to telecommuting. In addition, the study addressed the potential impact of telecommuting on transportation, energy consumption, pollution and telecommunications infrastructure in a large metropolitan area. The study was based on 57 telecommuters and 46 supervisors. The research for the study consisted of an initial survey that gathered baseline data such as how much time workers spend on commuting to work, how they feel about telecommuting, as well as worker demographics. A few months into the project, selected telecommuters were invited to a focus group meeting in which they discussed their experiences. At the end of the 10 month period, a follow-up survey was administered to the telecommuters and their supervisors.

There was a broad age range among the telecommuters along with good ethnic and racial representations. Women represented 60% of the telecommuting group; most of the telecommuters were married (76%); and

54% had dependents in their households. Education levels among the telecommuting group were high: 80% had at least a college degree. Household income was higher than the general population: 61% of households earned $50 000 per year or more.

A typical telecommuter in the study was a married woman with children at home. She and her husband both work. The typical telecommuter was a mid level manager who had been with her organisation for 10 years. Her motivations to become a telecommuter were to reduce the hassle of commuting, free up some time to be with her children and provide flexibility to attend school in the evening or deal with other everyday life chores.

Slightly more than half of the telecommuters (52%) used a car to commute to work. Among the telecommuters who used a car to commute, 97% drove alone and 3% participated in a car pool. They also reported that half of the drive to work (45%) was in congested traffic. They reported spending $11.08 per week for gasoline used in commuting. Table 22.1 shows these averages per trip.

Several months into the study, a series of focus group meetings were held with Bankers Trust, New York Telephone and NYC Department of Transportation employees who participated in the telecommuting project. The overall reactions by the telecommuters ranged from positive to very positive. No one expressed a negative attitude about their telecommuting experience. However, a number of problems and disadvantages were mentioned. Among the advantages mentioned, the most frequently cited issue was increased productivity. Focus group participants were nearly unanimous in their assessment that they got more work done at home compared to the office. The second most frequently mentioned issue was child care. Women indicated that they can better manage child care when they are telecommuting. For example, they do not have to drop off a child as early on a telecommuting day and can pick up the child earlier. Further, a few women mentioned that they were able to return to work earlier from maternity leave because of telecommuting.

In addition, many participants said that telecommuting leads to a better organisation of work and more accuracy in their work. Also, they feel that telecommuting leads workers to be more disciplined. Other participants indicated that there are fewer interruptions at home. There is less demand to attend meetings, answer phone calls and engage in office social inter-

Table 22.1   Average trip distance, travel time, cost and waiting time

Average total distance for trip	21.7 miles
Average total travel time for trip	59 min. 19 sec.
Average total cost per trip	$5.93
Average total waiting time per trip	6 min. 33 sec.
Average number of travel modes	2.6

action. So, a worker can concentrate on his or her work. Other advantages mentioned included saving money, e.g. reduced costs for travel, lunch and dry cleaning. Participants also indicated that with telecommuting, a worker has more free time. With a more concentrated work day, there is more time for household chores and parental duties.

Among the disadvantages mentioned were some resentment by other staff members. There was a feeling by many participants that some fellow staff members thought they were 'getting away with something'. The level of resentment was described as moderate, not high. Nonetheless, it was mentioned by many of the focus group participants. Other disadvantages mentioned included loss of money on monthly commuter tickets and increased air-conditioning bills in summer months. Some participants also mentioned that their telephone bills went up. A few participants said that they missed the comradeship of the office, though most felt that this was not a problem as long as telecommuting was one or two days a week. There was also some concern that they would not be as 'visible' as other workers and might in the long run not gain promotions as quickly as others.

When asked 'who is suited to telecommuting', focus group participants indicated that many workers are suited but not all. They said that tele-commuters need to be disciplined and to believe in the concept of telecommuting. People whose jobs require being in the office (e.g. a receptionist) could not telecommute and people who are *very* dependent on the social interactions in offices would probably not be suitable.

Discussing company policies about telecommuting, there was general agreement that companies and government agencies need to 'certify' telecommuting by developing policies that explicitly support telecom-muting activities. Second, many participants spoke about the positive messages that companies can send to workers if they develop telecom-muting policies. Telecommuting, they feel, indicates that a company or public agency is progressive and that it supports workers who can benefit from telecommuting, e.g. women who have been on maternity leave and want to return early to work. Also, they feel that a positive telecommuting policy will help to attract better workers and retain highly talented workers.

Survey findings

A survey at the end of the 10 month study provided findings about the impact of telecommuting on transportation patterns, energy and telecom-munications usage, work patterns and attitudes of both telecommuters and their supervisors.

Transportation

Telecommuting reduced the number of vehicle miles travelled by workers. The average telecommuter travelled 189 fewer miles per month due to telecommuting (Table 22.2).

Table 22.2 Estimated reductions in transportation associated with telecommuting

	Miles per month
Average reduction in vehicle miles travelled by all telecommuters:	189
Average reduction in vehicle miles travelled by all telecommuters who use a car on telecommuting day:	172
Average reduction in car miles travelled by all telecommuters who drive to work:	70
Average reduction in car miles travelled by all telecommuters who drive to work and use car on telecommuting day:	51
N=57	

This includes all forms of transportation—subway, bus, car. The reduction in vehicle miles travelled is based on the average number of miles travelled to and from work on a regular work day minus the average number of miles travelled on a telecommuting day during an average month with 3.7 telecommuting days. Among those who used a car to travel to work, there was an average reduction of 70 car miles per month. It appears that telecommuting can help reduce the strain on highways and other forms of transportation during rush hours.

Energy and other cost savings or increases

Telecommuters saved an average of $11.92 per day or $44.10 per month (based upon an average of 3.7 telecommuting days per month) due to lower costs for food, gasoline, subway fares, dry cleaning and other daily costs associated with work (Table 22.3).

Table 22.3 Cost savings on telecommuting days

Item	Number of telecommuters with increase or decrease	Average change
Gasoline	17	− $1.74
Tolls	6	−5.20
Food	35	−5.13
Dry cleaning	16	−2.34
Parking	7	−9.57
Day care	6	−28.67
Train	8	−2.75
Subway/bus	37	−3.70
Average net savings per day for all telecommuters:		$11.92
Average net savings per month for all telecommuters:		$44.10
N=57		

This includes all forms of transportation—subway, bus, car. The reduction in vehicle miles travelled is based on the average number of miles travelled to and from work on a regular work day minus the average number of miles travelled on a telecommuting day during an average month with 3.7 telecommuting days. Among those who used a car to travel to work, there was an average reduction of 70 car miles per month. It appears that telecommuting can help reduce the strain on highways and other forms of transportation during rush hours.

Telecommunications

Virtually all of the telecommuters relied on the telephone to conduct work and stay in touch with fellow workers. Further, many used fax machines, modems and other telecommunications equipment to support their work from home. In addition, 88% of the telecommuters believe that additional household telecommunications capacity in the form of a second telephone line is essential or very helpful to support telecommuting (Table 22.4). At the same time, some telecommuters indicated that they needed some technical assistance with the extra equipment and software they used for tele-commuting. This suggests that telecommuting is a strong potential application for an advanced telecommunications 'highway' to the home, but many workers will need assistance in setting up equipment to run on the highway.

Work patterns

Telecommuters indicated that they were able to keep work and home obligations separate when telecommuting. At the same time, there were a few noteworthy patterns about the days and hours when people do telecommuting. Most telecommuting occurred during mid-week days, not on Monday or Friday (Table 22.5).

Equally important, many telecommuters indicated that they were not able or preferred not to have a designated telecommuting day—the day of the week for telecommuting changed week by week or month by month. This suggests that flexibility is important for a telecommuting schedule.

While many worked the same schedule at home as in the office, others indicated that they started their work day earlier and ended it earlier compared to a regular work day. Presumably, this frees up time in late afternoon to spend with children, attend school or do household chores.

Table 22.4 Importance of a second telephone line

How important is a second telephone
line for you to do telecommuting work? Percentage of telecommuters

Essential	63
Helpful but not required	25
Not needed	9
Not applicable	4
N=57	

Table 22.5   Days of the week for telecommuting

Normal telecommuting day	% of telecommuters
Monday	2
Tuesday	12
Wednesday	28
Thursday	25
Friday	5
No set day	28
N=57	

Effects and attitudes

There is a strong perception by telecommuters that they are more productive in working at home (Table 22.6). Many supervisors of the telecommuters also saw improvements in productivity, although their perceptions were not as strong as for the telecommuters themselves. Telecommuters also indicated that telecommuting helped them to better manage their time and work at personal peak times (e.g. some people work best in early morning). Telecommuting also improved morale, attitudes about their job and attitudes about New York City Department of Transportation. In addition, 98% reported a favourable or very favourable attitude about telecommuting after 10 months of experience as a telecommuter (Tables 22.7 and 22.8), and 96% would like to continue as a telecommuter. The principal obstacle or problem associated with telecommuting as reported by telecommuters is the need for more equipment in homes (Table 22.9).

Conclusions and policy implications

Based upon this study, it is clear that telecommuting will influence urban activities but not eliminate traditional office functions. The attitudes of the telecommuters were particularly noteworthy. They liked telecommuting

Table 22.6   Effects of telecommuting on telecommuters

	Score on a 5 point scale (1=Strongly agree; 5=Strongly disagree)
Better manage time	1.7
Be more productive	1.5
Work at more convenient hours	1.8
Better coordinate work and family activities	2.1
Work at personal peak time	1.9
N=57	

Table 22.7   Telecommuter attitudes about telecommuting after 10 or more months of experience

Overall attitude	% of telecommuters
Very favourable	72
Favourable	26
Neutral/unsure	2
Unfavourable	0
Very unfavourable	0
N=57	

Table 22.8   Perceived benefits of telecommuting

	% of telecommuters*
Flexible work schedule	75
Less commuting time	68
Less aggravation of commuting	47
Save money	30
Spend more time with family	18
More productivity	14
Assist in child care	9
More time for household chores	7
N=57	

* Multiple answers permitted

Table 22.9   Perceived problems and obstacles in telecommuting

	% of telecommuters*
Need a lot of equipment in home	24
Increased cost of household utilities	19
Coworkers are sceptical	18
Family obligations interfere	16
Not enough space at home for work	14
Company policies not in place to really support telecommuting	11
Carrying work back and forth from office	9
N=57	

* Multiple answers permitted

and wanted to continue as telecommuters. Further, it boosted their morale and improved their outlook about their job and their employer. From an employer's point of view, it appears that telecommuting had a positive impact on productivity and that it can help companies as well as government agencies to meet the requirements of clean air legislation by reducing the number of autos during peak periods. Several policy recommendations follow from these findings.

Transportation and clean air policies

The findings in this study suggest that telecommuting can contribute to a reduction in vehicle miles travelled generally and car traffic specifically. In order to develop a model of potential savings, it is first necessary to estimate the percentage of the workforce that could telecommute. Estimates vary a great deal, from 15 to 50% of the workforce. In the model below, we utilise the conservative end of these estimates, i.e. 15% of workers, and develop a model for the New York metropolitan region. Each work day, there are seven million commuters in the New York metropolitan area. A high percentage of these commuters drive alone in a car to work: 53.6% according to the US Census Bureau. The potential savings in pollution and car traffic are substantial (Table 22.10).

Three recommendations follow from these findings:

- companies and public agencies should be encouraged to adopt tele-commuting programmes as a way to help meet the requirements of clean air legislation;
- telecommuting should be incorporated into general transportation policies as one means to help manage the strain on the transportation infrastructure during peak travel periods; and
- there is a need for a broad education programme to inform public and private organisations about the benefits of telecommuting as well as how to successfully implement telecommuting programmes.

Telecommunications policy

There has been much discussion about the need to build an advanced telecommunications infrastructure or 'information highway' linking homes, businesses, schools and public agencies. An advanced telecommunications infrastructure is linked in policy discussions to the creation of new jobs, better social services and new forms of entertainment. The study reported here indicates that telecommuting is a strong potential application for an advanced telecommunications infrastructure. Telecommuters rely heavily on telecommunications to do their work. Further, they expressed a need for

Table 22.10   Potential impact of telecommuting on motor vehicle traffic and pollution in the New York metropolitan area (based on 15% of workers telecommuting 3.7 times per month)

95 000	Fewer motor vehicles per day used in commuting
50 000 000	Fewer motor vehicle trips per year for commuting
15 000 000	Fewer gallons of gasoline per month
2.5	Fewer tons of toxic emissions per day

services that could be provided through a new telecommunications infrastructure, e.g. replace the physical shipment of large volumes of documents between office and home with electronic transfer of documents or allow workers at home to fully participate in meetings at the office through a video link between home and office. Some recommendations that follow from this general policy position include:

- encourage telephone companies and other groups to develop an advanced telecommunications infrastructure;
- encourage public and private agencies to experiment with new services that utilise existing telecommunications networks as well as new networks as they become available; and
- investigate and encourage new tariffs that promote enhanced use of existing telecommunications networks as well as new networks as they come on line. This might include ways that companies can pay directly for telephone lines and other telecommunications services in tele-commuter homes. It might also include new interactive video services that link telecommuter homes and business offices.

Energy policy

The study indicated that there is a trade-off between potential savings as a telecommuter for items such as gasoline and increased utility costs associated with air-conditioning in the summer and daytime heating in the winter as well as additional use of house lights. It was not clear whether there might be any savings in energy costs for organisations due to fewer personnel in offices. One indicator—reduced demand for office space—was negative. That is, supervisors did not generally feel that there was a reduced demand for office space as a result of telecommuting. While there was no apparent reduction in energy costs, there was a shift in energy use; that is, a reduction in gasoline consumption and a modest increase in electricity and gas consumption. Further research will be required to gain a more comprehensive assessment of the potential effects of telecommuting on energy consumption. Based upon this limited study, it appears that any effects are likely to be small.

Company or organisational policies about telecommuting

The study provided positive findings about the benefits of telecommuting for organisations and their employees. In addition, it suggested that telecommuting can help organisations to meet requirements of clean air legislation through a reduction in the use of cars by employees commuting to work. Organisations must also learn what types of management adjust-ments are necessary to support telecommuting and other requirements for a successful telecommuting effort. It is important that company policies support telecommuting and that an effort is undertaken to win the support of non-telecommuting coworkers.

Cities and telecommuting

This chapter highlights the potential benefits that telecommuting offers to workers and employers. However, there is little reason to believe that remote work with communications technology will eliminate the activities and interaction that occur within offices. Telecommuting may, however, allow the further dispersion of work to remote locations, by allowing workers to have greater residential choice and longer, but fewer, commutes to work. In addition, the growth of telecommuting may require offices to be reconfigured to provide space for social interaction and conferences that cannot be achieved through remote work. Cities may need to adapt to telecommuting rather than treat telecommuting as a threat to their economic viability.

PART 5 The Sustainable City

T HE IDEA OF *sustainability* has been implicit in much of the discussion engendered so far in the various contributions already presented in this book but we are at last in a position to address the idea directly. It is quite clear that the idea has evolved considerably during the last 20 years since it first became significant to city planning through studies of the 'limits to growth' and the 'first oil crisis'. Some of these changes to the concept have already been reflected here in notions concerning competition between cities, the changing hierarchy which reflects such competition and the changing basis of cities as productive systems, all set within the fascinating but perverse interaction between global and local scales. In short, our ideas concerning what is sustainable in terms of cities now reflect the complexity and diversity of locational behaviour, of energy use in transportation, of massive changes in information technologies, and of the limits to growth posed by the interaction between the developed, developing and underdeveloped worlds.

James Lovelock (1979) in his book, *Gaia*, demonstrates quite unequivocally that all systems are sustainable within general limits but it is the '*quality of life*' concerned with sustainability that is the crucial issue. Twenty years ago when the first studies of the interaction between energy and population on a global scale were begun, there was little emphasis upon the extent to which societies, cities and various production technologies provided a competitive mix which determined levels of development and responses to the evident growing mismatch between available non-renewable energy sources and levels of population. These early scenarios suggested a world catastrophe which would afflict all nations and cities by the middle of the 21st century.

Since then our understanding of these possibilities has been considerably changed by the emergence of *universal information technologies*, by the role of *political change*, and by the rapidity of *individual responses* to old problems and new ideas. The problem of developing sustainable urban futures which will not only guarantee a quality of life equivalent to that associated with the developed world but will raise standards for all, is ever more serious but at least we now have a better understanding of the factors which influence this future, and the resilience and competition between the various elements which comprise it. It is clear that the notion of global collapse in terms of the quality of life for all is too general a scenario to be contemplated in the face of massive and unimagined *technological change and adaptation*.

The very title of this book, *Cities in Competition*, belies the *paradigm shift* which is changing our view of sustainability, and several of the contributions hitherto have shown how limited conventional thinking about energy use, transportation and the spatial form of cities has been. There has been a shift in thinking from *macro*-analysis to *micro*, from the notion that cities are strong, collectively organised systems to ideas that cities are composed of many groups and individuals in competition, betraying great *diversity* but also great *adaptability*, acting locally but generating *organisation and order* which is manifest at more global scales through the urban hierarchy. Cities can no longer be characterised as simple, uniform, homogeneous systems, for the interaction of the local and the global which is best seen in the way world cities are structured, shows a degree of complexity which is becoming deeper and more convoluted as populations grow and as new technologies are evolved. Moreover, although our understanding of the forces which determine urbanisation is increasing, our ability to make predictions concerning these urban futures seems to be retreating in the face of this increasing complexity. Indeed, one of the themes which emerges from several of the previous contributions in this volume illustrates the poverty of our existing approaches to analysis and the need for new directions.

In this final part, we will broach these issues directly, changing our focus from one based on competition within and between cities to ways in which the urban future might unfold in terms of the relationship between *energy* and *population*. But as we will see, these contributions take as a starting point the assumptions and ideas of the previous chapters, for each of the authors assumes that any attempt to provide a sustainable urban future which maintains and improves the quality of life locally and globally must be based on the way cities and those who live in them work in competition. Chapter 23 by Harris begins by sketching the conventional problem of urban sustainability: that based on the agreed forecast that current demand for energy (on a per capita basis) cannot be maintained in terms of non-renewable sources in the light of increasing population. In short, unless there are technological miracles in the conversion of fossil fuels to useable energy or the use of hitherto untapped sources of renewable energy, the world in general and large cities in particular will face major limits on their energy use within two generations. These facts are largely incontrovertible.

What Harris succeeds in demonstrating, however, is that these impacts are likely to be massively uneven between First and Third Worlds, between industrial and post-industrial cities, between west and east and so on. Moreover, there is the possibility that these trends might be changed substantially by technology, perhaps by individuals and societies adapting to the change, but also by competition which, in its darkest form, involves *political instability*. It is never easy to generalise from recent events but some of these trends might be indicated by the current world recession in western nations and by the breakup of Europe and the political instabilities facing those societies moving from command to capitalist styles of economic organisation. Harris paints a frightening picture but one that is informed by much subtler arguments than those mooted some 20 years or so ago.

A more optimistic picture, however, is suggested by Meier (chapter 24) who, alone amongst all the authors writing here, provides a blueprint for one form of *sustainable urban ecosystem* which is currently feasible and could provide a model for contemporary urban living massively conserving resources. Just as Harris shows that the interaction between the First and Third Worlds is all-important to any discussion and forecast concerning the urban future, Meier shows how we might learn from both worlds in fashioning strategies for a sustainable urban future. Meier's blueprint only covers one feature of the urban system, albeit a crucial one, and that deals with the *recycling of waste products* and the *production of basic foodstuffs*. But he does show how his model of an idealised urban system relates to other social and economic processes which are central to the way cities are organised and in this sense, his ideas are highly suggestive of how we might move in general to the sustainable city. The mix of technologies from first and Third World which combine old and new, alternative, conventional and cutting-edge, provides a refreshing insight into the way we might *invent the future*. Unlike many urban utopias, his proposals are no mere pie-in-the-sky but manifestly feasible to the point whereby he suggests how his vision might be implemented now in an area of west Manhattan. He does not extend this vision to the entire city or to systems of cities or to the global pattern of cities but his argument provides a strong basis for such speculation.

In the last chapter (25), many of these threads are pulled together and abstracted by Batty who attempts to assess ways in which our approaches to analysis and forecasting need to change if we are to make use of the substantive insights into the functioning of cities which form the essential arguments of this book. His emphasis is upon the paradigm shift which we referred to above, from macro to micro, from global to local and back again, and from the collective, aggregative view of cities to more individualistic, disaggregate approaches. In essence, these shifts reflect the notion of cities in competition. Batty argues that we need to grasp the notion that cities are intrinsically complex, that our treatment must emphasise their dynamics, and that our ideas not only of what but how we might predict and manage the urban future must change to accommodate these insights.

All this is easier said than done but some rudimentary ideas for such forecasting are presented with an emphasis on how we might predict the interaction between global and local levels within the hierarchy of world cities. Sustainability, Batty argues, must be modelled in this way if *local and global interactions* and their surprising consequences are to be understood and anticipated. The various themes in this section interact with one another in diverse ways and will be picked up once again in the Epilogue which gives suggestions for how we might develop these ideas further in the next decade.

23 The Nature of Sustainable Urban Development

Britton Harris

Introduction

I begin by providing a broad and systemic description of 'sustainability' as it might be applied to virtually any major social and technological problem like urban growth. I then develop the view that the continuing growth of cities of the world is an inevitable and indeed essential feature of continued balanced development, and of finding a sustainable path for that development. The concentration of populations in cities is needed to provide relief from pressure on the land and the environment, and a setting in which education and socialisation will lead to a decline in birth rates. This growth of cities is most pressing in presently underdeveloped nations, and will result in greatly increased demands for capital investment and the application of energy. However, the world cannot sustain energy uses on a world wide basis at the present level of urbanisation and the per capita energy consumption which obtains in the West; neither can it politically sustain continued deprivation in the Third World. The development of an equitable, urbanised and prosperous world ultimately demands a technology for urban production and urban living which is less energy intensive than the best which is currently available.

Fortunately, the micro-technologies of information processing and transmission, computing, and biological manipulation can all be environmentally benign and are still in a period of rapid and fruitful development. Their expanding use to meet this challenge does not have any immediately foreseeable limits, but its course is not predictable on the basis of the scope of our experience to date. This direction will require many major changes in lifestyles and in modes of social interaction, not all of which are humanistically appealing when judged by historical standards. There are many potential pitfalls as well. Some of the new directions which are implied, and their implications, are developed in the balance of the chapter.

What is sustainable?

The idea of sustainability implies that if some process is continued into the future, the conditions necessary to support that process will not be impaired. In particular, it suggests that the process itself will not undermine the conditions which sustain it. Here are some of the conditions which I consider necessary for continued urban development, and I feel strongly that they are also necessary for continued human progress:

1 A sustainable process cannot exhaust non-reproducible resources on which it directly or indirectly depends. For people on earth, these resources include fossil fuels, ambient temperature and climate, flows of usable water, and other aspects of the environment.
2 Such a process should not impair the means by which the supporting environment can adapt to changing conditions and continue its support; this implies, among other things, a continuation of terrestrial species diversity.
3 Conditions of international tension and political instability have implications for the abuse of the environment and of human interrelationships which are in the long run inimical to the sustainability of the environment and of human social adaptation.
4 A politically realistic and morally sensitive view of the human condition leads to the conclusion that deprivation, ignorance and oppression are not sustainable. Not only are they repugnant, but they become a source of social conflict and instability.
5 These views then lead to the conclusion that many human social processes, including urban development, are not sustainable without improvements in the human condition. Further analysis and reflection leads to the view that in the long run, the human population on earth cannot indefinitely increase, and that its growth must ultimately come to a halt.

These conditions are not altogether scientifically established, and may never be because of their moral content. They are however widely believed and increasingly supported. I do not propose to argue their correctness here, but to offer them as a framework for addressing the issue of sustainable urban development. Others are free to make use of different descriptions of sustainability, always remembering that sustainability must extend into the indefinite future. From this standpoint, then, I now investigate the sustainability of urban development, imagining and evaluating the indefinite future on the basis of current circumstances and foreseeable trends.

Energy setting for the problem

Cities are centres of large-scale production of goods and services, and the habitat and cultural medium for a growing proportion of the world's growing human population. We ask what forms of cities are capable of sustained growth, and we take the idea of sustainability to refer to many different aspects of cities—moral, social, political, economic and environmental.

In dealing with a problem of this magnitude, we have to take a large-scale system view of matters. Urban growth is related to rural overpopulation, which is in turn related to population growth, environmental degradation and potential shortages of food supplies. These in turn are related to the use of energy, especially fossil energy whose use releases carbon dioxide which increases global warming, and other oxides which increase acid rain. The depletion of natural resources will ultimately greatly increase the price of energy as a monopoly rent or as a means of conservation, and possibly also the cost as a result of more laborious modes of extraction.

In the main, I intend to take these trends in energy prices and costs as given, although there are technological scenarios under which some of these difficulties could be mitigated or postponed. Such scenarios include successful atomic fusion, which might, however, increase nuclear contamination, and new means to capture solar energy. Either of these would greatly change the impact of the growth of population and settlements which I foresee, and which I examine in what follows. Since these technological fixes are not yet seen to be clearly in the realms of possibility, I will assume for this discussion that they are unlikely to be available.

The population setting for urban growth

Advanced Western countries like Britain and the USA are highly urbanised, with less than 10% of their populations living in rural conditions, and less than that engaged in agriculture. Even so, countries like this can produce agricultural surpluses, but partly through energy-intensive methods of cultivation. On the other hand, more than three-quarters of the population of most of the world is in rural areas. As a result, there is a population perhaps three times as large as the presently urbanised world population, which is in the process of slowly (and in many cases rapidly) moving into cities. The prospect of such a huge increase in urban population is one dangerous aspect of the immediate prospect, because of the impending pressure which it will put on world resources.

The continued increase of population itself exacerbates these pressures and generates additional ones. The Malthusian pressure of a rapidly growing population increases the danger of famine. In the rural areas this population growth increases pressure on the land, brings marginal new land under cultivation with consequent environmental damage, and increases rural under-employment, which leads to migration to cities. Under conditions of poverty, urban birth rates remain high, so that in the Third World, urbanisation does not lead to a decline in population growth at all comparable to what has occurred in Europe, America and Japan.

From the point of view of population pressure alone, there is an imperative for a level of education, welfare and prosperity which will lead to voluntary control of population growth like that which has occurred in the West. Under proper economic conditions, this possibility is a major advantage offered by the process of urbanisation. It remains to be seen how this will play itself out under some of the further conditions which are discussed below.

The large-scale political setting for development

Given the trend toward urbanisation, the resources which it will consume and the concomitant urge to reduce poverty, economic development is an overwhelming imperative on many different dimensions. In the light of this imperative and the historic shortfalls in meeting it, the political importance of development is perhaps far greater than is commonly recognised. Various scenarios of possible trends in development are easily projected in the light of post-war and more recent events.

Resource imbalances and the breakdown of development efforts lead to anger, frustration and sporadic violence. If these increase on a worldwide basis, then different outcomes may eventuate. Pressures for migration to resource-rich and less densely populated areas may increase. This is already evident on a small scale in Europe and the US. In these situations, both the need for migration and organised resistance to it are politically destabilising.

Similarly, these frustrations amplify the danger of internal and border conflicts, which may escalate into war and civil war. In the event that the pressures have already led to internal instability, there may be an impetus to new forms of political adventurism with unforeseeable consequences; Ethiopia, Somalia, Yugoslavia and even the antics of the radical right in Germany and elsewhere are recent examples.

More serious collapse may lead to widespread famine and, independently perhaps, to disease on scales not observed in recent years. Under conditions of urban overcrowding and widespread poverty, a major international pandemic is by no means impossible, and its impacts would not be confined to poor nations. Viruses like influenza and HIV are good examples of the mutability and dangerous potential of new and existing diseases, and we cannot guarantee that new pestilences will not arise.

In all of these situations, there are excellent opportunities for the implicit political instability to lead to the formation of new movements and the fomenting of new world-scale conflicts. Religious extremism, political crusades and ethnic conflicts at that scale are by no means ultimately impossible.

Scenarios for development

There are thus several composite scenarios for the future development of the Third World. It is possible, but to my mind unlikely, that many nations will stay at low levels of development, but that none of these dangers may reach a runaway level of intensity. This implies high poverty coupled with a decline in population growth and great docility. Continued low levels of development, with the sporadic emergence of new powers and the rise of dissatisfaction and conflict, is a strong possibility which the developed world should not tolerate.

On the other hand, such intolerance might give rise to an effort to maintain the status quo by Draconian measures. This is not viable in moral, political or military terms. Modest development on a wide scale might prevent much of the unrest and epidemic dangers sketched above, but it would probably not resolve the problems of population growth and resource pressure. In any event, moderate development seems inherently unstable, leading either to future lapses or to bursts of speed.

In the end, therefore, the most practical expectation is that world survival is best served by fairly rapid economic growth, of the kind that is now occurring on the Western Pacific Rim, in the style of Japan, South Korea, Taiwan, Hong Kong and Singapore. If such development reached all the people of China, India, Indonesia, and the Philippines, almost half the present world population would be affected, and the impact would be multiplied by the continued population growth in these countries. We now turn to an examination of the implications of such a scenario, as to sustainability under the present state of world resources.

The resource demands of large-scale urban growth

As I have indicated, the potential growth of population, combined with the movement of population to cities, poses an enormous challenge on many dimensions. It is not hard to believe that in 30–50 years the urban population of the world will be as much as, or more than, five times what it is today. If such growth is not achieved with increasing economic prosperity, then the potential political and social difficulties are enormous. But prosperity implies the increasing utilisation, both for investment and for consumption, of non-renewable resources. The potential rate at which such exhaustion of resources might proceed is virtually unimaginable. I discuss this in very general terms, in order to sketch the nature of one set of future problems.

The world economy has experienced premonitory shocks as the result of occasional resource shortages. In the early 1970s the OPEC boycott of Western oil markets threatened a runaway inflation, and not only of oil prices; the rise of relative prices for oil has been avoided in a manner which has perhaps lulled us into a false sense of security. From time to time, natural disasters damage or destroy some crop in heavy demand in the US, with a resulting rise in price which is offset by consumer substitution. Most recently, lumber prices in the US have risen sharply, doubling in spot instances, partly as a reaction to political pressure against the over-exploitation of timber resources in this country.

The increase in lumber prices is a particularly indicative example. World resources of timber are shrinking everywhere, not only from commercial exploitation, but owing to the growth of slash-and-burn agriculture and the use of wood lots for domestic cooking fuel. Yet wood is a principal material for the construction of housing and other structures in the growing cities of the Third World. Substitutes for wood such as steel and cement are made with very substantial energy consumption. Thus, the expansion of cities and their material infrastructures will bear heavily on the non-renewable resources of the world simply for their construction, let alone for their operation.

This operation itself is costly of energy, and produces numerous forms of pollution, some of which have been mentioned above. In addition to contamination of the air and water, cities produce vast quantities of trash and garbage, especially in the West. Recycling all of these residues is becoming increasingly arduous and complex, and complete recycling is at present very labour-intensive. We see some of the cost of this labour being passed back to the consumers.

We thus see a future in which there may be an enormous steady rise in pollution and in the prices of non-renewable resources—most notably energy. We need to place permanently on our agenda the systematic exploration of alternative developments which may forestall the worst of these possibilities. Some of the potentials in this direction are explored in the next section.

Seeking the future of urban investment and consumption

Some years ago the distinguished economist William Baumol (1961) published an economic fantasy, in which he projected that the relative costs of material goods, especially manufactured products, would fall asymptotically to zero, while the costs of services would ultimately consume the entire income of the world. We imagine a similar scenario, but with certain modifications, for the future of world urbanism and consequently for the future of the world economy.

The decline of manufacturing costs has already begun, but unevenly affects different branches of the economy. The general restructuring of the world economy which is now in progress really consists of three different but related trends. First, manufacturing costs are indeed falling, as is in part signalled by the reduction in the workforce taking place throughout industry; but it is still difficult to separate this trend from the general depression in world production. Second, since this trend is uneven, there is a shift to the consumption of goods whose relative cost is falling most rapidly. Third, the release of labour and purchasing power induces a shift to service industries, just as Baumol predicted, but with some emphasis on those services which are best supported by cheaper manufactures (i.e. by electronic means), and on health care services.

The uneven decline in manufacturing costs is worth special comment. A striking argument suggests that if the cost of services provided by automobiles had declined since 1970 as rapidly as the cost of services provided by home computers, the average family sedan would cost less than $100. Similarly, construction costs have continued to rise in real terms, largely because of their dependence on manufactured materials, and in spite of many technical improvements. The most rapid declines in manufacturing costs are in the field of electronics and communications, including the instrumentation of health care; the slowest declines are in the costs of goods whose production requires large inputs of energy—steel, chemicals, building materials, and the like.

It is obviously impractical, indeed impossible, to do away completely with the use of manufactured products which are made with the consumption of considerable energy. At the same time, it appears that many of these goods are in the nature of durable investments: public, corporate and individual. For these types of goods, a programme to improve the durability and prevent the obsolescence of infrastructure, structures, machines and appliances will be imperative. This applies especially to preventing the use of inferior materials and to forestalling the introduction of planned obsolescence. There is an obvious difficulty in the case of genuine obsolescence, but if the goods were more durable and better cared for, then there could be more filtering of machines and appliances, just as there is filtering of housing. Indeed, movable equipment has a larger secondary market than housing, because it is not geographically constrained.

Both durable and non-durable goods must be designed and packaged so that their residues can be completely, easily and inexpensively recycled; and their production would need to use a minimum of non-renewable resources. Since there is substitution all the way from the use of irrigation water to produce natural fibres to the end point where products are recycled for the 'last' time, the calculus of minimum resource use is difficult, complex and technologically dependent.

Ordinarily, the use of resources in consumption is even more significant than their use in producing durable goods. At the outset, we recognise heating, cooling and the movement of goods and people as the major direct consumers of energy. Cooking is a major user of fuel in Third World economies, and as economies advance, the use of energy to operate machines and implements becomes significant as well. In all of these direct uses of energy, the design of the equipment which uses it is of utmost importance. Very large decreases in energy consumption are possible through the redesign of buildings, automobiles, and heating, lighting and ventilating systems, and cooling equipment.

We can imagine that all of these rather direct approaches, based on present patterns in the developed nations, could reduce energy consumption at a given level of living by as much as 50%. Foreseeable technology is not apt to push this much further. Given the continued exhaustion of resources and the potential enormous escalation of population and income, this is not enough. It is apparent that major changes in lifestyles will be required, shifting demand away from energy-intensive consumption and in the direction of energy-benign consumption. Such a shift will require structural changes in the economy, in consumption and even in culture—changes of which we are beginning to become aware, but which have not yet played themselves out. It also appears that these forces may require major changes in the organisation of cities. We will first review some of these possible shifts, and then point to some dangers and difficulties.

Services, communications and urban form

The trend toward the increased importance of services seems in a sense to contradict Baumol's predictions, since a decrease in the cost of manufactured goods should stimulate their consumption. As a result, people would consume more goods and less services, measured in (say) 1970 prices. There are several reasons why the shift which is occurring does not seem to follow this pattern exactly.

First, the labour cost of manufactures has fallen while the resource cost has not. As a result, the figures on employment by sector reflect this trend, while the weight of consumption may not, partly because the unit costs of manufactures reflect their resources embodiment, which generate rents and not wages.

Second, many services are delivered partly through the use of machines, with a double implication. In many cases, the labour cost of services has not risen relatively as fast as Baumol anticipated. For an example, consider the average US retail food store, which is highly mechanised, and where the consumer invests his own time and movement expense to take partial advantage of the entrepreneur's economies of scale. At the same time, services in the entertainment and information fields in particular have virtually zero marginal costs, so that the average cost is low, and especially so in labour, while the income pays for monopoly rents and mechanised delivery.

Third, prosperous consumers take much of their consumption in the form of improved housing, space consumption in the suburbs, and mobility amongst opportunities in the urban region. This is a form of service consumption which is capital- and energy-intensive, and hence does not fall entirely into the pattern foreseen by Baumol.

When Baumol's ideas first appeared, I used to imagine a scenario of life in a prosperous world under the conditions he projected. Material poverty could be abolished since material goods would be very low in cost. Most people would earn their living performing services, and would have the option of greatly reducing their hours of work in favour of increased leisure. Since some day-to-day services would be mechanised, much of people's time could be devoted to recreation, sports and amateur arts such as writing, performing, woodworking, the invention of games, and writing computer programs. Indeed, some people would engage in these activities for pay for the enjoyment of others, and some would devote their leisure time and income to them. The dirty work of the world might be performed by young people under a programme of compulsory community service.

This somewhat utopian scenario founders on the reality of scarce energy resources and pressing worldwide poverty and population growth. There is also, so to speak, another worm in this apple of wisdom: recent gains in productivity are not being translated into reduced hours but into unemployment, and the cultivation of leisure time and the production of services is suffering through the virtues of economies of scale. Entertainment is packaged for the attraction of large viewing audiences on TV, for the production of blockbuster movies, and for the employment of multi-millionaire sports stars and rock artists. The bulk of service employment is not creative, but directed to tedious repetitive tasks which provide an interface between the public and computerised facilities and might better be computerised themselves.

At the same time, the obvious success and the future promise of electronic communications, information processing and dissemination, and enormous computational capacity is giving rise to a different image of utopia, sketched on a different dimension. We can now see the possibility that cities and their production can be decentralised, employees will work at home or in nearby office centres, goods will be produced by automata, and the movement of goods and services will also be automated. Consumers and producers will not need to travel nearly as much, so that congestion will not be a problem and energy consumption and pollution will be greatly reduced.

This scenario has its own difficulties. Personally I find it repugnant because it assumes an enormous reduction in the variety of human contact which people enjoy on the average, it basically restricts travel, and it may be in danger of recreating in a new form, what Marx called 'the idiocy of rural life'. The scenario does offer great possibilities for lowering the volume of future urban construction and the costs in energy and pollution for continued interaction on the large scales which have seemed inevitable in urban metropolises around the world. Finally the shape of this future is clouded with uncertainty, because the supporting electronic technology is still in its infancy, and the form in which the actuality might be realised is in contention and in continuous flux.

Because the scenario has both great attractions and great dangers, and because it is grounded on a view of technology which is virtually certain to be realised in some form, I think that the hypothesised future provides a good framework for research and design regarding the future of world urbanism. The next and final section of this chapter raises some of the issues which will arise in this research and design.

Major issues in guiding future development

It is apparent from the outset that developing this scenario cannot proceed entirely under the direction of market forces. Both urban development and the growth of the electronic industries which underlie this future entail large economies of scale and substantial externalities. Current decisions, wise or unwise, can capture these externalities in a way which pre-empts future paths of growth, and closes off others which might be superior. The embodiment of these decisions is costly and difficult to change or reverse, whether it is in the form of buildings and infrastructure or in the bodily and mental development of later generations.

Similar considerations apply to the development of another important technology which we have not discussed in detail—biological and genetic engineering. This technology promises to revolutionise many fields, from health care through manufacturing to waste disposal. Here the externalities and their dangers are even more readily apparent.

There is another side to the need for controlled and directed growth in these technologies and in the urban arrangements which they will influence. The need for control arises out of the very different time scales of technological change and the change in urban arrangements. The pace of technological development is forced by entrepreneurial competition, and without this competition (exercised either in the market place or in the realm of academic research), we should expect bureaucracy, stagnation, and frustration. Most particularly, the exploration of new alternatives will be inadequate under excessive controls, and irresponsible in their development under minimal supervision. The nature of this delicate balancing act defines a major area for research.

A particular aspect of technological development (which I discussed in a previous paper in this series of conferences (Harris 1991)) is the danger of monopoly in the intellectual development of new technology, as against the advantages of temporary monopolies granted by patents and copyrights in stimulating invention. The enormous scope of these technologies will make this dilemma a pressing matter in public policy research.

With respect to urban arrangements themselves there are many areas of research and planning requiring attention. Sound planning is based in part on a prediction of the outcomes of plans and an evaluation of their impact on the future welfare of people and societies for whom they are devised. In the matters which we are now considering, our knowledge of the behaviours which will be induced by the emerging technologies, and the consequent utilities which they will confer on the users, is largely based on speculation— speculation which is often due to the proponents of the new technologies rather than to informed judgment. Behavioural research in this area is a vital necessity, based on present indications and perhaps a better understanding of basic psychology and social psychology.

This research will go beyond the behaviour of individuals, households and organisations, and will extend to the socially desired forms of future interaction, mobility, education, choice of jobs and opportunities for recreation. As to urban form itself, the issues to be ultimately decided have a wide range. For example, compact settlements may be more energy efficient than dispersed ones solely on the basis of interaction costs, while compact structures may be more energy efficient than free standing homes. The extension of present patterns of space consumption may be technologically feasible but socially undesirable for a variety of reasons. In some countries there may be a conflict between the expansion of space consumption and the preservation of agriculture. A de-emphasis on the movement of people assumes that there will be a decreased need for the assembly of work forces, a decline in the importance of public assemblies, and a restricted circle of family recreational and social activities. Some of these trends are already apparent, but their extent, their limits and their social implications are not apparent.

In brief, therefore, I foresee a long period of rapid technological change and turbulent development in the possibility of new modes of development and new lifestyles. Huge fortunes will be made and lost by entrepreneurs; gigantic new systems of electronic interaction will appear and disappear; presently unimaginable changes will take place in cities—some of them enormously successful and others equally disastrous. While the need to plan and anticipate such changes is all too evident, we are in a position where we have not established an adequate framework for a discourse, either academic or public, on the issues involved. A major part of our responsibility is to establish such a framework, and to avoid generating new solutions and panaceas on the basis of present and immediately foreseeable conditions. I believe that such efforts, which come all too easily to us, should repeatedly provide a basis for new evaluations of the scope of our outlook, and for an appreciation of the limits of incrementalism in an era of great technological, economic and political discontinuities.

24 Sustainable Urban Ecosystems: Working Models and Computer Simulations for Basic Education

Richard L. Meier

Introduction

'SUSTAINABILITY' is an ecological word, while 'development' belongs to all the social and biological sciences. Thus 'sustainable development' means that a path of development is to be taken which can overcome all the known kinds of catastrophes originating from the environment and from internal defects. Another feature is assumed with respect to cities—the quality of life in a city should not fall back to subsistence in order to survive. That experience is defined as a catastrophe. So we plan for a resilient outcome over the long run. The greatest hazards are those of overshooting the supply of resources available; such scarcities are often accompanied by destructive conflict within the urban community. Long-lived communities have stable populations and steady state interactions with the environment.

These aims are a new kind of goal for planning and management. Instead of futures five years hence, or a generation or two, the horizon becomes indefinitely extended, transcending the disappearance of non-renewable resources. It is particularly challenging to think about large modern cities in this fashion, since they consume such large quantities of energy, water, food and materials, while polluting the surroundings with their wastes[1].

Schools of urban ecological research evolved in Poland, China, Australia and the United States, each of them incomplete in covering the necessary inputs or the totality of populations included in the community unit. A comprehensive, systematic synthesis resulted from their convergence at a World Congress of Urban and Peri-Urban Ecology near Beijing in 1988. Unfortunately, the proceedings remain in the archives of the Chinese Academy of Sciences, due to complications inside the government created by the incident with the students in Tiananmen Square months later. The comprehensive diagrammatic checklist, mutually arrived at, has been preserved in the notes of some attendees. Its application to the issues of sustainability over the long run are invaluable, because it calls attention to the full range of possible trade-offs between inputs, outputs and resident populations of actors. Because everything is measurable or countable, a system of ecological accounting is suggested. It permits the calculation of efficiencies with respect to stocks and flows of such inputs as energy, water, urban land, human time and attention, and information/knowledge, whichever is truly scarce.

The strategy for planning for a global metropolis like New York City is most easily illuminated in terms of this diagrammatic model (Figure 24.1).

The community is shown as having a highly permeable *boundary*, as befits an open system, and the interaction is carried on so as to obtain win-win outcomes for transactions. The *yin–yang* symbol was insisted upon by the Asians as an indicator that institutions for the maintenance of harmony with the environment, and between actors, need to be in place. Then the outcomes are win-win-win, and the environment is sustained.

A metropolitan community must export goods and services in order to arrange exchange for inputs. Its wastes are a concern, because they can stifle an urban community very quickly. The waste energy warms cooling water, or is radiated to outer space as low-grade heat. The water is evaporated to some degree, but mostly it flushes away sewage. The population of people *emigrating* should be roughly the number *immigrating*, as adjusted by birth/death rates. Individuals leaving are expected to be better educated or informed than those who are entering. The *materials of construction* go into artefacts and the assembly of the built environment, but what is left over is dumped into holes, or left on the ground as a contribution to 'soil'. In this manner the base of the city rises above the plain at the rate of about 30 cm per century.

The biomass of the city is predominantly human, but there are quite large numbers of animals, domesticated and wild, and *plants*, both cultivated and weeds. They enter into some of the transactions. Structures are inserted to provide *homes* and *support services* for the living populations, although squatters who occupy unplanned sites are common in all these categories.

Vehicles are prominent in the modern city. About a quarter of all ground space is given over to their needs. *Machines* of all kinds are proliferating, but the information processing equipment is currently infiltrating the built environment. The *automata*, called that by philosophers who had to have a place for makers of very simple decisions, were hardly noticed in the 1970s, but now their proficiency in completing transactions is rising rapidly. The most practical means of taking a census of automata is to count the number of pieces of software in actual use.

Organisations are created to expedite transactions. They mobilise members of the other populations, using 20–30% of the time and attention of humans, to speed up the life of the city. They have names, addresses and standing in law as if they were people; their life cycles are both shorter and longer than humans' expectations for survival. Organisations often distil information from their transactions, which is then digested, tested and recorded as *knowledge*. Since knowledge can now be miniaturised, it costs very little energy, space and time to accumulate. Therefore knowledge is growing at rates of 1–3% per year—a rate that cannot be tolerated in the

COMMUNITY ECOSYSTEM

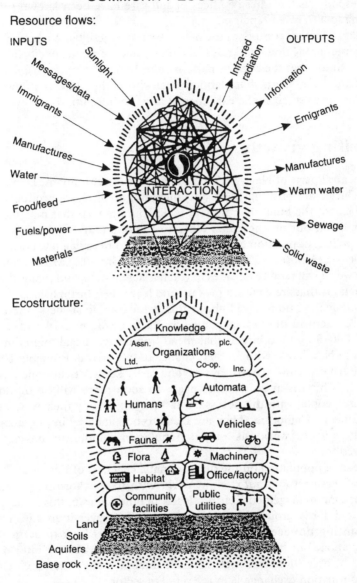

Figure 24.1 Living systems with essential flows and structure.
This is a comprehensive list for what has been learned about urban
communities. It can be used as a check for assessing the full range of the
impacts of a programme or project, and as an indicator of requirements
for a sustainable future.

other populations very long before crowding sets in to inhibit the growth.
(Software engineers will debate this last point, because their product is a
kind of knowledge, and they can set automata to tend to automata, and the

hierarchy will theoretically make no greater demands upon human time. My response is that these edifices of software have been known to crash, leading to a frenzied repair effort.)

That statement completes the necessary introductions. We have a cast of freely interacting flows and roles which play out the drama of city life. What are the limits that the present patterns of action place upon sustainability? What are the ailments that planners deal with in their interventions? What can they *do* to stabilise the existence of large communities?

Quelling growth

The central parts of New York have fewer and fewer humans. Demands of the vehicles are pushing them out, including their waste products. To a lesser extent the built environment is deteriorating, so that many look for better living out on the edge, close to the boundary. The overall total metropolitan population expansion has dropped to only a few per cent per decade—very close to ZPG (Zero Population Growth). Most other cities in the world will require decades to reach a steady state, and many also have populations that are eager to pick up and leave their birthplaces.

While feeling quite good about this demographic success, there is an ominous item in the newspaper reporting from the underworld. It costs $US5 000–$8 000 to hire a 'snakehead' to manage illegal migration from China to New York; with an 80% chance of safe arrival. Probably 100 000 people have already arrived, but no one really knows. Economic development in China is at an all-time peak, so soon there will be millions of Chinese people, with the necessary cash, trying to buy their way into the Big Apple or California. They are likely to be joined by immigrants from elsewhere in the world, as long as population equilibrium has not been achieved there.

Ahead of population on the diagrammatic checklist of Figure 24.1 is the flow of *messages and data*. New York, with its extraordinary financial markets and management centres, leads all other cities in this respect. The estimated 4–6% growth rate in information flow gives it an advantage in pinpointing unnecessary growth. No one knows how to stem illegal immigration, but New York, with its capacity for acquiring information, should be able to defend itself better than almost any other city. Most of this information exchange is in the services sector.

The demand for *manufactures* is not difficult to meet over the foreseeable future. The households in the suburbs and the offices downtown can afford the best quality available.

The prospects for *water* supply are somewhat problematic. The aquifers are tapped close to capacity, and the reservoirs in the hinterland are gradually filling. If the city reaches further out it encounters local residents, who protest vociferously because they would lose their environment. No longer will money buy them out. So New York must learn to conserve water in dry years and recycle its waste waters.

The supplies of *food* for New York seem not to be in jeopardy, because it can draw upon parts of the world in surplus through its rail connections and harbour. However, the Green Revolution is running out of momentum, so there are chances of a world food shortage occurring before the 21st century is a decade old, and the odds are increasing over time. Is it proper for New Yorkers to bid up the markets and deprive others? Moral concerns may impel the City to reduce waste, set up internal stockpiles, improve the efficiency of food processing technologies, and eventually eat less meat and other wasteful foods.

The future *energy* picture is also one of global shortage as the other cities increase their uptake of the cheap non-renewable fossil sources to achieve a quality of life high enough for their citizens to be willing to stay at home. New York has barely started the management of energy demand that should reduce fuel use by at least 30%, probably more when rationalising its fertiliser fixed nitrogen demand and when removing archaic regulation (Schipper & Meyers 1992).[2] Also, it is increasingly dependent upon hydro-electric power, a renewable fuel.

When looking at the populations in the Ecostructure, there is one that is troublesome. The moving of New Yorkers to suburbs has made them dependent upon automobiles. Families move out to the edges, often at great inconvenience, 'for the children's sake', but then the children become addicted to the personal automobile. They cannot imagine a free lifestyle without 'my car.' Strong environmentalists are willing to be hypocrites, if challenged to give up their cars. A great deal of enlightenment regarding convenient substitutes and ecological ethics will be needed to achieve a sustainable metropolis energy wise.

The American bathroom syndrome

Most of New York is already equipped with a classical American bathroom, because the zoning codes demand it. As with the automobile, an addiction became a necessity. American city dwellers see nothing wrong with this, but they do not realise that these features of lifestyle are infectious. The cities elsewhere in the world, which must accommodate another 5–8 billion population, will not be able to obtain the water or the energy. It sounds silly to say this, but students overseas who are deprived of the American-type bathroom and way of life that goes along with it, feel justified to conspire to get to America and overstay their visa.

Americans are not conscious of the fact that they have adopted an unwritten taboo that underlies the bathroom. Other societies have taboos against eating pork, or beef or shellfish, but none of these is as deep, or as wasteful, as the American. The taboo against *shit* is extended to the metaphors and many closely associated images. Breaking a taboo induces nausea, and dark humour in defence, with many other cultural complications. This would not be so serious if the taboos were not written into codes, regulations and standards and exported to organisations building potable water and sewage treatment infrastructure in the other cities of the world. American and European environmental engineers virtually universally recommend treating urban sewage with activated sludge processes, despite their energy requirements and the loss of organic phosphorus and nitrogen. Almost all the World Bank loans for sanitation are channelled into this unsustainable technology, because the alternatives are 'not documented.' The engineers need teaching along with the public. The solar-energised, less capitalised approaches pioneered in Asia should be demonstrated in a place like New York. Interweaving with advanced Western technology can make them safer against breakdown.

Thus the logic of sustainability of urban ecosystems brings us to the need for a working demonstration model of the Asian technology for converting the organic matter in sewage directly into food, while preventing the dissemination of waterborne disease.

Whatever is done will have to have the very highest backing, city, state and federal, because it will be violating code, and would be virtually uninsurable. It would have to demonstrate feasibility, in America and elsewhere, from the technological, political, economic, social and cultural point of view. It should be designed to solve some of New York's special problems as well. It should mobilise at least as many allies as any other City project at the proposal stage. Because it would involve re-education of both professionals and youth, it has to be not only interesting and technically accurate, but dramatic and exciting. It is not possible to review all the considerations required to undercut a solidly established cultural *no-no*, so I will jump to a plausible outline proposal which illustrates what can be done for teaching the implications of sustainable urban ecosystems. It undertakes a 'walkthrough' that an urban designer uses to help translate function into spatial form and pattern.[3]

A working model

The West Side of Manhattan has a hole in it that should be filled in with residences and business compatible with a projected future as a global city. A disused railroad yard adjacent to the Hudson River has opened up a space, facing south to south-west. The original developer's proposals were delayed by controversy ('too much shadowy canyon'), and then the financing fell apart, so new high value schemes are eligible.

The incoming population to New York City is no longer made up of various European ethnic entities, but of Asians, Chinese, Indians, Koreans, Muslims and ex-Communists, mostly belonging to the technical/professional classes, who are able to pay the rents in new apartment houses. A proposal has already been formulated that emphasises the energy economies inherent in their traditional space-conserving lifeways and cuisines. It can be made to fit the local building codes, so there should be minimal delay (Meier & Crane 1978). It could also be organised to teach people to occupy older parts of the City, introducing energy economy. It is an opportunity that fits this site particularly well, and is highly complementary to what comes later, but it is not a prerequisite, since one feature of this part of New York is that it is poorly served by sewage treatment capacity. The load from 8–10 000 apartments and small businesses will need to be met.

The natural place to put the park that serves this population and its semi-deprived neighbours is 9 ha (approx. 21.5 acres) adjacent to the shoreline. Unfortunately, it is crossed by a raised freeway, an artery that is probably too expensive to move. It is proposed that this park contain a working model that is a major tourist attraction (Figure 24.2).

Working Model and Computer Simulations
NEW YORK CONSORTIUM OF MUSEUMS

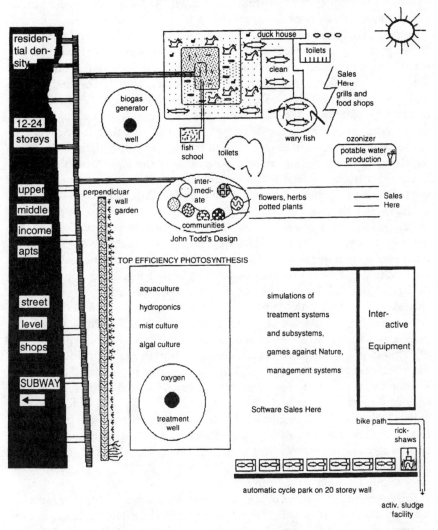

Figure 24.2   A working model and computer simulation of a sustainable urban ecosystem

The publicity blurb says that it is designed and managed by a consortium of leading city museums to be as much as possible a hands-on facility, like the Exploratorium in San Francisco and Chicago's Museum of Science and Industry. You cannot get there by taxi or car. The approach starts at the subway station at 66th and Broadway. From there one either rents a bicycle (several varieties are available), hires a cycle rickshaw, waits in a queue for an electric van, or walks a kilometre. Explanations are given that this design extracts experience from all over the world, in order to gain the greatest economic yield for a city.

Once there one sees a colourful helical pond (Korea, Berkeley), from which fountains occasionally spurt to help oxygenate the effluent, over-arched by plastic and glass transparent promenades (from the *Futurist Monthly*). Several species of ducks and geese (Peking) are on the pond, which changes in colour from grey to dull green as the flow moves outward into wider, shallower channels. There the activity of fish becomes evident. Feeding time is announced with a bell, the fish stand on their tails in the water waiting. Visitors fling out pills of feed they bought from a slot machine. Golden and silver carp jump into the air, sometimes a half metre high, to catch the pill on the fly (Taiwan). The ducks try to get their share, but profit mostly from the pills that are missed. The water churns. This is the epitome of feeding frenzy. The package explains that the pellets are made from selected restaurant waste to provide a healthy diet for the fish, which in between feeding times nibble away at the algae in the effluent. Since the amount of feed allowed, and the number of visitors, are both variable, the price charged for the privilege of feeding is also variable.

High above the pond at representative points are large faces for pollution reading instruments (New Jersey), which show how fast the sunlight cleans out the sewage (two to four days depending on weather). At the end it reaches a level purer than river water, where the fish are netted and the ducks and geese go off to their nests. The netted fish are dropped into a pool bubbling with air for a few days to become 'purified'.

Beyond is a pool with 20 positions around it, from which people of all ages are trying to catch fish. This is no mean task, because these are intelligent fish. They have been trained in their youth to avoid typical kinds of bait and lures, and have been tagged accordingly (Japan). The fish that have been caught can be grilled immediately, but if you are Japanese or Korean you may wish to have the fish you catch sliced directly to eat it raw with sauce. At the next kiosk you could order a duck egg omelette or a curried slice of goose egg, with fresh vegetable trimmings (Indonesia, India). Around the corner is the 'fish school' where the fingerlings are being trained to avoid lures, receiving a small electric shock from a transparent grid when they get too close (Japan). A trained fish could run a gauntlet of lures without setting off the buzzer.

Adjacent are exhibitions of alternative designs. John Todd, for example, developed a process by which higher plants can clean out the sewage. The effluent runs through a sequence of tubs (6–12), each with a more advanced ecosystem and a recording system showing pollution level, the last of them yielding water fit to drink. This is for people in north-east North America where residents are highly suspicious and have queasy stomachs. They are willing, however, to buy flowers and plants to take home. An ozoniser would guarantee the purity of the water in the drinking fountains, without creating chlorinated hydrocarbon traces as by-products

The toilets are out on the edge. They have posters requesting a contribution to keep the ecosystem going. Users are informed that this overall system for sewage treatment has been retarded when subjected to overflow crowds, which have left the district stinking. The threatened overflow in this case is delivered to a deep well treatment system (United Kingdom) that is very economical of this high-valued land. A similar deep well system should be able to produce biogas.

Over there, against the 10–20 metre cliff, is a perpendicular garden. The pattern of vegetables is a piece of art. The south west exposure produces vegetables and small fruits earlier than any normal garden. Window washing equipment has been borrowed from a nearby skyscraper to take on the harvesting chore. Seeds with fertiliser are embedded in foamed polyurethane sheets according to the prescribed design, and are then hung up on the cliffside with pond water trickling over them (Canada). The vegies are sold on the spot.

Adjacent are demonstrations of the most efficient ways a city can produce food in the long-term future, given the high price of land (and, therefore, sun). Hydroponics (California) can produce a variety of vegetables, and mist culture (Israel) will yield a little more. Aquaculture of shellfish (Japan) could add different kinds of delicacies with high efficiency. The algal culture could supply a green sherbet, or a yogurt (Japan).

The kids are hanging out around the gaming tables, playing games with Nature, redesigning simulations of ecosystems, and working out ecological strategies to fight off pests. Most will buy software, rather than picture books. The Japanese and American designers of computer games have very much downplayed this theme in favour of trivia. They are here a by-product of the computer simulations needed to design the original facility with its built-in automatic controls.

One cannot forget the bicycles in this scene. Paths segregate them from pedestrians, but where can they be parked? The land is too valuable to install standard parking. The best suggestion is to entrust them to a facility like the moving rack at a cleaners, but make it perpendicular. We need something that can go 10 storeys high to take care of the demand, storing thousands of bikes, and scores of cycle rickshaws for all kinds of weather.

Near the parking facility is a directional sign for a side trip along the river. A couple of miles away a standard old-fashioned sewage treatment plant employing the activated sludge process stands out into the river. In Vienna the artist Hundertwasser repainted the ugliest downtown infrastructure in very bright colours and surprising patterns. New York has not yet found an artist to make their mark.[4]

The not infrequent spills of partially treated sewage remain undocumented, because it was too uninteresting for eager environmentalists to keep watching. Visitors would get the practical idea, however, for encroaching upon the Hudson River surface to expand the ecosystem in novel ways out from the shoreline, so new things can happen every year.

What happens next?

We see here a well elaborated, high impact concept that illustrates the principles of urban ecosystems and aims at overcoming a cluster of attitudes that obstruct progress toward sustainability. The concept should be linked to some of the more immediate goals of the city planners, such as waterfront development. Then a team is assembled from the mayor's office, and from both the museums and the planners which would work out the sizes, the connections and the layout, and set up a Request for Proposals (RFP) for consultants or internal design teams that leads to a specific design. If the Planning Commission and the mayor's office are really enthusiastic, the RFP would lay out the procedure for a Design Competition which would include preliminary costing.[5]

There is a very good chance that, if credits are given for the marginal cost of providing standard sewage treatment, and every possible means is used for collecting revenue during peak periods, this working model can quickly make a profit, and thus pay rent for the public park on a sliding scale. It could be leased to a private corporation that promises to franchise designs for cities with other climates and cultural combinations. Sustainability of urban ecosystems depends upon a high rate of acceleration and multiplication of this facility, along with the development of many variants that fit local circumstances elsewhere in the world.

A very large share of the new urbanism in the developing world should be equipped with this kind of waste disposal. It may not be designed to provide high drama, but it should be associated with communities that process foodstuffs and prepare delicatessen items for the metropolis. People can live and work there in clean surroundings that allow mixing traditional ways with the very modern procedures for producing packaged foods.

Making urban space—the aquatic frontier

As great cities try to find their way to a sustainable future they encounter another grave shortage, which is a major impediment to resource-conserving urban development. They need land suited for waste handling and sewage treatment facilities, international airports, container-handling docks, residences simultaneously accessible to the central business district and to recreation, and many miscellaneous purposes. In the 19th and early 20th centuries, cities filled adjacent tideland with construction debris, dredgings and shaved-off hills to make space for commerce, shipbuilding, warehousing and some related manufacturing, but the space that is economical to fill has been developed, and the remaining coastal areas have been claimed by the environmentalists.

Open space lies only a few hundred metres offshore, the only claimant being navigation lanes. The latter could be re-routed more narrowly with better instrumentation. Technologies are known for building, or floating, platforms, and for constructing marinas for servicing houseboats and apartments-on-barges. Insurance charges would be higher, but recent bombings in New York and London will bring them closer together. A result of the scarcity of urban land is an excessive price for sites, especially those that are just inshore. New York could use space at the edge, and just beyond, to good advantage, but it is not truly essential because its growth is topping out.

Aquatic urban space appears to be crucial in any growth model that lays out a future for the expanding metropolises of the world. Jakarta and Surabaya were used as surrogates for such metropolises as Guangzhou-Hong Kong, Shanghai, Dhaka, Manila, Bombay, Karachi, Pusan, Bangkok, Singapore-Johore-Bataam, Alexandria-Cairo and Ho Chi Minh City to explore possible advances to the sea (Meier 1974; 1976; 1980).

A pent-up demand for life on the water is discovered by exploring the coastline from the water side, where in every great estuarine city there are visible niches for thousands of illegal residents. They are, in effect, squatters, although they hold respectable technical and professional positions in the city.

In no place has an organised experiment been undertaken as yet for the modern settlement of aquatic surfaces.[6] Each culture has a different taboo which has roots in its history. The future need is evident, the opportunity exists in many places, but who will take the lead?

Education for sustainability

All the reviews of fertility reduction in the poorer parts of the world suggest that it is not happening fast enough, except possibly in China (where the measures taken may be too extreme, thus incurring a major backlash when political discipline is relaxed). There is new evidence suggesting that limitation of family size occurs very rapidly when girls and young mothers reach secondary education (Caldwell *et al.* 1992; Levine *et al.* 1991).

The possibilities for setting up a working model for the accelerated education of girls and young mothers must depend upon some independent growth process in communications in the poor and isolated parts of the world. To be significant, the educational program must expand rapidly regardless of politics, economics and culture. For the first time in human history there is such a change under way, and its potential is not yet being used to real advantage for development.

The decade of the 1990s will see the majority of the world's poor gain access to television, just as the 1970s opened up the era when radio became broadly accessible. By the turn of the century, the television manufacturing industry is of the opinion that two-thirds of India and China will have access, and quite a few of these viewers will be from districts not yet electrified. Television becomes an obsession for the poor in virtually all the societies that it invades—unless the message becomes too authoritarian.

Until recently the program content was censored, but the censors are stymied by the ability of relatively free societies to present images of high quality across international boundaries. A majority of TV sets now sold are displaying colour, despite the extra cost. Quite a few of them are from districts that have not yet been electrified, and so are forced to use batteries. Television in rural areas is believed to increase the confidence of villagers that they can survive in the city. Soon many of them proceed to work out means of immigrating. So the future residents of the city will very often in the future have been already introduced to television. The medium should be used consciously to make the transition easier, particularly for young women.

It should be noted that television of the future will be more than it is today. The personal computer, starting in 1994, will able to manage the television screen, bringing with it a keyboard, sketchpad and simple computing. Tapes and disks are already copied and smuggled all over the developing world. Object-oriented programming is much more easily comprehended by the less literate portions of the population. Yet the cost will be about the same as at present.

Here the planners may expect to have access for the first time in history. What can be done with it? The experiments should be designed soon, and be ready to proceed on a hundred million dollar a year scale by the turn of the century, expanding by 10–20% per year, in order to have the needed impact.

The primary goal is rapid introduction of young women into the educated stratum that exists within walking distance by now virtually everywhere in the world. The highest payout for everyone in the next generations is low fertility, but for now it should generate excitement and HOPE.

At this very early stage what can we envision as the steps to the grand experiment?

1 For a hundred or so subcultures pick a set of ten stages of personal advancement, starting from simple numeracy going all the way to composing a written report on an incident in community affairs and its contacts with the outside world.
2 Create short scenarios that illuminate problem-solving processes, and a set of problems of apparently equal difficulty; they would replicate standard school problems presented in abstract or literary form within the cultural context.
3 Choose groups of five, plus or minus one, to work with the new TV set, starting from roughly equal backgrounds. They gain confidence in their own capacities, starting from being a part-time operator and a full-time

participant. Social problem solving is included (e.g. how to reduce the likelihood of wife and child battering, how to win a game against the weather in growing a crop, responding to infants in such a manner as to enhance communication and future educability).

4 Although collaborative problem solving predominates, print-outs can be obtained for individuals as a substitute for report cards. By this means each student should acquire an updated portfolio that shows qualitatively the level of achievement.

5 Allowance should be made for fast learners to accelerate their development by joining more advanced groups. They should also be granted opportunities of formal schooling.

6 A very delicate issue must be faced: similar opportunities must be offered to the much smaller number of boys who are deprived of education. That is a community level decision which is likely to have many different resolutions within the same culture.

Experiments of this kind need discussion, but there is little relevant experience due to the newness of the technology.

Conclusions

Some essential transitions in technology and social organisation to reach sustainable futures for cities are held back by strongly entrenched taboos and prejudices. A promising means of overcoming them is to dramatise the fact that there is nothing to fear. The next generation can play with the new technology without inhibition, and learn how to take best advantage of it.

A celebrated public conversion of sewage into food using solar-stimulated bioprocesses would save water, some energy and scarce materials, such as fixed nitrogen and materials and soluble phosphates, and require very little of the scarce urban space.

A means of expanding urban space so as to reduce the costs of crowding, is available for cities adjoining bodies of water. Although the technology allowing people to move out on the water surface has been available for decades, it has been prevented by archaic regulations regarding shipping and taboos left over from centuries ago reinforced by recent preconceptions of environmentalists. Therefore this opening into the neighbouring empty spaces has not yet been realised. The price of downtown and coastal sites has risen to such a high level that experiments in moving out on the water should save expenditures in infrastructure on the part of the city, so all citizens would profit.

The standard procedures for expanding educational systems which teach numeracy, literacy, absorb knowledge from a library, and to write a simple report are too expensive, and, especially, too slow. We need to invent new ways of transmitting the experience and information needed for women to participate fully in the urban community. Then we can be confident that the family size will drop to levels close to replacement, as it has already in about a third of the world. Brave and promising attempts have been made in most cultures—they make up a series of dramatic and heart-warming stories—but even if multiplied tenfold, and backed with resources as large as the World Bank, planners cannot see a sufficient impact to reach sustainability for urban ecosystems. Meadows *et al.* (1992) have to invoke a miracle in population control in order reach sustainable conditions, but they offered no strategy to bring it off. We have to look around for the leverage that would be a 'substitute for schooling'.

NOTES

[1] Calls for expanding the outlook of the planning profession in this direction were made in a series of contributions entitled 'The Coming Global Metropolis' in the *Journal of the American Planning Association* (Winter 1991). An attempt is made here to offer more explicit proposals that overcome critical barriers that lie between cities of today and a sustainable future.

[2] Fixed nitrogen is a key input for steady state urbanism, both in foodstuffs and materials. At the present time about half the nitrogen is fixed by sunlight and soil bacteria, and the other half by chemical processes using electric power. (China is already dependent upon fossil fuels for 60% of its fertiliser nitrogen (Smil, 1991).) It is lost in agriculture due to evaporation of ammonia and to decomposition to gaseous nitrogen. The activated sludge process puts 70–90% out of the cycle for reuse, whereas it is expected that the working model using solar technology would lose no more than 20%. This is one of the best energy savings in prospect for the long run.

[3] The basic idea here is borrowed from the fishermen of Muhally in Calcutta. In 1961 a group of riverine fishermen acquainted with the sewage fisheries of Calcutta rented 70 ha of waterlogged land. In 1980 they undertook a redesign based upon an ecological appraisal by Dr Dhrubajyoti Ghosh, and later, under supervision of Dr Sumita Sen, documented the inputs and outputs as in Figure 23.1. The Muhally Fisherman's Co-operative added waterfowl, planted 120 species of higher plants to stabilise the bunds, and even imported deer to browse the plants they preferred. Without public subsidy, they built facilities for teaching school children about ecology and for recreation on their ponds, calling it an Eco-Park. The initial design for a facility in New York may depend upon data they collect.

[4] Hundertwasser is a colourful, energetic architect-artist-urban designer who has made a striking impact upon Vienna. His style for highlighting essential urban infrastructure, like power plants, old factories and gasworks, is seen in the handbills for city tours. His irreverence for refinement in standard urban facades is recorded in *Das Haus Hundertwasser*, Hundertwasser, Osterreichisher Bundesverlag and Compress Verlag, Vienna, 1985.

[5] A world city has a plethora of talent on hand, all seeking prestige. A consortium of museum directors (from, say, the United Nations, the American Museum of Natural History, the New York Botanical Garden, the New York Aquarium, and the Museum of Modern Art, which are world-class institutions) could take on responsibility for planning and implementation. Willoughby (this volume) has noted that, despite congestion and high rents, projects in biotechnology are expedited quickly in downtown New York, and the total cost is frequently less.

[6] It is necessary to differentiate this outlook from the colourful macro-engineered sea-cities published by the engineering companies of Japan and reviewed in *The Futurist* (Conway, 1993). Those designs are not tested for cost-effectiveness or marketability. Occasionally their capacities are put to test, such as for floating runways for a round-the-clock international airport to decongest the airspace over Japan, but usually they overshoot the low- to middle-tech needs for urban space.

25 Cities and Complexity: Implications for Modelling Sustainability

Michael Batty

Introduction: sustainability and complexity

In her seminal book *The Death and Life of Great American Cities*, Jane Jacobs (1961, pp.428, 433) wrote:

> Merely to think about cities and get somewhere, one of the main things to know is what kind of problems cities pose, for not all problems can be thought about in the same way...Cities happen to be problems in organised complexity...They present 'situations in which half a dozen or even several dozen quantities are all varying simultaneously and in subtly interconnected ways'.

Over 30 years later, we are still grappling with the enormity of urban problems and the invention of an appropriate logic for their understanding and management. Throughout this book, the difficulties of inferring causes and effects from aggregate analysis have loomed large and the ambiguities of unravelling the impact of spatial competition upon energy and travel patterns, on local and global economic development, have been demonstrated time and again. Frequently, aggregate analysis conducted on static cross-sectional data has yielded contradictory conclusions with respect to the patterns and processes so necessary in the understanding of urban problems. Moreover, much of our urban analysis and theory is still predicated on ideas which ignore the fact that cities adapt continuously and rapidly and display a level of balance, diversity and resilience which indicates deep and subtle but organised, hence understandable, complexity. The idea that we might be able to steer cities towards some 'more sustainable' future by employing the blunt physical instruments of the past flies in the face of what we are slowly but painfully learning about the nature of social systems in general, and cities in particular. In this chapter, we will attempt to highlight some of these dilemmas referred to in this book and map out possibilities for their better resolution.

In essence, we will present a critique of contemporary approaches to sustainability and spatial competition which, we will argue, are embodied in a paradigm which defies our emergent knowledge of how cities adapt and change. In the last 30 years, there has been a remarkable shift in viewpoint within systems theory and the study of system dynamics. Traditional analogies of social as physical systems have been replaced by analogies based on biological systems, in this sense mirroring the shift from problems of disorganised to organised complexity first mooted in 1948 by Warren Weaver (1967) and referred to so presciently by Jane Jacobs (1961). This change in approach has profound implications for how we might develop our understanding of cities; it suggests that we must have much greater concern for the definition of our system of interest, of the way it interacts with its environment, the way it is organised hierarchically, and the way elements compete with one another in time as well as space. Traditionally our understanding of cities has been based on aggregate observations, on macro and global properties pertaining to distribution and location, but it is increasingly clear that those aggregate patterns are consistent with many different types of local structure, that the way those aggregate patterns arise is but one manifestation of the underlying mechanisms and processes at work, and that for any successful study of how cities change, compete and adapt, more micro concerns must predominate.

We are painting an extreme picture in one sense, for our thesis here will not deny the usefulness of contemporary and traditional aggregate analysis; in any case, this is our starting point. Nevertheless, a refocussing on local processes which give rise to the aggregates we observe imply new approaches to organised complexity which are beginning to have an impact on many sciences, particularly economics (Anderson *et al.* 1988). These new ways of approaching systems suggest that problems such as the emergence and impact of new technologies on cities, the way novelty and innovation pervade the urban system, and the role of historical determinism, can now be managed within the explanations we seek. And to this end, ideas about sustainable futures based on lowering energy use, denying growth, seeking alternative technologies and reverting to past, seemingly simpler ways of organisation, no longer appear credible, no matter how laudable their motive. Here, we will suggest ways in which new varieties of analysis and model might be developed for explaining and forecasting the spatial structure of urban systems.

We will, of course, sketch out the rudiments of this evolving systems approach but before we do so, it is worth noting some very striking conclusions from various contributions already presented in this book which give fuel to our fire. Much of the research reported here involves the crucial relationships between energy use, travel patterns and commuting, urban densities and the suburbanisation-decentralisation of urban populations from monocentric to polycentric city forms. The general notion of the sustainable city in this context is one in which energy use might be lowered, primarily through the reduction of trip making and/or the use of different modes of transport. The age-old notion of effecting such change is one of making cities more compact, of reinforcing the central city, which is built on the view that when people spread out in space, they use more energy for travel, and habitation generally. But much of the evidence produced here (Breheny, chapter 21 this volume; Cervero, chapter 17; Gordon &Richardson, chapter18; amongst others) is mixed, if not contradictory of this idea.

Moreover, most of these analyses have been conducted in a fairly narrow context in which income levels with respect to location, the substitution of different energy uses for one another, increasing levels of interaction, changes in the fuel effectiveness of various transport technologies, and a host of other relevant factors, cannot be accounted for. There is, in fact, strong evidence that by changing the system of interest to embrace a wider range of factors, the conclusions concerning energy use and hence sustainability change radically. And this is without considering more global issues relating to competition between cities, the substitution of information for energy-based technologies, and rapid rises in productivity being occasioned by organisational as well as technological change. In short, there is increasing evidence that where we define the system of interest in traditional ways, we face an ecological fallacy in which our conclusions can be contradicted by quite small changes in definition. This suggests the need for new approaches of which those associated with the 'new systems theory'—the science of complexity—are designed to address.

To this end, we will first sketch traditional ways of understanding cities using systems theory and then focus on the difficulties and dilemmas in developing theory which is truly dynamic in temporal terms. This is not only because of the difficulties in observing and recording urban data in time but also because of intrinsic difficulties in theorising about appropriate processes. We will introduce the biological analogy which now represents the dominant paradigm of systems theory, linking this to the emergent science of complexity from which we draw new insights concerning sustainability. We will particularly emphasise the notion of important events in urban change as being transitions at the boundaries of systems and environments and then discuss how these new theories might represent the local-global nexus which is one of the most perverse themes in understanding the changing urban system. Lastly we will briefly sketch some of our own work where we are developing a global (but spatial) economic-demographic model in which the world cities feature as one of the core organising concepts. We are not well advanced in the development of this model but we are able to sketch its design and show how it draws on some of these new principles of systems theory as well as new types of world database and visualisation media in which this work can be accomplished. In short, this example shows the importance of defining the system of interest in terms of its temporal dynamics as well as its spatial structure for understanding how world cities compete with one another and rise and fall in the global urban hierarchy.

The traditions of general systems theory

Systems theory had emerged as a rudimentary body of formal thought by the mid-20th century, largely motivated by the need to apply scientific principles to problems which could not be cast, in Weaver's (1967) terms, as either those of physical simplicity or 'statistical' complexity. In fact, many of its early adherents claimed that the most relevant of scientific problems were those which only the new theory could inform. Biology did influence the emergent theory but this was the biology of classification and the more speculative mathematical variety pioneered by von Bertalanffy (1968). The biology of genetics and evolutionary adaptation which has dominated the field for the last 40 years was in its infancy and it is only now that systems theory is being rewritten in its terms. Thus the theory of the mid-century was much more like classical and statistical physics than modern biology and this determined the analogies it provided for social systems. The theory which emerged was quite widely applied to the softer social sciences, particularly social psychology and political science; and with respect to urban phenomena generally and cities in particular, location theory emanating from quantitative economic geography and optimisation theory from operations research provided two of its most important constructs (Harris 1961; Berry 1964).

Systems were conceived as having structure and behaviour, equivalent in economics to the idea of stocks and flows. However in urban and economic applications, the emphasis was on understanding, modelling and predicting their structure rather than behaviour. The evolution of such systems in terms of their temporal dynamics was implicit at best, largely due to the sheer difficulties of both recording and observing such change as well as the more perverse problems of identifying and articulating the relevant processes of urban change. In turn, structure was articulated in terms of hierarchies of functions, the most coherent of which involved the notion of subsystems whose interaction both within and between levels provided an elaboration of the framework of the system. The relationship between the system and its wider environment was also of some import in forcing recognition of the wider context.

The way in which such ideas were developed for urban systems was based on a much reduced version of this general theory. By and large, the models built were static, comparative static at best in which system structures were compared at two distinct and often distant points in time. There was little emphasis on how the transition from one state to another was made in time and this belied a complete dearth of theory concerning urban processes. This has always been problematic; such analysis was usually predicated on the basis that such models might be used predictively, to evaluate the future state of the system and to be used in association with policy making and planning based on the realisation of public goals, such as urban sustainability. But this type of urban analysis, developed over the last 40 years, is inconsistent with theorising about urban change, for the mechanisms of such change are nowhere addressed in the models which have been developed. This problem is apparent from the lack of success in practical use of such analysis although it should be noted that where dynamic models have been attempted too, the experience has been equally salutary.

The emphasis on statics in existing approaches has also been problematic due to the asymmetry between system and environment. The environment has been regarded as a 'catchall', akin to the modeller's garbage can in which extraneous and other difficult to simulate system features have been placed. This is graphically portrayed in input-output models where such environments are termed 'rest-of-the world' sectors. In fact, the system environment has often been assumed to be both physically and temporally distant from the system although in practice this is rarely the case, having equal but different importance within the entire system nexus. One element of this emerging science of complexity is that the real action in any system takes place between the system and its environment, something which has not been the focus in the previous and current generations of urban systems models. The idea of an hierarchically organised system has rarely been followed through either, apart from some organisation of systems into subsystems at fixed levels of resolution or aggregation. Where hierarchy has been used, it has been as a mathematical convenience for simplification or as a means of taking account of missing data. Even where there are well-articulated principles of hierarchical organisation such as in the retailing system, most models which have been built concentrate on flows of activity at a fixed level of hierarchy and make little attempt to generate the kind of diversity and competition which exists across the hierarchy.

There are many important reasons for these limitations of systems theory in urban analysis and simulation. These relate to our difficulty in observing and measuring urban phenomena, in observing processes which take place at the individual or micro level but only manifest themselves at the macro, even in articulating the configurations of factors which determine how the critical decisions are made which change the urban system and its spatial structure in time. And there is also only now a dawning realisation as to how complex this phenomena actually is and how ineffective the approaches of the past have been in unravelling explanations of spatial structure. This is clear from the usual types of analysis which seek to explain urban change which is comparatively static in nature, for example, based on comparing the spatial structure of cities and regions between census dates. Processes of change have to be inferred from such comparisons, and assumptions that the implied trends are continuous in some sense are usually made. If discontinuities are detected when more than two snapshots in time are available, then the assumption is one of a trend reversing, rather than of any comparative discontinuity in system behaviour taking place. In short, the process of analysis is confounded by the aggregation problem in which dominant trends clearly mask the diversity and continuous variations at more micro levels. The fact that surprising events may be discerned from macro trends is unlikely; there may well be surprising things occurring in urban systems but their appearance at the more aggregate levels is likely to be as much determined by the vagaries of aggregation and smoothing as by actual changes in the underlying dynamics of the system.

When it comes to the development of policy associated with such static aggregations of the urban system, there are profound problems of both identifying appropriate instruments and in their implementation. A cardinal principle of systems theory to which only lip-service has been paid in the past is Ashby's (1956) 'principle of requisite variety'. In essence, Ashby argues that to control or plan some system, that is to steer it towards explicit goals, then there must be as much variety or detail in the controller as there is in the system itself. This suggests that before any policy making is initiated, the designer must represent the system at its appropriate level of aggregation so that appropriate policies might be devised. If the representation is pitched at too aggregate a level and in static terms only, then policies pitched at the same level are likely to be ineffective as the system will not be working towards real goals, only those which appear to pertain to the aggregative, hence unrealistic, level. In urban planning, it is very clear that attempting to impose aggregate policy instruments such as measures to contain growth which are not intrinsic to the way the system operates in terms of locational choice and decision making, is not only inimical to the actors so affected but is also likely to exacerbate the problems confronting the system. This is over and above the problem that the goals of the system are often conceived in terms of long-term static solutions or blueprints to be reached at some future state using instruments tantamount to brute force. The traditional instruments of city planning fall into this realm, e.g. those such as green belts, new towns, certain types of zoning control and some types of local economic development.

The debate over the development of sustainable cities is largely predicated in these terms. Sustainability, say Meadows *et al.* (1992; p.209), implies a system (state of society) 'that can persist over generations, one that is far-seeing enough, flexible enough, and wise enough not to undermine either its physical or its social systems of support'. This definition translates into a concern for measuring the performance of the global system in terms of population, capital and technology, and the resources necessary to sustain the system within acceptable limits. In city planning, a concern with efficiency and equity has been broadened to embrace ecology and energy. For obvious reasons, with a style of analysis and policy making largely pitched at the aggregate level, problems of effective planning to realise sustainable cities in these terms are increased above those which presently characterise the activity. In fact, as we implied earlier, sustainability is mainly predicated in terms of the energy demands associated with different spatial structures. All the evidence suggests not that energy is unimportant with respect to different settlement structures but that the level of analysis at which this type of research takes place is inappropriate in picking up the detail of the phenomena in question.

If the system is changed in even quite small ways, the conclusions also change. For example, in much of the work on examining the energy consequences of different patterns of trip making, there has not been any analysis of the different ways in which energy demands vary with respect to the time-budgeting of trips. Nor has there been any examination of the changing fuel efficiency of automobiles, or the effect of substituting information flows for physical movement, all set in the wider context of massively increasing demands for interaction associated with greater prosperity and opportunities for new types of communication. There is a distinct possibility that if compact cities were to be developed according to the ideas of the sustainability movement, these would have much the same energy demands as contemporary urban forms. It would be virtually impossible to change the demand for interaction, short of massive investment in alternative forms of energy-efficient transit, severe controls on the private car, and draconian policies on energy pricing. Doubtless urban location patterns could be changed if these types of control were instituted, but it is yet a further comment on the poverty of our analysis that we have little idea as to how cities would be changed in the face of these policies. All this suggests the need for a change in approach to more micro, more dynamic conceptions which underlie our analysis.

Urban equilibrium and dynamics

The idea of equilibrium is deeply ingrained throughout the physical and social sciences and is a product of the search for an ordered, stable, and hence understandable world. Such a world at any scale or level appears increasingly illusory as we learn more about its structure and behaviour, and there is a sense now in which we must move beyond the idea of equilibrium to make progress. Equilibrium, it seems, is often confused with stability, resilience and permanence, which are the more likely properties of the city systems that we observe. Cities rarely blow themselves apart although they continually change and admit innovation. Moreover, the emphasis on equilibrium brings with it the notion that change is smooth, and continuous, that the system manifests itself in terms of underlying trends which if we can validate over sufficiently many time periods, can be used to predict or chart the course of the changing equilibrium. In this sense, we associate the idea of equilibrium with convergence.

All these assumptions are embodied within state-of-the-art urban analysis. For some years however, there has been substantial critique of this dominant approach (Harris 1970). Discontinuities in aggregate trends, or rather changes in the direction of trends, can sometimes be discerned, although in the urban domain these are muted. Some cities, for example, have achieved economic renaissance against all the odds so it might seem, some show counter-intuitive changes in the patronage of transit systems, some reveal surprising pockets of affluence and poverty in places where theory would suggest they should not be. But these are the exception rather than the rule. Discontinuities have been easier to observe in terms of the way growth occurs within cities, not simply by constant accretion but by rapid and lumpy development as in the emergence of 'edge' cities. To handle these notions, analysts have turned to deeper and richer models of dynamics such as those based on catastrophe and bifurcation theory, although these have still been developed in mainly aggregative terms (Wilson 1981; Dendrinos & Mullally 1985; Dendrinos 1992).

The equilibrium perspective is as much confounded by the treatment of aggregates as it is by the lack of appropriate frameworks for dynamics.

1 Whenever we aggregate detail, we smooth variation. Things thus look more regular and smoother, and manifest greater continuity than they actually show at the more micro level where the processes which determine their form actually operate.
2 Whenever we examine cross-sections in time, we intuitively interpolate continuously, assuming smooth trends.
3 Whenever we examine structure rather than behaviour, we unwittingly give more weight to historically determined stocks and locations. In most western cities which had strong industrial bases and which are now rapidly transforming to the post-industrial context, urban forms may appear to be dominated by their history although their dynamic is now quite different from that which determined earlier spatial structures. For example, even in Melbourne which by American standards is still strongly monocentric, only 21% of employment is located in the central core city. Most is in what we ubiquitously refer to as suburbs (Brotchie, Anderson & McNamara, chapter 20 this volume). This suggests we should be differentiating the suburbs much more finely than present analysis employs. The more general point however is that we must find ways of dealing with the historical persistence of spatial structures that are more closely in tune with the dynamics that we now know from casual experience and observation are in place but which we also know did not create those structures in the first place. This again implies the need for a much finer level of analysis at the level of the basic system dynamics. On aggregation, such analysis would provide the kinds of trends that we observe but also imply a rich and diverse set of processes accounting for the complexity we know is present.

There are several types of dynamics which we know take place in urban systems but which are difficult, if not impossible, to observe in macro trends. There is purely random change in which locational decision making is incidental to the real processes at work. More important are innovations which are surprising or novel and these often show themselves in locational change which is rapid, perhaps discontinuous, where new types of centres and organisations suddenly reveal themselves, and which we find hard to explain from the macro view. If the analysis is sufficiently rich and detailed, then these types of change may not be quite so surprising. Or at least, the dynamics required to generate them need to be explicit so that their emergence is possible, likely, but not necessarily predictable in time or space. The problem in any analysis is defining the system of interest, and focusing upon the critical variables and parameters which enable us to make sense of the phenomena and generate meaningful explanations and responses.

There is also technological change which is hard to account for in the evolution of urban systems. Of course such change has to be assumed but its impact should be predictable once the technology is known. That is, the models we work with should be sufficiently rich to enable us to generate the impact of new technologies which show themselves in terms of rapid changes in spatial structure. The impact of telecoms on location is perhaps one of the best examples, where profound changes in media and hardware as well as in the regulatory environment are clearly having an enormous impact on the way people make routine decisions concerning work and leisure. But we have little idea as to their impact on spatial structure because the models and analyses we use simply do not operate at the level of detail in space and time that is necessary in the construction of this bigger picture. Other dynamic phenomena such as divergence or bifurcation are hard to identify from aggregate data too. These are clearly important to spatial decisions in that initial decisions to invest can generate lock-in effects which continue to be reinforced, notwithstanding other trends (Arthur 1988). Finally, the whole notion of many units, individuals and agencies continuously adapting to each other but within the broader structural constraints and opportunities characterising any particular urban system, provides the sort of seedbed in which all kinds of unusual activities might be spawned. Until we have this kind of image of the city, our analysis will remain quite limited; for the diversity existing at a level which captures this requisite variety is unlikely to show itself in the kinds of macro and aggregate data which dominates urban analysis at present.

There is another consequence of working in a static and aggregate framework and this involves the concept of equifinality—the notion that very different processes can give rise to the same system state or structure. In an urban context, this might mean that the same city form could be generated by entirely different conditions and interactions of key processes. This is tantamount to arriving at the same place by different paths but it is a very real problem when aggregate data and aggregate trends are being used to explain local processes. Cheshire & Gordon (chapter 5 this volume) add a corollary when they say: 'the same forces across advanced cities may generate quite different patterns' while it is clear that the same patterns may be generated by different forces, once again across the same domain of cities. Excellent examples of such equifinality exist in trip-making patterns where it is possible to construct the same pattern from many different sets of individuals. Indeed in the 1960s, the very models used to represent such trip-making patterns were derived on the basis that such equifinality was desirable for modelling aggregates where trip volumes on highways were the central focus of interest.

Where the focus changes to micro-behaviour, what was once required becomes problematic. For example, in the case of energy use, different sets of individuals will have access to different sets of resources. Although the total distance travelled with respect to a trip pattern might be constant, its energy use may vary considerably because of differential access to energy resources, different types of vehicle, different travel time budgets and so on. In short, one might envisage a spatial structure with a given travel pattern that might have associated with it energy usage which could vary more widely (dependent upon which individuals formed the pattern) than very different travel patterns associated with, say, dispersed and compact city forms. These kinds of issue must go some way in explaining why the evidence on energy use and settlement dispersion is so ambiguous.

Somewhat ironically perhaps, the idea of urban sustainability comes from a concern for how systems are changing and heading towards perceived catastrophe, but proposed solutions invariably are cast in terms of reimposing a kind of equilibrium which is quite uncharacteristic of the ecological and biological systems from which the notions originate. In one sense, of course, James Lovelock's (1979) concept of Gaia lies beyond this in that it assumes that the world system is infinitely sustainable but that the components of sustainability change and that sustainability itself is a concept that always pertains to some explicit group or goal. In this sense then, what is required in the study of city systems are methods and models which are rich enough to admit many types of spatial structure and which show how cities in general rise and fall in both local and global hierarchies, illustrating how they can be transformed slowly or rapidly, thereby giving us the possibility of evaluating the impact of changes in technology, energy use, information flow and so on. To this end, we require a change in approach based on the ways in which competition, conflict, growth and change in biological systems are conceived, rather than those based on static equilibrium which form the conventional wisdom of the social sciences.

The biological analogy

Analogies between cities and organisms have been made from prehistory but the impact of evolutionary biology on the social sciences is more recent. In the 19th century, Herbert Spencer blazed the way but these analogies have held little more than casual comparisons until quite recently (Steadman 1979). Darwinism, for example, has made little impact on the modelling of social dynamics, possibly because the original theories were based on reproduction and survival of the fittest through random mutation and this was clearly against any sense of how social systems actually evolve. Yet in the last 30 or so years, much more selective Darwinism has developed. No longer is the reproduction of populations seen as being a random affair. It appears that selection and then reproduction within biological populations is exercised in proportion to the degree of fitness of those involved and, in this way, populations evolve towards greater fitness much more quickly than would be the case with random mutation. In fact, the inability of random mutations to lead to the development of the kinds of species that populate the earth today in feasible time has been the main factor in the construction of contemporary Darwinian theory (Dawkins 1986).

A much richer and diverse dynamics based on punctuated equilibria, rapid and slow periods of evolution, evolution of the fitness function itself, now characterises the modern theory. The attraction of this biological model to social evolution and change relates to the fact that populations change through the mating of individuals who are effectively blind to the process of optimisation but who nevertheless reproduce in proportion to their degree of fitness. In this sense, the overall fitness of the population—the global property—rapidly increases through random interactions within the population at the local level. In this way, fitness increases exponentially and new qualities of system can emerge and diffuse quickly. In some senses, change in the system in its local context can be seen as embodying a kind of rationality which, in the large, increases the overall fitness, but in the small can appear as both rational and random.

There have been two broad developments of this kind of theory in modelling social systems. First there have been many successful applications of the 'genetic algorithm' (just sketched above) to problems of optimisation which have withstood solution because of the nature of their solution space. Problems in which variables are indivisible, of integer type, objectives which are non-linear and discontinuous, these are the types of problem for which these algorithms work best by sampling the space for solutions and then 'reproducing' new (and usually better) solutions from this same population (Goldberg 1988). However, the application of these ideas to social systems is more speculative because the notion of 'reproduction' is somewhat foreign in the biological sense to the way such systems develop. Nevertheless, it is possible to associate births of new behaving units to this type of theory. Operating within a given historical pattern in which institutions and behaviours are already established, it is possible to see how such models might generate local behaviours which manifest global order of the kind seen in cities.

We will speculate on how this analogy might be developed for cities in the next section and ultimately use pieces from it in assembling ideas for our world cities model. However the analogy suggests that to generate global order from local 'non-order' or apparent randomness, very detailed simulations of systems are required at the most basic or 'atomic' level of the behaving units. What the design of such models would ensure however is the possibility of extensive interactions. For such systems to be resilient, fairly sparse patterns of interaction are usually the norm, while such disaggregate modelling would produce very strong opportunities for adaptation at the individual level in the presence of various fitness functions which enable the system to improve in general or degenerate. In this sense, systems with optimising or pathological behaviour might be constructed.

We are a very long way from building models of this kind although important steps are being made in economics, in game theory, in artificial intelligence and like areas where this paradigm is being developed. The value of bringing this into the debate concerning sustainability and competition is in suggesting a change in focus, in attempting to persuade a change in viewpoint from aggregative to disaggregate, from static to dynamic, from macro to micro. In fact, in one of the main areas of concern in this book—the impact of changing locational patterns in cities on trip making and energy use—this shift in viewpoint is immediately valuable. If disaggregate behaviour is essential in showing how travel patterns and energy demands are impacting on spatial structure, selective sampling of different socio-economic groups across the city could be immensely more valuable than the sorts of aggregate analysis which provide the usual approach. From samples, one can generalise but even if this is not possible, the setting of aggregate against some disaggregate data in situations where conclusions are impossible to draw from the aggregate only, could be extremely useful.

In the next section, we will pull these threads together and sketch the emerging science of complexity and focus upon some of its critical aspects. But at this point it is worth summarising the issues so far. It is important to see cities as being in continual flux, as being resilient as they clearly are, but never being in any form of equilibrium in the classical sense and probably not converging on any future state that might be imagined or desirable. If we cannot predict our own evolution, and most of us are careful not to, then how do we have the audacity to predict the future of our own institutions, cities especially? By shifting our focus away from the stable state, we will have a much greater chance of handling the emergence of new qualities and artefacts in cities, of understanding the way cities compete across the hierarchy, of how local change generates global order, and of the ways in which resilience and stability are transmitted across space and time. In the next section, we will dwell on the idea of phase transitions as being the essential levers of urban change which we need to identify, transitions across boundaries which are both spatial and non-spatial. And we will indicate how a shift in viewpoint might enable us to make better policy prescriptions or at least prescriptions which have a better chance of being implemented. Before we outline our world cities model, we will return to one of the themes of this book concerning energy and transportation and speculate on how these ideas might help generate new insights there.

Order from chaos: complexity and transition

This chapter might seem a little remote from many of the contributions which precede it although the concerns expressed are of extremely wide import throughout the social sciences, particularly in areas where quantitative and formal analysis is the norm. For example, the entire edifice of economic equilibrium theory is under scrutiny for precisely the reasons noted here; the inability of the theory to enable predictions any better than the simple extrapolation of past trends, and the inability to deal with the emergence of new and unanticipated events. Of course no theory will ever be able to predict the intrinsically unpredictable, but the problem with economic and urban theory is that the unpredictable cannot even be accounted for within the frameworks used. In this sense, unpredictable and counter-intuitive conclusions are often treated as suspect. Economic equilibrium theory shows how an economy composed of many actors holds itself together and generates global order from local goals, but such theory is frozen in time and unable to handle even marginal change. In short, the very dynamics of reaching local goals are defined away within the theory and thus the processes of changing those goals and generating a different global order is simply not possible within the given framework.

In the last decade, from the development of chaos theory which in turn grew from a concern for the properties of highly non-linear systems in physics, and from the development of new types of geometry involving cellular and fractal growth, has come a general theory entitled the 'science of complexity'. The theory has gained widespread popularity of late largely due to its iconoclastic and colourful expositors, although the fact that it brings together so many different disciplinary viewpoints has added to its appeal (Waldrop 1992; Lewin 1992). We have already sketched the rudiments of the theory—its emphasis on basic dynamics, multiple interactions, and many behaving units—although one of its central features is the old concern for the relationship of the system to its environment. In many guises, complexity theory suggests that the real focus of analysis should be on the relationships or boundaries between the system and its environment where most energy interactions take place. More generally, the idea that the system has many such boundaries or edges has developed and the explanation of the basic mechanisms of such systems, it is argued, are best understood by identifying these boundaries. This, in itself, is seldom easy for it depends on the explanation sought as well as the fact that these boundaries are forever changing.

An urban example is in order. Typically, the degree to which spatial change is taking place in the city is highly variable with respect to location. Inner cities and outer suburbs are usually the most active areas in terms of changing development and in this strict sense, these are the areas it is important to study in terms of spatial change. However the idea of physical boundaries is simply one manifestation of the concept of critical edges. Consider the development of new transit lines. These may have much greater impact not on the centre or edge of the city but on established suburbs where locations are comparatively stable and where dramatic changes in travel demand might emerge due to the availability of new modes. This in turn might have important implications for travel densities although not for residential movements. This type of example can be repeated *ad nauseam* for cities, and thus a central quest is to identify such 'boundaries' or points of articulation which represent the critical thresholds of urban change.

In fact, it is never likely to be possible to identify such boundaries in anything other than cursory terms. What complexity theory implies is that the model of the system must be rich enough to enable such boundaries or points of articulation to be represented implicitly through the overall structure of the simulated system and its dynamics. Extraneous changes which lie beyond the ability of the system's predictive capacity *per se*, when fed into such a model would generate changes in terms of these implicit thresholds. In short, the entire system might be conceived of as a set of boundaries between its elements and the environment, those of importance varying according to the local and global system dynamic. A clear consequence of such a representation would be the persistence of some structures in the city, and change in others, hopefully in a way that explained some of the puzzles that presently beset the kinds of aggregate observations we are dealing with in analyses of competition between cities.

What this viewpoint also presumes is that nothing in the city or in its simulated representation is fixed. Everything is in flux, boundaries change, interactions change, new behaviours emerge. The trick is to design a structure sufficiently rich to enable such changes to be accommodated. This implies systems composed of many behaving units, highly differentiated in terms of their attributes, multiple interactions, and a detailed dynamics in the temporal sense, probably construed in terms of months rather than years. Such representations sound far-fetched but computation is no longer a problem and the structure would be such that the interaction of these units would be routine and repetitive. It would not be necessary to fit each unit to some pre-specified data attributes; this might be achieved in terms of averages. In some senses, this sounds like models of 'disorganised complexity' in Weaver's (1967) terms, although the types of interactions and attributes would be rather specific, and in this sense the model would be more highly organised. It would not be chaotic in the statistical sense, although it would contain much random processing of behaviour. Good models of this genre would in fact put together all these elements and their interactions in such a way that the structural conditions which yield various observed behaviours, both in local and global terms, are used to dimension the model.

Other features of this approach relate to the fact that as in chaos theory, the specific trajectories of the behaving units in the large and the small cannot be predicted. In this sense, the system is sensitive to its initial conditions, and the way its trajectories might be generated would mirror the kind of random bifurcation behaviours which have been employed by Allen (1982) and his colleagues in their models of urban structure. The emphasis on identifying system and environment and the focus on boundaries is predicated on the idea that these boundaries are the points at which 'phase transitions' occur, qualitative as well as quantitative changes which display properties long associated with innovation and the emergence of new structures. In the context of the world cities model that we will sketch in the next section, such transitions would be generated by the structural positioning of cities in the global hierarchy illustrating their rapidly changing role from local to world scale, and vice versa.

With respect to models of individual cities, then these approaches have not been employed so far, although Arthur (1988) has demonstrated how some of these ideas might be used to explain the location and growth of cities in general. In fact, the idea of models based on many behaving units suggests that one such approach might conceive these as basic cells, building on new developments in cellular automata theory. In this way, cities would develop spatially from small to large numbers of cells, and their complexity might be represented as some function of their spread across space, for example as n^2 where n is the number of units. In fact, in spatial terms, this power of n is likely to be less than 2, say D where D lies between 1 and 2, thus nicely connecting up to fractal theories of cities where D is the dimension. There is a strong link between the value of D and the morphology of the city, linear cities having $D = 1$, compact cities $D = 2$, and there are immediate links to density theory, and onwards to ideas about travel and energy. This is getting beyond our discussion, but the point is there are many such straws in the wind. Models of cities as cellular automata are promising as Tobler (1979) and Couclelis (1985) in a number of papers have speculated upon, and recent work on these ideas under the rubric of simulating artificial life is also suggestive (Langton 1992).

The world cities model

We are attempting to develop several of these ideas in a model of the world's cities which is designed to explain and predict the rise and fall of the global city in the context of world spatial economic development in general. This is no simple task. The project is highly speculative and is not geared to making specific predictions in the way the world econometric and input–output models such as LINK and GLOBUS are. However the approach we are taking does depend upon very rich and detailed data from a variety of sources but using many 'commonsense' but 'non-calibrated' relationships akin to those used in the systems dynamics models of Meadows et al. (1992). The particular features of the model which make it very different from other global models mainly relate to its emphasis on spatial location and interaction, its emphasis on cities per se and its focus on predicting flows or rates of change rather than stocks or structures.

The model makes a very sharp and clear distinction between system and environment, with the environment in one sense acting as the motor which drives the economic dynamic of the model. The environment is the spatial economic backcloth composed of different nation states for which data is readily available and which reflects political organisation, while the system is composed of many cities, from the city size distributions for the various nation states. Our favourite image of the model is one in which the cities are the seeds which grow and die in the economic soil which forms the environment for change or the backcloth on which all dynamics take place. The way the environment is linked to the system in terms of cities as its elements is through the idea that cities 'feed' off their immediate environment in terms of growth and change, that competition is not directly between the cities as such but that it is through the nation states which comprise the backcloth. As yet we are assuming that we will use nation states as the basic variety of the environment although it is possible that we may make some explicit regionalisation of the data to reflect the fact that some countries must be partitioned and/or aggregated differently from their national boundaries. Although the model is able to generate the emergence of new global cities, it is less able to generate new country or national groupings, notwithstanding the fact that such new groupings can be input to the model and their impact assessed.

The key structure of the model is based on demographic–economic sectors which compete spatially and which are tied together in terms of the backcloth through three types of spatial interaction:

• trade in goods;
• information flows; and
• population migration.

This is at the level of the environment and the structure of the global system is captured at any point in time through these sets of flows organised as input-output and migration matrices. Trade in goods and migration flows are easy to compile but flows of information are much more difficult and we will be developing these from international telecoms data and from the physical representation of related networks. The way the cities relate to this backcloth is through their competition for population and economic growth/resources which reflect the share of national economies, rural-urban migration and so on. We are also linking the economic sector to the use of physical resources and to the development of physical infrastructure while two key indicators which spin off from the economic-demographic dynamic are pollution and the ubiquitous index of prosperity/wealth which represents some catchall measure of social and economic performance. The economic and demographic sectors are also driven by pure births and deaths of activities as well as differential interactions and movements, and these in turn influence the economic structure at any point in time.

In essence, the model is driven not by predicting stocks of activity but by flows, rates of change such as births, deaths, migration and trade. In this sense, stocks simply represent that state of the system at a given point based on an accumulation of the effect of changing rates. This is the way in which global properties such as the distributions of cities in terms of the rank-size relation, the pattern of their location, the levels of pollution and wealth etc. can be predicted or at least represented. The way in which cities compete with one another and with their resource pools is important. The basic redistribution of fixed resources at any point in time is operationalised at the level of the environment through changing trade, population and such like interactions. These in turn can be a function of the distribution and types of cities at the system level. However the real growth and decline in the system (which is not fixed exogenously) is a function of the way cities compete for resources from the resource pool which in turn is increased or decreased dependent upon the critical indicators of pollution, prosperity and changes in these rates. In short, the model is able to generate absolute growth or decline which is dependent upon the configuration and state of the world cities and the economic backcloth over any period of time. Thus the way that the environment and system interpenetrate and influence each other is complex and there are many feedback loops which enable relative and absolute changes in these rates of growth.

All these elements of the model are tied together in terms of the stocks of activity or spatial structure of the system—in terms of the location and size of activities and the patterns of interaction. Basic accounting of the input-output variety is used extensively as the frame to link sectors and places together, while the key relations which drive the changes in rates are constructed either as econometrically estimated functions or as simple causal relations specified on *a priori* grounds. It is virtually impossible in advance to anticipate how this model will perform and thus as a precursor to its full development, a theoretical pilot model concentrating on the representation and the relationships for a much simplified environment and set of cities is being developed. This will provide some sense of how the various components operate together before the full model is run and will provide opportunities for its key causal structure to be manipulated.

To date all that has been done is to begin to explore the backcloth on which the model is to be built. We are using three global data sources of an intrinsically spatial kind. World Bank and UN data used for the LINK and input–output models is available by country and we have already begun to explore the spatial relationships inherent in this data. A unique set of physical and statistical data is now available in the ARC-World package which has been culled from several sources such as the CIA World Data Bank, US Bureau of the Census (international data source) and so on. Thirdly the Digital Chart of the World (DCW) provides extremely good physical data at a scale of 1:1 million and contains much data which we can use to dimension the physical geographic elements in the model. All of our work will be developed using various GISs associated with the global databases in ARC-World and DCW.

It is onto this backcloth that we will plant our cities and watch them grow, change, decline, or prosper, and from which we will assess the differential growth of the world economy in terms of urbanisation. It is from this that we hope we will be able to generate a model with enough competition between its parts to provide an accurate reflection of the changing global hierarchy of cities and of the emergence of new foci and the decline of existing ones. We will attempt to dimension our model to fit known trends, but at the same time the model should be able to generate novel and surprising change. The fact that we have little idea as to how it might perform is both a good and bad sign in that it suggests a sufficiently radical departure from past practice to enable new concepts to be handled, but it also suggests the need for caution. The way we hope to presume the effect of political structure is through the present configuration of nation states although we have yet to study how strictly this will limit the model's ability to predict and generate new alignments. Finally in terms of the sustainability argument, we feel that we will have a much stronger handle on the spatial pattern of energy use and on the comparative resilience or otherwise of the global economy in these terms.

Conclusions: next steps

This has been a somewhat speculative chapter concluding this book but it is motivated by a general concern for developing analytical approaches which might inform us better about urban sustainability and particularly the influence of energy and trip-making patterns on urban form and the impact of local change on global order. The problem that we see is not that static and aggregative analysis is no longer suggestive, it surely is, but that it is incapable of enabling us to detect the real import of relevant changes which we know must have an impact in some way on location. This has always been a problem with existing analysis but it is much exacerbated when the domain of interest is broadened beyond issues of spatial efficiency and equity to include energy and ecology. Moreover, because the major transition which is sweeping the world at present involves the substitution of energy for information, or rather the transformation of energy into information, the problem is further compounded when we seek to disentangle energy use from this broader context.

What we suggest here is not new, for the ideas were known and discussed a generation or more ago. But the issues now seem more important. We have had much experience with aggregate and comparative static analysis in the last 30 years and it now appears that its conclusions must be viewed with greater caution than we previously thought. To this end, we advocate a shift in viewpoint towards more micro, dynamic concerns. Of course, it is never possible to make this transition immediately and it is not our purpose to suggest that aggregate analysis should be abandoned. Nor is it our purpose to give the impression that these ideas are the prerogative of the very few. There is a groundswell of opinion throughout the social and physical sciences which is suggesting the same types of shift, from ideas more in tune with classical physics to those more in tune with contemporary biology; from models of disorganised complexity to those of organised complexity in Warren Weaver's (1967) terms. The illustration we have provided, which is no more than a sketch of the model we are beginning to build, draws on these ideas. In a global context, this should provide us with a clearer view about the future sustainability of the world economy, the use of energy and associated patterns of urbanisation, all cast in a framework which will provide a richer and more insightful mode of analysis than that available hitherto.

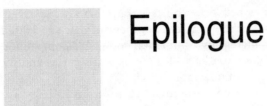

Epilogue

The study

THIS VOLUME IS the latest from an international study of technological change and its effect on cities. The study began a decade ago and has focused on the continuing development of new information, communication and transport technologies and their increasing effects on cities and their industries and on urban living and working. It is convened by the International Council for Building Research as its Working Commission 72. The three previous volumes—*The Future of Urban Form*, 1985; *The Spatial Impact of Technological Change*, 1987; and *Cities of the 21st Century*, 1991—identified, chronicled and interpreted these changes. These include the continuing development and dissemination of information and communication technologies and their use in combination, the spread of fast rail networks and their increasing travel speeds, the spatial development patterns of high technology industries including biotechnologies, the market development of these in 'silicon landscapes' and their planned inducement in high technology parks, precincts and new cities. Also included are the development of automation and flexible specialisation in manufacturing, of office automation, the emergence of global networks, global markets and global cities, and of new industrial processes and alliances based on these technologies, and their impacts on our daily lives.

Telecommunications, computing and fast transport are considered to be both the primary agents of this change and the critical infrastructures for economic development over the next decade. They are creating another industrial revolution, ushering in the transition to an information-based economy. In this new economy, information and knowledge are emerging as the key factors of production for both goods and services, partly displacing materials, energy and labour. These new technologies are being increasingly linked with production of goods and services in complex geographical patterns embracing the entire globe to form the new engines of the economy.

Global telecommunications networks and fast transport are increasing the spatial scale of markets, enabling the development of new specialised services and goods. There is increasing continental and global competition among cities for a share of these markets—for goods and services and for the development and attraction of new industries.

The changes occurring as a consequence are profound. They are affecting our daily activities, where we work and live, the nature of our work and leisure activities and their environments. These and other forces are combining to induce further changes to our cities and our daily lives.

Changing forces operating on cities

The range of forces operating to change our cities and the relationships between them are not only technological but also economic, social, political, regulatory and environmental.

Technological

The increasing capabilities of information technologies—the computer and microchip, telecommunications networks of expanding size and bandwidth, and information recording devices—are having a profound effect on our cities and their activities. The confluence and networking of these technologies will create a second and perhaps larger impact. The inclusion of artificial intelligence will cause a further and still larger effect. Already information technology has increased the locational choices available to urban industries and service activities. This has resulted in a shift in employment distribution from the central city to the suburbs with only about one-third of jobs remaining in the central region of major cities. This increased choice of location has affected various industrial activities differently. Large manufacturing firms have generally moved to the periphery of development, retail and service industries and various back-office functions have moved to the suburbs, but company headquarters and finance, insurance, real estate, legal and advertising (i.e. producer services) have generally remained in the central city.

The demand for employment skills has also shifted away from manual and craft towards the processing of information- and knowledge-based skills, enabling a wider work force participation—of females as well as males—in an improved workplace environment. This new nature of work has facilitated its closer integration with other urban activities, both residential and recreational. New opportunities resulting from the combination of information technologies may include a further widening of choice of work and residential location and even a merging of the two. Teleworking, from home or from a neighbourhood office, may be easier with broadband telecommunications—including multimedia transmission and video conferencing—where joint decision making or negotiation is required. Its increasing economic viability—a function of declining telecommunications costs—could also see a change in attitude to make teleworking more socially and organisationally acceptable (Moss and Carey, chapter 22).

The global expansion of these networks has resulted in expansion of markets to national and global levels, causing greater competition among cities at a regional and global level. The major cities located at network nodes have gained a special advantageous position as information centres and market places, as headquarters for multinational firms, and as centres for highly specialised producer services and finance. These producer services appear to be the most locationally mobile and are highly concentrated in global and national centres.

High-speed transport modes, particularly air and fast rail, but also expressways, further influence the competitiveness, prosperity and development of cities. Expressways, generally at urban peripheries, have attracted manufacturing industries by providing easier access to suppliers and markets and to cheaper production space. This allows better integration of production and distribution, including just-in-time manufacturing and production—and distribution-on-demand. Air and fast rail have had a similar impact on information-based industries. Whereas information can be transmitted along electronic networks, associated knowledge or expertise still moves in people's heads, requiring access to fast transport networks.

The Shinkansen network in Japan has been shown to have a major impact on location of information industries. High speed ground transport has also linked urban markets, and allowed migration of people and firms along these corridors—creating mega conurbations.

At a larger scale, the existing high speed rail network in Europe and its planned extension to link Birmingham, London, Brussels, Cologne, Frankfurt, Stuttgart, Zurich and Milan—the so-called 'blue banana'—is increasing the desirability and prosperity of these cities. The Tokaido line in Japan, linking Tokyo and Osaka, has had a similar effect on the larger cities along this corridor.

Economic

The globalisation of markets has increased the competition among cities for the provision of services and goods, and for the attraction or development of new economic activities. The global networks creating these markets provide particular advantage to market places at their nodes. Cities located both at electronic communication nodes and fast transport nodes will gain particular advantage.

The transition to an information economy has increased locational options for existing industries, enabling them to choose locations with low factor costs. This transition has also changed the relative importance of these factors by substituting information technology for labour. It has fostered the growth of new industries based on these technologies or on more specialised information services. Global markets and increasingly integrated technologies have also created the need for strategic alliances between firms to bring together the wider range of skills and technologies required. Within any country, these new technologies are developing in largely new regions, while older technologies and their regions may stagnate.

The increasing integration of production, distribution, communication and computing is further changing the industrial and urban landscape. The imperatives of just-in-time production, the compression of time in the production process and the greater integration of individual client preferences into products, are causing further spatial and temporal concentrations of activity. Production-on-demand with an increasingly wide range of options makes the maintenance of large inventories prohibitive; but the compression of production time—to a few days for a house or car in Japan, or to an hour for a pocket pager in the US—obviates the need for inventories, but requires a concentration of suppliers close (in time) to the production scene.

Concentration of ownership, the merging of companies within and between countries, joint ventures, partnerships and other alliances, including international alliances, are further consequences of these changes, and are further forces for industrial concentration within these alliances. The new information and communication technologies also provide the networks or media for these larger organisational structures.

The inclusion of artificial intelligence as a further element in these economic engines will enable further quality improvements and increased competitiveness for those firms and nations which embrace it early and effectively. It will also enhance the quality of working and living environments and increase even further their locational choices.

Future growth in global population and economic activities will occur largely in its metro areas as urbanisation continues, particularly in the developing economies, emphasising the need for their efficient planning and sustainable development. There will also be major changes in the share of global economic output over the next few decades if the nations of south and east Asia continue their present rates of economic growth.

Social

Social forces include a greater participation of women in the work force as information and knowledge become key factors of production and as family sizes decrease. Many women prefer part-time jobs and shorter commuting trips, and hence suburban job locations. There is also a wider range of skill levels required over the work force as a whole as many routine jobs are deskilled or eliminated, while the need for specialist services, innovation and R&D increases. Income differences increase as a consequence, causing social equity problems. The suburbanisation of jobs means less of them can be accessed by public transport, adding to these equity concerns.

Decreased security of employment, less stability in marital relationships and de-institutionalisation are further factors removing security of income, accommodation and supporting services from many of those most in need—dramatically increasing urban social problems and emphasising a darker side to urban living.

The ageing of populations in the developed world will result in further economic and social change and further contrast with the developing world.

Political

The formation of multi-nation trading blocs and regional alliances is providing further impetus for change. It is diminishing the significance of national barriers and the need for firms to be located in each individual country of a trading bloc, further increasing competition among the cities of that bloc for the location of regional centres for these firms and for the bloc as a whole.

Regulatory

The move towards deregulation in a number of industries—transport and travel, finance and trade, and in urban and regional planning—is providing further forces of change. Deregulation in transport has reduced the costs of travel and freight movement and increased economies of distribution, allowing centralisation or concentration of production. Inventory management using information technology is reducing the need for warehouses and wholesalers. The cleaner, new information industries are reducing the need for separate zones, allowing a merging of commercial, residential and recreational activities. Privatisation is complementing deregulation and effectively extending it to other activities such as urban infrastructure provision, travel, communications and services. This is further increasing locational choices, but also reinforcing the larger centres where network hubs are forming, and higher levels of service are viable, reducing cross-subsidies to other service locations.

Environmental

Environmental sustainability is an emerging constraint on urban activities and their locations. There is an increasing awareness that the further growth and development of this planet will be largely through urbanisation and will be constrained not only by resources but by receptors for its wastes—the atmosphere, land and waters. This second constraint may be more critical, influencing the location of development as well as its form and density. It could also constrain the flows of goods and people between activities and the modes adopted for those flows, particularly those using fossil fuels. Cleaner, leaner and more environmentally compatible forms and locations of development will be sought.

Urban impacts

These forces are impacting on the nature and spatial distribution of urban activities at a metropolitan, regional and global level.

Global

At a global level, these forces are reinforcing the largest and most productive cities as world centres and global network nodes as outlined above. They are also increasing the distinction between levels in this hierarchy. The forces of change are creating increasing competition among cities for these higher level roles and for the development and attraction of new industries. They are resulting in global and regional production systems and the concentration of suppliers around some production centres as inventories are reduced and as time between placement of an order and its delivery is compressed.

Flexible specialisation is enabling a wide range of product options, and even personalisation of the products, but requires the further compression of time for production through re-engineering of the process to avoid inventory stocks. Spatial concentration of activities and suppliers is a further consequence; and a wider spatial organisation of marketing and distribution provides economies of scale and scope.

The development of larger market sizes, including global, is increasing the level of specialisation viable in service activities and hence the quantity and diversity of specialist services available. Flexible specialisation is facilitating a similarly wider range of products, enabling personalisation on demand. Time is becoming the critical factor in these processes as diversity prevents a full range of stocks and production must be linked to demand. Time is displacing mass as the factor requiring agglomeration economies or spatial concentration.

Regional

A further effect of this wider scale of organisation via telecommunications and, to some extent, of policies for income redistribution at a national level, is a reduction in the difference in mean per capita incomes between a nation's cities by extending city benefits to other parts of a region—creating a larger virtual city and reducing apparent economies of scale of larger cities in this sense. At the same time, suburbanisation of industries is constraining the rate of growth of urban congestion and the transport task, thereby reducing the diseconomies of scale. The net effect is that larger cities retain and increase their competitive advantage. Further scale and agglomeration economies include better access to global networks, particularly broadband telecommunications, and to higher level producer services.

Larger cities also nurture innovation centres and R&D. The patterns of innovative activity are themselves concentrated (Goddard, chapter 6). There are ten such concentrations or islands of innovation in Europe, for example, and they are located in the major urban regions such as London, Paris, Amsterdam and Dortmund. The concentration of corporate control is similarly located in a few key centres, e.g. the 'blue banana' centres in Europe. These headquarters and R&D centres are largely located on high-speed passenger routes, enabling the easy movement of expertise and knowledge between such centres, providing further scale or agglomeration economies. The need for regional competitiveness is increasing with this concentration.

Paradoxically, the increasing capabilities of telecommunications and fast travel could spread the benefits of major cities to other urban areas by linking them into larger virtual city activities—possibly changing the European development 'banana' to something more like a bunch of grapes (Goddard, Wegener chapters 6, 7).

Metropolitan

The increasing location options for industry resulting from improvements in communication and information technologies is causing even larger relative changes at the metropolitan level, but in this case the changes are in the opposite direction—to deconcentration or dispersal of employment into suburban centres, edge cities and beyond the urban fringe. Routine production is seeking locations with low factor costs and this generally means cheaper land outside the central city and better access to suppliers and markets. Back-office functions can move to the suburbs. The less routine activities such as innovation and R&D are following people and skills into higher amenity areas.

Thus the growth or prosperity of global and regional cities is not necessarily in the city centres but largely in their suburbs or beyond, but still retained essentially within their daily activity patterns. However, key elements of that prosperity in the form of corporate control and specialised producer services may well remain in the centre or key subcentres.

New forms of planned cities are emerging (Hall, chapter 1). Commercial subcentres are being developed on reclaimed urban land (e.g. London Docklands). Metro centres are being constructed at public transport nodes, sometimes serving as downtowns (e.g. Croydon, Shinjuku). Edge cities are developing, some with reliance on the automobile (e.g. Tysons Corner) and others (e.g. Stamford, Kawasaki, Reading) at public transport nodes. The science- and technology-based city or Technopolis in Japan provides another form in which the whole city is planned as a technology incubator, with export potential, e.g. Hsinchu, Taiwan (Blakely chapter 15) and the planned MFP in Australia (Hamnett chapter 16). Growth corridors based on high-speed transport links (e.g. East Thames corridor, UK) represent a further planned development form.

This decentralisation of employment in the post-industrial city represents a reversal of the pattern in the industrial city where industry was concentrated at the city centre. This, in turn, was a reversal of the pattern for the pre-industrial city where economic activity was largely dispersed in cottage industries and in the surrounding fields. Commuting patterns, networks and modes changed accordingly, as illustrated in Table 26.1.

A comparison of employment dispersal and commuting trip length characteristics over selected US, UK and Australian cities indicates the ratio of employment dispersal to residential dispersal is similar for each, with the US cities furthest dispersed. On the other hand, the ratio of mean commuting trip length to what it would be with all jobs concentrated at the city centre varies substantially, particularly among US cities. The mean for

Table 26.1 Urban employment and commuting pattern reversals over time

Period	Employment pattern	Pattern	Commuting Network	Mode
Pre-industrial	Dispersed	Dispersed	Minimal grid	Private, walk
Industrial	Central	Focussed	Radial rail	Public, rail
Post-industrial	Dispersed	Dispersed	Extensive grid	Private, car

the sample of US cities is 1.0 but individual cities vary from 0.6 to 1.5. For UK and Australian cities the ratio is lower—between about 0.55 and 0.85, indicating higher travel energy costs and perhaps more congested roads. Travel times appear to be relatively stable over time, giving some support to the concept of constant travel time budgets, and travel distances changing with transport technologies.

Breheny has argued the infeasibility of containment of urban activities generally and shown the ineffectiveness of such containment in the UK. Containment policies may drive activities to more accommodating cities, providing competition among neighbouring cities for these activities, to the detriment of the city so contained.

Another feature of the new spatial structure of metropolitan areas is their robustness in the face of future uncertainties. Their dispersal of employment enables a potential redistribution of trip destinations or origins to provide shorter commuting and shopping trips, lower energy using patterns and increasing self-sufficiency or self-containment of suburban centres. Formation of additional centres would enable even lower energy use. However, this lower energy use would require behavioural changes—or the imposition of higher energy costs—for realisation.

Fast rail and road links are providing corridors for further decentralisation of urban activities, creating larger conurbations as urban centres extend and merge.

In the US, manufacturing, retail, back-office functions, R&D and construction are dominant activities in the suburbs and in edge cities. Central cities largely retain corporate headquarters, producer services, FIRE (finance, insurance and real estate) and legal services. They are more specialised, including advertising, accounting, money markets, stock exchanges and travel services, and depend on suburban commuters who are largely better skilled and better paid. These central activities are generally in competition with those of other cities. Producer services in particular are highly centralised and located in the largest US cities, particularly New York.

The decentralised concentration of production and its suppliers is a result of flexible specialisation, just-in-time processing, re-engineering and production-on-demand. These new processes are interdependent and with all of the product options offered, would require large inventories if demand and production were not integrated and if the process of production was not dramatically compressed in time. New industrial structures and spatial concentrations in decentralised centres are the result.

Conclusion

The transition from an industrial to an information society—brought about by the impacts of information and communication technologies and fast travel—has substantially changed the form of our cities and the relationships among them. Further and more fundamental waves of change may be expected from the confluence of information and communication technologies, from their further development and the incorporation of artificial intelligence, and from the development and expansion of fast transport such as fast rail networks and air transport.

The transport technology within our cities has been said to determine their scale—given human instinct or behavioural characteristics, including essentially fixed budgets for daily travel time. The faster the transport technology, the larger the distance which can be covered in that time and the larger the city which can accommodate these trips. Fast intercity travel can facilitate the extension of cities along these intercity routes and their eventual merging into larger conurbations.

The expanding capabilities of telecommunication and information technology, including multimedia teleconferencing, may substitute for some face-to-face contact in consultation, in negotiation, in joint decision making and certainly in document processing to enable ever larger urban systems than those encompassing daily movement patterns. Many firms, of course, have already expanded beyond individual cities. These capabilities may also affect the competition developing between present cities—perhaps reinforcing the larger ones at those network nodes which are first connected to broadband services.

There is close analogy between the development of motorised transport systems and the development of information and communication technologies. The development of a high quality highway system realised the full potential of the motor vehicle, and its urban impact, assisting the development of suburban living, large suburban shopping centres and decentralised production. The confluence of information technologies and broadband telecommunication networks will similarly increase their impacts on urban metropolitan development and particularly on residential and service activities.

The shift from energy and materials to information and knowledge as key factors of production is not only increasing the productivity of industries and cities but also increasing quality of goods and services by allowing their quality control and specialisation or personalisation to individual preferences and needs.

This shift to information and knowledge is also the key to sustainability of our urban systems as energy and material inputs per product are reduced over the range of goods and services produced. Transport energy needs can also be reduced with improvements in the bandwidth and capabilities of information and communication technologies—allowing further movement of jobs towards residences and their incorporation in these residences for an increasing range of activities. The better integration of commercial, residential and recreational activities can also reduce recreational and shopping trips in a future environment in which travel energy is constrained.

The holistic planning of urban areas by considering their various elements, natural and built, as parts of an integrated system, enables further sustainability. For example, greater retention of stormwater for gardens and lakes requires less water supply and less drainage provision, reducing infrastructure needs and providing better local recreational opportunities. Walking and cycling paths to local shops and facilities will further enhance this sustainability and amenity. Concentrated activity centres, combining information-oriented jobs and consumer services, could be grouped around strong public transport nodes with interchange between buses and/or rail, thus facilitating a polycentric urban structure, especially in the larger metropolitan areas. As well as intercepting suburban commuters who would otherwise need to make long journeys to downtown areas, these nodes would also provide for reverse commuters coming out from the inner urban areas and also for circumferential commuters, who might require the provision of new services, most likely provided by bus or paratransit running on reserved high-occupancy lanes on freeways. These same nodes would logically incorporate workcentres for telecommuters needing facilities for regular contact by telecommunications and also occasional face-to-face contact; these would depend on the provision of high-level broadband communication networks which would permit the ready exchange not merely of voice and written data but also of multiple media. Teleworking is an increasingly viable option for some workers and for some part of the week. Providing a wide range of choice in housing types, densities and environments will better match people's needs, and perhaps creating these in each suburb will enable closer linking of home and work, and shorter commuting trips by lower energy modes. It is also comforting to note that the higher levels of human needs are not physical but social and psychological and perhaps even transcendental, and thus not based on consumption of physical resources but more compatible with the information economy.

Thus, the factors making cities more competitive, more efficient and more productive can also make them more robust, potentially sustainable and more livable. These benefits, however, are not uniformly shared. Technological change is creating both winners and losers, among people and places. New technologies are requiring new skills from some, particularly in innovation, research and development, and increasingly specialised services, but deskilling or displacement of others in more routine production operations. New technology industries are growing in some areas and leaving other areas with older technologies to stagnate. The benefits of change, if they are to be fully enjoyed, will need to be extended to all urban citizens, and this perhaps is both the promise and the challenge for cities of the 21st century.

Bibliography

ABC Air Guide 1946, Thomas Skinner, London.

ABC *World Airways Guide*, Reed Publishing (Various years).

Adrian, C. 1984, *Urban Impacts of Foreign and Local Investment in Australia*, Australian Institute of Urban Studies, Publication No. 119, Canberra.

Allen, P.M. 1982, 'Evolution, modelling and design in a complex world', *Environment and Planning B*, 9, 95–111.

Amano, K., Toda, T. & Nakagawa, D. 1991, 'The rapid transportation system and the socioeconomic restructuring of Japan', in *Cities of the 21st Century*, eds J.F. Brotchie, M. Batty, P. Hall, & P.W. Newton, Longman Cheshire, Melbourne.

Amin, A., Charles, D. & Howells, J. 1992, 'Corporate restructuring and cohesion in the new Europe', *Regional Studies*, 26, 319–32.

Amin, A. & Goddard, J.B. 1986, *Technological Change, Industrial Restructuring and Regional Development*, Allen and Unwin, London.

Amrhein, C.G. & Harrington, J.W., Jr 1988, 'Location, technical change and labour migration in a heterogeneous industry', *Regional Studies*, 22(6), 515–29.

Anderson, M., Roy, J.R. & Brotchie, J.F. 1986. 'On the adaptability of alternative urban configurations', *Environment and Planning B*, 13, 305–18.

Anderson, P.W., Arrow, K.J. & Pines, D. (eds) 1988, *The Economy as an Evolving Complex System*, Addison-Wesley, Reading, Massachusetts.

Andersson, Å.E. 1985, *Kreativitet: StorStadens Framtid*, Prisma, Stockholm.

Andersson, Å.E. 1986, 'The four logistical revolutions', *Papers of the Regional Science Association*, 59, 1–12.

Andersson, Å.E. & Batten, D. 1988, 'Creative nodes, logistic frameworks and the future of the metropolis', *Transportation*, 14, 281–93.

Argyle, M. 1990, 'The social psychology of home working', *Working Environment News*, February/March.

Arizona Department of Commerce 1991, *State of Arizona Telecommuting Pilot Six Month Evaluation*, Arizona Energy Office and AT&T, Phoenix.

Armington, C., Harris, C. & Odle, M. 1979, *Formation and Growth in High Technology Firms: A Regional Assessment*, Brookings Institution, Washington, DC.

Arthur, W.B. 1988, 'Urban systems and historical path dependence', in *Cities and Their Vital Systems: Infrastructure, Past, Present, and Future*, eds J.H. Ausubel & R. Herman, National Academy Press, Washington DC, 85–97.

Ashby, W.R. 1956, *Introduction to Cybernetics*, Methuen, London.

Ashworth, G.J. & Voogd, H. 1990, *Selling the City: Marketing Approaches in Public Sector Urban Planning*, Belhaven Press, London and New York.

Atlanta Regional Commission 1985, *Transportation Problems and Strategies for Major Activity Centers in the Atlanta Region*, Atlanta Regional Commission.

Australian Bureau of Statistics 1985, *Travel to Work, School and Shops, Victoria, October 1984*, Catalogue No. 9201.2, ABS, Melbourne.

Aydalot, P. (ed.) 1986, *Milieux Innovateurs en Europe*, GREMI, Paris.

Aydalot, P. & Keeble, D. (eds) 1988a, *High Technology Industry and Innovative Environments: The European Experience*, Routledge, London.

Aydalot, P. & Keeble, D. 1988b, 'An overview', in *High Technology Industry and Innovative Environments: The European Experience*, eds P. Aydalot & D. Keeble, Routledge, London.

Bachtler, J. & Clement, K. 1990, T*he Impact of the SEM on Foreign Direct Investment in the UK*, HMSO, London.

Bae, C. 1993, 'Air quality and travel behavior: untying the knot', *Journal of the American Planning Association*, 59(1), 65–75.

Bairoch, P. 1988, *Cities and Economic Development*, The University of Chicago Press.

Banister, D. 1980, *Transport Mobility and Deprivation in Inter Urban Areas*, Saxon House, Farnborough.

Banister, D. 1992, 'Energy use, transport and settlement patterns', in *Sustainable Development and Urban Form*, ed. M. Breheny, Pion, London.

Bar, F. 1989, 'Telecommunications networks: exploring new infrastructure concepts', Presented to '*Regulation, Innovation and Spatial Development*' *Symposium, Cardiff*, 13–15 September.

Barlow, J. 1989, 'Regionalisation or geographical segmentation? Developments in London and south east housing markets', in *Growth and Change in a Core Region: The Case of South East England*, eds M. Breheny & P. Congdon, Pion, London.

Bateman, M. 1985, *Office Development: A Geographical Analysis*, Croom Helm, London.

Batten, D.F. 1992, 'The fourth logistical revolution: implications for Australian cities', *Urban Policy and Research*, 9(4), 233–37.

Batty, M. 1988, 'Home computers and regional development: an exploratory analysis of the spatial market for home computers in Britain', in *Informatics and Regional Development*, eds M. Giaoutzi & P. Nijkamp, Avebury, Aldershot, 147–65.

Baumol, W. J. 1961, *Economic Theory and Operations Analysis*, Prentice Hall, Englewood Cliffs, NJ.

Bell, M. 1992, *Internal Migration in Australia, 1981–1986*, Bureau of Immigration Research, Australian Government Publishing Service, Canberra.

Beniger, J.R., 1986, *The Control Revolution: Technological and Economic Origins of the Information Society*, Harvard University Press, Cambridge, Massachusetts.

Benington, J. 1986, 'Local economic strategies: paradigms for a planned economy', *Local Economy*, 1,7–24.

Bennett, R.J. & Krebs, G. 1991, *Local Economic Development: Public-Private Partnership Initiatives in Britain and Germany*, Belhaven Press, London.

Berry, B.J.L. 1964, 'Cities as systems within systems of cities', in *Regional Planning and Development: A Reader*, eds J. Friedman & W. Alonso, MIT Press, Cambridge, Massachusetts, 116–37.

Berry, B.J.L. 1991, *Long-Wave Rhythms in Economic Development and Political Behavior*, Johns Hopkins UP, Baltimore.

Berry, B. & Gillard, Q. 1977, *The Changing Shape of Metropolitan America*, Ballinger, Cambridge, Massachusetts.

Bibby, P. & Shepherd, J. 1991, *Rates of Urbanisation in England 1981–2001*, Pion, London.

Billi, A. 1992, *An Evaluation of Network Theories: Towards a New Theory*, Honours Dissertation, Department of Management, The University of Western Australia.

Blakely, E.J. 1987, 'Innovation in national industrial policy', *International Journal of Technology Management*, 2(5/6), 731.

Blakely, E.J. 1992, *Shaping the American Dream: Land Use Choices for America's Future*, Working Paper No. 588, Institute for Urban and Regional Development, University of California, Berkeley.

Blakely, E.J. & Fagan, R.H. 1988, *Metropolitan Strategy in Sydney: Employment Distribution and Policy Issues*, Monograph No. 36, Institute of Urban and Regional Development, University of California, Berkeley.

Blakely, E.J. & Nishikawa, N. 1989, *The Search for a New Golden Goose: State Strategies for the Biotechnology Industry*, Working Paper No. 500, Institute of Urban and Regional Development, University of California, Berkeley.

Blakely, E.J. & Shapira, P. 1984, 'Industrial restructuring: public policies for investment in advanced industrial society', *Annals of the American Academy of Political and Social Sciences*, 475, 96–109.

Blakely, E.J. & Stimson, R.J. 1992, *The New City of the Pacific Rim*, Monograph No. 43, Institute for Urban and Regional Development, University of California, Berkeley.

Blakely, E.J. & Willoughby, K.W. 1990, 'Transfer or generation? Biotechnology and local industry development', *Journal of Technology Transfer*, 15(4), 31–38.

Bluestone, B. & Harrison, B. 1982, *The Deindustrialization of America: Plant Closings, Community Abandonment, and the Dismantling of Basic Industry*, Basic Books, New York.

Boddy, M., Lovering, J. & Bassett, K. 1986, *Sunbelt City? A Study of Economic Change in Britain's M4 Growth Corridor*, Clarendon Press, Oxford.

Boeing Company 1990, *Current Market Outlook*, Boeing Commercial Airplane Group, Seattle.

Boeing Company 1992, *Current Market Outlook*, Boeing Commercial Airplane Group, Seattle.

Boghani, A., Kimble, A. & Spencer, E. 1991, *Can Telecommunications Help Solve America's Transportation Problems?* Arthur D. Little, Cambridge, MA.

Bookout, L. 1992, 'Neotraditional town planning: cars, pedestrians, and transit', *Urban Land*, 51(2), 10–15.

Bowman, G. & Davis, J. 1990, *Telecommuting Pilot Study: Final Report*, County of San Diego Department of Public Works, San Diego, CA.

Brander, J. 1987, 'Rationales for strategic trade and industrial policy', in *Strategic Trade Policy and the New International Economics*, ed. P. Krugman, MIT Press, Cambridge, MA..

Breheny, M.J. 1990, 'Strategic planning and urban sustainability', in *Proceedings of the Town and Country Planning Association Conference on Planning for Sustainable Development*, TCPA, London.

Breheny, M.J. 1992, 'The contradictions of the compact city: a review, in sustainable development and urban form', in *Sustainable Development and Urban Form (European Research in Regional Science, 2)*, ed. M.J. Breheny, Pion, London, 138–59.

Breheny, M. 1993, 'Counterurbanization and sustainable urban form', this volume.

Breheny, M., Gent, T. & Lock, D. 1992, 'Alternative development patterns: new settlements', draft report to Department of the Environment, London.

Breheny, M.J. & McQuaid, R.W. 1987, *The Development of High Technology Industries: An International Survey*, Croom Helm, London.

Brooks, A. 1991, 'Omaha: thriving through both boom and bust', *New York Times*, 21 July.

Brotchie, J.F. 1984, 'Technological change and urban form', *Environment and Planning A*, 16, 583–96.

Brotchie, J. F. 1991, 'Fast rail networks and socioeconomic impacts,' in *Cities of the 21st Century*, eds J.F. Brotchie, M. Batty, P. Hall, & P.W. Newton, Longman Cheshire, Melbourne.

Brotchie, J.F. 1992, 'The changing structure of cities', *Urban Futures*, Special Edition No. 5, 13–26.

Brotchie, J.F., Anderson, M. & McNamara, C. 1992, 'Developments in urban transport', Urban Transport Conference, IIR, Sydney, April.

Brotchie, J.F., Hall, P. & Newton, P.W. (eds) 1987, *The Spatial Impact of Technological Change*, Croom Helm, London.

Brotchie, J.F., Newton, P.W., Hall, P. & Nijkamp, P. (eds) 1985, *The Future of Urban Form: The Impact of New Technology*, Nichols, New York.

Brown, R. 1989, 'A DITAC view of the multi-function polis proposal', *Australian Planner*, 27(2), 8–13.

Bruinsma, F. & Rietveld, P. 1992, *Stedelijke Agglomeraties in Europese Infra-struktuurnetwerken*, Werkstukken 27, Stedelijke Netwerken, Amsterdam.

Brunet, R. 1989, *Les Villes Européennes*, Rapport pour la DATAR, RECLUS, Montpellier.

Buck, N., Drennan, M. & Newton, K. 1992, 'Dynamics of the metropolitan economy' in S. Fainstein, I. Gordon & M. Harloe (eds) *Divided Cities: New York & London in the Contemporary World*, Blackwell, Cambridge, MA.

Buck, N., Gordon, I., Young, K., Ermisch, J. & Mills, L. 1986, *The London Employment Problem*, Oxford University Press, London.

Budd, L. & Whimster, S. (eds) 1992, *Global Finance and Urban Living: A Study of Metropolitan Change*, Routledge, London.

Bulusu, L. 1990, 'Internal migration in the United Kingdom', *Population Trends*, 62, 33–6.

Bunker, R. 1991, 'Adelaide: a multi-functional metropolis', *Australian Journal of Public Administration*, 50(2), 145–53.

Burnley, I.H. 1980, *The Australian Urban System*, Longman Cheshire, Melbourne.

Burnley, I.H. 1988, 'Population turnaround and the peopling of the countryside? Migration from Sydney to country districts of New South Wales', *Australian Geographer*, 19, 268–82.

Business Week 1992, 'Mr. Sam's experiment is alive and well', 20 April, 39.

Business Week 1993a, 'Intel: What a tease—and what a strategy', 22 February, 40.

Business Week 1993b, 'Sweating out the HDTV contest', 22 February, 92–93.

Caldwell, J.C., Orubuloye, I.O. & Caldwell, P. 1992, 'Fertility decline in Africa: a new type of transition?', *Population and Development Review*, 18, 211–42.

CALUS 1983, *Property and Information Technology—the Future of the Offices Market*, College of Estate Management, Reading.

Cameron, G., Moore, B., Nicholls, D., Rhodes, J. & Tyler, P. 1990 (eds), *Cambridge Regional Economic Review: The Economic Outlook for the Regions and Counties of the United Kingdom in the 1990's*, Department of Land Economy, University of Cambridge, Cambridge.

Carter, C. 1981, *Industrial Policy and Innovation*, Heinemann, London.

Carvel, J. 1993, 'Changing face of EC refugees', *The Guardian (London)*, 1 June.

Castells, M. 1984, *Towards the Informational City? High Technology, Economic Change, and Spatial Structure: Some Exploratory Hypotheses*, Working Paper No. 430, Institute of Urban and Regional Development, University of California, Berkeley.

Castells, M. 1985, 'High technology, economic restructuring, and the urban-regional process in the United States', in *High Technology, Space and Society*, ed. M. Castells, Sage Publications, Beverly Hills.

Castells, M. 1986, *High Technology, Economic Policies and World Development*, Working Paper No. 18, Berkeley Roundtable on the International Economy, University of California, Berkeley.

Castells, M. 1991, *The Informational City: Information Technology, Economic Restructuring and the Urban-Regional Process*, Basil Blackwell, Oxford.

Castells, M. & Hall, P. 1993, *Technopoles of the World: The Making of 21st-Century Industrial Complexes*, Routledge, London.

Cederlund, K., Erlandsson, U. & Törnqvist, G. 1991, *Swedish Contact Routes in the European Urban Landscape*, Department of Social and Economic Geography, University of Lund, Lund.

Centre for Technology and Social Change 1990, *Strategic Alliances in the Internationalisation of Australian Industry*, Australian Government Publishing Service, Canberra.

Cervero, R. 1986a, *Suburban Gridlock*, Center for Urban Policy Research, Rutgers University, New Brunswick, New Jersey.

Cervero, R. 1986b, *Jobs-Housing Imbalances as a Transportation Problem*, Institute of Transportation Studies Research Report 86–9, University of California, Berkeley.

Cervero, R. 1989a, *America's Suburban Centers: The Land Use-Transportation Link*, Unwin Hyman, Boston.

Cervero, R. 1989b, 'Jobs-housing balancing and regional mobility', *Journal of the American Planning Association*, 55(2), 136–50.

Cervero, R. 1991a, 'Jobs-housing balance as public policy', *Urban Land*, 50(10), 10–14.

Cervero, R. 1991b, 'Congestion relief: the land use alternative', *Journal of Planning Education and Research*, 10(2), 119–29.

Cervero, R. & Landis, J. 1992, 'Suburbanization of jobs and the journey to work: a submarket analysis of commuting in the San Francisco Bay Area', *Journal of Advanced Transportation*, 26(3), 275–97.

Champion A.G. 1987, 'Momentous revival in London's population', *Town and Country Planning*, 56, 80–2.

Champion, A.G. (ed.) 1989, *Counterurbanization: The Changing Pace and Nature of Population Deconcentration*, Edward Arnold, London.

Champion, A. & Congdon, P. 1988, 'An analysis of the recovery of London's population change rate', *Built Environment*, 13(4), 193–211.

Chapman, K. & Humphrys, G. 1987, *Technical Change and Industrial Policy*, Basil Blackwell, Oxford.

Charles, D. & Feng, L. 1993, *Lean Production, Supply Chain Management and New Industrial Dynamics: Logistics in the Automobile Industry*, PICT Research Report, CURDS, University of Newcastle-upon-Tyne.

Cheshire, P.C. 1990, 'Explaining the recent performance of the major urban regions of the European Community', *Urban Studies*, 27(3), 311–33.

Cheshire, P.C. 1992, *European Integration and Regional Response*, CeSAER Working Paper No. 1, Faculty of Urban & Regional Studies, University of Reading.

Cheshire, P.C., Camagni, R. & Gaudemar, J.P. 1991, '1957 to 1992, towards a Europe of regions and regional policy', in *Industrial Change and Regional Economic Transformation: the experience of Western Europe*, eds L. Rodwin & H. Sazanami, Harper Collins Academic, London.

Cheshire, P.C., D'Arcy, E. & Giussani, B. 1992, 'Purpose built for failure? Local, regional and national government in Britain', *Environment and Planning C: Government and Policy*, 10, 355–69.

Cheshire P.C. & Hay, D.G. 1986, 'The development of the European urban system', in *The Future of the Metropolis*, ed. H.-J. Ewers, Walter de Gruyer and Co., Berlin.

Cheshire, P.C. & Hay, D.G. 1989, *Urban Problems in Western Europe: An Economic Analysis*, Unwin Hyman, London.

Clark, C. 1940, *The Conditions of Economic Progress*, Macmillan, London.

Clark, C. 1951, 'Urban population densities', *Journal of the Royal Statistical Society*, A, 114, 490–96.

Clark, C. 1957, 'Transport: maker and breaker of cities', *Town Planning Review*, 28, 237–50.

Clark, C. 1967, *Population Growth and Land Use*, Macmillan, London.

Clark, G.L. 1992, 'Global interdependence and regional development', Presented to international meeting on *Regional Economies and Global Interdependence: Strengthening Business Linkages in the Western Pacific*, UNCRD, Wakayama, November.

Clark, W. & Burt, J. 1980, 'The impact of workplace on residential location', *Annals of the Association of American Geographers*, 70(1), 59–67.

Cohen, J. & Simmie, J. 1991, 'Regional policy, innovation and technopolis planning in Britain and France', University College, London (mimeo).

Cohen, R. 1987, *The New Helots: Migrants in the International Division of Labour*, Avebury, London.

Cohen, S.S. & Zysman, J. 1987, *Manufacturing Matters—The Myth of the Post Industrial Economy*, Basic Books, New York.

Commission of the European Community 1987, *Third Periodic Report on the Regions of Europe*, CEC, Luxembourg.

Commission of the European Communities 1990, *Green Paper on the Urban Environment*, European Commission, Brussels.

Commission of the European Communities 1991, *Europe 2000: Outlook for the Development of the Community's Territory*, Office of Official Publications of the EC, Luxembourg.

Commission of the European Communities 1992a, *Urbanization and the Function of Cities in the European Community*, A research report prepared by the European Institute of Urban Affairs, Liverpool John Moores University, Directorate General for Regional Policies, Brussels.

Commission of the European Community 1992b, *Fourth Periodic Report on the Regions of Europe*, CEC, Luxembourg.

Commission of the European Communities 1992c, *The Impact of Transport on the Urban Environment*, European Commission, Brussels.

Conway, M. 1993, 'Tomorrow's supercities', *The Futurist*, 27, May–June, 27–33.

Coopers and Lybrand Consultants 1992, *National Report on Population Growth Ranking in Australia*, June.

Coopers & Lybrand Deloitte 1991, *London: World City Moving into the 21st Century*, a research project sponsored by the London Planning Advisory Committee *et al.*, HMSO, London.

Corporation of London 1989, *City of London Local Plan Second Monitoring Report July 1987–June 1988*, Corporation of London, London.

Couclelis, H. 1985, 'Cellular worlds: A framework for modeling macro dynamics', *Environment and Planning A*, 17, 585–96.

Cowan, P., Fine, D., Ireland, J., Jordan, C., Mercer, D. and Sears, A. 1968, *The Office: A Facet of Urban Growth*, Heinemann, London.

Crawford, J.R., Ioannou, S., Katz, M., Penford, P., Newton, P.W., Cavill, M. & Connor, J. 1991, *Pharmacies GIS Study for ISDN: Report 1—Melbourne; Report 2—Sydney*, Telecom Research Laboratories, Melbourne, June.

Cresswell, P. & Thomas, R. 1972, 'Employment and population balance', in *New Towns: The British Experience*, ed. H. Evans, Charles Knight and Co. for the Town and Country Planning Association, London.

Critchley, J. 1992, 'The change in global route networks. A case study of Qantas 1965–1990', BA Honours Thesis, Department of Geography and Environmental Science, Monash University, Clayton, Victoria.

Cropper, M. & Gordon, P. 1991, 'Wasteful commuting: a re-examination', *Journal of Urban Economics*, 29, 2–13.

Cross, D. 1990, *Counterurbanisation in England and Wales*, Avebury, Aldershot.

Daly, M.T. & Stimson, R.J. 1991, 'Australian cities and property development in the international context: the challenges', *Pacific Rim Council on Urban Development, Conference Proceedings*, Hong Kong, October.

Daly, M.T. & Stimson, R.J. 1992: 'Sydney: Australia's gateway and financial capital', Ch. 18, in *New Cities of the Pacific Rim*, eds E.J. Blakely & R.J. Stimson, Monograph No. 43, Institute for Urban and Regional Development, University of California, Berkeley.

Daly, M.T. & Stimson, R.J. (forthcoming), 'Dependency in the modern global economy: Australia and the changing face of Asian finance', *Environment and Planning A*.

Daniels, P.W. 1972, 'Transport changes generated by decentralized offices', *Regional Studies*, 6(3), 273–89.

Daniels, P.W. 1981, 'Transport changes generated by decentralized offices: a second survey', *Regional Studies*, 15(6), 507–20.

Daniels, P.W. 1991, 'Producer services and the development of the space economy', in *The Changing Geography of Advanced Producer Services*, eds P.W. Daniels & F. Moulaert, Belhaven Press, London & New York.

Daniels, P.W. 1993, *Service Industries in the World Economy*, Blackwell, Oxford.

Daniels, P.W. & Bobe, J.M. 1990, *Planning for Office Development in the City of London and Adjacent Boroughs: A Review of Recent Policy*, Service Industries Research Centre Working Papers, No. 1, University of Portsmouth.

Daniels, P.W., O'Connor, K. & Hutton, T.A. 1991, 'The planning response to urban services growth: an international comparison', *Growth and Change*, Fall, 3–26.

Dasgupta, M., Frost, M. & Spence, N. 1989, *Mode Choice in Travel to Work in British Cities 1971–1981*, Research undertaken for the Transport and Road Research Laboratory, Crowthorne, Berkshire. Research Paper, Department of Geography, London School of Economics and Political Science.

Dasgupta, M., Frost, M. & Spence, N. 1990a, *Employment Patterns and Travel Distances in British Cities: 1981*, Research undertaken for the Transport and Road Research Laboratory, Crowthorne, Berkshire, Research Paper, Department of Geography, London School of Economics and Political Science.

Dasgupta, M., Frost, M. & Spence, N. 1990b, *Journey to Work Trends in British Cities 1971–1981*, Research undertaken for the Transport and Road Research Laboratory, Crowthorne, Berkshire, Research Paper, Department of Geography, London School of Economics and Political Science.

Davelaar, E.J. & Nijkamp, P. 1989, 'The role of the metropolitan milieu as an incubation centre for technological innovations: a Dutch case study', *Urban Studies*, 26, 517–25.

Davies, R.E.G. 1992, 'Technology: history, evolution and future of aircraft. Future trends in air transport', *Lessons from History, Papers of Conference on Airlines, Airports and Aviation*, Institute for Advanced Legal Studies, University of Denver, May, 105–14.

Dawkins, R. 1986, *The Blind Watchmaker*, Longmans Scientific and Technical, London.

DEET 1991, *Australia's Workforce in the Year 2000*, Department of Employment, Education and Training, Australian Government Publishing Service, Canberra.

Dempsey, P.S. 1992, 'The state of the airline industry: deregulation and concentration', *Lessons from History, Papers of Conference on Airlines, Airports and Aviation*, Institute for Advanced Legal Studies, University of Denver, May, 1–64.

Dendrinos, D.S. 1992, *The Dynamics of Cities: Ecological Determinism, Dualism, and Chaos*, Routledge, London

Dendrinos, D.S. & Mullally, H. 1985, *Urban Evolution: Studies in the Mathematical Ecology of Cities*, Oxford University Press, New York.

Department of the Environment (DOE) 1989, *Draft Strategic Planning Guidance for London*, Department of the Environment, London.

Department of the Environment 1990, *This Common Inheritance*, HMSO, London.

Department of the Environment 1991, *Household Projections, England 1989–2011*, Department of the Environment, London.

Department of the Environment 1992, *Development Plans and Regional Planning Guidance, PPG12*, HMSO, London.

Department of the Environment 1993, *Transport, Draft PPG13*, Department of the Environment, London.

Department of Transport 1988, *National Travel Survey 1985/86 Report*, HMSO, London.

Dicken, P. 1986, *Global Shift: Industrial Change in a Turbulent World*, Harper and Row, London.

Dicken, P. 1992a, *Global Shift: The Internationalisation of Economic Activity*, The Guilford Press, New York.

Dicken, P. 1992b, *Global Shift*, 2nd Edn, Paul Chapman, London.

Dillon, D., Weiss, S. & Hait, P. 1989, 'Supersuburbs', *Planning*, 55, 7–21.

Donnelly, S.B. 1991, 'The west: mixing business and faith', *Time*, 29 July.

Dosi, G. 1984, *Technical Change and Industrial Transformation*, St Martin's Press, New York.

Dosi, G., Freeman, C., Nelson, R., Silverberg, G. & Soete, L. (eds) 1988, *Technological Change and Economic Theory*, Pinter Publishers, London.

Douglas, B. 1992, 'Comparison of commuting trends between downtown, suburban centers, and suburban campuses in the Washington metropolitan area', Parsons-Brinckerhoff-Quade-Douglas, Washington, DC (mimeo).

Downs, A. 1962, 'The law of peak-hour expressway congestion', *Traffic Quarterly*, 16, 393–409.

Downs, A. 1992, *Stuck in Traffic: Coping with Peak-Hour Traffic Congestion*, The Brookings Institution, Washington, DC.

Drennan, M.P., 1989, 'Information intensive industries in metropolitan areas of the United States of America', *Environment and Planning A*, 21, 1603–18.

Drennan, M.P. 1992, 'Gateway cities: the metropolitan sources of US producer service exports', *Urban Studies*, 29(2), 217–35.

Dubin, R. 1991, 'Commuting patterns and firm decentralization', *Land Economics*, 67(1), 15–29.

Duffy, F. 1986, 'Exploding the myths of New York', *The Banker*, 136(727), September, 109.

Duffy, F. & Henney, A. 1989, *The Changing City*, Bulstrode Press, London.

Dunning, J.H. & Norman G. 1987, 'The location and choice of offices of international companies', *Environment and Planning A*, 19, 613–31.

Eco, U. 1987, *Travels in Hyper-Reality*, Picador, London.

Ecologically Sustainable Development 1991a, *Greenhouse Report*, ESD Secretariat, Australian Government Publishing Service, Canberra, ACT.

Ecologically Sustainable Development 1991b, *Report by Transport Working Group on Ecologically Sustainable Development*, Department of Prime Minister and Cabinet, Australian Government Publishing Service, Canberra, ACT.

ECOTEC 1993, *Reducing Transport Emissions Through Planning*, HMSO, London.

Edwards, D.W., Elam, J.J. & Mason, R.O. 1989, *Securing an Urban Advantage: How US Cities Compete Through IT*, International Centre for Information Technologies, Washington, DC.

Egerton-Smith, G., Moore, J. & Webster, S. 1985, 'Office accommodation costs', *Estates Gazette*, 19 January, 31–4.

Elkin, T, McLaren, D. & Hillman, M. 1991, *Reviving the City: Towards Sustainable Urban Development*, Friends of the Earth, London.

Elwood, S. 1988, 'Refurbishing office blocks: re-paving paradise', *Estates Gazette*, 10 December, 65–92.

Emerson, M. 1988, *The Economics of 1992*, Oxford University Press, Oxford.

Enderwick, P. (ed.) 1989, *Multinational Service Firms*, Routledge, London.

Engelbrecht, H.J. 1986, 'Analysis of structural change using an information sector perspective', *Asian Economies*, 58, 22–46.

EPAC 1991, *A Joint Approach to Tourism—The Main Issues*, Discussion Paper 90/04, Economic Planning Advisory Council, Australian Government Publishing Service, Canberra, June.

Evans, A.E. 1973, 'The location of the headquarters of industrial companies', *Urban Studies*, 10, 387–95.

Fainstein, S.S., Gordon, I.R. & Harloe, M. 1992a, *Divided Cities: Economic Restructuring and Social Change in London and New York*, Basil Blackwell, London.

Fainstein, S.S., Gordon, I.R. & Harloe, M. (eds) 1992b, *Divided Cities: New York and London in the Contemporary World*, Blackwell, Cambridge, MA and Oxford.

Feder, B.J. 1991, 'Omaha: talk, talk, talk of telemarketing', *New York Times*, 20 July.

Federal Highway Administration 1991, *1990 National Personal Transportation Study: Early Results*, Office of Highway Information Management, FHWA, Washington, DC.

Feldman, M. 1983, 'Biotechnology and local economic growth: the American pattern', *Built Environment*, 9(1), 40–50.

Felsenstein, D. & Shachar, A. 1988, 'Locational and organizational determinants of R&D employment in high technology firms', *Regional Studies*, 22(6), 477–86.

Fielding, A. 1991, *Migration to and from south-east England*, report to Department of the Environment, London.

Findlay, C.C. 1985, *The Flying Kangaroo: An Endangered Species?*, Allen and Unwin, Sydney.

Fisher, J. & Mitchelson, R. 1981, 'Extended and internal commuting in the transformation of the intermetropolitan periphery', *Economic Geography*, 57(3), 189–207.

Flatow, I. 1992, *They All Laughed...*, Harper Collins, New York.

Flood, J. 1992, 'Internal migration in Australia: who gains, who loses', *Urban Futures*, Special Issue 5, February, 44–53.

Flood, J,. Maher, C.A., Newton, P.W. & Roy, J.R. 1991, *The Determinants of Internal Migration in Australia*, Indicative Planning Council for the Housing Industry (IPC), Canberra.

Florida, R.L. & Kenney, M. 1988, 'Venture capital, high technology and regional development', *Regional Studies*, 22(1), 33–48.

Florida, R.L. & Kenney, M. 1990, 'High-technology restructuring in the USA and Japan', *Environment and Planning A*, 22(3), 233–52.

Forester, T. 1992, 'The electronic cottage revisited', *Urban Futures*, Special Issue 5, February, 27–33.

Fortune Magazine 1992, 'Unfriendly skies', *Fortune*, 2 November, 32–3.

Fothergill, S. & Gudgin, G. 1982, *Unequal Growth: Urban and Regional Employment Change in the U.K.*, Heinemann Educational, London.

Freeman, C. 1987, *Technology Policy and Economic Performance: Lessons from Japan*, Pinter Publishers, London.

Friedmann, J. & Wolff, G. 1982, 'World city formation: an agenda for research and action', *International Journal of Urban and Regional Research*, 6(3), 309–44.

Friedrichs, J. 1985, *Stadtentwicklungen in West- und Osteuropa*, de Gruyter, Berlin.

Frost, M. & Spence, N. 1981, 'Employment and worktravel in a selection of English inner cities', *Geoforum* 12, 107–60.

Frost, M. & Spence, N. 1991a, 'Employment changes in Central London in the 1980s: the record of the 1980s', Geographical Journal, 157, 1–12.

Frost, M. & Spence, N. 1991b, 'Understanding recent forces for change and future development constraints', *Geographical Journal*, 157, 125–35.

Frost, M. & Spence, N. 1993, 'Global city characteristics and Central London's employment', *Urban Studies*, 30(3), 547–58.

Fulton, W. 1990, 'The long distance commute', *Planning*, 4, 6–12.

Furino, A. 1988, *Cooperation and Competition in the Global Economy: Issues and Strategies*, Ballinger, Cambridge, Massachusetts.

Gappert, G. 1987, 'Urban issues in an advanced industrial society', in *The Spatial Impact of Technological Change*, eds J. F. Brotchie, P. Hall & P. W. Newton, Croom Helm, London.

Garreau, J. 1991, *Edge City: Life on the New Frontier*, Doubleday, New York.

Garrison, W.L. & Deakin, E. 1988, 'Travel, work and telecommunications: a view of the electronics revolution and its potential impacts', *Transportation Research*, 22A, 239–45.

Gershuny, J. & Miles, I. 1983, *The New Service Economy: The Transformation of Employment in Industrial Societies*, Frances Pinter, London.

Gillespie, A. & Robins, K. 1989, 'Geographical equalities: the spatial bias of new communications technology', *Journal of Communications*, 39, 7–18.

Giuliano, G. 1991, 'Is jobs-housing balance a transportation issue?', *Transportation Research Record*, 1305, 305–12.

Giuliano, G. & Small, K. 1991, 'Subcenters in the Los Angeles region,' *Regional Science and Urban Economics*, 5, 163–82.

Giuliano, G. & Small, K. 1992, *Is the Journey to Work Explained by Urban Structure?*, University of California Transportation Center Working Paper 107, Berkeley.

Glaeser, E.L., Kallal, H.D., Scheinkman, J.A. & Shleifer, A. 1992, 'Growth in cities', *Journal of Political Economy*, 100(6), 1126–52

Glasmeier, A. 1988a, 'Factors governing the development of high tech industry agglomerations: a tale of three cities', *Regional Studies*, 22(4), 287–301.

Glasmeier, A. 1988b, 'The Japanese technopolis programme: high tech development policy or industrial policy in disguise?', *International Journal of Urban and Regional Research*, 12, 6–8.

Glasmeier, A., Hall, P. & Markusen, A. 1984, 'Metropolitan high-technology industry growth in the mid 1970s: can everyone have a slice of the high-tech pie?', *Berkeley Planning Journal*, 1(1), 30–142.

Goddard, J.B. 1989, 'Preface', in *Geography of the Information Economy*, ed. M.E. Hepworth, Belhaven Press, London.

Goddard, J.B. 1990, *The Geography of the Information Economy*, PICT Policy Research Paper 11, CURDS, November.

Goddard, J.B. & Gillespie, A.E. 1987, 'Advanced telecommunications and regional economic development', in *Managing the City: The Aims and Impacts of Urban Policy*, ed. B. Robson, Croom Helm, London, 84–109.

Goddard, J.B. & Gillespie, A.E. 1988, 'Advanced telecommunications and regional economic development', in *Informatics and Regional Development*, eds M. Giaoutzi & P. Nijkamp, Avebury, Aldershot, 121–46.

Goetz, A. 1992, 'Air passenger transportation and growth in the US urban system', *Growth and Change*, 23, 217–38.

Goldberg, D. 1988, *Genetic Algorithms in Search, Optimization and Machine Learning*, Addison-Wesley, Reading, Massachusetts.

Gordon, I.R. 1988, 'Resurrecting counter-urbanisation: housing market influences on migration fluctuations from London', *Built Environment*, 13(4), 193–211.

Gordon, I.R. 1990, 'Regional policy and national politics in Britain', *Environment and Planning C: Government and Policy*, 8, 427–38.

Gordon, I. & Jayet, H. 1991, 'Territorial policy between cooperation and competition', University of Reading (mimeo).

Gordon, P., Kumar, A. & Richardson, H. 1988, 'Beyond the journey to work', *Transportation Research A*, 22A(6), 419–26.

Gordon, P., Kumar, A. & Richardson, H.W. 1989a, 'Congestion, changing metropolitan structure, and city size in the US', *International Regional Science Review*, 12(1), 45–56.

Gordon, P., Kumar, A. & Richardson, H.W. 1989b, 'The influence of metropolitan spatial structure on commuting time', *Journal of Urban Economics*, 26, 128–51.

Gordon, P., Kumar, A. & Richardson, H.W. 1990, 'Peak-spreading: how much?', *Transportation Research A*, 24A, 165–75.

Gordon, P. & Richardson, H.W. 1989a, 'Notes from the underground: the failure of urban mass transit', *The Public Interest*, 94, 77–86.

Gordon, P. & Richardson, H. 1989b, 'Gasoline consumption and cities—a reply', *Journal of the American Planning Association*, 55(3), 342–45.

Gordon, P. & Richardson, H.W. 1991a, 'L.A. lost and found', Presented at the *38th North American Meeting of the Regional Science Association*, New Orleans.

Gordon, P. & Richardson, H. 1991b, 'Anti-planning', presented to the Planning Transatlantic Conference, Oxford Polytechnic, Oxford.

Gordon, P., Richardson, H.W. & Jun, M.J. 1990, *The Commuting Paradox— Evidence from the Top Twenty*, Working Paper, School of Urban and Regional Planning, USC, Los Angeles.

Gordon, P., Richardson, H.W. & Jun, M.J. 1991, 'The commuting paradox: evidence from the top twenty', *Journal of the American Planning Association*, 57(4), 416–20.

Gordon, P., Richardson, H. & Wong, H. 1986, 'The distribution of population and employment in a polycentric city: the case of Los Angeles', *Environment and Planning A*, 18, 161–73.

Gordon, R. & Kimball, L. 1986, *Industrial Structure and the Changing Global Dynamics of Location in High Technology Industry*, Working Paper No. 3, Silicon Valley Research Group, University of California, Santa Cruz.

Graham, S.D. 1992, 'Electronic infrastructures in the city: some emerging municipal policy roles in the UK', *Urban Studies*, 29, 755–81.

Granstrand, O. 1982, *Technology, Management and Markets*, Frances Pinter, London.

Great Britain, Department of the Environment and Department of Transport 1993, *Reducing Transport Emissions Through Planning* (ECOTEC Research and Consulting Ltd in Association with Transportation Planning Associates), HMSO, London.

Greater London Council 1985, *The London Industrial Strategy*, GLC, London.

Green, R. & Holliday, J. 1991, *Country Planning—A Time for Action*, Town and Country Planning Association, London.

Grubler, A. 1990, *The Rise and Fall of Infrastructures*, Physica-Verlag, Heidelberg, Germany.

Gunn, J. 1988, *The Defeat of Distance: Qantas 1919–1939*, Queensland University Press, Brisbane.

Gunn, J. 1990a, *Challenging Horizons: Qantas 1939–1954*, University of Queensland Press, Brisbane.

Gunn, J. 1990b, *High Corridors: Qantas 1955–1970*, University of Queensland Press, Brisbane.

Gurstein, P. 1990, *Working at Home in the Live-in Office: Computers, Space, and the Social Life of Households*, PhD dissertation, Department of Architecture, University of California, Berkeley.

Hacking, A.J. 1986, *Economic Aspects of Biotechnology*, Cambridge University Press, Cambridge, Mass.

Hall, P. 1966, *The World Cities*, McGraw-Hill, New York.

Hall, P. 1987, 'The anatomy of job creation: nations, regions and cities in the 1960s and 1970s', *Regional Studies*, 21, 95–106.

Hall, P. 1988, 'The intellectual history of long waves', in *The Rhythms of Society*, eds M. Young & T. Schuller, Routledge, London and New York.

Hall, P. 1989a, 'The rise and fall of great cities: economic forces and population responses', in *The Rise and Fall of Great Cities: Aspects of Urbanization in the Western World*, ed. R. Lawton, Belhaven, London, 20–31.

Hall, P. 1989b, *London 2001*, Unwin Hyman, London.

Hall, P. 1990, *The Generation of Innovative Milieux: An Essay in Theoretical Synthesis*, Working Paper No. 505, Institute of Urban and Regional Development, University of California, Berkeley.

Hall, P. 1991a, 'Structural transformation in the regions of the United Kingdom', in *Industrial Change and Regional Economic Transformation: The Experience of Western Europe*, eds L. Rodwin & H. Sazanami, Unwin Hyman, London, 39–69.

Hall, P. 1991b, 'Moving information: a tale of four technologies', in *Cities of the 21st Century: New Technologies and Spatial Systems*, eds J. Brotchie, M. Batty, P. Hall & P. Newton, Longman Cheshire, Melbourne, 1–21.

Hall, P., Bornstein, L., Grier, R. & Webber, M. 1988, *Biotechnology: The Next Industrial Frontier*, Working Paper No. 474, Institute of Urban and Regional Development, University of California, Berkeley.

Hall, P., Breheny, M., McQuaid, R. & Hart, D. 1987, *Western Sunrise: The Genesis and Growth of Britain's Major High Tech Corridor*, Allen and Unwin, London.

Hall, P. & Hay, D. 1980, *Growth Centres in the European Urban System*, Heinemann Educational Books, London.

Hall, P. & Markusen, A. 1985, *Silicon Landscapes*, Allen and Unwin, Boston.

Hall, P. & Preston, P. 1988, *The Carrier Wave: New Information Technology and the Geography of Innovation, 1846–2003*, Unwin Hyman, London.

Hall, P., Sands, B. & Streeter, W. 1993, *Managing the Suburban Commute: A Cross-National Comparison*, University of California, Institute of Urban and Regional Development, Working Paper WP 5xx, Berkeley.

Hall, P., Thomas, R., Gracey, H. & Drewett, R. 1973, *The Containment of Urban England*, George Allen and Unwin, London.

Hamilton, B.W. 1982, 'Wasteful commuting', *Journal of Political Economy*, 90, 1035–53.

Hamilton, F.E.I. 1987, *Industrial Change in Advanced Economies*, Croom Helm, London.

Hamilton, W. 1991, *Serendipity City: Australia, Japan and the Multifunction Polis*, ABC Books, Crows Nest, NSW.

Hanks, J. & Lomax, T. 1991, 'Roadway congestion in major urban areas: 1982 to 1988', *Transportation Research Record*, 1205, 177–89.

Harding, A. 1991, 'Growth machines–UK style', *Environment and Planning C: Government and Policy*, 9, 295-316.

Harris, B. 1961, 'Some problems in the theory of intra-urban location', *Operations Research*, 9, 695–721.

Harris, B. 1970, *Change and Equilibrium in the Urban System*, Institute of Environmental Studies, University of Pennsylvania, Philadelphia, Pennsylvania.

Harris, B. 1991, 'What price knowledge in the electronic age?', in *Cities of the 21st Century: New Technologies and Spatial Systems*, eds J. Brotchie, M. Batty, P. Hall, & P. Newton, Longman Cheshire, Melbourne, pp. 339–52.

Harris, R. 1991a, 'The geography of employment and residence in New York Since 1950', in *Dual City: Restructuring New York*, eds J.H. Mollenkopf & M. Castells, Russell Sage Foundation, New York.

Harris, R.L.G. 1991b, 'The central London office market: a study of the structure of demand', Unpublished PhD Thesis, University of Bristol.

Harris, R.L.G. & Thrift, N. 1987, 'Internationalisation of demand', *World Property*, April, 65–7.

Hårsman, B. 1992, 'Strategies for the Stockholm-Mälar region', Paper to conference on Europe and its Regions, Summer University of Southern Stockholm/K.T.H., Stockholm.

Hartshorn, T.A. & Muller, P.O. 1986, *Suburban Business Centers: Employment Expectations*, Final Report for Department of Commerce, Economic Development Administration, Washington, DC, November.

Harvey, D. 1989, *The Condition of Postmodernity*, Basil Blackwell, Oxford.

Heikkila, E., Gordon, P., Kim, J., Peiser, R., Richardson, H. & Dale-Johnson, D. 1989, 'What happened to the CBD-distance gradient?: land values in a polycentric city', *Environment and Planning A*, 21, 221–32.

Hennings, G. & Kunzmann, K.R. 1990, 'Priority to local economic development: industrial restructuring and local development responses in the Ruhr area—the case of Dortmund', in *Global Challenge and Local Response—Initiatives for Economic Regeneration in Contemporary Europe*, ed. W.B. Stühr, Mansell, London, pp. 199–223.

Hennings, G. & Kunzmann, K.R. 1993, 'Local economic development in a traditional industrial area: the case of the Ruhrgebiet', in *Comparative Studies in Local Economic Development*, ed. P. Meyer, Greenwood Press, London, pp. 35–54.

Hepworth, M.E. 1987, 'The information city', *Cities*, August, 253–62.

Hepworth, M.E. 1989, *Geography of the Information Economy*, Belhaven Press, London.

Herbig, P. & Golden, J.E. 1993, 'How to keep that innovative spirit alive: an examination of evolving innovative hot spots', *Technological Forecasting and Social Change*, 43(1), 75–90.

Hill, C.T. & Utterback, J.M. 1979, *Technological Innovation for a Dynamic Economy*, Pergamon Press, New York.

Hirst, P. & Zeitlin, J. 1990, 'Flexible specialization versus post Fordism', presented to the *Conference on Pathways to Industrialization and Regional Development in the 1990s*, Lake Arrowhead, California, March.

Hoare, A.G. 1974, 'International airports as growth poles: a case study of Heathrow airport', *Transactions, Institute of British Geographers*, 63, 75–96.

Hodson, N. 1992, *The Economics of Teleworking*, UK: British Telecommunications, Ispwich.

Holmes, J.H. 1983, 'Telephone traffic dispersion and nodal regionalisation in the Australian States', *Australian Geographical Studies*, 21, 231–50.

Holtzclaw, J. 1990, 'Manhattanization versus sprawl: how urban density impacts auto use comparing five Bay Area communities', *Proceedings of the Eleventh International Pedestrian Conference*, 11, City of Boulder, Colorado, 99–106.

Holzer, H. 1991, 'The spatial mismatch hypothesis: what has the evidence shown?', *Urban Studies*, 28, 105–22.

Hopkins, N. 1992, 'MFP which way now', *Adelaide Advertiser*, 1 August.

Hoyle, B.S. 1988, 'Development dynamics at the port-city interface', in *Revitalizing the Waterfront: International Dimensions of Dockland Redevelopment*, eds B.S. Hoyle, D.A. Pinder, D.A. & M.S. Husain, Belhaven Press, London, 3–19.

Hu, P. & Young, J. 1992, *Summary of Travel Trends: 1990 Nationwide Personal Transportation Survey*, Federal Highway Administration, Washington, DC.

Hughes, M. 1992, 'Regional economics and edge cities', in *Edge City and ISTEA— Examining the Transportation Implications of Suburban Development Patterns*, Federal Highway Administration, Washington, DC.

Hugo, G.J. 1989, 'Australia: the spatial concentration of the turnaround', in *Counterurbanization: The Changing Pace and Nature of Population Decentralization*, ed. A.G. Champion, Edward Arnold, London.

Hugo, G.J. & Smailes, P.J. 1985, 'Urban-rural migration in Australia: a process view of the turnaround', *Journal of Rural Studies*, 1, 11–30.

Huxley, M. 1990, 'The multifunction polis: the issues', *Arena*, 90, 43–9.

Ihlandfeldt, K. & Sjoquist, D. 1990, 'Job accessibility and racial differences in youth employment rates', *American Economic Review*, 80(1), 267–76.

Inkster, I. 1991, *The Clever City: Japan, Australia and the Multifunction Polis*, Sydney University Press in association with Oxford University Press Australia.

International Civil Aviation Organisation (ICAO) 1989, *The Economic Situation of Air Transport, Review and Outlook 1978–2000*, Circular 272, ICAO, Montreal.

International Civil Aviation Organisation (ICAO) 1992, *World Wide Air Transport Colloquium: Exploring the Future of International Air Transport Regulation*, ICAO, Montreal, April.

Itaki, M. & Waterson, M. 1990, *European Multinationals and 1992*, Discussion Papers in Economics, Series B, III, No. 141, University of Reading, Reading.

Jacobs, J. 1961, *The Death and Life of Great American Cities*, Random House, New York.

Jacobsen, R. 1987, 'Shaping the information age policy agenda: the California experience', in *The Ideology of the Information Age*, eds J.D. Slack & F. Fejes, Ablex, Norwood, New Jersey, 170–78.

James, Paul (ed.) 1990, *Technocratic Dreaming: Of Very Fast Trains and Japanese Designer Cities*, Left Book Club, Melbourne.

Jarvie, W. 1989, 'Changes in internal migration in Australia: population or employment led', in *Regional Structural Change: Experience and Prospects in two Mature Economies*, eds L.J. Gibson & R.J. Stimson, Regional Science Research Institute, Peace Dale, Rhode Island, 47–60.

Jencks, C. & Mayer, S. 1990, 'Residential segregation, job proximity, and black job opportunities', in *Inner-City Poverty in the United States*, eds L. Lynn & M. McGeary, National Academy Press, Washington, DC.

Jencks, C. & Peterson, P.E. eds 1991, *The Urban Underclass*, Brookings Institution, Washington DC.

Johnson, D. 1991, 'Prosperity must make room for diversity in Utah', *New York Times*, 25 August.

Johnson, W.R. 1991, 'Anything, anytime, anywhere: the future of networking', in *Technology 2001: The Future of Computing and Communications*, ed. D. Leebaert, MIT Press, Cambridge, Massachusetts.

Joint Steering Committee 1990, *Multifunction Polis Feasibility Study*, Report to the Australian and Japanese Governments.

Jones, B. 1989, 'The multi-function polis: Australia's newest new city proposal: introduction', *Australian Planner*, 27(2), 6–7.

Jones Lang Wootton, 1990, *Business Space in Perspective: A Review of 'High Tech' Development in Australia*, JLW Property Research Pty Ltd, Sydney.

Karunaratne, N.D. 1986, 'Analytics of information and empirics of the information economy', *The Information Society*, 4, 313–31.

Kasarda, J. 1991, 'Global air industrial complexes as development tools', *Economic Development Quarterly*, 5(3), August.

Keeble, D. 1976, *Industrial Location and Planning in the United Kingdom*, Methuen, London.

Keeble, D., Offord, J. & Walker, S. 1988, *Peripheral Regions in a Community of Twelve Member States*, Final Report for the Commission of the European Communities, Office of Official Publications of the EC, Luxembourg.

Keen, P.G.W. 1986, *Competing in Time: Using Telecommunications for Competitive Advantage*, Ballinger, Cambridge, Massachusetts.

Keen, P.G.W. 1991, *Shaping the Future*, Harvard Business School Press, Cambridge, Massachusetts.

Keil, R. & Ronneberger, K. 1992, 'Going up the country: internationalization and urbanization on Frankfurt's northern fringe', Presented at the UCLA International Sociological Association, Research Committee 29. *A New Urban and Regional Hierarchy? Impacts of Modernization, Restructuring and the End of Bipolarity*, April 24–26, 1992.

Kellerman, A. 1984, 'Telecommunications and the geography of metropolitan areas', *Progress in Human Geography*, 8(2), 222–46.

Kennedy, P. 1987, *The Rise and Fall of Great Powers*, Random House, New York.

Kenney, M. 1986, *Biotechnology: The University-Industrial Complex*, Yale University Press, New Haven.

King, A.D. 1990, *Global Cities, Post Imperialism and the Internationalisation of London*, Routledge, London.

Knight, R.V. 1987, 'City development in advanced industrial countries', in *Cities in the Twenty First Century*, eds G. Gappert & R.V. Knight, vol. 23, Urban Affairs Annual Reviews, Sage Publications, Beverly Hills.

Knight, R.V. 1989, 'City building in a global society', in *Cities in a Global Society*, eds R.V. Knight & G. Gappert, vol. 25, Urban Affairs Annual Review, Sage Publications, Newbury Park, 326–34.

Kokudocho 1991, *Daitoshiken no Seibi*, (The affairs of metropolitan area), National Land Agency.

Koushki, P.A. 1991, 'Auto fuel elasticity in a rapidly developing urban area', *Transportation Research A*, 25A(6), 399–405.

Krugman, P.R. 1986, *Strategic Trade Policy and the New International Economics*, MIT Press, Cambridge, Massachusetts.

Kunzmann, K.R. 1990, 'Das regionale Kreativitätspotential: Standortfaktor für Innovationen', in *Innovationschancen für Mittelstädte. Heidenheimer Schriften zur Regionalwissenschaft*, ed. R. Funk, Verlag Stadt Heideuheim au der Brenz, Heidenheim, 10, 52–72.

Kunzmann, K. & Wegener, M. 1991, *The Pattern of Urbanisation in Western Europe 1960–1990*, Berichte aus dem Institut für Raumplanung 32, Institut für Raumplanung, Universität Dortmund, Dortmund.

Lamberton, D.M. 1987, 'The Australian information economy: a sectoral analysis', in *Challenges and Change: Australia's Information Society*, ed. T. Barr, Oxford University Press, Melbourne.

Lamberton, D.M. 1988, 'Secondary sector analysis: methods and data requirements', in *The Cost of Thinking*, eds M. Jussawalla *et al.*, Ablex Publishing Corporation, Norwood.

Langdale, J.V. 1991, *Internationalisation of Australia's Services Industries*, Department of Industry, Technology and Commerce, Australian Government Publishing Service, Canberra.

Langton, C.G. 1992, 'Life at the edge of chaos', in *Artificial Life II*, eds C.G. Langton, C. Taylor, J.D. Farmer & S. Rasmussen, Addison-Wesley, Reading, Massachusetts, 41–91.

Lash, S. & Urry, J. 1987, *The End of Organized Capitalism*, Polity Press, Cambridge.

Lazak, D. 1987, 'Evolution of teleports and telematic cities', in *Information Networks and Data Communication*, ed. D. Khakhar, Elsevier, Amsterdam.

LBS (London Business School) 1992a, *The City Research Project: Key Issues for the Square Mile*, Interim Report, Corporation of London, London.

LBS (London Business School) 1992b, *London's Competitive Position*, Interim Report, Corporation of London, London.

Lee, R. 1992, 'Travel demand and transportation policy beyond the edge', University of California, Department of City and Regional Planning, Berkeley (mimeo).

LEEL, 1992, *European City Regions: Establishing Edinburgh's Competitive Position*, a Research Brief, Lothian and Edinburgh Enterprise Ltd., Edinburgh.

Lever, W. & Moore, C. eds 1986, *The City in Transition: Policies and Agencies for the Economic Regeneration of Clydeside*, Oxford University Press, Oxford.

Levine, R.A., Levine, S.E., Richman, A., Uribe, F.M.T., Sunderland, C.C. & Miller, P.M. 1991, 'Women's schooling and child care in the demographic transition: a Mexican case study', *Population and Development Review*, 17, 459–96.

Lewin, R. 1992, *Complexity: Life at the Edge of Chaos*, Macmillan Publishing Company, New York.

Ley, D. 1985, 'Work-residence relationships for head office employees in an inflating housing market', *Urban Studies*, 22(1), 21–38.

Leyshon, A., Daniels, P. & Thrift, N. 1987a, *Large Accountancy Firms in the U.K.: Spatial Development*, Working Paper, St David's University College, Lampter, UK, and University of Liverpool.

Leyshon, A., Thrift, N.J. & Daniels, P.W. 1987b, *'Sexy Greedy': The New International Financial System, the City of London and the South East of England*, Working Papers on Producer Services, No. 8, University of Bristol and Service Industries Research Centre, University of Portsmouth.

Liao, Y. 1993, 'Trip chaining in urban travel', Presented at the University of Southern California, Urban Economics Group Seminar.

Lieb, R.C. & Miller, R.A. 1988, 'JIT and corporate transportation requirements', *Transportation Journal*, Spring, 5–10.

Linzie, M. & Boman, D. 1991, *Mälarregionen i ett gränslöst Europe*, Allmänna Förlaget, Stockholm.

Lion, C.P. & Van De Mark, G. 1990, 'Los Angeles', in *The New Urban Infrastructure*, eds J. Schmandt, F. Williams, R.H. Wilson & S. Strover, Praeger, New York.

Little, A.D. 1992, *New Directions for South Australia's Economy*, Final Report of the Economic Development Strategy Study, August.

Lobb, O. 1979, *1976 Urban and Rural Travel Survey Summary of Findings: Travel Data, 4*, Report prepared for the Southern California Association of Governments.

Loe, E. 1987, 'What is not an intelligent building', *Estates Gazette*, 4 July, 2-21.

Logan, J.R. & Molotch, H.L. 1987, *Urban Fortunes: the Political Economy of Place*, University of California Press, Berkeley.

London Planning Advisory Committee 1991, *London: World City*, LPAC, Romford.

Los Angeles County Transportation Commission 1992, *LACTC Proposed 30-Year Integrated Transportation Plan*, LACTC, Los Angeles.

Loveland, J. 1990, 'Minneapolis', in *The New Urban Infrastructure*, eds J. Schmandt, F. Williams, R.H. Wilson & S. Strover, Praeger, New York.

Lovelock, J.E. 1979, *Gaia: A New Look at Life on Earth*, Oxford University Press, London.

Lowry, I.S. 1988, 'Planning for urban sprawl', in *Transportation Research Board Special Report 221: Looking Ahead: The Year 2000*, Transportation Research Board, Washington, DC.

LPAC, 1990, *London: World City*, a Research Brief, London Planning Advisory Committee, London.

Lueck, T. 1992, 'Taking the 8:15 to the suburbs: commuting transformed by growth outside New York City', *The New York Times*, 23 December.

Luger, M. & Goldstein, G. 1992, *Technology in the Garden: Research Parks and Regional Economic Development*, The University of North Carolina Press, Chapel Hill.

MacKay, R. 1992, '1992 and the relations with the EEC', in *Regional Development in the 1990s: The British Isles in Transition*, eds P. Townwroe & R. Martin, Regional Studies Association, London.

Maillat, D. 1982, *Technology: A Key Factor for Regional Development*, Georgi Publishing Company, Saint-Saphorin.

Maillat, D. 1990, 'Innovation and local dynamism: the role of the milieu', *Sociologica Internationalis*, 28, 147–59.

Maillat, D. 1991, 'Local dynamism, milieu and innovative enterprises', in *Cities of the 21st Century: New Technologies and Spatial Systems*, eds J. Brotchie, M. Batty, P. Hall & P. Newton, Longman Cheshire, Melbourne, 265–74.

Malech, A. & Grodsky, M. 1991, *The Managerial Implications of Telecommuting: Final Report*, University of Maryland Center for the Study of Management & Organizations.

Malecki, E.J. 1981, 'Science, technology, and regional development: review and prospects', *Research Policy*, 10, 312–34.

Mandeville, T. 1988, 'A 'multi-function polis' for Australia', *Prometheus*, 6(1), 94–106.

Mandeville, T. 1991, 'The multifunction polis: an information age institution', in *Cities of the 21st Century*, eds J. Brotchie, M. Batty, P. Hall & P. Newton, Longman Cheshire, Melbourne.

Marchetti, C. 1992, *Anthropological Invariants in Travel Behaviour*, International Institute of Applied Systems Analysis, Laxenburg, Austria.

Markusen, A. 1985, *Profit Cycles, Oliogopoly and Regional Development*, MIT Press, Cambridge, Massachusetts.

Markusen, A. & Gwiasda, V. 1991, 'Multi-polarity and the layering of functions in the world cities: New York City's struggle to stay on top', presented in Tokyo, Japan, at the *Conference 'New York, Tokyo & Paris'*, October, Revised for publication 1993.

Markusen, A., Hall, P., Campbell, S. & Deitrick, S. 1991, *The Rise of the Gunbelt: The Military Mapping of Industrial America*, Oxford University Press, New York.

Markusen, A., Hall, P. & Glasmeier, A. 1986, *High Tech America: The What, How, Where, and Why of the Sunrise Industries*, Allen and Unwin, Boston.

Marshall, A. 1993, 'Refugees pay a high price as Europe raises the drawbridge', *The Independent (London)*, 28 May.

Marshall, J.N. *et al.* 1986, 'Uneven development in the service economy: understanding the location and role of producer services', Report of the Producer Services Working Party, Institute of British Geographers and the ESRC.

Marshall, M. 1987, *Long Waves of Regional Development*, Macmillan, London.

Mascarenhas, A. 1992, Presentation to *World Wide Transport Colloquium*, Paper WATC 1.17, ICAO, Montreal.

Masser, I. 1989, 'Technology and regional development policy: a review of Japan's technopolis programme', Presented to *Annual Conference of the American Collegiate Schools of Planning*, Portland, Oregon.

Masser, I., Sviden, O. & Wegener, M. 1992, *The Geography of Europe's Futures*, Belhaven Press, London.

Massey, D. & Allen, J. 1988, *Uneven Re-Development: Cities and Regions in Transition*, Hodder and Stoughton, London.

Matteaccioli, A. & Peyrache, V. 1989, 'Milieux et réseaux innovateurs: synthése sous l'angle de la complexité', *Cahiers du C3E*, 78, 1–25.

McCormack, G. 1990, 'Metamorphosis of the MFP', *Australian Society*, 9(9), 20–1.

McCormack, G. (ed.) 1991a, *Bonsai Australia Banzai. Multifunction Polis and the Making of a Special Relationship with Japan*, Pluto Press, Leichhardt, NSW.

McCormack, G. 1991b, 'Coping with Japan: The MFP proposal and the Australian response', in *Bonsai Australia Banzai*, ed. G. McCormack, Pluto Press, Leichhardt, NSW.

McLaren, D. 1992, 'Compact or dispersed?—Dilution is no solution', *Built Environment*, 18(4), 268–84.

Meadows, D.H., Meadows, D.L. & Randers, J. 1992, *Beyond the Limits: Confronting Global Collapse, Envisioning a Sustainable Future*, Chelsea Green Publishing Company, Post Mills, Vermont.

Meier, R.L. 1974, *Planning for an Urban World: Design of Resource-Conserving Cities*, MIT Press, Cambridge, Massachusetts.

Meier, R.L. 1976, 'A stable urban ecosystem: its evolution within densely populated societies', *Science*, 192, 962–68.

Meier, R.L. 1980, 'A stable urban ecosystem: anticipations for the third world', *Urban Ecology*, 4.

Meier, R.L. & Crane, J. 1978, 'Less is more in megalopolis', *Society, March/April 8*, 55–63.

MFP Adelaide Management Board 1991, *MFP-Adelaide: Design Concept Development and Core Site Assessment*, Vol. 1, Report by Kinhill-Delfin Joint Venture, May.

Ministry of Planning and Environment 1988, *A Population and Household Forecast for Metropolitan Melbourne*, Information Bulletin No. 2, Ministry of Planning and Environment, Melbourne.

Mokhtarian, P. 1990, 'A typology of relationships between telecommunications and transportation', *Transportation Research*, 24A(3), 231–42.

Mokhtarian, P.L. 1991, 'Telecommuting and travel: state of the practice, state of the art', *Transportation*, 18, 319–42.

Mollenkopf, J. & Castells, M. eds 1991, *Dual City: Restructuring New York*, Russell Sage Foundation, New York.

Molotch, H. 1976, 'The city as a growth machine', *American Journal of Sociology*, 82, 309–30.

Moodie, D. 1992, 'MFP: dream or nightmare', *Charter*, 63(8), 10–5.

Moravec, H. 1988, *Brain Children*, Harvard University Press, Cambridge, Massachusetts.

Morgan, K. 1992, *The Challenge of Lean Production in German Industry*, Regional Research Report No. 12, Department of City & Regional Planning, University of Wales College of Cardiff.

Morgan, K. & Sayer, A. 1988, *Microcircuits of Capital: 'Sunrise' Industry and Uneven Development*, Westview Press, Boulder.

Moss, M. 1987, 'Telecommunications and international financial centres', in *The Spatial Impact of Technological Change*, eds J.F. Brotchie, P. Hall & P.W. Newton, Croom Helm, London.

Moss, M. 1988, 'Telecommunications: shaping the future', in *America's New Market Geography*, eds G. Sternlieb & J. Hughes, Center for Urban Policy Research, Rutgers State University, New Jersey.

Mouer, R.E. & Sugimoto, Y. (eds) 1990, *The MFP Debate: A Background Reader*, Latrobe University Press, Australia.

Mulgan, G.J. 1991, *Communication and Control: Networks and the New Economics of Communication*, The Guilford Press, New York.

Myrdal, G. 1957, *Competing in Time: Using Telecommunications for Competitive Advantage*, Ballinger, Cambridge, Massachusetts.

Nakamura, H. & Ueda, T. 1989, 'The impacts of the Shinkansen on regional development', presented to *Fifth World Conference on Transport Research*, Yokohama, 16 pp.

Nakicenovic, N. 1989, 'Expanding territories: transport systems past and future', in *Transportation for the Future*, eds D. Batten & R. Thord, Springer-Verlag, Berlin.

Nance, J.J. 1984, *A Slash of Colours: The Self Destruction of Braniff International*, William Morrow & Co, New York.

National Capital Planning Authority 1990, *MFP Urban Development Concept*, Report to the Department of Industry, Technology and Commerce, Canberra.

National Physical Planning Agency (The Netherlands) 1991, *Summary of the Fourth Report Extra on Physical Planning*, Ministry of Housing, Physical Planning and Environment, The Hague.

Newman, P. 1992, 'The compact city—An Australian perspective', *Built Environment*, 18(4), 285–300.

Newman, P.W.G. & Kenworthy, J.R. 1989a, *Cities and Automobile Dependence: A Sourcebook*, Gower, Aldershot.

Newman, P.G.W. & Kenworthy, J. 1989b, 'Gasoline consumption and cities—A comparison of US cities with a global survey', *Journal of the American Planning Association*, 55(1), 24–37.

Newman, P.W.G. & Kenworthy, J.R. 1992, 'Is there a role for physical planners?', *Journal of the American Planning Association*, 58, 353–62.

Newstead, A. 1989, 'Future information cities: Japan's vision', *Futures*, 21, 263–76.

Newton, P.W. 1991, 'Telematic underpinnings of the information economy', in *Cities of the 21st Century: New Technologies and Spatial Systems*, eds J.F. Brotchie, M. Batty, P. Hall & P.W. Newton, Longman Cheshire, Melbourne.

Newton, P.W. 1992, 'The new urban infrastructure: telecommunications and the urban economy', *Urban Futures*, Special Issue 5, February, 54–75.

Newton, P.W. 1993, 'Changing places? Firms, households and urban hierarchies in the information age', this volume.

Newton, P.W., Crawford, J.R., Ioannou, S., Bouchier, P. & Katz-Even, M. 1993a, 'GIS supports network analysis', *Business Geographics*, 1(1), 40–42.

Newton, P.W., Wilson, B.G., Crawford, J.R. & Tucker, S.N. 1993b, 'Networking construction: electronic integration of distributed information', in *Management of Information Technology for Construction*, eds K.S. Mathur, M.P. Betts & K.W. Tham, World Scientific and Global Publications Services, Singapore.

Nilles, J.M. 1991, 'Telecommuting and urban sprawl: mitigator or inciter?', *Transportation*, 18, 411–31.

Nittim, Z. 1990, 'Of feminist space and multi-function polis', *Refractory Girl*, No. 35, 44–7.

Nolan, C. & Shirazi, E. 1991, *The Telecommuting Manual: Flexible Jobs for a Productive, Successful Future*, Pacific Bell.

Nowlan, D. & Stewart, G. 1991, 'Downtown population growth and commuting trips', *Journal of the American Planning Association*, 57(2), 165–82.

Noyelle, T.J. & Dutka, A.B. 1988, *International Trade in Business Services: Accounting, Advertising*, Law and Management Consulting, Ballinger Publishing, Cambridge, Massachusetts.

Noyelle, T.J. & Peace, P. 1988, *The Information Industries: New York's New Export Base*, Columbia University, New York.

Noyelle, T.J. & Peace, P. 1991, 'Information industries; New York's new export base', in *Services and Metropolitan Development: International Perspectives*, ed. P.W. Daniels, 285–304, Routledge, London.

Noyelle, T.J. & Stanback, T.M. 1984, *The Economic Transformation of American Cities*, Rowman and Allanheld, Totowa, NJ.

O'Connor, K. 1980, 'The analysis of journey to work patterns in human geography', *Progress in Human Geography*, 4(4), 477–99.

O'Connor, K. 1990, *State of Australia*, National Centre for Australian Studies, Monash University, Clayton.

O'Connor, K. 1991, 'Creativity and metropolitan development: a study of media and advertising in Australia', *Australian Journal of Regional Studies*, 6, December, 1–14.

O'Connor, K. 1992, 'Economic activity in Australian cities: national and local trends and policy', *Urban Futures*, Special Issue 5, February, 86–95.

O'Connor, K. 1993, 'The geography of research and development activity in Australia', *Urban Policy and Research*.

O'Connor, K. & Edgington, D. 1991, 'Provider services and metropolitan development in Australia', in *Services and Metropolitan Development: International Perspectives*, ed. P. Daniels, Routledge, London.

O'Connor, K. & Scott, A. 1992, 'Airline services and metropolitan areas in the Asia–Pacific region', *Review of Urban and Regional Development Studies*, 4, 240–53.

O'Connor, W.E. 1971, *Economic Regulation of the World's Airlines*, Praeger, New York.

Ó hUallacháin, B. & Reid, N. 1992, 'The intrametropolitan location of services in the United States', *Urban Geography*, 13(4), 334–54.

Oakey, R. 1984, *High Technology Small Firms: Innovation and Regional Development in Britain and the United States*, Frances Pinter, London.

Oakey, R., Rothwell, R. & Cooper, S. 1988, *The Management of Innovation in High-Technology Small Firms: Innovation and Regional Development in Britain and the United States*, Pinter Publishers, London.

Office of Population Censuses and Surveys 1992a, *1991 Census—Preliminary Report for England and Wales*, OPCS, London.

Office of Population Censuses and Surveys 1992b, *OPCS Monitor PP1 92/1 Provisional Mid-1991 Population Estimates for England and Wales and Constituent Local and Health Authorities Based on 1991 Census Results*, OPCS, London.

Office of Technology Assessment 1984, *Technology, Innovation and Regional Economic Development*, United States Congress, US Government Printing Office, Washington, DC.

Office of Technology Assessment (OTA) 1990, *Critical Connections: Communications for the Future*, OTA, Washington, DC.

Olsen, M. 1965, *The Logic of Collective Action: Public Goods and the Theory of Groups*, Harvard University Press, Cambridge, MA.

Orlowska, A.M., Lister, A.M. & Fogg, I. 1992, 'Decentralising spatial databases', in *Networking Spatial Information Systems*, eds P.W. Newton, P.R. Zwart & M.E. Cavill, Belhaven Press, London.

Orrskog, L. & Snickars, F. 1992, 'On the sustainability of urban and regional structures', in *Sustainability Development and Urban Form*, ed. M. Breheny, Pion, London.

Owens, S.E. 1984, 'Spatial structure and energy demand', in *Energy Policy and Land Use Planning*, eds D.R. Cope, P.R. Hills & P. James, Pergamon, Oxford, 215–40.

Owens, S.E. 1986, *Energy, Planning and Urban Form*, Pion, London.

Owens, S.E. 1990, 'Land-use planning for energy efficiency', in *Energy, Land and Public Policy*, ed. J.B. Cullingworth, Transactions Publishers, Newark, Delaware, 53–98.

Owens, S. 1991, *Energy Conscious Planning*, Council for the Protection of Rural England, London.

Owens, S.E. 1992a, 'Energy, environmental sustainability and land-use planning', in *Sustainable Development and Urban Form*, ed. M.J. Breheny, Pion, London.

Owens, S.E. 1992b, 'Land-use planning for energy efficiency', *Applied Energy*, 43, 81–114.

Page, A. & Brain, D. 1992, 'Teleworking and decentralization: current realities and future trends', *Interlink 2000*, Chase Communications, Cornwall, UK.

Paris, C.T. 1992, 'The local context of international tourism urbanisation', *A New Urban Regional Hierarchy? Impacts of Modernization, Restructuring and The End of Bipolarity*, Conference at Lewis Centre for Regional Policy Studies, University of California, Los Angeles, April.

Parry, D.C. & Watkins, A.J. 1977, *The Rise of Sunbelt Cities*, Sage Publications, Beverly Hills.

Pavitt, K. 1980, *Technical Innovation and British Economic Performance*, Macmillan, London.

Peace, A. 1992, 'Development, democracy and the Australian MFP', *Chain Reaction*, 65, 23–6.

Peach, W. 1985, 'Changing patterns in the city', *Estates Gazette*, 19 January, 33–4.

Peach, W. & Jones, S. 1988, '1989—are you sitting comfortably?', *Estates Gazette*, 26 November, 86.

Pearce, W. 1992, 'Toxic fear to delay MFP start', *Adelaide Advertiser*, 28 November.

Pentecost, G. & Love, G. 1985 'Trends in office development', *Estates Gazette*, 23 November, 886–87.

Perpich, J.G. 1986, *Biotechnology in Society: Private Initiatives and Public Oversight*, Pergamon Press, New York.

Pickrell, D.H. 1992, 'A desire named streetcar', *Journal of the American Planning Association*, 58, 158–76.

Piore, M. & Sabel, C. 1984, *The Second Industrial Divide: Possibilities for Prosperity*, Basic Books, New York.

Pisarski, A.S. 1987, *Commuting in America*, Eno Foundation for Transportation, Westport, Connecticut.

Pisarski, A.S. 1992, *New Perspectives on Commuting*, Federal Highway Administration, Washington, DC.

Pisarski, A.S. 1992, *Travel Behavior Issues in the 90s*, US Department of Transportation, Federal Highway Administration, Office of Highway Information, Washington, DC.

Pivo, G. 1990, 'The net of beads: suburban office development in six metropolitan areas', *Journal of the American Planning Association*, 56(4), 457–69.

Polunin, I. 1989, 'Japanese travel boom', *Tourism Management*, March, 4–8.

Porat, M. 1977, *The Information Economy: Definitions and Measurement*, Special Publication 77–12(1), Office of Telecommunications, US Department of Commerce, Washington, DC.

PPK Consultants 1992, *MFP Australia. Gillman/Dry Creek Urban Development Proposal: Draft Environmental Impact Statement*, February.

Pred, A. 1975, 'On the spatial structure of organisations and the complexity of metropolitan interdependence', *Papers of the Regional Science Association*, 35, 115–40.

Prentis, S. 1984, *Biotechnology: A New Industrial Revolution*, George Brazillier, New York.

Prism Research Ltd. 1991, *Mobility of Industry from South East England*, report for Department of the Environment, London.

Pumain, D. 1992, 'Urban networks or urban hierarchies?', *Environment and Planning A*, 24(10), 1377–80.

Pumain, D. & Rozenblat, C. 1991, 'Multinational firms and the restructuring of the European urban system', Paper presented to North American Regional Science Congress, New Orleans.

Purvis, C. 1992, *County-to-County Commute Patterns in the San Francisco Bay Area*, Working Paper 3, Metropolitan Transportation Commission, Oakland, California.

Pushkarev, B. & Zupan, J. 1977, *Public Transportation and Land Use Policy*, Indiana University Press, Bloomington.

Quigley, J. & Weinberg, D. 1977, 'Intraurban residential mobility: a review and synthesis', *International Regional Science Review*, 1, 41–66.

Rees, J. 1986, *Technology, Regions, and Policy*, Rowman and Littlefield, Totowa, NJ.

Rees, J. & Stafford, H. 1986, 'Theories of regional growth and industrial location: their relevance for understanding high technology complexes', in *Technology Regions and Policy*, eds J. Rees, Rowman and Littlefield, Totowa, NJ.

Rice Center 1987, *Houston's Major Activity Centers and Worker Travel Behavior*, Rice Joint Center for Urban Mobility Research, Houston.

Rice Center 1989, *Suburban Activity Centers: Private Sector Participation*, Rice Joint Center for Urban Mobility Research, Houston.

Richardson, C. 1991, 'Some desirable extensions of economic theory', *Economic Papers*, 10(4), 59–69.

Richardson, H.W. 1989, 'Urban development issues in the Pacific Rim', *Conference on Urban Development of the Pacific Rim*, PRCUD, Los Angeles, August.

Richardson, H.W. & Gordon, P. 1989, 'Counting nonwork trips: the missing link in transportation, land use and urban policy', *Urban Land*, 48(9), 6–18.

Richardson, H.W. & Gordon, P. 1993, 'New data and old models in urban economics', Presented at the January meetings of the *American Economics Association*, Anaheim.

Richardson, R. 1993, 'Telecommunications, customer care and competitive advantage: social and economic issues arising from new telemediated services', PICT, CURDS, University of Newcastle-upon-Tyne (mimeo).

Rickaby, P. 1987, 'Six settlement patterns compared', *Environment and Planning B, Planning and Design*, 14, 193–223.

Rickaby, P. 1991, 'Energy and urban development in an archetypal English town', *Environment and Planning B, Planning and Design*, 18(2), 153–76.

Rickaby, P. & de la Barra, T. 1989, 'A theoretical comparison of strategic spatial options for city-regional development using the TRANUS model', in *Spatial Energy Analysis*, eds L. Lundqvist, L.-G. Mattsson & E. Eriksson, Avebury, Aldershot.

Rickaby, P., Steadman, P. & Barrett, M. 1992, 'Patterns of land use in English towns: implications for energy use and carbon dioxide emissions', in *Sustainable Development and Urban Form*, ed. M. Breheny, Pion, London.

Rimmer, P.J. 1988, 'Japanese construction contractors and the Australian states: another round of interstate rivalry', *International Journal of Urban and Regional Research*, 12(3), 404–24.

Rimmer, P.J. 1989, 'Putting multi-function polis into context: MITI's search for a place in the sun?', *Australian Planner*, 27(2)15–21.

Rimmer, P.J. 1991, 'The global intelligence corps and world cities: engineering consultancies on the move', in ed. P.W. Daniel, *Services and Metropolitan Development: International Perspectives*, Routledge, London and New York, 66–106.

Robertson, J. 1990, 'Alternative futures for cities', in *The Living City: Towards a Sustainable Future*, eds D. Cadman & G. Payne, Routledge, London.

Robinson, S. 1986, 'Analysing the information economy: tools and techniques', *Information Processing and Management*, 22, 183–202.

Rosenberg, N. 1982, *Inside the Black Box: Technology and Economics*, Cambridge University Press, Cambridge.

Rothwell, R. & Zegveld, W. 1983, *Innovation and the Small and Medium Sized Firm*, Frances Pinter, London.

Rothwell, R. & Zegveld, W. 1985, *Reindustrialization and Technology*, Longman, Harlow.

Roy, J.R. 1992, 'Transport efficiency in cities with sub centres', Invited paper to *Sixth World Conference of Transport Research*, Lyon.

Ruthven, P. 1992, 'Myths that hold us back', *Business Review Weekly*, 18 December, 15–21.

Saffo, P. 1990, 'Concepts take 20 years to become overnight success', *Infoworld*, 5 November, 66.

Saloman, I. 1984, 'Telecommuting: promises and reality', *Transport Reviews*, 4(1), 103–13.

Sassen, S. 1991, *The Global City: London, New York, Tokyo*, Princeton University Press, Princeton,

Sassen, S. 1994, *Cities in a World Economy*, Pine Forge/Sage, California (in press).

Sassen-Koob, S. 1981, 'Towards a conceptualization of immigrant labor', *Social Problems* 29 (October), 65–85.

Savel, F. & Rabin, G. 1992, *T.G.V.: Aménagement du Territorie et Environnement*, Bipe Conseil, for L'Association des Villes Européens T.G.V., Issy-les-Moulineaux.

Sawers, L. & Tabb, W.K. 1984, *Sunbelt/Snowbelt: Urban Development and Regional Restructuring*, Oxford University Press, New York.

Saxenian, A. 1983, 'The genesis of Silicon Valley', *Built Environment*, 9(1), 7–17.

Saxenian, A. 1989a, *The Cheshire Cat's Grin: Innovation, Regional Development, and the Cambridge Case*, Working Paper No. 497, Institute of Urban and Regional Development, University of California, Berkeley.

Saxenian, A. 1989b, 'Local area networks: industrial adaptation in Silicon Valley', Presented to *Third International Conference on Innovation, Technological Change and Spatial Impacts*, Selwyn College, Cambridge, UK.

Saxenian, A. 1990, 'Regional networks and the resurgence of Silicon Valley', *California Management Review*, 33, 89–112.

Saxenian, A.L. 1991, 'The origins and dynamics of production networks in Silicon Valley', *Research Policy*, 20, 423–37.

Saxenian, A. 1992, 'Contrasting patterns of business organization in Silicon Valley', *Environment and Planning D, Society & Space*, 10, 377–91.

Saxenian, A. 1993, *Regional Networks: Industrial Adaptation in Silicon Valley and Route 128*, Harvard University Press, Cambridge, Massachusetts.

Schipper, L. & Meyers, S. 1992, *Energy Efficiency and Human Activity*, Cambridge, Cambridge.

Schmandt, J. & Wilson, R. 1987, *Promoting High Technology Industry: Initiatives and Policies for State Governments*, Westview Press, Boulder.

Schnore, L. 1959, 'The timing of metropolitan decentralization: a contribution to the debate', *Journal of the American Institute of Planners*, 25(4), 200–6.

Schwartz, A. 1992, *Cities, Suburbs, and the Geography of Corporate Service Provision*, Working Paper No. 39, Center for Urban Policy Research, Rutgers University, New Brunswick, New Jersey.

Scott, A.J. 1983a, 'Industrial organization and the logic of intra-metropolitan location: I. theoretical considerations', *Economic Geography*, 59(3), 233–50.

Scott, A. 1983b, 'Industrial organization and the logic of intra-metropolitan location: the printed circuit industry', *Economic Geography*, 59(4), 343–67.

Scott, A. 1985, 'Location processes, urbanization, and territorial development', *Environment and Planning A*, 17, 479–501.

Scott, A.J. 1988a, *Metropolis: from the Division of Labor to Urban Form*, University of California Press, Berkeley.

Scott, A.J. 1988b, 'Flexible production systems and regional development: the rise of new industrial spaces in North America and western Europe', *International Journal of Urban and Regional Research*, 12, 171–86.

Scott, A.J. 1988c, *New Industrial Spaces: Flexible Production, Organization and Regional Development in North America and Europe*, Pion, London.

Scott, A.J. 1989, *The Technopoles of Southern California*, UCLA, Department of Geography, Research Papers in Economic and Urban Geography, 1, Los Angeles.

Scott, A.J. 1992, 'The collective order of flexible production agglomerations: lessons for local economic development policy and strategic choice', *Economic Geography*, 68(3), 219–33.

Scott, A.J. & Angel, D.P. 1987, 'The US semiconductor industry: a locational analysis', *Environment and Planning A*, 19, 875–912.

Scott, A.J. & Paul, A.S. 1990, 'Collective order and economic coordination in industrial agglomerations: the technopoles of Southern California', *Environment and Planning C: Government and Policy*, 8, 179–93.

Seek, N.H. & Dickenson, D. 1990, 'Australian commercial property markets: trends and opportunities', *Pacific Rim Council on Urban Development*, Hong Kong, October.

Segal Quince Wicksteed 1985, *The Cambridge Phenomenon: The Growth of High Technology Industry in a University Town*, Segal Quince Wicksteed, Cambridge, UK.

Self, P. 1989, 'International lessons for an Australian new city', *Australian Planner*, 27(2), 28–32.

Sellers, P. 1990, 'The best cities for business', *Fortune*, 22 October.

Shand, E. 1992, 'A big job for small business', *Business Review Weekly*, 13 December, 58.

Sharp, M. & Shearman, C. 1987, *European Technological Collaboration*, Routledge and Kegan Paul, London.

Shaw, R.P. 1985, *Intermetropolitan Migration in Canada: Changing Determinants Over Three Decades*, NC Press, Toronto.

Shaw, S. 1982, *Air Transport: A Marketing Perspective*, Pitman Publishers, New York.

Sherlock, H. 1990, *Cities are Good for Us*, Transport 2000, London.

Siebel, W. 1984, 'Krisenphänomene der Stadtentwicklung', *arch+ 75/76*, 67–70.

Small, K.A. and Song, S. 1992, 'Wasteful commuting: a resolution', *Journal of Political Economy*, 100, 4, 888-98.

Smil, V. 1991, 'Population growth and nitrogen: an exploration of a critical existential link', *Population and Development Review*, 17(4), 569–602.

Song, S. 1992, *Spatial Structure and Urban Commuting*, The University of California Transportation Center, Working Paper.

South Australian Government 1990, *Adelaide*, A Submission to the MFP Joint Secretariat by the South Australian Government, May.

Spiekermann, K. & Wegener, M. 1993, *Zeitkarten für die Raumplanung*, Arbeitspapier 117, Institut für Raumplanung, Universität Dortmund, Dortmund.

Stanback, T.M. Jr 1985, 'The changing fortunes of metropolitan economies', in *High Technology, Space and Society*, ed. M. Castells, (*Urban Affairs Annual Reviews, 28*), Sage, Beverly Hills, 122–42.

Stanback, T.M., Jr 1991, *The New Suburbanization: Challenge to the Central City*, Westview Press, Boulder, Colorado.

Stanback, T.M. Jr & Noyelle, T.J. 1982, *Cities in Transition: Changing Job Structures in Atlanta, Denver, Buffalo, Phoenix, Columbus (Ohio), Nashville, Charlotte*, Allenheld, Osmun, New Jersey.

Statistisches Bundesamt 1988, *Länderbericht EG-Staaten 1988*, Kohlhammer Verlag, Wiesbaden.

Steadman, P. 1979, *The Evolution of Designs: Biological Analogy in Architecture and the Applied Arts*, Cambridge University Press, Cambridge, UK.

Steadman, P. & Barrett, M. 1991, *The Potential Role of Town and Country Planning in Reducing Carbon Dioxide Emissions*, report to Department of the Environment, London.

Stevens, M. 1992, 'Sagasco braces for future battles', *Business Review Weekly*, 27 November, 20–3.

Stimson, R.J. 1991, *Brisbane–Magnet City: Strategies for the Development, Growth and Management of Brisbane into the 21st Century*, Brisbane City Council.

Stimson, R.J. 1992, 'Challenges for Australian cities in an increasingly competitive and uncertain world', *Australian and New Zealand Association for the Advancement of Science, 61st Congress*, Brisbane, September.

Storper, M. & Walker, R. 1989, *The Capitalist Imperative: Territory, Technology, and Industrial Growth*, Basil Blackwell, Oxford.

Suh, S. 1990, 'Wasteful commuting: an alternative approach', *Journal of Urban Economics*, 28, 277–86.

Sweeney, G.P. 1987, *Innovation, Entrepreneurs and Regional Development*, Frances Pinter, London.

Swyngedouw, E. 1989, 'The heart of the place: the resurrection of locality in an age of hyperspace', *Geografiska Annaler*, 71(B), 31–42.

Tarr, J.A. & Dupuy, G. 1988, *Technology and the Rise of the Networked City in Europe and America*, Temple University Press, Philadelphia.

Tatsuno, S. 1986, *The Technopolis Strategy: Japan, High Technology and the Control of the Twenty-first Century*, Prentice Hall, New York.

Taylor, M. & Thrift, N. (eds) 1986, *Multinationals and the Restructuring of the World Economy*, Croom Helm, London.

Teece, D.J. 1987, *The Competitive Challenge: Strategies for Industrial Innovation and Renewal*, Harper and Row, New York.

Thompson, C. 1992, *What A Great Idea!*, Harper Perennial, New York, p. xvii.

Thornton, R.L. 1970, *International Airlines and Politics*, Michigan University Press, Lansing, Michigan.

Thrift, N. 1987, 'The fixers: the urban geography of international commercial capital', in *Global Restructuring and Territorial Development*, eds J. Henderson & M. Castells, Sage, London.

Thwaites, A.T. & Oakey, R.P. 1985, *The Regional Economic Impact of Technological Change*, Frances Pinter, London.

Time 1993, 'The next magic box?', 18 January, 47.

Tobler, W.R. 1979, 'Cellular geography', in *Philosophy and Geography*, eds S. Gale & G. Olsson, D. Reidel Publishing Company, Boston, Massachusetts, 379–86.

Tornquist, G. 1983, 'Creativity and the renewal of regional life', in *Creativity and Context*, ed. A. Buttimer, Lund Series in Geography (B) No. 50, GWK Gleerup, Lund.

United Nations Committee on Transnational Corporations 1990, *Directory of the World's Largest Service Companies*, United Nations, New York.

Urban Land Institute 1987, *Market Profiles 1987*, Urban Land Institute, Washington, DC.

Urey, G. 1991, 'Information in space and time: a planning exploration', in *Proceedings, Second International Conference on Computers in Urban Planning and Urban Management*, ed. R.E. Klosterman, Vol. 1, Institute for Computer-Aided Planning, The University of Akron, Ohio.

US Bureau of the Census 1992, *Statistical Abstract of the United States: 1992*, US Government Printing Office, Washington, DC.

van den Berg, L., Klaassen, L.H. & van der Meer, J. 1990, *Marketing Metropolitan Regions*, European Institute for Comparative Urban Research, Erasmus University, Rotterdam.

Vedel, T. 1987, 'Local policies for wiring in France', in *Wired Cities: Shaping the Future of Communications*, eds W.H. Dutton, J.G. Blumler, & K.L. Kraemer, G.K. Hall & Co., Boston, Massachusetts, 255–78.

Vickerman, R.W. 1984, 'Urban and regional change, migration and commuting—the dynamics of workplace, residence and transport choice', *Urban Studies*, 21, 15–29.

von Bertalanffy, L. 1968, *General System Theory: Foundations, Developments, Applications*, Braziller, New York.

Wabe, J. 1967, 'Dispersal of employment and the journey to work', *Journal of Transport Economics and Policy*, 1(3), 345–61.

Wachs, M., Taylor, B., Levine, N. & Ong, P. 1992, 'The changing commute: a case study of the jobs/housing relationship over time', University of California, Graduate School of Urban and Regional Planning, Los Angeles (mimeo).

Waldrop, M. 1992, *Complexity: The Emerging Science at the Edge of Order and Chaos*, Simon and Schuster, New York.

Walker, R.A. 1985, 'Technological determination and determinism: industrial growth and location', in *High Technology, Space and Society*, ed. M. Castells, Sage, Beverly Hills.

Wassenbergh, H.A. 1992, 'Aspects of the civil aviation relation of the EEC after January 1 1993', *Lessons from History, Papers of Conference on Airlines, Airports and Aviation*, Institute for Advanced Legal Studies, University of Denver, May, 117–26.

Weaver, W. 1967, *Science and the Imagination*, Basic Books, New York.

Weber, A. 1929, *Theory of the Location of Industry*, University of Chicago Press, Chicago.

Webster, P. 1989, 'An article of associations', *Estates Gazette*, 2 December, 92–4.

Weinstein, B.L. & Firestine, R.E. 1978, *Regional Growth and Decline in the U.S.: The Rise of the Sunbelt and the Decline of the Northeast*, Praeger, New York.

White, M. 1988, 'Urban journeys are not "wasteful"', *Journal of Political Economy*, 96, 1097–110.

White, M., Braczyk, H.-J., Ghobadian, A. & Niebuhr, J. 1988, *Small Firms' Innovation: Why Regions Differ*, Policy Studies Institute, London.

Whittington, D. 1985, *High Hopes for High Tech*, The University of North Carolina Press, Chapel Hill.

Willoughby, K.W. 1990, *Technology Choice*, Westview Press, Boulder.

Willoughby, K.W. 1993, *Technology and the Competitive Advantage of Regions: A Study of the Biotechnology Industry in New York*, Monograph No. 44, Institute of Urban and Regional Development, University of California, Berkeley.

Willoughby, K.W. & Blakely, E.J. 1989, *Making Money from Microbes: Finance and the California Biotechnology Industry*, Working Paper No. 89–166, Center for Real Estate and Urban Economics, Institute of Business and Economic Research, University of California, Berkeley.

Willoughby, K.W. & Blakely, E.J. 1990, *The Economic Geography of Biotechnology in California*, Working Paper No. 90–176, Center for Real Estate and Urban Economics, Institute of Business and Economic Research, University of California, Berkeley.

Wilmoth, D. 1987, 'Metropolitan planning for Sydney', in *Urban Australian Planning Issues and Polices*, eds D. Hamnett & R. Bunker, Mansell Publications, 158–84.

Wilson, A.G. 1981, *Catastrophe Theory and Bifurcation: Applications to Urban and Regional Systems*, Croom Helm, London.

Wilson, W.J. 1987, *The Truly Disadvantaged: The Inner City, the Underclass, and Public Policy*, University of Chicago Press, Chicago.

Winchester, S. 1990, 'Leviathans of the sky', *The Atlantic Monthly*, 107–18.

Woerth, D.E. 1992, 'International aviation: cabotage, foreign ownership and international marketing alliances', *Lessons from History, Papers of Conference on Airlines, Airports and Aviation*, Institute for Advanced Legal Studies, University of Denver, May, 147–67.

Wren, C. 1990, *Regional Policy in the 1980s*, Discussion Papers in Economics, Series C, No. 52, University of Reading, Reading.

Wu, C.T. & C. Leung, K. 1992, 'High tech industries in Taiwan and S. Korea: a comparison', Paper presented to the *International Geographic Congress*, Washington, DC.

Wulff, M.G., Flood, J. & Newton, P.W. 1993, *Population Movements and Social Justice: An Exploration of Issues, Trends and Implications*, Bureau of Immigration Research and Department of the Prime Minister and Cabinet, Canberra.

Yanarella, E. & Levine, R. 1992, 'The sustainable cities manifesto: pretext, text and post-text', *Built Environment*, 18(4), 301–13.

Yasuda, Y. 1991, *40 Years, 20 Million Ideas*, Productivity Press, Cambridge, Massachusetts, p. 73.

Yencken, D. 1989, 'The multi-function polis: social and strategic issues', *Australian Planner*, 27(2), 22–5.

Yencken, D. 1990, 'Erewhon and Gnihton: in search of the multifunction polis', in *The MFP Debate: A Background Reader*, eds R.E. Mouer & Y. Sugimoto, Latrobe University Press, Australia.

Young, J. 1993, 'Ranks of telecommuters grow', *The Christian Science Monitor*, 25 February.

Zamanillo, D.G. 1985, 'Connection of facsimile terminals in an ISDN field trial environment', in *Data Communication in the ISDN Era*, ed. Y. Perry, Elsevier, Amsterdam.

Zysman, J. & Tyson, L. 1983, *American Industry in International Competition: Government Policies and Corporate Strategies*, Cornell University Press, Ithaca.

Index